TECTONIC AND EUSTATIC CONTROLS ON SEDIMENTARY CYCLES

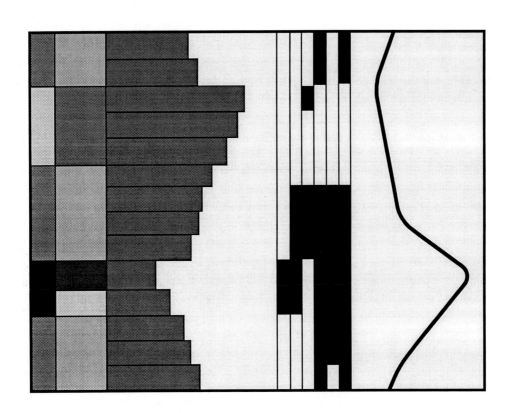

Edited by:
John M. Dennison, University of North Carolina at Chapel Hill, Chapel Hill, North Carolina
and
Frank R. Ettensohn, University of Kentucky, Lexington, Kentucky

Copyright 1994 by
SEPM (Society for Sedimentary Geology)

Peter A. Scholle, Editor of Special Publications
Concepts in Sedimentology and Paleontology Volume 4

Tulsa, Oklahoma, U.S.A.   *September, 1994*

A Publication of
SEPM (Society for Sedimentary Geology)

ISBN 1-56576-017-4
© 1994 by
SEPM (Society for Sedimentary Geology)
P. O. Box 4756
Tulsa, Oklahoma 74131
Printed in the United States of America

TABLE OF CONTENTS

PREFACE

This symposium was first suggested by Richard Diecchio, then president of Eastern Section of Society of Economic Paleontologists and Mineralogists, as a special session for the Eastern SEPM meeting in Baltimore, Maryland, on March 16, 1991. The gathering was in conjunction with the combined Northeast-Southeast Section meeting of the Geological Society of America.

The original symposium topic was "Sedimentary Cycle Control: Tectonic vs. Eustasy". We were asked to Co-Chair the session to represent two different emphases in our own published research, and thus we sought a broad national base of speakers, in addition to presentations from the Eastern Section area of SEPM. Response to the call for abstracts was so great that several papers could not be accommodated in the symposium, but all appeared somewhere in the general meeting in talks or poster-session format. Exchange in the all-day session was at times heated, and the audience of generally more than 100 people heard these ideas to the end.

This Concepts in Sedimentology and Paleontology Series Publication 4 documents the final Eastern SEPM symposium before SEPM reorganized into separate Eastern and Southeastern Sections in 1992, with plans for these separate SEPM entities to meet annually with the respective Northeastern and Southeastern regional sessions of the Geological Society of America. Also, this is the first publication in this series to represent a complete symposium done by desk-top publishing. We appreciate the cooperation of Barbara Lidz and Peter Scholle, who successively as Special Publications Editor, patiently awaited our learning how to integrate so many authors into such an effort. Dana Ulmer-Scholle facilitated in many ways the final preparation, serving as SEPM Special Publications Technical Editor. We especially acknowledge her pleasant skill in bringing together the publication details.

The collected volume begins with a brief perspective by one of the conveners, followed by articles in order of increasing stratigraphic age. Several papers presented for the symposium are not published here in their entirety, because other plans for publication were made before this volume was confirmed. However, abstracts of all papers submitted for the symposium are included at the end of this volume, if they were not developed here to full-length papers.

Change from the original title of the symposium to the title of this volume reflects the sharing and integration of ideas that went on both during and after the symposium. For some, interpretations of tectonism versus eustasy were no longer mutually exclusive. As a result, authors became less polarized, though still holding onto favored views, and several situations were recognized by various authors in which both eustasy and basin tectonism combined to produce the preserved stratigraphic record. Based on our later discussions with authors during the reviewing and editing of papers, it is apparent that the symposium proved to be a growing and sharing experience for most participants, and we hope that our readers will benefit similarly from this exchange of ideas.

John M. Dennison

Frank R. Ettensohn

ASSESSING THE ROLES OF TECTONISM AND EUSTASY

JOHN M. DENNISON

Geology Department, University of North Carolina at Chapel Hill, Chapel Hill, North Carolina 27599-3315

ABSTRACT: Eustatic sea-level changes and tectonic warpings of basins are competing mechanisms for explaining many stratigraphic patterns. The model for sea-level changes should be developed first for a basin, since it is allocyclic and leads to a series of time bands in the strata. The residual effects should then be modeled for tectonic patterns affecting the depositional processes. Doing the reverse limits time constraints on the tectonic warping models and will blur the resolution of detailed time surfaces in the strata. Case histories of situations with both tectonic warping and time surfaces marked by sea-level events will lead to improved interpretations of earth history.

HISTORICAL BASIS

One fundamental theme of geology is understanding earth history. The flow of changing events leaves an imprint on the rocks, and the historical timing is often important in determining rock characteristics and optimal development of economic resources.

Once a flow of history is established, then causal mechanisms can be better evaluated. These are not always easy to establish for stratigraphic deposition. First, there is the problem of uncertainty in correlating minute detail relative to the size of the region being studied. Second, facies relations and unconformities can result from a variety of causes—such as tectonic uplift and erosion of a source area, exposures of varying rock types in the source area, climatic fluctuations, biologic additions to the sedimentary accumulation, changes in sea level, and tectonic activity of the depositional basin.

The arguments are old and go back to the beginning of stratigraphy. Do the drift deposits of northern Europe represent a sea-level change (the Flood) or glacial deposits reaching out from mountainous tectonic uplifts? Many of the original systemic boundaries were angular unconformities related to tectonism, but others were related to marine transgressions over parallel substrates. At a time when orogeny was believed globally episodic, it was easy for Ulrich (1911) to divide North American stratigraphy into a new set of systems bounded by regional paleontologic breaks interpreted as unconformities, for Grabau (1940) to envision an episodic stratigraphy, and for Umbgrove (1942) to sense a pulse of the earth.

The coming of plate tectonics changed all of that. First, Gilluly (1949) challenged the idea of global tectonic pulsations. Next, drifting continents vigorously reappeared in geologic thought after a lapse of nearly three decades, followed by the concept of oceanic spreading centers, and now we generally accept moving lithospheric plates. With the plate-tectonic mechanism powered by radioactivity deep in the earth and with slow rates of convection and cooling contraction of spreading plates, it is difficult to envision a detailed global synchroneity of tectonic controls on sedimentation. The plate-tectonic paradigm is now mature, so its tectonic effects on basin history are well understood today.

Recognition of the effects of climatic episodes on sea level and on erosional and depositional mechanisms permits expectation of chronostratigraphy in a depositional basin, and perhaps on a global scale. These climatostratigraphic markers include some sediment color banding, the alternation between coals and semiarid or arid conditions, and sea-level changes related to glaciation or to water stored in arid inland lake basins or within the pore space of soil and rock.

Local structural weaknesses often become reactivated during the history of a basin. The effects of basin flexures related to advancing thrusts, sediment loading, and erosion of hinterland source terrane are now being mathematically modeled using the principles of engineering physics.

RESOLVING DIFFERENCES OF OPINION

Needless to say, any specialist views the whole from the basis of personal experience and expertise. There is an inherent human tendency to explain the known facts by the mechanisms one understands best. My early work on Appalachian basin stratigraphy (lithostratigraphy and biostratigraphy, aided by some sedimentologic detail) led to a sea-level curve for the mid-Paleozoic in that basin (Dennison and Head, 1975). Later that sea-level curve was expanded to include the entire Paleozoic preserved in the Appalachian basin (Dennison, 1984, 1989). Because my effort studied only one basin, I could not claim eustasy as the cause, even though I strongly suspected (and hoped) it was the cause. Only comparison with other Paleozoic basins in North America and other continents can lead to recognition of truly eustatic sea-level events. The largest sea-level changes will be recognized first on an intercontinental scale, and the seven sequences of Sloss (1963) seem to accomplish this. Global mechanisms, whether tectonic or climatic, affecting transgressions and causing changes of sea level can lead to a powerful correlation tool and exploration aid for mineral/hydrocarbon resources. The "Vail Revolution" of seismic stratigraphy (Vail and others, 1977) has focused mostly on Mesozoic and Cenozoic strata, with very fruitful results integrating it all into modern sequence stratigraphy. Many old understandings have been greatly refined, accompanied by a plethora of new terminology.

Another powerful influence on stratigraphic thought is application of mathematical laws of engineering physics, which has resulted in increasingly useful models for flexural tectonics in depositional basins. These models have been fine-tuned to represent time, position, thickness, water depth, and land height with considerable detail. The temptation is great to assume that if the model appears to work (that is, produces results similar to what we understand to be the flow of geologic history), then Nature used that mathematical formula to produce the rocks. It is now evident that mathematical models can be adjusted to fit about any situation in a basin. We are fast getting to the position that we can call out (or have faith that a computer specialist can call out) a mathematical model that will explain any water-depth change we want to place in a depositional basin. This perhaps leads to a simplistic faith that because warping *can* be used to explain nearly everything, then basin flexure *did cause* everything. Alternative procedures in the real world (such as sea-level changes and tectonic warping) can produce very similar results in the rocks.

RESOLVING INTERACTIONS

This situation of alternative geologic choices leads us to the reason behind this symposium. We have two opposing paradigms which claim to be the dominant factor in basin development. Which one is right? Are they both important factors? How do these two causes interact? How do we separate the effects of one factor from another? The symposium call for papers went out hoping to attract strongly polarized views. This was even true for the conveners, with Dennison (1989) strongly favoring eustasy as the dominant factor, and Ettensohn (1991; this volume) strongly favoring flexural warping as dominant. We drew a widespread opinion, sometimes even heated argument in search of the truth.

I personally believe it is important to push the sea-level model first for a basin, and then to model the residual for tectonic effects, whether they be basin tilting, flexural bulges and foredeep basins, or more localized basement fractures and flexures. It matters little whether the sea-level change is for a whole basin (such as general continental uplift or subsidence of a cratonic basin) or is eustatic for the whole earth. It matters little if a truly eustatic change is one caused by mysterious global pulses of the earth's heat engine, by global cooling of the crust as total spreading slows, by assembly, disassembly, or rotation of super-continents, or by climatic changes involving glaciation or aridity. The sea-level model is allocyclic, so that fluctuations will affect the entire marine basin, albeit manifested variously in different depositional situations. Pulses of sea-level change provide theoretical time surfaces, but in detail they are blurred by differences in catch-up response of sedimentation, by the varying effects of different clastic source areas, by tectonic warping and faulting affecting the depositional basin, and perhaps even by climatic differences within a large basin. Once the tracing of sea-level changes in a basin is extended to its ultimate limit of temporal detail and geographic extent, then the residual stratigraphic effects can be analyzed to separate out the other causes.

Of course, the sea-level model will probably be traced in too much detail and too far geographically in the end, and that

model may produce blind faith in the layer-cake tracing of time lines. Still though, such errors are likely to be less than the errors from following first the results of those other causes of stratigraphic variation to their ultimate expression in the strata.

There is no way to predict in advance when and where peripheral bulges and sub-basins will develop, when siliciclastic source areas will arise, or even when climatic changes will occur, unless some sort of cyclicity is assumed, like Milankovitch cyclicity of climate. Because changes in sea level, certainly, and changes in climate, probably, will affect the entire basin, these have the greatest potential as time lines within the basin, to supplement instantaneous time surfaces like volcanic ash or sedimentologic or geochemical signatures related to bolide impacts. Sea-level changes have the greatest likelihood of being recorded in marine sediments, and wet-dry climate changes of moderate extremes will more likely be recorded in nonmarine strata.

A peripheral bulge, an antiperipheral bulge, or a foredeep basin implies a moving event in response to a transient tectonic situation. These features grow, move, and diminish in configuration. Chronostratigraphy, by whatever means, is essential to prove and track their history. These bulges and sub-basins can best be documented as autocyclic deviations from the time patterns established by allocyclic phenomena, bentonite stratigraphy, or carefully refined biostratigraphic chronology.

More localized tectonics (as distinguished from peripheral bulges, antiperipheral bulges, and foredeep basins) are simpler to recognize. Correlation is usually not much of a problem on a local scale: key beds or fossil zones are simply thinned, thickened, or absent. Very local facies (such as a chain of reefs) or abrupt facies changes suggest a declivity. Care must be taken to recognize possible channels (which have two margins rather than one topographic declivity). The time of local tectonism is easily dated, confined between the last "normal" sedimentation pattern and the return to "normal" more layer-cake sedimentation. Furthermore, local tectonic features commonly are reactivated at nearly the same location at various times throughout geologic history, with effects vertically stacked within the stratigraphic record.

We are mostly talking about two entirely different mechanisms when we consider eustasy and basin tectonics. Both occurred in the Pleistocene through Holocene time span, and presumably both occurred in the more distant past. One of these may be so dominant that the other is masked. During the Pleistocene, the effects of sea-level changes clearly dominated the effects of tectonics on the southeastern trailing edge of the United States. On the tectonically active Pacific coast, local tectonism masks the sedimentation effects of eustatic change. Distinguishing sea-level changes during the Pleistocene and Holocene near the non-glaciated coast of California and southern Oregon will be exceedingly complex in that area of very active tectonism along the Pacific basin margin. On the passive rocky coast extending from Connecticut to Labrador, the stratigraphic record of the Pleistocene will be preserved as deeply scoured valleys with unusually coarse and poorly sorted sediments. Unconformities may dominate the stratigraphic record of that region because of repeated glaciation, and the results of most sea-level changes will be largely missing from the stratigraphic record there. The area from Connecticut to Labrador will be further confused by numerous short-lived, glacially related peripheral bulges and subsidences that produced many tens of meters of apparent sea-level changes. Perhaps these geologically young examples are in an unusual setting within geologic history, but they serve to remind us that tracing ancient eustatic sea-level changes over large areas will probably also be difficult.

FORWARD FROM HERE

This symposium brought people with different emphases in experience together to discuss often conflicting conclusions. Listening to each other clarified our thought processes, made us more attentive to the arguments of others, and reminded us of the often forgotten fact that in most long-lasting controversies both sides are commonly right in part.

A real sign of progress is that many original presentations mellowed their views in written form, a response to careful hearing of opposing perspectives. Some papers present convincing evidence of simultaneous sea-level change *and* tectonic warping. Their criteria are especially instructive in teaching us how to resolve detail in the true history of the rocks.

I personally learned that my original hypothesis of a mid-Ordovician sea-level drop, as presented in this volume in abstract form (Dennison, 1994), cannot be clearly evaluated without much more field and laboratory data. Haynes (1989) has traced by careful petrography the occurrence of the distinctive Deicke and Millbrig K-bentonites along the Cincinnati arch and in the southern Valley and Ridge Province of the Appalachians. These K-bentonites occur at the approximate time of the sea-level drop I proposed, but are not distinctly identified in my data base. Once the Deicke and Millbrig K-bentonites are traced over the entire Appalachian basin, then there will be a time frame to use for rigorous evaluation of whether my proposal of a basin-wide (eustatic?) sea-level drop just before the accumulation of this pair of distinctive ash beds is correct, or if Ettensohn's migrating Ordovician peripheral-bulge and antiperipheral bulge model is correct. Use of this bentonite chronology should provide the best opportunity to trace precise migration through time of such flexures during an episode in geologic history when orogeny was clearly happening in the hinterland of the southern Appalachians. Certainly my suggested criteria for shallowing water have not been examined throughout the basin. I leave my idea in abstract form, as probably premature to evaluate the detailed hypothesis with the data I now have available.

Frank Ettensohn, on the other hand, has presented a bold, basin-wide model, as the final paper of this symposium volume, which can be matched against independent measures of time. The mathematical manipulations of the computer modelers will provide a realistic check on what is likely to be physically possible in space and response time. These factors set the stage

for another advance in stratigraphic resolution. Perhaps other workers will help resolve the conflict between our two approaches by developing more case studies with careful stratigraphic, sedimentologic, and paleontologic documentation.

REFERENCES

DENNISON, J. M., 1984, Geology of the Eastern Overthrust: Kentucky Geological Survey, Series XI, Proceedings of the Technical Sessions, Kentucky Oil and Gas Association Forty-fifth Annual Meeting June 10-12, 1981, Special Publication 11, p. 22-36.

DENNISON, J. M., compiler, 1989, Paleozoic sea-level changes in the Appalachian basin: Washington, D. C., American Geophysical Union, 28th International Geological Congress Field Trip Guidebook T354, 56 p.

DENNISON, J. M., 1994, SEA-LEVEL DROP CONTRASTED WITH PERIPHERAL BULGE MODEL FOR APPALACHIAN BASIN DURING MID-ORDOVICIAN, IN DENNISON, J. M., AND ETTENSOHN, F. R., EDS., TECTONIC AND EUSTATIC CONTROLS ON SEDIMENTARY CYCLES: TULSA, SEPM CONCEPTS IN SEDIMENTOLOGY AND PALEOTOLOGY, V. 4, P. 249.

DENNISON, J. M., AND HEAD, J. W., III, 1975, Sea level variations interpreted from the Appalachian basin Silurian and Devonian: American Journal of Science, v. 275, p. 1089-1120.

ETTENSOHN, F. R., 1991, Flexural interpretation of relationships between Ordovician tectonism and stratigraphic sequences, central and southern Appalachians, U. S. A., in Barnes, C. R., and Williams, S. H., eds., Advances in Ordovician Geology: Ottawa, Geological Survey of Canada Paper 90-9, p. 213-224.

ETTENSOHN, F. R., 1994, Tectonic control on formation and cyclicity of major Appalachian unconformities and associated stratigraphic sequences, in Dennison, J. M., and Ettensohn, F. R., eds., Tectonic and Eustatic Controls on Sedimentary Cycles, SEPM (Society for Sedimentary Geology) Concepts in Sedimentology and Paleontology, v. 4, p. 217-242.

GILLULY, J., 1949, Distribution of mountain building in geologic time: Geological Society of America Bulletin, v. 60, p. 561-590.

GRABAU, A. W., 1940, The Rhythm of the Ages: Peking, China, Henri Vetch, 561 p. (Reprinted 1978, Huntington, New York, Robert E. Krieger Publishing Company.)

HAYNES, J. T., 1989, The mineralogy and stratigraphic setting of the Rocklandian (Upper Ordovician) Deicke and Millbrig K-bentonite beds along the Cincinnati arch and in the southern Valley and Ridge: Unpublished Ph. D. Dissertation, University of Cincinnati, Cincinnati, 237 p.

SLOSS, L. L., 1963, Sequences in the cratonic interior of North America: Geological Society of America Bulletin, v. 74, p. 93-113.

SLOSS, L. L., 1991, The tectonic factor in sea level change; a countervailing view: Journal of Geophysical Research, v. 96, p. 6609-6617.

ULRICH, E. O., 1911, Revision of the Paleozoic Systems: Geological Society of America Bulletin, v. 22, p. 281-680.

UMBGROVE, J. H. F., 1942, The Pulse of the Earth: The Hague, M. Nijhoff, 179 p.

VAIL, P. R., MITCHUM, R. M., JR., AND THOMPSON, S., III, 1977, Seismic stratigraphy and global changes of sea level, Part 4: Global cycles of relative changes of sea level, in Payton, C. E., ed., Seismic Stratigraphy–Applications to Hydrocarbon Exploration: Tulsa, American Association of Petroleum Geologists Memoir 26, p. 83-97.

CYCLES IN LAKE BEDS OF THE TRIASSIC SANFORD SUB-BASIN OF NORTH CAROLINA

LISA N. HU AND DANIEL A. TEXTORIS

Geology Department, The University of North Carolina, Chapel Hill, North Carolina 27599

ABSTRACT: Five wells were examined for evidence of cycles in the Middle Carnian Cumnock Formation. Strata in three wells, located in the lake depocenter, show strong periodicities which may be related to astronomical climate forcing as indicated by time series analyses using a modified Cooley-Tukey Fast Fourier Transform of gamma-ray logs from the three wells. The Butler well has strong signals at thicknesses of 4.2, 19.4, and 62 m. The Groce well has strong signals at 6.0, 25.6, and 61.5 m. The Hall well shows strong signals at 6.3 and 51m. The thicknesses and ratios between them correspond to Van Houten cycles in outcrops of other Newark Supergroup strata. Periodicities may represent present-day 21,700-, 109,000-, and 412,000-year astronomical cycles.

Lithofacies sequences and petrology suggest expansion and contraction of the lake, and possibly correlate with the present-day 21,700- and 109,000-year cycles shown in the power spectra. If the 4.2-, 6.0-, and 6.3-m thicknesses represent the present-day 21,700-year precession cycle, the sedimentation rates of strata in the three wells range from 0.19 to 0.29 mm of rock/yr, west to east. The life span of the lake was at least 1.2 million years. Renewed tectonic activity along the Jonesboro fault system to the east caused an increase in sedimentation rates in that part of the basin, masked lake cycles, and eventually eliminated the Cumnock lake.

Two other wells, located in the basin perimeter, do not display obvious cycles. The Dummit Palmer well, to the northwest, did not penetrate the entire Cumnock Formation and is affected by a diabase intrusive, faults, several coal beds and related basin-edge complications. The Gregson well, to the southeast, contains only minor lake-margin strata, and they were not amenable to any of our cycle analyses.

INTRODUCTION

Events leading to the breakup of Pangea and the evolution of the Atlantic passive margin are recorded in the rock record of more than 40 offshore and onshore Upper Triassic to Lower Jurassic synrift basins that formed on the Variscan-Alleghanian orogen. These rift basins of eastern North America are a series of grabens and half-grabens that reflect components of strike-slip movement along transform faults (Gore, 1986) and wrench faults forming rhomb-shaped half-grabens or pull-apart basins (Bain and Harvey, 1977). They trend southwest to northeast and are present from northern Florida to Nova Scotia. Rocks within these basins comprise the Newark Supergroup (Luttrell, 1989). The rocks include coarse siliciclastic sediments representing alluvial fans along faulted basin margins, to sand-mud siliciclastics representing channels and floodplains, fine-grained, organic-rich sediments representing deeper lake deposits, and evaporites from playa deposits (Gore, 1986). Basaltic flows and intrusives occur in some of the sequences and were emplaced during Early Jurassic time (Smoot and others, 1988). Paleontologic (mainly palynologic), radiometric, paleomagnetic, and geochemical studies show that sedimentation began in the southernmost basins and that sediments became younger to the north (Traverse, 1986; Bain and Harvey, 1977).

Local Geologic Setting

The Deep River basin is the southernmost, major exposed Triassic basin. It is located along the eastern edge of the North Carolina Piedmont, where it is enclosed by Upper Proterozoic to Upper Paleozoic metasedimentary, metavolcanic and intrusive rocks of the Carolina slate belt and the Raleigh belt (Fig. 1). The basin is approximately 260 km long and 9 to 25 km wide, extending from Granville County, North Carolina, to Chester-field County, South Carolina (Gore, 1986). The Deep River basin is divided into three interconnected sub-basins, the Durham sub-basin to the north, the Sanford sub-basin in the middle, and the Wadesboro sub-basin to the south. Based on palynological studies, the three connected half-graben basins developed in central North Carolina during Carnian time, and are filled with strata defined as the Chatham Group of the Newark Supergroup (Smoot and others, 1988; Luttrell, 1989). The Durham and Sanford sub-basins are separated by the Colon cross-structure. The Sanford and Wadesboro sub-basins are separated by the Pekin cross-structure. These cross-structures are interpreted as basin constrictions and basement highs (Gore, 1986).

The Deep River basin is bounded on the east by a series of lystric, normal, low- and high-angle faults (Davis and others, 1991; Bain and Harvey, 1977; Brown, 1988) known as the Jonesboro fault system. The basin is bounded by post-sedimentation normal faults and unconformities on the west; thus forming a half-graben. Coastal Plain strata cover the southern portion of the Jonesboro fault system (Fig. 1).

Stratigraphy

The strata in the Sanford sub-basin dip approximately 10 degrees east and southeast, toward the Jonesboro fault system. The rocks found in the sub-basin are known as the Newark Supergroup, Chatham Group (Smoot and others, 1988; Luttrell, 1989), and are subdivided into the Pekin, Cumnock and Sanford Formations (Campbell and Kimball, 1923) (Fig. 2). These formations are vertically gradational and partially facies equivalent (Reinemund, 1955; Bain and Harvey, 1977). The three formations comprise a 2,800-m thick continental sedimentary sequence at the depocenter, dominated by poorly-sorted conglomerate, arkosic and lithic sandstone, siltstone, black shale, carbonate, and coal (Gore, 1986). The sub-basin also contains a Jurassic diabase in the form of dikes and sills (Bain and Brown, 1981).

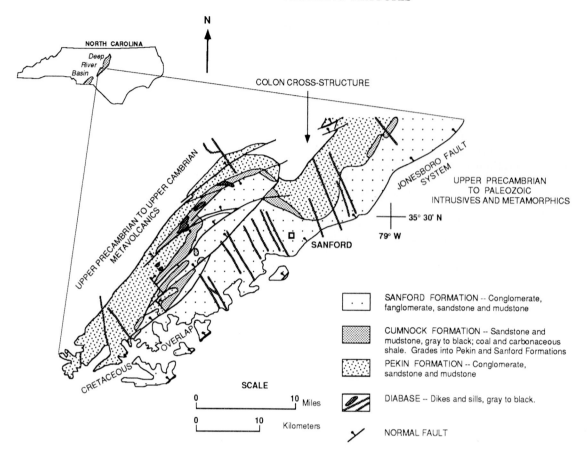

FIG. 1.—Geologic map of the Sanford sub-basin (modified after Textoris and Gore, 1992).

Pekin Formation.—

Named after the town of Pekin in the Wadesboro sub-basin, the Pekin Formation is composed of red-brown or gray fluvial sandstone, conglomerate, and siltstone (Gore, 1986). The Pekin Formation was deposited during Early and Middle Carnian time (Smoot and others, 1988; Luttrell, 1989). It is 542 to 1240 m thick, and at its base in many locations on the western passive basin margin is a distinctive gray, quartz-rich conglomerate known as the "millstone grit" (Reinemund, 1955).

Cumnock Formation.—

Named after the Cumnock Coal Mine (Campbell and Kimball, 1923), the Cumnock Formation is a potential source for oil, natural gas, coal, sulfate, phosphates, and iron (Reinemund, 1955). The Cumnock Formation was deposited during upper Middle to lower Late Carnian time (Smoot and others, 1988; Luttrell, 1989). It is 230 to 250 m thick and, black to gray carbonaceous, calcareous shale and mudstone, black to gray calcareous siltstone, fine-grained gray sandstone, coal and some carbonate predominate. There are two major coal beds in the lower half of the Cumnock Formation. The lower Gulf coal consists of one bed ranging in thickness from a few centimeters to nearly 1 m, and the upper Cumnock coal consists of three beds, each ranging from 1 to 3 m (Robbins and Textoris, 1986; Textoris and Robbins, 1988).

The shales and mudstones of the Cumnock Formation represent the evolution of a deep, meromictic to oligomictic lake which was fed primarily by southward-flowing axial rivers from the northern Durham sub-basin, and by major rivers from the west and northwest (Textoris and others, 1989). The two major coal beds are interpreted as having formed in shoreline swamp systems during wet conditions (Robbins and others, 1988). The lake and swamp facies are especially well developed in the northwestern part of the Sanford sub-basin. The siltstones and the fine sandstones are interpreted to represent lake-margin, deltaic, and shoreline facies. Vertebrate fossils in the black shales of the Cumnock Formation include fishes, amphibians, reptiles, and mammal-like reptiles. Common invertebrate fossils are ostracods and conchostracans (Gore, 1986).

Sanford Formation.—

The Cumnock Formation grades upward into the red-brown, coarser grained rocks of the Sanford Formation, deposited during Late Carnian time (Smoot and others, 1988; Luttrell, 1989). Named after the city of Sanford, located in the eastern portion of the sub-basin, the Sanford Formation consists of 930 to 1240 m of red-brown, alluvial, coarse-grained, arkosic sandstone and conglomerate, with associated red-brown/gray, fluvial claystone, siltstone, and fine-grained sandstone (Reinemund, 1955). The Sanford Formation is thickest along the southeast-

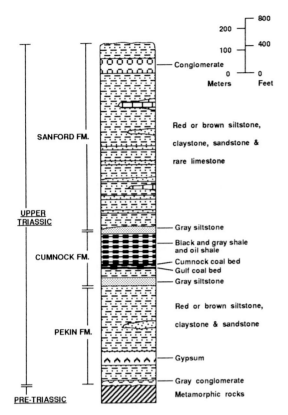

Fig. 2.—Generalized stratigraphic column of the Sanford sub-basin (modified after Reinemund, 1955 and Textoris and Gore, 1992).

ern portion of the sub-basin, adjacent to the Jonesboro fault system (Reinemund, 1955).

Previous Investigations

General.—

The Sanford sub-basin has been studied for the past 160 years. Early investigations dealt primarily with coals in the Cumnock Formation. Olmstead (1824) was the first to describe the Triassic sedimentary rocks with which the coal is associated. Emmons (1856) made the first comprehensive report on the Sanford sub-basin and noted that the rocks could be divided into three informal units. The three units were later formally named by Campbell and Kimball (1923) as the Pekin, Cumnock, and the Sanford Formations. Pollen, spore, and vertebrate assemblages suggest that the oldest rocks in the Sanford sub-basin are as old as Early Carnian in age (Smoot and others, 1988).

The most extensive research in the Sanford sub-basin was done by Reinemund (1955), who concentrated on cores and outcrops that contain the coals in the Cumnock Formation in the western part of the sub-basin. Based on the presence of megaflora in the upper Pekin Formation, Hope (1975) and Gensel (1986) concluded that the siltstones are swamp and floodplain sediments, deposited during a warm and moist climate. During sedimentation of the Cumnock Formation, more moisture and less tectonic activity allowed lake-fringe swamps to extend into

an anoxic lake where gray and black siliciclastic muds were deposited (Robbins and others, 1988; Robbins and Textoris, 1988a, 1988b). Reinemund (1955) believed that the lake in the sub-basin was caused by drainage blockage along the Colon cross-structure in the north and along the Pekin cross-structure in the south.

Cumnock Formation.—

Gore (1989) interpreted the Cumnock Formation to represent a hydrologically-open lake deposit. The thick sequence of black shale is suggestive of constant water level and shoreline stability in a hydrologically open lake. The thick sequence of lacustrine black shale overlying the coal probably represents a profundal (deep-lake) deposit, and it was apparently uninterrupted by major transgressions and regressions, subaerial exposure, paleosol development, or fluvial-colluvial deposition (Gore, 1989). The open-basin model is also based on the absence of evaporites and the presence of strata-bound siderite concretions which form in low-sulfate, freshwater lakes.

Climate during Cumnock time.—

During Middle Carnian time, the Sanford sub-basin lay at 10 degrees north latitude and experienced a warm, humid climate (Textoris and others, 1989), enabling paludal and lacustrine environments to form.

Olsen (1986) suggested that many of the Newark basins had experienced a monsoonal climate along with orographic effects that modified the magnitude and expression of the climate change. Changes in heating intensity of North America caused directly by precession of the equinoxes and modulated by the eccentricity of the earth's orbit alter the intensity of low-pressure cells that drive the monsoonal winds near the equator (Rossignol-Strick, 1985). According to the Rossignol-Strick model, the obliquity cycle has little direct effect on insolation in low latitudes where monsoonal conditions, such as these represented by the Cumnock Formation, predominate.

During Late Carnian time, the North American plate progressively migrated northward from the warm and humid region to a drier climate zone. Traverse (1986) noted that the pollen in the sub-basin show an overall drying trend during Carnian time. Local climate variations were probably due to periodic tectonic activities along the Jonesboro fault system, causing periodic blockage of rain-bearing easterly winds (Textoris and Gore, 1992).

Lacustrine cycles.—

Lacustrine facies commonly exhibit repetitive sequences known as Van Houten cycles. Van Houten cycles were first reported from the Lockatong Formation in the Newark basin by Van Houten (1962, 1964). Many other Newark Supergroup lacustrine deposits also exhibit cyclic sequences similar to those described by Van Houten (Olsen, 1980a, 1980b, 1984, 1986, 1988; Olsen and others, 1989b, 1991a, 1991b; Smoot, 1991).

Van Houten suggested that the vertical sequence of bed types in the cycles reflected the expansion and contraction of lakes, governed by changes in precipitation. The Van Houten cycles consist of two end members, one detrital and the other chemical. Detrital cycles were produced by lakes that were deep enough to have reached their outlets and were hence through-flowing or hydrologically open. Chemical cycles were produced by lakes that never reached their outlets and therefore have high concentrations of salts allowing the precipitation of minerals such as analcime, dolomite, or evaporites (Van Houten, 1962; Olsen, 1984). We have recognized only the detrital end member in the Sanford sub-basin.

According to Van Houten, each detrital cycle in the Lockatong Formation can be divided into three lithologically distinct divisions, in ascending order: division 1 contains platy to massive gray mudstone deposited during lake transgression; division 2 is a fine, calcareous, black mudstone often platy to microlaminated and organic-rich, deposited during lake highstand; and division 3 consists of thicker beds of platy to massive, gray mudstone deposited during lake regression, commonly with evidence of subaerial exposure (Van Houten, 1962, 1964; Olsen, 1984). Van Houten and Olsen suggested that the sedimentary fabrics in each detrital cycle can be used to interpret relative lake depth. Olsen (1984) developed a classification of sedimentary fabrics with seven categories, ranked from 0 to 6, 0 being the shallowest and 6 being the deepest. Olsen (1980a, 1980b, and 1984) used criteria such as presence of vertebrate and invertebrate fossils, organic carbon content, presence of footprints, plant material, and coprolites to define relative lake depth.

Power spectra of depth rank curves constructed from four Newark Basin sections in the Lockatong and Passaic Formations were interpreted by Olsen (1986) to exhibit cyclicity. The ratios of Olsen's shortest cycles to his longer cycles correspond closely to the ratios of the present-day orbital periodicities that appear to influence climate. The periodicities of thicknesses of sedimentary sequences in other Early Mesozoic Newark Supergroup basins are suggestive of Milankovitch-type cyclic controls on climate caused by astronomical forcings (Olsen, 1986). A full spectrum of Milankovitch cycles is present, including the "21,700"-year precession cycle, the "41,000"-year obliquity cycle, and the "109,000-" and "412,100"-year eccentricity cycles (Olsen, 1986; Olsen and others, 1989a). Recent core data from more than 6 km of Newark lake beds in New York, New Jersey, and Pennsylvania suggest also the existence of 2,000,000- and 6,000,000-year cycles within a 30,000,000-year record (Olsen and others, 1991a).

The Cumnock Formation lacks the microlaminated facies typical of most other Newark lacustrine sequences (Smoot, 1991). From preliminary examination of the cores used by Reinemund (1955), it is not certain whether they contain Van Houten cycles, or whether they represent continuous deep-water deposition (Gore, 1989). However, Liggon (1972), Olsen (1980b), and Smoot (1991) have observed that the Cumnock Formation appears to contain some type of cyclic sedimentary sequence. Liggon (1972) reported fining-upward sequences in one of the cores. Smoot (1991) suggested that the cycles are not rhythmic or hierarchical and may be related to deltaic sedimen-

tation. Olsen (1980b) reported repetitive alterations of shallower and deep-water sediments in an outcrop. He suggested that some sort of cyclicity affected lake depth or shoreline aggradational sequences. These alternations may reflect the climatic fluctuations responsible for the Van Houten cycles in other basins (Olsen, 1986; Olsen and others, 1989a, 1991b). Hu and others (1990) and Hu and Textoris (1991) reported strong periodicities in some of the exploratory wells in the Sanford sub-basin using time series analysis of gamma-ray logs and depth rank curves of lithologic logs.

Current research indicates that precession and obliquity cycles may have been shorter during pre-Quaternary time (Klein, 1991). Berger and others (1989) argued that the shortening of the earth-moon distance and of day length back in time induces a shortening of the fundamental astronomical periods for obliquity and precession. Estimated precession and obliquity periods during Triassic time are 20,000 and 37,000 years respectively with apparently no variation in the eccentricity periods (Berger and Loutre, 1989).

METHODS

Well cuttings, core, and gamma-ray logs from five exploratory wells were examined for this study: Dummit Palmer No. 1 (core), Butler No. 1 (well cuttings), Groce No. 1 (well cuttings), Hall No. 1 (well cuttings), and Gregson No. 1 (well cuttings). The five wells are located perpendicular to the general strike of the Sanford sub-basin, providing cross-basinal control (Fig. 3). Clay minerals were identified by XRD.

Petrologic Analysis

For the petrologic study and facies interpretation, 365 samples of well cuttings and core were used. Well cuttings were sampled every 3 m. Two hundred sixty-nine thin sections were used for the petrologic study. Point counts of 200 each were used to determine mineralogy and provenance on 40 representative sandstone thin sections.

Gamma-ray Log Analysis

Lithofacies analysis.—
Gamma-ray and spectral gamma-ray logs of the five wells were examined to help determine the lithologies present in the Cumnock Formation. Because of excellent resolution of lithologic changes in siliciclastic sequences, gamma-ray log patterns were used to aid in the identification and interpretation of the lithofacies.

Time series analysis.—
Gamma-ray logs from the five wells were used to determine cyclicity. The logs were digitized to obtain an American Petroleum Institute (API) reading at 0.3-m intervals. These numerical values versus depth (m) were used for the Fast Fourier analysis.

Time series analyses using a modified Cooley-Tukey Fast

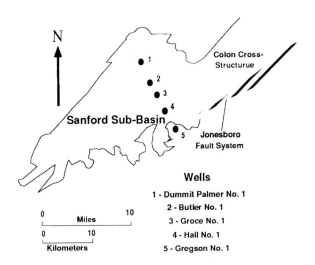

FIG. 3.—Well locations in the Sanford sub-basin.

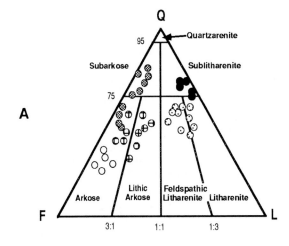

Fourier Transform of the gamma-ray values were performed with Systat 3.0 (Systat, 1985).

PETROLOGY

Siltstone and Sandstone Petrology

Five types of coarse siltstone or sandstone are present in the Cumnock Formation of the Sanford sub-basin. The compositions of the sandstones are displayed as ternary plots (Fig. 4A), and reflect provenance terrane and source rock (Fig. 4B).

Dummit Palmer No. 1 well.—

The siliciclastics in the Dummit Palmer well range from coarse siltstones to fine- to medium-grained sandstones: gray and dark gray, sub-mature to immature argillaceous sublitharenites. Quartz is the most abundant component. All quartz types are present, with schistose quartz and monocrystalline quartz the most common. Rock fragments include fine-grained quartz-mica schists and fine-grained mica phyllites. Matrix is composed of muscovite, chlorite, illite, and quartz. Authigenic pyrite and siderite are present. Cements include calcite, dolomite, and silica.

Butler No. 1 well.—

The Butler well contains medium to coarse siltstones and fine- to coarse-grained sandstones: gray and reddish brown, submature to immature argillaceous litharenites to feldspathic litharenites. Quartz is the most abundant constituent grain, and all quartz types are present. Schistose quartz and monocrystalline quartz are the most common varieties. Fine-grained quartz-mica schists and fine-grained mica phyllites are the dominant rock fragments with minor amounts of granitic fragments. Silt-size to fine sand-size potassium and calcium feldspar are present. The matrix is composed of muscovite, chlorite, kaolinite, illite, and quartz silt. Cements include hematite, calcite, dolomite, and silica. Authigenic pyrite is present in the gray siltstones.

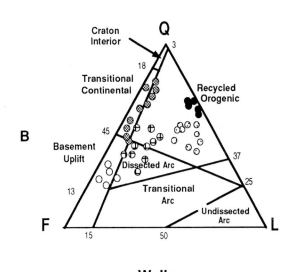

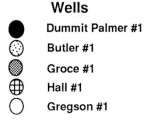

Wells

Dummit Palmer #1
Butler #1
Groce #1
Hall #1
Gregson #1

FIG. 4.—(A) Quartz-feldspar-lithic diagram (Folk, 1980). (B) QFL ternary plot of sandstones in the Cumnock Formation using Dickinson and others (1983) classification scheme (Q=poly-crystalline and monocrystalline quartz, F=potassium and plagioclase feldspars, L=fine- and coarse-grained lithic fragments).

Groce No. 1 well.—

The Groce well contains fine- to coarse-grained sandstones: gray, greenish gray, and reddish-brown, immature argillaceous subarkoses to arkoses. Monocrystalline quartz and schistose quartz are the most abundant quartz types. Medium to coarse grains of potassium and calcium feldspars are common. Minor amounts of coarse-grained quartz-mica schists, mica phyllites, and rare granite are also present. The matrix is composed of muscovite, illite, kaolinite, chlorite, feldspar, and quartz. Calcite and minor amounts of siderite are present as cements and replacements. The subarkoses and arkoses are cemented with feldspar, calcite, quartz, and hematite. Minor amounts of authigenic pyrite are present in the gray siltstones and sandstones.

Hall No. 1 well.—

The Hall well contains medium- to coarse-grained sandstones: gray and reddish brown, immature lithic arkoses to arkoses. Monocrystalline quartz is the dominant quartz type with minor amounts of schistose quartz. Medium to coarse grains of potassium and calcium feldspar are abundant. Rock fragments are composed of coarse-grained quartz mica schists and granite. Muscovite, illite, kaolinite, chlorite, feldspar, and quartz constitute the matrix. The dominant cements are silica and calcite. Some lithic arkoses are also cemented by hematite. Minor amounts of authigenic pyrite are found in the gray siltstones and sandstones.

Gregson No. 1 well.—

The Gregson well contains coarse sandstones: reddish brown and maroon, immature hematitic arkoses. Monocrystalline quartz, schistose quartz, and volcanic quartz are the quartz types. Medium to coarse grains of potassium and calcium feldspar are abundant. Rock fragments consist of quartz mica schists and granite. The matrix is composed of illite, kaolinite, and quartz silt. The arkoses are cemented by hematite, clay minerals, quartz, and calcite.

Shale and Mudrock Petrology

Dummit Palmer No. 1 well.—

The shales and mudstones in the Dummit Palmer well are dominantly dark gray to black, calcareous, carbonaceous, fossiliferous, illite-, chlorite-, phosphate-, pyrite-, and siderite-bearing, micaceous shales, silty shales, and mudstones. The rocks are cemented by quartz, dolomite, siderite, and calcite. Authigenic minerals consist of ammonium illite, calcite, dolomite, siderite, apatite, and pyrite (Krohn and others, 1988; Robbins and Textoris, 1988a, 1988b). Fossils include fish fragments, ostracods, conchostracans, fecal pellets, burrows, and root casts.

Butler No. 1 well.—

The shales and mudstones in the Butler well are gray and dark gray to black, calcareous, fossiliferous, pyrite-bearing, illite/chlorite-bearing, micaceous shales, silty shales, mudstones and silty mudstones. The well also contains reddish-brown, slightly calcareous, nonfossiliferous, hematitic, micaceous, illite- and kaolinite-bearing silty shales and silty mudstones. Fossils include fish fragments, ostracods, and conchostracans, which are present in the gray to black shales.

Groce No. 1 and Hall No. 1 wells.—

Shales and mudstones in these wells are commonly gray and reddish-brown, slightly calcareous, silty shales and silty mudstones. There are some black, calcareous, slightly fossiliferous, pyrite-, illite-, and chlorite-bearing silty shales and mudstones. Fossils include a small number of ostracods and conchostracans, present only in the gray rocks.

Gregson No. 1 well.—

The Gregson well contains reddish-brown, slightly calcareous, nonfossiliferous, hematitic, illite-, kaolinite-, and smectite-bearing silty shales and silty mudstones.

PROVENANCE

According to the tectonic model of Dickinson and others (1983), sandstones of rift belts such as the Newark basins in the eastern United States generally should plot in the transitional continental provenance area. The sandstones from the five wells of the Cumnock Formation in the Sanford sub-basin plot in the recycled orogenic area, dissected arc area, and basement-uplift categories, as well as in the transitional continental category (Fig. 4B).

Dummit Palmer No. 1 and Butler No. 1 Wells

The litharenites of these two wells plot in the recycled orogenic area on the Dickinson and others (1983) format (Fig. 4B). The arenites contain an abundance of quartzose and fine lithic rock fragments from metasedimentary and metavolcanic rocks of the Carolina slate belt. The feldspathic rocks of the Butler well indicate some type of granitic source, probably from the east where the Raleigh belt consists of high-grade metamorphic and intrusive igneous rocks.

Groce No. 1 Well

The subarkosic and arkosic rocks of the Groce well plot in the transitional continental area on the format of Dickinson and others (1983) (Fig. 4B). The subarkoses include both a feldspathic suite from granitic complexes and high-grade metamorphic rocks of the Raleigh belt from the east, as well as a quartzose and fine lithic suite from recycled orogenic rocks of the Carolina slate belt to the west.

Hall No. 1 Well

The lithic arkosic to arkosic rocks of the Hall well plot in the dissected-arc and recycled-orogenic categories of Dickinson and others (1983) (Fig. 4B). The arkoses contain framework grains of the feldspathic suite of granitic and gneissic rocks from the Raleigh belt.

Gregson No. 1 Well

The arkoses in the Gregson well plot in the basement uplift category of Dickinson and others (1983) (Fig. 4B). The arkoses are derived from the granitic and gneissic rocks of the Raleigh belt.

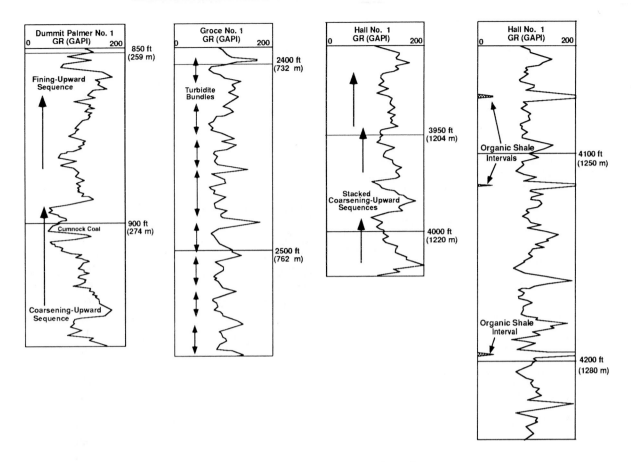

FIG. 5.—Gamma-ray signatures of depositional patterns recognized in the Cumnock Formation.

Lithologies recognized in this study are from one core and well cuttings from the five wells in the Sanford sub-basin. Because of mixing, small size, and sample interval of the well cuttings (3 m), most structures and some textures of the rocks are not apparent; therefore, gamma-ray patterns were used to aid in the lithofacies interpretation of the rock units. The lithologies were divided into five lithofacies: lacustrine, turbidite, deltaic, paludal, and basin-margin sand (Figs. 5, 6).

Lacustrine Lithofacies

Description.—
Two lithologies constitute the lacustrine lithofacies. The major one is finely laminated to non-laminated, black to dark gray, organic, calcareous, fossiliferous, siderite/pyrite-bearing, ammonium illite (Textoris and others, 1989), chlorite shales, silty shales, and mudstones. This lithofacies is best represented in the Dummit Palmer and Butler wells. Reinemund (1955) defined as "blackbands" the ferruginous shales associated with the coal beds in the Cumnock Formation containing limonite-, siderite-, ammonium-, and phosphate-rich nodules with an iron content up to 15%. Fossils are abundant but diversity is low.

Ostracods and conchostracans are abundant in the shales and mudstones. Fossil copopod (aquatic microcrustacean) fecal pellets are also abundant (Porter and Robbins, 1981). Redfieldiids and coelacanth scales are scarce (Olsen and others, 1982). Pollen and spores are also common in the shales (Traverse, 1986; Robbins and Textoris, 1986).

High gamma-ray readings are associated with this shaly sequence. The shaly zones are occasionally interrupted by silty layers represented by lower gamma-ray signatures (Fig. 5).

The other lithology present in this lithofacies is a carbonate, identified as a dolomicrite. It is rare and has only been recognized in the Dummit Palmer well.

Interpretation.—
The texture and the fossils indicate an offshore lacustrine depositional environment in a large meromictic to oligomictic, perennially stratified lake (Gore, 1986, 1989). The "blackband" siderite deposits indicate at least interstitially anoxic, low sulfate waters (Berner, 1981). The lack of evaporites suggests low-salinity waters. The fauna in this sequence suggest an open-water, shallow lake environment (Olsen, 1988). Smoot (1991) interpreted the facies as representing a relatively deep perennial lake.

LITHOFACIES	ROCK DESCRIPTION
LACUSTRINE	Finely laminated to non-laminated, black to dark gray, organic, calcareous, fossiliferous, pyrite-, siderite-bearing, illitic, chloritic shales, silty shales, mudstones, finely crystalline carbonates.
TURBIDITE	Alternating black shales/mudstones and dark-gray to gray fine- to medium-grained immature siltstones and fine-grained sandstones.
DELTAIC	Coarsening-upward sequences of black and gray silty mudstones and shales grading to illitic, kaolinitic gray to reddish-brown siltstones and sandstones. Presence of root casts and coals at top of sequence.
PALUDAL	Gulf and Cumnock coals at top of deltaic sequences.
BASIN MARGIN SAND	Fining-upward sequences of reddish-brown sandstones and siltstones to reddish-brown silty shales and silty mudstones.

LITHOFACIES	GAMMA-RAY SIGNATURES
LACUSTRINE	Shaly sequences indicated by moderate to high gamma-ray readings.
TURBIDITE	Erratic patterns with curve swinging to and fro with a a large amplitude.
DELTAIC	Coarsening-upward sequences indicated by decrease in gamma-ray reading.
PALUDAL	Low radiation with symmetrical, blocky signatures.
BASIN MARGIN SAND	Fining-upward sequences indicated by increase in gamma-ray readings and blocky signatures.

FIG. 6.—Rock descriptions and gamma-ray signatures of the lithofacies found in the Cumnock Formation (see Fig. 5 for gamma-ray signature graphics).

Turbidite Lithofacies

Description.—

This lithofacies is characterized by alternating black shales and mudstones with dark gray, fine- to medium-grained, illite-chlorite, immature siltstones. It is best represented in the Butler and the Groce wells. The gamma-ray logs show a very erratic pattern with the curve swinging to and fro with a large amplitude (Fig. 5). The vertical resolution of the gamma-ray logs and the vertical spacing of the well cuttings do not show individual graded beds but show discrete shale and siltstone bundles 3 to 7 m thick.

Interpretation.—

The erratic nature of each discrete bundle reflects the inter-bedding of siltstones and shales. Contraction and expansion of the lake caused by the rise and fall of lake levels may have caused instabilities along the slopes of the Cumnock lake resulting in the production of turbidity currents. Earthquakes and storms may also have produced instabilities which initiated formation of turbidite flows.

Deltaic Lithofacies

Description.—

This lithofacies consists of coarsening-upward sequences of black and gray silty mudstones and shales, grading to gray

siltstones topped by gray to reddish-brown, fine- to medium-grained, carbonaceous litharenites to arkoses. Coal is rarely associated near the top of this sequence. This lithofacies is present in all wells except the Gregson well. Coarser litharenites and arkoses are present in some sequences. Some of the sequences are capped by coal beds. Root casts and bioturbation are present in the shales and siltstones of the Dummit Palmer well. The gamma-ray logs exhibit coarsening-upward sequences with minor upward-shaling sequences (Fig. 5).

Seismic profiles of the Sanford sub-basin show large deltaic lobes prograding into the sub-basin from the west (J. W. S. Davis, Jr., pers. commun., 1991).

Interpretation.—

Deltaic sequences are typically coarsening-upward sequences or stacked coarsening-upward sequences which may be capped by coal beds. Mudstones, shales, and siltstones are prodelta sediments. Overlying them are coarser sandstones that fine upward to reddish-brown shales and siltstones. Because the coal beds are directly overlain by lacustrine siliciclastics, rising lake levels and accompanying fine clastics may have drowned each swamp community (Robbins and Textoris, 1988b).

Paludal Lithofacies

Description.—

Coal beds are present near the base of the Cumnock Formation. The upper and the most extensive coal in the Sanford sub-basin is the Cumnock Coal, and it is recognized in the Dummit Palmer, Butler, and Groce wells. The Cumnock Coal is best developed as three beds in the northwestern part of the sub-basin where the middle bed can be 3 m thick (Reinemund, 1955). The lower coal is the Gulf Coal, found in the Dummit Palmer well, as one bed that can be as much as 1 m thick (Robbins and others, 1988). It is also present in the Butler well. The coals are mainly bituminous A in rank, and fall in high-volatile to medium-volatile categories (Robbins and others, 1988).

Textoris and others (1989) reported that wood cells, gelified wood, and palynomorphs of ferns and seed ferns predominate in the coals. In carbonaceous shales associated with the coals, wood cells and palynomorphs of horsetails and lycopods predominate.

Interpretation.—

Coal in the Cumnock Formation is evidence for a tropical paleoclimate of high precipitation or high humidity in a lake-fringing swamp environment where clastic sedimentation was slow (Hope, 1975; Gensel 1986; and Textoris and others, 1989).

Basin Margin Sand Lithofacies

Description.—

Fining-upward sequences of reddish-brown, medium- to coarse-grained, illite/kaolinite lithic arkoses, arkoses, siltstones, and shales characterize this lithofacies. These sequences are common in the Gregson well and at the top and base of the Hall well (Fig. 5).

Reinemund (1955) reported that the Cumnock Formation thins and coarsens to the southeast and southwest, grading into strata resembling the Pekin and Sanford Formations.

Interpretation.—

Fining-upward sequences of reddish-brown, coarse- grained, nonfossiliferous silty shales to sandstones suggest oxidizing environments associated with channel or beach deposition in either a deltaic or fluvial environment. Sandstones associated with the coals may represent delta topsets (Smoot, 1991). The exact environment in which the sands were deposited is uncertain. Therefore, the term "basin-margin sands" is used in this study.

DEPTH-RANK CURVES

Using Olsen's (1984) format, the relative lake depths for the Cumnock lake are based on fish and ostracod preservation and content, pelletoid content, amount of burrowing, presence of root casts, presence of coal, rock color, grain size, and organic content (Fig. 7). Depth ranks 0 to 4 were used in this study. Depth rank 4 represents lake high stand, and depth rank 0 represents lake low stand.

When the depth ranks are plotted against the measured Cumnock Formation, the depth-rank curve represents relative lake level, and hence local climate through time (Figs. 8 to 12).

Cycles from the Depth Rank Curves

The Dummit Palmer well depth-rank curve displays two distinct cycles. Each cycle is approximately 40 m thick (Fig. 9).

The Butler well depth-rank curve shows 12 cycles. These cycles are 18 to 21 m thick (Fig. 10). The Groce well depth-rank curve contains at least 13 cycles. Each cycle is approximately 6 m thick (Fig. 11). The 13 cycles cover approximately the central third of the Cumnock Formation. Cycles may also be present in the upper and lower parts of the formation, but due to the small size of the well cuttings from the upper and lower parts of the section depth rank interpretations could not be made. Only the middle third of the section, which had coarser cuttings, was used for depth-rank interpretation (Fig. 11).

The Hall well depth-rank curve indicates two major cycles. Each cycle is approximately 45 m thick (Fig. 12).

TIME SERIES ANALYSIS
Introduction

Time series analysis is a collection of numerical observations arranged in a natural time progression. Each observation is associated with an instant of time or interval of time, and it is this that provides the ordering (Bloomfield, 1976).

Fourier analysis of a time series is a decomposition of the series into a sum of sinusoidal components. Fourier analysis reveals any persistent sinusoidal components in the data (Bloomfield, 1976).

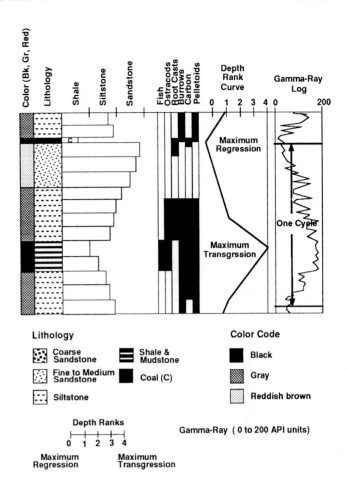

FIG. 7.—An example of an idealized lithologic log with its depth rank curve and gamma-ray log displaying one full cycle (using the format of Olsen, 1984).

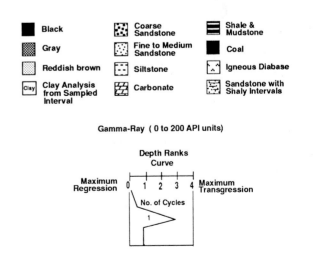

FIG. 8.—Legend for Figures 9 to 12.

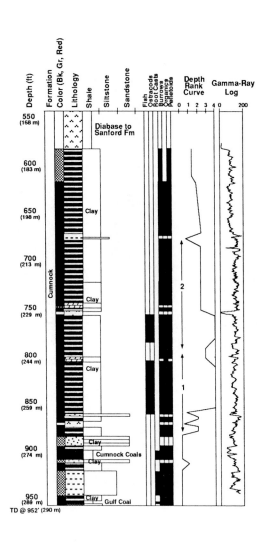

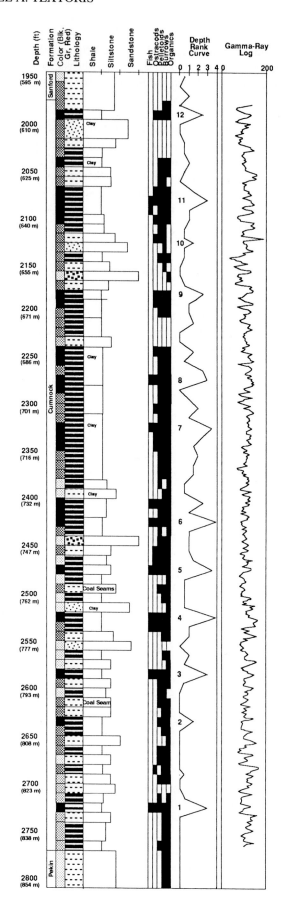

FIG. 9.—Lithologic log, depth rank curve with two possible cycles, and gamma-ray log of the Cumnock Formation from the Dummit Palmer No. 1 well.

Fourier transforms are time consuming because they involve numerous trigonometric functions. Cooley and Tukey (1965) developed a fast algorithm that significantly reduces the computation time. Time series analyses using the Cooley and Tukey Fast Fourier Transform (FFT) has made possible the transformation of large sets of data quickly (Systat, 1985). The frequencies and their intensities from the FFT are displayed in power spectra diagrams.

Gamma-ray Logs

Gamma-ray logs record the radioactivity of the rock as a scintillation device is passed down the borehole. The gamma-ray log device is sensitive to mineralogy and organic content and the readings are the amount of potassium (illite) in the rock. Indirectly, the gamma-ray logs provide information on the elemental composition and grain size of the strata. The logs provide practically continuous, objective, and quantified information that can be interpreted.

Due to poor outcrops and the large sampling interval of well cuttings, outcrops and well cuttings were not very useful for recognizing and deciphering cycles in the Cumnock Formation. Because the major portion of the lake phase contains highly organic shales and siltstones, gamma-ray logs were used to identify cycles. The gamma-ray logs display well-defined signatures of high resolution which are usually not evident in the well cuttings or outcrops.

Cycles from the Gamma-ray Logs

Time series analyses using a modified Cooley-Tukey Fast Fourier Transform were performed on the gamma-ray logs from the Dummit Palmer, Butler, Groce, Hall, and Gregson wells. The power spectra show the frequencies in depositional cycles (m) and time (1,000 years) versus intensity (Fig. 13). Figure 14 illustrates the steps involved in the conversion of the rough power spectra frequencies to cycle thicknesses and time periods.

Dummit Palmer No. 1 well.—
Time series analysis of the Dummit Palmer gamma-ray log shows strong frequencies of 2.8 in the 110 m of Cumnock Formation. This corresponds to depositional cycles of 38.8 m possibly representing the 412,100-year period (Fig. 13).

Butler No. 1 well.—
Time series analysis of the Butler gamma-ray log shows strong frequencies of 56, 30.3, 12, and 3.8 in 236 m of Cumnock Formation. These frequencies correspond to depositional cycles of 4.2 m or the 21,700-year period, 7.7 m or the 39,783-year period, 19.4 m or the 100,233-year period, and 62 m or the 320,333-year period.

Groce No. 1 well.—
Time series analysis of the Groce gamma-ray log shows strong frequencies of 39.5, 24.5, 9.8, and 3.8 in the 238 m of Cumnock Formation. These frequencies correspond to depositional cycles of 6.0 m or the 21,700 year period, 9.8 m or the 35,443-year period, 25.6 m or the 92,587-year period, and 61.5 m or the 222,425-year period.

Hall No. 1 well.—
Time series analysis of the Hall gamma-ray log shows strong frequencies of 18, 10.5, and 2.3 in the 113 m of Cumnock Formation. The frequencies correspond to depositional cycles of 6.3 m or the 21,700-year period, 10.9 m or the 37,544-year period, and 51 m or the 175,667-year period.

Gregson No. 1 well.—
Time series analysis of the Gregson well does not show obvious cyclicity due to poorly developed lacustrine facies.

COMPARISON WITH MODERN ORBITAL PERIODS AND OTHER NEWARK BASINS

The thicknesses of the highest frequency cycles in the Cumnock Formation are similar to thicknesses of Olsen's (1986) Van Houten cycles in the Newark Supergroup of other basins (Table 1).

We assume that the shortest cycles in the Cumnock Formation are Van Houten detrital cycles which represent either present-day (Olsen, 1986) or estimated Triassic (Berger and Loutre, 1989) precession periods of 21,700 and 20,000 years respectively. The ratios of the thicknesses of these shortest Cumnock Van Houten cycles with the longer Cumnock cycles resemble the ratios of present-day and estimated Triassic precession and obliquity orbital terms that appear to influence climate (Table 1). The Cumnock ratios for the eccentricity orbital terms do not correlate well with present-day or estimated Triassic eccentricity ratios. Our 109,100-year eccentricity cycle average ratio of 4.4 compares favorably with Olsen's (1986) at 4.3. However, our 412,100-year eccentricity cycle average ratio of 11.1 does not compare favorably with Olsen's (1986) of 16.3.

Orbital Forcing

The periods seen in the time series analysis of the Cumnock Formation are similar to those predicted by the orbital theory of climate change. The 4.2-, 6.0-, and 6.3-m cycles of the Butler, Groce and Hall wells respectively correspond to the cycle of the precession of the equinoxes or Van Houten cycles (Table 1). The 7.7-, 9.8-, and 10.9-m cycles of the Butler, Groce, and Hall wells respectively correspond to the obliquity cycle (Table 1). The 19.4 and 25.6 m cycles of the Butler and Groce wells respectively correspond to the eccentricity of the earth's orbit (Table 1). Other thicker cycles appear strong but do not correlate well with any cycle predicted by the orbital theory. Longer cycles may represent episodes of tectonic activity.

COMPARISON OF DEPTH RANK CURVES WITH THE TIME SERIES ANALYSES

The thickness of the two cycles seen in the Dummit Palmer depth rank curve (Fig. 9) correspond approximately to the two 38.8 m periods indicated in the time series analysis of the gamma-ray logs. The two periods may be related to the 412,100-year or 109,000-year eccentricity of the earth's orbit.

The cycles and their thicknesses seen in the depth-rank curves of the Butler and Hall wells (Figs. 10 and 12) correlate with periods of about 100,233 and 175,667 years seen in the power spectra. The precession and obliquity cycles are too small

FIG. 10.—Lithologic log, depth rank curve with 12 possible cycles, and gamma-ray log of the Cumnock Formation from the Butler No. 1 well.

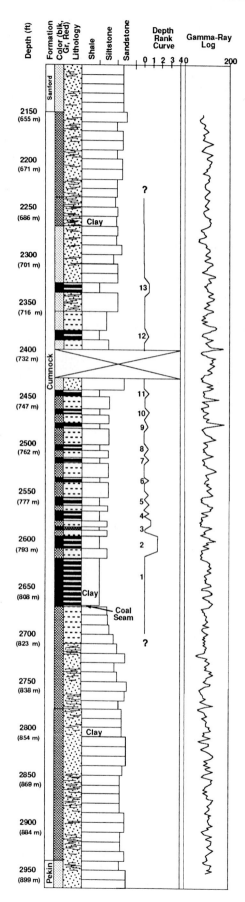

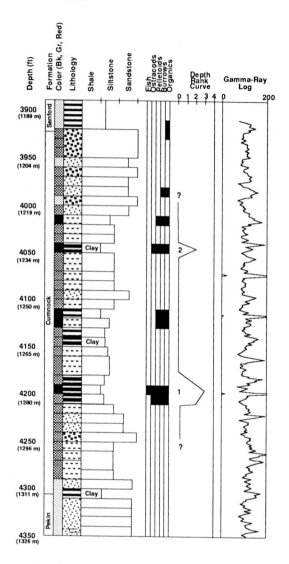

FIG. 12.—Lithologic log, depth-rank curve with two possible cycles, and gamma-ray log of the Cumnock Formation from the Hall No. 1 well.

to be detected by the examination of well cuttings. The well cuttings were sampled at ten-foot (3-m) intervals and do not have enough resolution to exhibit the precession and obliquity periods.

The Groce well depth-rank curve (Fig. 11) displays cycles that have similar thicknesses as the precession-cycle thicknesses given by the time series analysis. Because thirteen 6.0 m cycles were interpreted from approximately a third of the Groce depth rank curve, thirty-nine 6.0-m cycles could represent the entire Cumnock section. The approximate thirty-nine 6.0-m cycles interpreted from the depth-rank curve would then correspond to the 39.5 frequencies obtained from the time series analysis for the 6.0-m depositional cycles.

FIG. 11.—Lithologic log, depth rank curve with 13 possible cycles, and gamma-ray log of the Cumnock Formation from the Groce No. 1 well.

TABLE 1.—RATIOS OF CUMNOCK PRECESSION CYCLE THICKNESSES TO THE OTHER CUMNOCK CYCLE THICKNESSES COMPARED TO THE PRESENT-DAY AND LATE TRIASSIC ORBITAL RATIOS AND OTHER NEWARK SUPERGROUP ORBITAL RATIOS

Item	Precession	Obliquity	Eccentricity	Eccentricity
Present-day periodicities (x 100 yrs) (Olsen, 1986)	21.7	41.0	109.0	412.1
Ratios of present-day precession periodicity to other present-day orbital periodicities (Olsen, 1986)	1.0	1.9	5.1	19.0
Late Triassic periodicities (x 1000 yrs) (Berger and Loutre, 1989)	20.0	37.0	109.0	413.0
Ratios of Late Triassic precession periodicity to other Late Triassic orbital periodicities (Olsen, 1986)	1.0	1.9	5.5	20.6
Thickness of Cumnock cycles (m)				
Dummit Palmer No. 1 Well	NA	NA	NA	38.8
Butler No. 1 Well	4.2	7.7	19.4	62.0
Groce No. 1 Well	6.0	9.8	25.6	61.5
Hall No. 1 Well	6.3	10.9	NA	51.0
Ratios of the Cumnock precession thickness to other Cumnock orbital thickness				
Dummit Palmer No. 1 Well	NA	NA	NA	NA
Butler No. 1 Well	1.0	1.8	4.6	14.8
Groce No. 1 Well	1.0	1.6	4.3	10.3
Hall No. 1 Well	1.0	1.7	NA	8.1
Average ratios of Cumnock precession thickness to other Cumnock orbital thicknesses	1.0	1.7	4.4	11.1
Ratios of Van Houten cycles to other orbital cycles in the Newark Supergroup (Olsen, 1986)				
Eureka Section	1.0	1.7	4.3	NA
Gwynned Section	1.0	1.9	3.9	15.5
Delaware River Section	1.0	1.7	4.1	16.6
Ottsville Section	1.0	NA	4.7	NA
Average ratios of Van Houten cycles to other orbital cycles in the Newark Supergroup (Olsen, 1986)	1.0	1.8	4.3	16.3

SEDIMENTATION RATE AND LAKE DURATION

The estimated sedimentation rates in the Butler, Groce, and Hall wells, using Berger and Loutre's (1989) estimated Triassic precession period, are 0.21 mm/year (rock), 0.30 mm/year (rock), and 0.32 mm/year (rock) respectively (Table 2). The sedimentation rates in the Butler, Groce, and Hall wells, using Olsen's (1986) estimated present-day precession period, are 0.19 mm/year (rock), 0.28 mm/year (rock), and 0.29 mm/year (rock) respectively (Table 3). The calibrations in Tables 2 and 3 depend on the assumption of a constant sedimentation rate. If the sediments were compacted on average to 50% of their original thickness, the true sedimentation rates, in the Butler, the Groce, and the Hall wells, are respectively 0.42 or 0.38 mm/year, 0.60 or 0.56 mm/year, and 0.64 or 0.58 mm/year (Tables 2 and 3). The lake duration can be determined by multiplying the number of precession cycles by the present-day precession period (21,700 years) or by the estimated Triassic precession period (20,000 years) in the Butler, the Groce, and the Hall wells (Tables 2 and 3). The calibrations show a lake duration of 1,215,000 or 1,120,000 years for the Butler well, 857,000 or 790,000 years for the Groce well, and 391,000 or 360,000 years for the Hall well respectively. Because the Butler well is nearest to the lake depocenter, lake duration was at least one million years. The lake duration progressively decreased toward the sub-basin margins due to sedimentologic complications.

Using published dates from Cowie and Bassett (1989), Palmer (1983), and Olsen and others (1989a), the duration of Carnian sedimentation is approximately eight million years. Assuming 2 million years as the duration for the upper portion of Middle Carnian time during which lake sedimentation occurred, Hu and others (1990) calculated the sedimentation rate in the Cumnock lake as 0.236 mm/year (0.118 mm/year of rock).

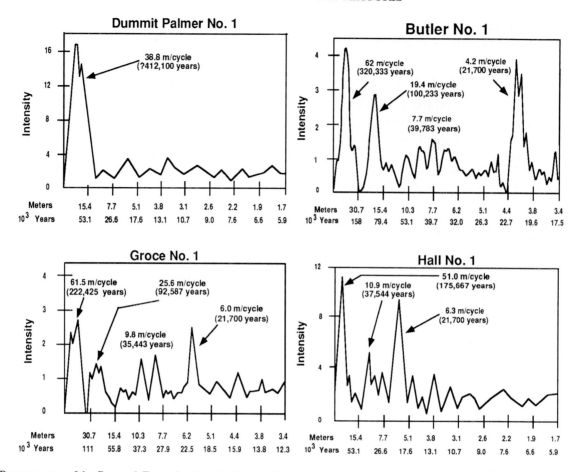

FIG. 13.—Power spectra of the Cumnock Formation from the Dummit Palmer, Butler, Groce, and Hall wells in the Sanford sub-basin. The reader should round off the cycle years to the nearest one thousand for a more realistic number.

CUMNOCK DEPOSITION

The major controls on the deposition of the Cumnock Formation were probably climate and tectonic activity along the Jonesboro fault system along the eastern border of the Sanford sub-basin (Reinemund, 1955; Robbins and others, 1988). The humid, warm conditions, along with tectonic quiescence, were appropriate for the development of a lacustrine environment with accompanying peripheral deltas, swamps, streams, and beaches.

Lithologic logs, depth-rank curves, and gamma-ray logs indicate that the depocenter of the Cumnock lake was on the western side of the sub-basin, between the Butler and the Dummit Palmer wells. The Butler well contains the thickest sequence of black shales and the greatest number of sedimentary cycles (Figs. 15 and 16). Also, seismic profiles of the Sanford sub-basin show the Cumnock lake depocenter to have been located between the Dummit Palmer and the Butler wells (Davis and others, 1991). The overall basin depocenter, which contains 2800 m of strata, however, is located closer to the eastern border, between the Hall and the Gregson wells (Textoris and Gore, 1992).

Expansion and contraction of a hydrologically open lake

resulting from changes in precipitation related to astronomical forces and local topographical controls produced wetter and drier climate cycles, but the climate was not dry enough to produce red evaporitic subaerial cycles as in the northern basins which produced typical Van Houten chemical cycles.

Renewed tectonic activity along the eastern Jonesboro fault system caused an increase in sedimentation rates in the eastern part of the sub-basin. Basin-margin deposits from the east eventually masked the lake cycles near the lake's depocenter and eliminated the Cumnock lake (Hu and others, 1990; Hu and Textoris, 1991).

CONCLUSIONS

Four of the five wells in the Sanford sub-basin contain recognizable Cumnock Formation. Based on petrologic analyses and gamma-ray signatures, the Cumnock Formation consists of five lithofacies: lacustrine, turbidite, deltaic, paludal, and basin-margin sand. The easternmost well (Gregson) contains a lateral facies equivalent to the Cumnock Formation, consisting mostly of red basin-margin sand deposits.

Five sandstone suites are present throughout the sub-basin: litharenite, feldspathic litharenite, subarkose, lithic arkose, and arkose. These provide evidence that provenance was a quartz-

(Continued on p. 20)

TABLE 2.—SEDIMENTATION RATES AND LAKE DURATIONS FOR THE BUTLER, GROCE, AND HALL WELLS.

Well	Total Cumnock Thickness (m)	Number of Precession Cycles	Thickness of Prec. Cycle (m) (Total Cumnock/ No. of Prec. Cycles)	Sedimentation Rate (mm/yr/rock) (Thickness of Prec. Cycles/21,700 yrs)	Lake Duration (yrs)* (21,700 yrs x No. of Prec. Cycles)
Butler No. 1	236.0	56.0	4.2	0.21	1,120,000
Groce No. 1	238.0	39.5	6.0	0.30	790,000
Nall No. 1	113.0	18.0	6.3	0.32	360,000

*Late Triassic precession periodicity (20,000 years) from Berger and Loutre (1989).

TABLE 3.—SEDIMENTATION RATES AND LAKE DURATIONS FOR THE BUTLER, GROCE, AND HALL WELLS

Well	Total Cumnock Thickness (m)	Number of Precession Cycles	Thickness of Prec. Cycle (m) (Total Cumnock/ No. of Prec. Cycles)	Sedimentation Rate (mm/yr/rock) (Thickness of Prec. Cycles/21,700 yrs)	Lake Duration (yrs)* (21,700 yrs x No. of Prec. Cycles)
Butler No. 1	236.0	56.0	4.2	0.19	1,215,000
Groce No. 1	238.0	39.5	6.0	0.28	857,000
Nall No. 1	113.0	18.0	6.3	0.29	391,000

*Present-day precession periodicity (21,700 years) from Olsen (1986).

Power Spectra from Time Series Analyses Yield All Frequencies (Cycles) in the Cumnock Formation

↓

Total Thicknesses of Cumnock / No. of Cycles = Cycle Thicknesses in m/cycle

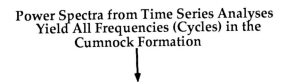

> **e.g., Butler No. 1 Well**
> 236 m / 56 cycles = 4.2 m/cycle
> 236 m / 30.3 cycles = 7.7 m/cycle
> 236 m / 12.0 cycles = 19.4 m/cycle
> 236 m / 3.8 cycles = 62 m/cycle

↓

Assumption: The Highest Frequencies or the Thinnest Cycles = Precession Cycles (21,700 years)

↓

(Longer Cycle Thicknesses / Precession Cycle Thicknesses) X 21,700 years = Longer Cycle Time Periods

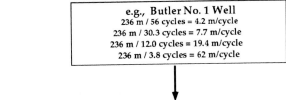

> **e.g., Butler No. 1 Well**
> (7.7 m / 4.2 m) X 21,700 years = 39,783 years
> (19.4 m / 4.2 m) X 21,700 years = 100,233 years
> (62.0 m / 4.2 m) X 21,700 years = 320,333 years

FIG. 14.—Example of the conversion of power spectra to cycle thickness and present-day time periods. The reader should round off the cycle years to the nearest one thousand for a more realistic number.

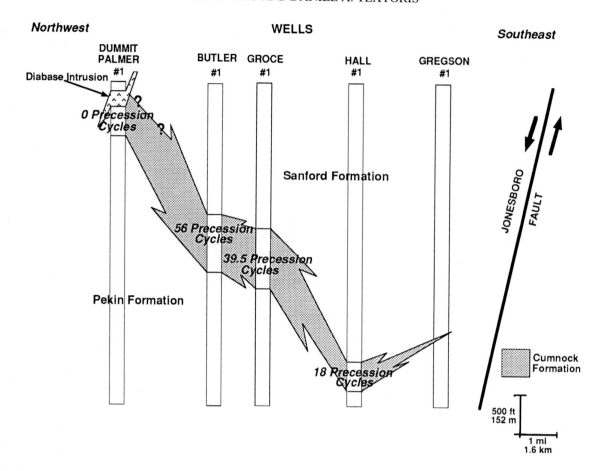

Fig. 15.—Number of precession cycles in the Cumnock Formation across the Sanford sub-basin.

rich metamorphic source from a recycled orogenic area in the Carolina slate belt to the west, and feldspathic sources from transitional-continent, dissected-arc, and basement-uplift areas in the Raleigh belt to the east.

Time series analysis using modified Cooley-Tukey Fast Fourier Transform of gamma-ray logs from wells indicate the presence of Milankovitch-type cycles. Their periodicities and their ratios resemble periodicities and ratios of northern lacustrine sections of the Newark Supergroup.

Assuming that the shortest cycle periods are the duration of either present-day or estimated Triassic precession cycles, the sedimentation rates in the three central sub-basin wells range from 0.19 to 0.32 mm of sediment/year, west to east. Lake duration during a portion of Middle Carnian time was at least one million years.

Depth-rank curves based on petrologic depth indicators show cycles that correlate best with the eccentricity (109,000 year) cycles displayed in the power spectra from time series analyses. For the Groce well, the depth rank cycles correlate with the 21,700-year precession cycle.

The precession cycles are similar to Van Houten's detrital cycles. The Cumnock lake was large and through-flowing, not suitable for the development of Van Houten's chemical cycles which would have resulted in the precipitation of evaporites.

The cycles in the Cumnock Formation were probably produced by climate-controlled expansion and contraction of the lake. Climate was controlled by astronomical forcing.

Tectonic activity along the eastern border Jonesboro fault system caused an increase in the sedimentation rates across the sub-basin eventually masking the lake cycles and eliminating the Cumnock lake by the deposition of the dominantly fluvial Sanford Formation.

ACKNOWLEDGMENTS

Financial assistance was provided in part by the Geological Society of America— Southeastern Section, the American Association of Petroleum Geologists, Sigma Xi, and the North Carolina Department of Environment, Health, and Natural Resources. Samples and geophysical logs were generously provided by the North Carolina Geologic Survey. We thank Pamela Gore, Paul Thayer, John Dennison, and Frank Ettensohn for thorough readings of our manuscript and excellent constructive criticism.

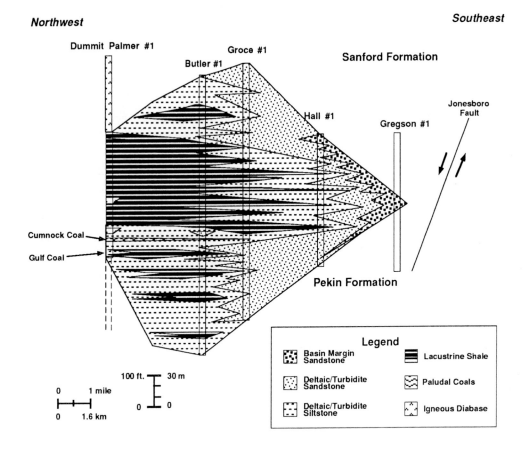

FIG. 16.—Lithofacies cross-section of the Cumnock Formation across Sanford sub-basin.

REFERENCES

BAIN, G. L., AND BROWN, C. E., 1981, Evaluation of the Durham Triassic basin of North Carolina and techniques used to characterize its waste storage potential: Washington, D. C., United States Geological Survey, Open-File Report 80-1295, 138 p.

BAIN, G. L., AND HARVEY, B. W., 1977, Field Guide to the Geology of the Durham Triassic Basin: Raleigh, Carolina Geological Society, Fortieth Anniversary Meeting, North Carolina Department of Natural Resources and Community Development, 84 p.

BERGER, A., AND LOUTRE, M. F., 1989, Pre-Quaternary Milankovitch frequencies: Nature, v. 342, p. 133.

BERGER, A., LOUTRE, M. F., AND DEHANT, V., 1989, Influence of the changing lunar orbit on the astronomical frequencies of pre-Quaternary insolation patterns: Paleoceanography, v. 4, p. 555-564.

BERNER, R. A., 1981, A new geochemical classification of sedimentary environments: Journal of Sedimentary Petrology, v. 51, p. 359-365.

BLOOMFIELD, P., 1976, Fourier Analysis of Time Series: An Introduction: New York, John Wiley & Son, p. 1-9.

BROWN, C. E., 1988, Determination of rock properties by borehole-geophysical and physical-testing techniques and groundwater quality and movement in the Durham Triassic basin, North Carolina: Washington, D. C., United States Geological Survey Professional Paper 1432, 29 p.

CAMPBELL, M. R., AND KIMBALL, K. W., 1923, The Deep River coal field of North Carolina: Raleigh, North Carolina Department of Conservation and Development Bulletin 33, 95 p.

COOLEY, J. W., AND TUKEY, J. W., 1965, An algorithm for the machine computation of complex Fourier series: Mathematical Computation, v. 19, p. 297-301.

COWIE, J. W., AND BASSETT, M. G., 1989, Global stratigraphic chart with geochronometric and magnetostratigraphic calibrations: Episodes, v. 12, Supplement.

DAVIS, J. W. S., JR., TEXTORIS, D. A., AND PAULL, C. K., 1991, Seismic observations of the geometry and origin of the Upper Triassic Deep River basin, North Carolina (abs.): Geological Society of America Abstracts with Programs, v. 23, p. 20.

DICKINSON, W. R., AND OTHERS, 1983, Provenance of North American Phanerozoic sandstones in relation to tectonic setting: Geological Society of America Bulletin, v. 94, p. 222-235.

EMMONS, E., 1856, Geological Report of the Midland Counties of North Carolina: Raleigh, George Putnam and Company, 352 p.

FOLK, R. L., 1980, Petrology of Sedimentary Rocks: Austin, Hemphill's, 182 p.

GENSEL, P. G., 1986, Plant fossils of the Upper Triassic Deep River basin, in Textoris, D. A., ed., Society of Economic Paleontologists and Mineralogists Field Guidebooks, Southeastern United States, Third Annual Midyear Meeting: Tulsa, Society of Economic Paleontologists and Mineralogists, p. 82- 86.

GORE, P. J. W., 1986, Depositional framework of a Triassic Rift basin: the Durham and Sanford sub-basins of the Deep River basin, North

Carolina, *in* Textoris, D. A., ed., Society of Economic Paleontologists and Mineralogists Field Guidebooks, Southeastern United States, Third Annual Midyear Meeting: Tulsa, Society of Economic Paleontologists and Mineralogists p. 55-115.

GORE, P. J. W., 1989, Toward a model for open and closed basin deposition in ancient lacustrine sequences: the Newark Supergroup (Triassic-Jurassic), Eastern North America: Palaeogeography, Palaeoclimatology, Palaeoecology, v. 70, p. 29-51.

HOPE, R. C., 1975, A paleobotanical analysis of the Sanford Triassic Basin, North Carolina: Unpublished Ph.D. Dissertation, University of South Carolina, Columbia, 74 p.

HU, L. N., AND TEXTORIS, D. A., 1991, Possible Milankovitch cycles in lake beds of the Triassic Sanford sub-basin of North Carolina (abs.): Geological Society of America Abstracts with Programs, v. 23, p. 47.

HU, L. N., TEXTORIS, D. A., AND FILER, J. K., 1990, Cyclostratigraphy from gamma-ray logs, Upper Triassic lake beds (Middle Carnian), North Carolina, USA (abs.): Nottingham, 13th International Sedimentological Congress Abstracts, p. 548-549.

KLEIN, G. Dev., 1991, Reply on "Pennsylvanian time scales and cycle periods": Geology, v. 19, p. 407-408.

KROHN, M. D., EVANS, J., AND ROBINSON, G. R., JR., 1988, Mineral-bound ammonium in black shales of the Triassic Cumnock Formation, Deep River basin, North Carolina, *in* Froelich, A. J., and Robinson, G. R. Jr., eds., Studies of the Early Mesozoic basins of the Eastern United States: Washington, D. C., United States Geological Survey Bulletin 1776, p. 86-98.

LIGGON, G. H., 1972, Geology of some non-marine Triassic sediments of North Carolina: Unpublished M.S. Thesis, North Carolina State University, Raleigh, 99 p.

LUTTRELL, G. W., 1989, Stratigraphic nomenclature of the Newark Supergroup of eastern North America: Washington, D. C., United States Geological Survey Bulletin 1572, Plate 1.

OLMSTEAD, D., 1824, Report on the geology of North Carolina conducted under the direction of the Board of Agriculture: Raleigh, J. S. Gales & Son, 44 p.

OLSEN, P. E., 1980a, Fossil great lakes of the Newark Supergroup in New Jersey, *in* Manspeizer, W., ed., Field Studies of New Jersey Geology and Guide to Field Trips: Newark, 52nd Annual Meeting of the New York State Geological Association, Rutgers University, p. 352-398.

OLSEN, P. E., 1980b, The latest Triassic and Early Jurassic formations of the Newark basin (eastern North America, Newark Supergroup): stratigraphy, structure, and correlation: New Jersey Academy of Science Bulletin, v. 25, p. 25-51.

OLSEN, P. E., 1984, Periodicity of lake-level cycles in the Late Triassic Lockatong Formation of the Newark Basin (Newark Supergroup, New Jersey and Pennsylvania), *in* Berger, A. L., and others, eds., Milankovitch and Climate, Part I: New York, Reidel Publishing, p. 129-146.

OLSEN, P. E., 1986, A 40-million-year lake record of early Mesozoic orbital climatic forcing: Science, v. 234, p. 842-848.

OLSEN, P. E., 1988, Paleontology and paleoecology of the Newark Supergroup (Early Mesozoic, Eastern North America), *in* Manspeizer, W., ed., Triassic-Jurassic Rifting: Continental Breakup and the Origin of the Atlantic Ocean and Passive Margins, Part A: Amsterdam, Elsevier, p. 185-213.

OLSEN, P. E., FOWELL, S. J., CORNET, B., AND WITTE, W. K., 1989a, Calibration of Late Triassic-Early Jurassic time scale based on orbitally induced lake cycles: Washington, D. C., 28th International Geological Congress Abstracts, v. 2, p. 547.

OLSEN, P. E., FROELICH, A. J., DANIELS, D. L., SMOOT, J. P., AND GORE, P. J. W., 1991a, Rift basins of Early Mesozoic Age, *in* Horton, J. W., Jr., and Zullo, V. A., eds., The Geology of the Carolinas: Carolina Geological Society 50th Anniversary Volume, University of Tennessee Press, p. 142-170.

OLSEN, P. E., KENT, D. V., AND CORNET, B., 1991b, Thirty million year record of tropical orbitally-forced climate change from continental coring of the Newark Early Mesozoic rift basin: EOS Supplement, p. 269.

OLSEN, P. E., McCUNE, A. R., AND THOMPSON, K. S., 1982, Correlation of the Early Mesozoic Newark Supergroup by vertebrates, principally fishes: American Journal of Science, v. 282, p. 1-44.

OLSEN, P. E., SCHLISHE, R. W., AND GORE, P. J. W., eds., 1989b, Tectonic, Depositional, and Paleoecological History of Early Mesozoic Rift Basins, Eastern North America: Washington, D. C., International Geological Congress Field Guidebook T351, American Geophysical Union, 174 p.

PALMER, A. R., 1983, The Decade of North American Geology 1983 Geologic Time Scale: Geology, v. 11, p. 503-504.

PORTER, K. G., AND ROBBINS, E. I., 1981, Zooplankton fecal pellets link fossil fuel and phosphate deposits: Science, v. 212, p. 931-933.

REINEMUND, J. A., 1955, Geology of the Deep River coal field, North Carolina: Washington, D. C., United States Geological Survey Professional Paper 246, p. 11-119.

ROBBINS, E. I., AND TEXTORIS, D. A., 1986, Fossil fuel potential in the Deep River basin, North Carolina, *in* Textoris, D. A., ed., Society of Economic Paleontologists and Mineralogists Field Guidebooks, Southeastern United States, Third Annual Midyear Meeting: Tulsa, Society of Economic Paleontologists and Mineralogists, p. 75-79.

ROBBINS, E. I., AND TEXTORIS, D. A., 1988a, Analysis of kerogen and biostratigraphy of core from the Dummit Palmer No. 1 well, Deep River basin, North Carolina: Washington, United States Geological Survey Open- File Report 88-670, 15 p.

ROBBINS, E. I., AND TEXTORIS, D. A., 1988b, Origin of Late Triassic coal in the Deep River basin of North Carolina (abs.): Beijing, International Association of Sedimentologists International Symposium on Sedimentology Related to Mineral Deposits Abstract, p. 219-220.

ROBBINS, E. I., WILKES, G. P., AND TEXTORIS, D. A., 1988, Coals of the Newark rift system, *in* Manspeizer, W., ed., Triassic-Jurassic Rifting: Continental Breakup and the Origin of the Atlantic Ocean and Passive Margins, Part B: Amsterdam, Elsevier, p. 649-678.

ROSSIGNOL-STRICK, M., 1985, Mediterranean Quaternary sapropels, an immediate response of the African monsoon to variation of insolation: Palaeogeography, Palaeoclimatology, Palaeoecology, v. 49, p. 237-263.

SMOOT, J. P., 1991, Sedimentary facies and depositional environments of Early Mesozoic Newark Supergroup basins, eastern North America: Palaeogeography, Palaeoclimatology, Palaeoecology, v. 84, p. 369-423.

SMOOT, J. P., FROELICH, A. J., AND LUTTRELL, G. W., 1988, Newark Supergroup correlation chart, *in* Froelich, A. J., and Robinson, G. R., eds., Studies of the Early Mesozoic Basins of the Eastern United States: Washington, D. C., United States Geological Survey Bulletin 1776, Plate 1.

SYSTAT, 1985, The System for Statistics: Evanston, Systat, Inc., p. 323-367.

TEXTORIS, D. A., AND ROBBINS, E. I., 1988, Coal resources of the Triassic Deep River Basin, North Carolina: Washington, D. C., United

States Geological Survey Open-File Report 88-683, 16 p.

TEXTORIS, D. A., AND ROBBINS, E. I., AND GORE, P. J. W., 1989, Origin of organic-rich strata in an Upper Triassic rift basin lake, eastern USA (abs.): Washington, D.C., 28th International Geological Congress Abstracts, v. 3, p. 229.

TRAVERSE, A., 1986, Palynology of the Deep River basin, North Carolina, *in* Textoris, D. A., ed., Society of Economic Paleontologists and Mineralogists Field Guidebooks, Southeastern United States, Third Annual Midyear Meeting: Tulsa, Society of Economic

Paleontologists and Mineralogists, p. 66-71.

VAN HOUTEN, F. B., 1962, Cyclic sedimentation and the origin of analcime-rich Upper Triassic Lockatong Formation, west-central New Jersey and adjacent Pennsylvania: American Journal of Science, v. 260, p. 561-576.

VAN HOUTEN, F. B., 1964, Cyclic lacustrine sedimentation, Upper Triassic Lockatong Formation, central New Jersey and adjacent Pennsylvania: Geological Survey of Kansas Bulletin 169, p. 497-531.

GLACIO-EUSTATIC ORIGIN OF PERMO-CARBONIFEROUS STRATIGRAPHIC CYCLES: EVIDENCE FROM THE SOUTHERN CORDILLERAN FORELAND REGION

WILLIAM R. DICKINSON, GERILYN S. SOREGHAN, AND KATHERINE A. GILES

Department of Geosciences, University of Arizona, Tucson, AZ 85721

ABSTRACT: Shallow-marine and marginal-marine depositional systems of foreland regions between cratons and orogens are sensitive to all the potential influences thought to control stratigraphic cyclicity: autogenic, climatic, tectonic, tectono-eustatic, and glacio-eustatic. In foreland settings, both clastic and carbonate depositional systems of shelf or epeiric seas and adjacent coastal plains are responsive to (a) shifts in shoreline position caused by local or regional variations in sediment supply, and (b) fluctuations in accommodation space controlled by either global eustasy or tectonic flexure that changes relative sea level. For half a century, many have attributed development of prominent Permo-Carboniferous foreland cyclothems to glacio-eustasy in response to Gondwanan glaciations, but dominance of other controls is not easy to exclude. Stratigraphic analysis of key sequences in the southern Cordilleran region supports a primarily glacio-eustatic origin for Permo-Carboniferous cyclothems. This is based on the following observations difficult to reconcile jointly with dominance of other controls: (a) ubiquitous development of stacked cyclothems in both the Permo-Carboniferous foreland of the Ouachita-Marathon orogenic system and in correlative non-foreland settings of Nevada and Utah; (b) absence of comparable cyclicity in older Antler and younger Sevier foreland successions; (c) distinctly diachronous stratigraphic records produced by foreland tectonic flexure and associated migratory forebulges; (d) basinwide distribution of multiple intrabasinal cyclothems in selected Ancestral Rockies basins; (e) provisional interbasinal correlation of individual cyclothems from the midcontinent region through the Ouachita-Marathon foreland; and (f) apparent cyclothem duration within the Milankovitch time band.

INTRODUCTION

Vertical cyclicity of stratal composition and bedding style is as inherent a part of stratigraphy as lateral facies change. Different types and scales of cyclicity imply varied controls for cycle development. We argue that the prominent cyclicity of Permo-Carboniferous strata in sedimentary basins of varied origin is most reasonably attributed primarily to glacio-eustatic changes in global sea level governed by fluctuations in the size of Gondwanan ice sheets. The special character of resulting depositional systems is reflected by the widespread recognition of repetitive stratigraphic cycles in typical sequences of Permo-Carboniferous age. The strata include both cyclothems, which embody multiple alternations of marine and nonmarine strata, and stacked shoaling-upward cycles capped by exposure surfaces or strandline deposits.

Our interpretation is neither original nor unique, for Wanless and Shepard (1936) long ago suggested the potential association of cyclothemic sedimentation with Gondwanaland glaciations, and many authors have followed their lead with vigor (for example, Heckel, 1986, 1990). For many years, however, doubt of the reality of continental drift, and consequent suspicion of the interpretation of massive glaciers in Gondwanaland, cast a cloud over their conclusions. Acceptance of continental drift, with its rationale for positioning the southern continents together over the pole, gave renewed impetus to their concepts (Crowell, 1978). Recent analysis of Gondwanan glacial sequences infers major ice volumes in Gondwanaland from Late Mississippian to mid-Early Permian time, corresponding closely to the interval of pronounced cyclothemic sedimentation in Euramerica (Veevers and Powell, 1987).

We derive independent arguments for the glacio-eustatic control of Permo-Carboniferous cyclothems and related cyclostratigraphic sequences by consideration of relationships within the southern Cordilleran region (Fig. 1). In our view, arguments that the classic cyclothems of Euramerica may have been controlled partly by tectonic effects in foreland settings is refuted by two observations: (a) in strata of Permo-Carboniferous age within the southern Cordilleran region, pronounced cyclicity is evident not only in foreland settings but in other tectonic environments as well, and (b) strata deposited in Cordilleran foreland basins of other ages (mid-Paleozoic and Cretaceous) do not display comparable cyclicity. Provisional correlation of late Pennsylvanian (Missourian-Virgilian) cycles of the southern Cordilleran region with those described for the midcontinent strengthens the inference of their synchroneity over wide areas. We also confirm previous conclusions (Heckel, 1986; Veevers and Powell, 1987) that durations of Permo-Carboniferous cycles lie within the Milankovitch band of frequencies thought to be controlled by orbital parameters of the earth.

FORELAND DEPOSITIONAL SYSTEMS

Depositional systems of low-lying foreland regions (Fig. 2) between cratons and orogens are notably sensitive to changes in relative sea level because they occupy broad expanses within the interiors of continental blocks. Relative highstands readily flood broad foreland basins with marine waters, but water depths remain modest even at maximum flooding. Drawdown during relative lowstands commonly induces widespread exposure and terrestrial sedimentation or formation of paleosols. It is important to observe, however, that this marked sensitivity to eustatic fluctuations in sea level does not apply to phases of continent or arc collision during which incipient subduction carries a continental margin into deep water where eustatic influences are indirect. Eustasy may affect deep environments by influencing sediment supply and the dynamics of turbidite systems, but

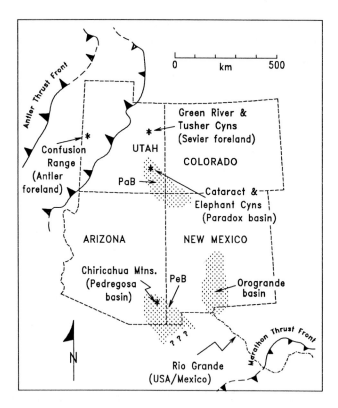

FIG. 1.—Sketch map of southern Cordilleran region showing positions of principal Phanerozoic thrust fronts (Devono-Mississippian Antler, Permo-Carboniferous Marathon, and Cretaceous Sevier), and locations of comparative stratigraphic columns (Confusion Range in Antler foreland, Green River/Tusher Canyons in Sevier foreland, Cataract/Elephant Canyons in Paradox basin of Colorado Plateau, and Chiricahua Mountains in Pedregosa basin of Ouachita-Marathon foreland region). Selected sedimentary basins: OgB—Orogrande basin; PaB—Paradox basin; PeB—Pedregosa basin.

fluctuation in sea level alone becomes a less significant factor for sedimentation as water depth increases.

Relative sea level in foreland basins can be envisioned as the resultant of interplay between three quasi-independent parameters (Fig. 3): subsidence, sediment supply, and global eustasy. In general, transgressive episodes can be triggered by any effective combination of increased subsidence rate, decreased sediment supply, and rise in global sea level (and regressive episodes by the reverse). At times when variations of these controlling parameters are conflicting in tendency, their net effect is inherently difficult to predict but, in principle, a guiding balance must still be struck because they jointly determine the available accommodation space for continued sedimentation.

None of the controlling parameters is truly independent, and treatment of each in isolation is artificial (Klein and Willard, 1989). For example, the same regional tectonic processes that govern subsidence of foreland basins in response to thrust loading also govern to some extent the relief of orogenic sediment sources and, thus, influence sediment supply. On the one hand, climatic factors may also modulate sediment supply by

influencing erosion rates and the nature of prevailing depositional systems with contrasting autogenic (internally driven) behavior. On the other hand, orogenic relief may, in turn, influence local climate through rain shadow effects. Climatic patterns may also be affected by fluctuations in global sea level and the resultant degree of continental emergence.

FORELAND STRATIGRAPHIC CYCLICITY

Cyclic variations in foreland sedimentation could stem from cyclic fluctuations in tectonic, climatic, or eustatic influences on foreland depositional systems. Expected amplitudes and frequencies for cycles controlled by tectonics and eustasy can be inferred from established concepts (Fig. 4). Temporal patterns of non-eustatic climatic cycles, which may independently influence sediment supply, probably mimic those of glacio-eustasy because of analogous response to orbital forcing parameters (Cecil, 1990). Episodic autogenic processes operating at different mean frequencies peculiar to particular depositional systems can also induce stratigraphic cyclicity of cyclothemic type (Moore, 1959).

As the chart of Figure 4 indicates diagrammatically, several different potential cycle controls operate within inferred frequency bands and amplitude ranges that overlap. Consequently, identifying cycle controls is inherently ambiguous from observations of thickness and age at individual outcrops. Supplementary information is required to make a logical choice among competing alternatives.

On Figure 4, the field shown in the upper right corner for a tectonic cycle represented by an entire basin fill is included only as an example of "cyclic" behavior at a frequency and amplitude entirely apart from the scale of events reflected by individual cyclothems. Similarly, cyclic tectono-eustasy, at least as we now understand it to be a function of changing spreading rates and continental configurations, involves a lower frequency of variation than the fluctuations in sedimentation recorded by typical cyclothems (Cloetingh and others, 1985).

On the other hand, at least four other types of processes might involve frequencies and amplitudes suitable for sequences of stacked cyclothems and correlative all-marine cycles: (1) structural movements on folds and faults affecting the relative elevation of a depositional site, (2) autogenic processes of depositional systems that display inherently unsteady behavior (such as switching of delta lobes, migration or avulsion of fluvial channels, shifting of local environments on carbonate platforms), (3) flexure of lithosphere beneath the depositional site as a result of changing intraplate stresses (including but not limited to those imposed by successive emplacement of nearby thrust sheets), and (4) repetitive glacio-eustatic fluctuations in global sea level and associated climatic changes.

Although such diverse types of processes could thus conceivably generate "cyclothems" having similar thicknesses, their lateral extents would be quite different. Structural or autogenic controls would produce cyclothems of only local extent, confined to just some affected part of a sedimentary

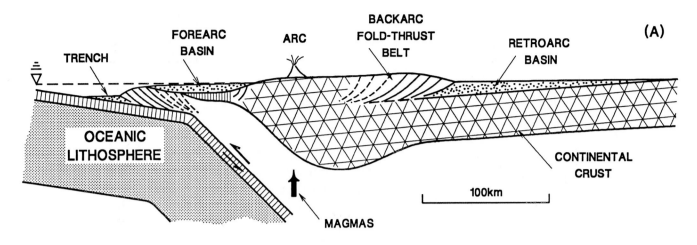

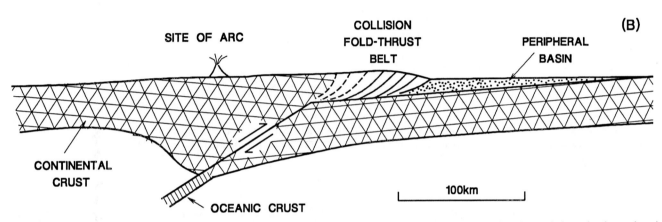

FIG. 2.—Schematic cross sections of asymmetric foreland basins formed by downflexure of continental lithosphere beneath thrust loads emplaced along flanks of adjacent orogenic systems: (A) retroarc foreland basin along rear flank of arc-trench system forming active continental margin, with thrusting antithetic to subduction at associated trench; (B) peripheral foreland basin along passive side of collisional orogen (dormant arc on active side), with thrusting synthetic to subduction at associated suture belt. In either case, intrabasinal structures may include uplifts either parallel or orthogonal to the trend of the orogen.

basin. By contrast, cyclothems of glacio-eustatic origin would be developed throughout a given sedimentary basin, and would be potentially correlated with synchronous cyclothems in other basins throughout the world.

Distinction between glacio-eustatic cyclothems and cyclothems produced by the flexural response of lithosphere to time-dependent variations in intraplate stress presents a more severe challenge to interpretation, particularly in foreland settings. Variable intraplate stresses generated by interplate stresses transmitted across subduction zones or suture belts, or the flexural stresses of overthrust loads emplaced episodically, could reasonably be expected to affect a large foreland region simultaneously. Accordingly, Klein and Willard (1989) have postulated that Permo-Carboniferous foreland cyclothems may reflect episodic thrust loading (combined with glacio-eustasy).

Fortunately, the areal relations of lithospheric flexure afford the means, in principle, to distinguish between eustatic and flexural effects. Lithospheric flexure, whatever the nature of the stresses causing it, has a characteristic wave length which varies with flexural rigidity (Beaumont, 1981). Wherever lithospheric flexure creates additional accommodation space for sedimentation, areas of downflexure are delimited by areas of upflexure (Fig. 5). The latter include uplifted basin margins rimming basin centers depressed by regional intraplate stresses (Cloetingh, 1988) and uplifted forebulges flanking foreland basins depressed by adjacent thrust loads (Jordan, 1981). Consequently, cyclothems produced by stress fluctuations of any origin are predicted to be out of phase between basin interiors and basin flanks. Moreover, nodal areas (Fig. 5) between domains of downflexure and upflexure should lack cyclothems of either phase. By contrast, cyclothems produced by glacio-eustasy will be in phase wherever preserved and should be present throughout a given basin except for complications produced by stratigraphic condensation.

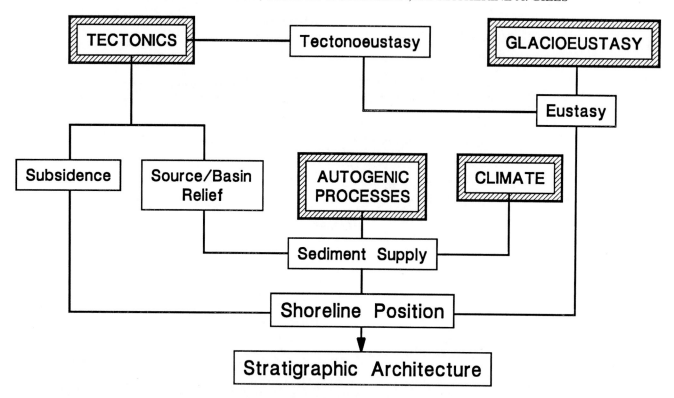

F<small>IG</small>. 3.—Diagram showing main inferred interactions of tectonic influences, autogenic processes, and climatic factors to produce shifts in shoreline position and consequent variations in stratigraphic architecture as generated by shallow-marine and marginal-marine depositional systems; modified after Galloway (1989).

OUACHITA-MARATHON FORELAND CYCLICITY

During late Paleozoic time, southern Arizona and New Mexico, and adjacent parts of Sonora and Chihuahua, lay northwest of the southwesterly extension of the Ouachita-Marathon thrust front in Texas (Fig. 1), and experienced complex foreland deformation (Armin, 1987) in a peripheral foreland setting (Fig. 2B). Within this foreland region, the Pedregosa and Orogrande basins were bounded by uplifts of the Ancestral Rockies system, which evolved in response to stresses associated with the continental collisions that assembled the supercontinent Pangea (Kluth and Coney, 1981). Pennsylvanian and Lower Permian strata of both basins display pervasive and ubiquitous stratigraphic cyclicity (Soreghan, 1990, 1991). The two basins afford a superb natural laboratory for distinguishing the effects of climatic (Armin, 1991), eustatic, tectonic, and autogenic influences on the development of carbonate (Pedregosa) and mixed carbonate/clastic (Orogrande) cyclic stratigraphy in Permo-Carboniferous foreland successions.

Preliminary results from our study of stratigraphic cyclicity in the Pedregosa basin (Fig. 1) suggest that the same 17 Missourian and Virgilian carbonate cycles are present at multiple sampling sites distributed widely through the basin. Apparently equivalent cycles are present within the Orogrande basin, but are more complex there because of more pronounced clastic influence on local sedimentation (Soreghan, 1991). The 17 identified

cycles span the stratigraphic interval from the base of the Missourian Stage to a mid-Virgilian horizon that is correlative with some upper part of the Shawnee Group (Heckel, 1986) in the midcontinent region. Intrabasinal correlation has been established by cycle diagrams (Fischer plots) supplemented by fusulinid collections (Soreghan, 1990). Each stratigraphic cycle is thinner at inner shelf sites around the margins of the basin than at outer shelf sites closer to the basin interior, but no sites have yet been found where any of the 17 cycles are missing. Each appears on outcrop as a shoaling-upward sequence of subtidal carbonate capped in some instances, commonly on the inner shelf but rarely on the outer shelf, by intertidal or supratidal algal laminites and oxidized or brecciated exposure surfaces. The stratigraphic cyclicity was clearly controlled by repetitive alternations in relative sea level that varied local water depth. Moreover, many individual cycles display incomplete shoaling of subtidal strata prior to terrestrial exposure. Such abrupt facies juxtapositions imply eustatic or tectonic rather than autogenic control.

The apparent correlation of the same array of Missourian and Virgilian stratigraphic cycles throughout the Pedregosa and Orogrande basins, at widely separated sites representing inner-shelf to outer-shelf environments, suggests eustatic control. Tectonic or autogenic controls would be expected to introduce local cycle variability that is not observed. We are struck, moreover, by the fact that the number of midcontinent Missou-

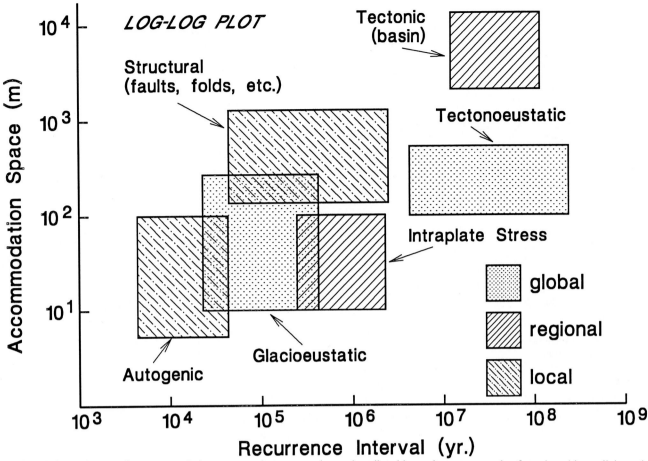

Fig. 4.—Estimated ranges in accommodation space and recurrence interval attributable to alternate controls of stratigraphic cyclicity, where accommodation space (per cyclic event) sets limits on thicknesses of individual cycles and recurrence interval (between initiation of successive cycles) governs characteristic frequency of observed cyclicity. The structural field for folds and faults is delimited to embrace the full growth of individual folds, or the integrated duration of activity and net offset along individual faults, rather than separate increments of episodic deformation. All fields shown are tentative and subject to revision or reinterpretation (tectonic box pertains to full basin evolution).

rian to mid-Virgilian cyclothems prominent enough to be correlated from Kansas to Texas (Boardman and Heckel, 1989) matches the number of cycles (17) that we observe in the Pedregosa basin. Although more paleontological work is needed to confirm the correlation in detail, we reach the provisional conclusion that the same 17 glacio-eustatic cycles can be correlated from the midcontinent region throughout the Ouachita-Marathon foreland. Independent work by Connolly (1991) also suggests the feasibility of correlating individual middle Pennsylvanian cycles (Desmoinesian and basal Missourian) from Kansas to Arizona.

EXTRA-FORELAND CYCLICITY

Pennsylvanian cyclic sedimentation was also pronounced at settings within the southern Cordilleran region far removed from the Ouachita-Marathon thrust front. Well known cyclic successions include the Bird Springs Formation in the Arrow Canyon Range of the miogeosynclinal belt in Nevada (Langenheim, 1991) and strata of the intracontinental Paradox basin (Fig. 1), located midway between Cordilleran and

Mesoamerican continental margins of North America in Paleozoic time. Detailed measured sections through Missourian and Virgilian strata of the Paradox basin in Utah (Loope and others, 1990) invite comparison with correlative cyclic strata of the Pedregosa basin in the Ouachita-Marathon foreland region (Fig. 6). The strata in the Paradox basin are composed of interstratified marine carbonate sequences and eolian dune deposits of unmistakably terrestrial origin. The nature of the lithologic alternation renders the identification of dominant cycle boundaries seemingly conclusive from measured sections alone, and pronounced changes in relative sea level were unquestionably associated with cycle development. Seventeen marine/nonmarine cycles are evident in these strata (Fig. 6). The overall scale of cyclicity in the Paradox basin is also comparable to that in the Pedregosa basin, but the paucity of diagnostic fusulinids in the former precludes unequivocal interbasinal correlation of the cycles observed. Comparison of the two key stratigraphic sections of Figure 6 permits the inference, however, that the same Missourian and Virgilian cycles delimited by our work in the Pedregosa basin may be present also in the Paradox basin.

If future work confirms that arrays of Pennsylvanian cycles

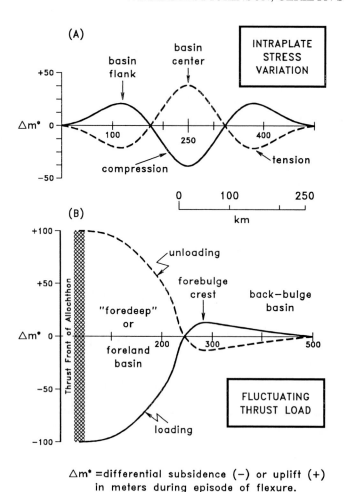

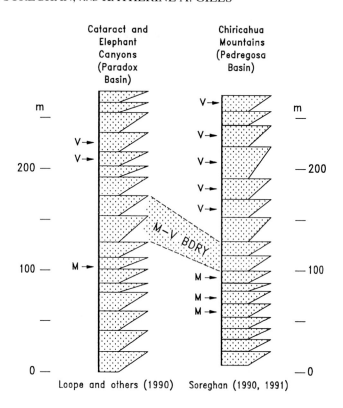

FIG. 5.—Diagram showing relations of incremental downflexure and coupled upflexure for (A) symmetric basin influenced by cyclic variations in lateral intraplate stresses (modified after Cloetingh, 1988), and (B) asymmetric basin influenced by cyclic fluctuations in adjacent thrust loads (modified after Goebel, 1991). Curves indicate schematically how hypothetical episodes of flexural subsidence in sediment-loaded basin center (A) or thrust-loaded basin keel (B) are paired with complementary episodes of uplift on basin margin (A) or forebulge (B), whereas episodic flexural rebound of basin center or keel is paired with episodic flexural collapse of basin margin or forebulge. Zero lines are normalized to basin configurations in existence prior to episodic flexure that occurs in response to changes in lateral intraplate stress or vertical thrust load. Scales are illustrative only, because vertical amplitudes of differential subsidence or uplift during postulated episodic events are uncertain, and lateral dimensions of flexural features are dependent upon the variable elastic thicknesses of lithospheric plates. See Karner (1986) for similar plots and for time-dependent effects on flexural rigidity (not shown).

in the southern Cordilleran region are not confined to foreland settings, but can be traced throughout multiple Permo-Carboniferous sedimentary basins regardless of tectonic setting, then no simple appeal to foreland flexure under variable lithospheric stresses can explain their occurrence. Moreover, the very existence of persistent and widespread stratigraphic cycles may

FIG. 6.—Comparison of Pennsylvanian cyclicity in the Paradox (left) and Pedregosa (right) basins of the Ancestral Rockies province. Each column shows 17 Missourian and Virgilian cyclothems (M and V denote occurrences of Missourian and Virgilian fusulinids, respectively) presumed tentatively to be correlative. Cycles in Pedregosa basin are shoaling-upward carbonate cycles whereas cycles in Paradox basin are cyclic alternations of marine carbonate and terrestrial eolianite. (Schematic representation depicts decrease in water depth and/or increased subaerial relief from left to right on columns). Scale in meters. See text for discussion of tentative correlation and Figure 1 for locations of Cataract/Elephant Canyons (Paradox basin) and Chiricahua Mountains (Pedregosa basin).

well afford the means to achieve stratigraphic correlation throughout the continent at a scale difficult to attempt with confidence using biostratigraphic criteria alone.

Any argument strengthening the interpretation that the cycles were controlled by global glacio-eustasy would enhance the attraction of this approach to stratigraphic correlation in detail.

CYCLE FREQUENCY

Such an argument is provided by the observed frequency of Permo-Carboniferous stratigraphic cycles in geologic time. For example, the average apparent duration of the 17 Missourian and Virgilian stratigraphic cycles can be calculated from estimates for the total time span of the base-Missourian to mid-Virgilian interval. Interpolation using the various time scales of Harland and others (1982, 1990), Ross and Ross (1987), and Klein (1990) yields a mean estimate of 6 to 8 my for the time span during which the 17 key cycles were deposited (the full range of uncertainty is perhaps 5 to 9 my). Our best estimate for the average duration of the 17 cycles is thus in the range of 353,000 to 471,000 yr, with a nominal median value of 412,000 yr. The latter figure is perhaps just fortuitously close to the periodicity of 413,000 years calculated for the so-called long eccentricity cycle from orbital parameters (Berger, 1977; Imbrie and Imbrie, 1980). In any case, however, given inherent uncertainties of the geologic time scale and of relevant orbital calculations, the cyclic sedimentation represented by the 17 Missourian and Virgilian cycles defines an apparent frequency indistinguishable from the long eccentricity cycle. There are sound theoretical reasons to suppose, moreover, that the frequency of the long eccentricity cycle has not varied since Carboniferous time (Maynard and Leeder, 1992).

Other authors have also called attention recently to near congruence of the inferred frequency of regionally developed Permo-Carboniferous cyclothems and analogous cycles with the periodicity of the long eccentricity cycle (Veevers and Powell, 1987; Algeo and Wilson, 1991; Chesnut, 1991; Langenheim, 1991). We are unsure why the shorter orbital cycles related to obliquity and precession have been less readily detected in the Permo-Carboniferous stratigraphic record, but some detailed studies suggest that they are recorded as well (Heckel, 1986; Boardman and Heckel, 1989; Maynard and Leeder, 1992). The apparent dominance of the long eccentricity cycle for Permo-Carboniferous sedimentation may be explained by the supposition that the amplitude of insolation variations associated with it were greater in late Paleozoic time than the analogous variations associated with other orbital parameters (Collier and others, 1990, Fig. 11).

The viewpoint that especially prominent stratigraphic cycles in Permo-Carboniferous foreland settings are a function of synchronous Gondwanan glaciations, rather than tectonic setting, predicts a lack of comparable stratigraphic cyclicity within foreland sequences of other ages, to which we now turn briefly.

ANTLER-SEVIER FORELAND SEQUENCES

Thrust belts with vergence toward the interior of the continent developed in association with both the Antler and Sevier orogenic belts (Fig. 1) along the Cordilleran margin. The Antler event of Devono-Mississippian age was most probably an arc-continent collision (Dickinson and others, 1983), which thrust the eugeoclinal Roberts Mountains allochthon across the Cordilleran miogeoclinal prism to produce a peripheral foreland basin best developed in eastern Nevada and western Utah. Clastic sediments of the Antler foreland basin are separated from underlying miogeoclinal shelf carbonates by a transitional interval including limestone and shale (Fig. 7). The Sevier event, of more prolonged and migratory backarc thrusting (Dickinson, 1976) during Cretaceous time, emplaced thrust sheets as far east as central Utah to produce a retroarc foreland basin that extended across Utah into Colorado. Clastic sediments of the Sevier foreland basin overlie terrestrial redbed successions of Permian and Mesozoic age now exposed over much of the Colorado Plateau.

If foreland stratigraphic cyclicity reflects episodic or pulsating tectonism related to stresses associated with emplacement of thrust sheets, stratigraphic sequences in both the Antler and Sevier forelands should display well developed cyclothems. We have assembled Figures 8 and 9 to show that this prediction is not borne out by field observations. The two figures compare key sequences from the Devono-Mississippian Antler (Fig. 8) and the Cretaceous Sevier (Fig. 9) forelands with the Missourian through mid-Virgilian Pennsylvanian succession of the Pedregosa basin. For comparison, we selected Antler and Sevier successions of approximately the same thickness as the Pedregosa succession, and deposited within comparable spans of time.

Much of the Antler foreland basin fill is composed of turbidites for which a clear glacio-eustatic record would not be expected in any case, and overlying deltaic deposits lack fossils adequate to establish time constraints closely. Consequently, the Antler succession of Figure 8 is composed mostly of transitional deposits (Fig. 7) represented by the Pilot Shale and Joana Limestone. These units record the passage, from west to east, of a migratory forebulge and accompanying back-bulge basin, which was located between the migrating forebulge and the undisturbed craton (Goebel, 1991). Conodont biostratigraphy documents the diachroneity of successive sedimentary environments related to changing tectonic setting. The structural crest of the forebulge stood high enough to nucleate the development of the Joana carbonate platform in shallow marine waters. As the position of the forebulge through time was controlled by lithospheric flexure in response to the advancing thrust front of the Roberts Mountains allochthon, any high-frequency fluctuations in lithospheric stresses associated with thrust emplacement would have caused changes in the amplitude and bathymetry of the forebulge that should have been adequate to induce cyclothemic sedimentation. The principal cyclicity observed in the Antler foreland (Fig. 8) is of a distinctly different character, however, from the pervasive cyclicity characteristic of the Pedregosa basin, and can be attributed to discrete tectono-eustatic events of a non-repetitive nature (Goebel, 1991). The high-frequency cycles present within the Joana Limestone (Fig. 8) involve alternation of oolitic to peloidal grainstones with intertidal to supratidal laminites. They reflect modest variations in water depth, from wave-washed shoals to peritidal lagoonal environments, within the carbonate platform that developed on the flexural forebulge. Contacts between these Joana cycles become gradational as the cycles thicken basinward.

In the Sevier foreland basin, fluvio-deltaic clastics prograded eastward into shallow marine waters (Fouch and others, 1983). Alternations of shelf, prodelta, shoreface, and delta plain deposits are characteristic. The resulting distribution of interbedded marine, marginal marine, and terrestrial facies was controlled by variations in local sediment supply in a regime of steady or incremental rise in relative sea level resulting from persistent foreland subsidence (Swift and others, 1987). In a schematic form, Figure 9 shows a measured section of the portion of the Sevier foreland-basin fill displaying the most pronounced cycli-

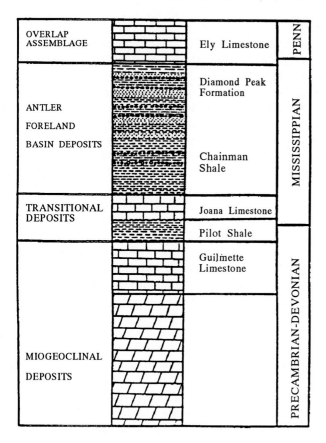

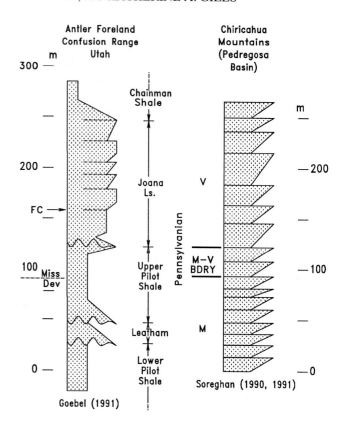

FIG. 7.—Upper Precambrian through lower Pennsylvanian stratigraphy of Antler foreland region in eastern Nevada and western Utah. Miogeoclinal deposits represent shelf sedimentation along a passive continental margin prior to the Antler orogenic event of latest Devonian and Early Mississippian time. Antler foreland basin deposits reflect the evolution of the peripheral foreland basin that developed adjacent to the Antler thrust front, along which the Roberts Mountains allochthon was emplaced above lateral equivalents of the subjacent miogeoclinal deposits. Transitional deposits (Fig. 8) record the development of the structural forebulge and coupled back-bulge basin (located between forebulge and craton) that migrated across the foreland region just prior to initial subsidence of the foreland basin. Pilot Shale records subsidence of the back-bulge basin, and Joana Limestone reflects subsequent deposition across the crest of the migratory forebulge.

FIG. 8.—Comparison of typical stratal succession of transitional deposits (Fig. 7) in the Devono-Mississippian Antler foreland region (left) to well developed stratigraphic cyclicity (Fig. 6) in Permo-Carboniferous strata of the Pedregosa basin (right) in the Ouachita-Marathon foreland region. Columns show marine sequences of comparable thickness (scale in meters) deposited in comparable time spans (6-8 my for Pedregosa basin, 10-12 my for Antler foreland). Widths of columns schematically depict decrease in water depth from left to right (wavy lines mark scour surfaces of subaerial unconformities). Leatham Member is part of Pilot Shale; "Miss/Dev" horizon marks gradation from Devonian to Mississippian strata; "FC" marks horizon within Joana Limestone inferred to record passage of migratory forebulge crest past the depositional site; dashed lines within Joana Limestone denote contacts between shoalwater limestone cycles inferred to reflect autogenic migration of depositional environments within a carbonate bank system (see text for discussion). See Figure 1 for locations of Confusion Range (Antler foreland) and Chiricahua Mountains (Pedregosa basin).

city where it is superbly exposed in the deep canyon of the Green River in central Utah (Lawton, 1983). In our view, most of the cyclicity present can be ascribed to a variety of autogenic processes intrinsic to the progradation and internal evolution of delta lobes and strandplains. Most cycles depicted coarsen upward from prodelta or lower-shoreface deposits to upper-shoreface or mouthbar deposits of delta fronts. The style of observed cyclicity does not contrast as strongly with that of the Pedregosa basin as is the case for the forebulge section from the Antler foreland, but still differs in important respects. Not only is the cycle frequency lower and less uniform, but the diverse shoaling-upward cycles present reflect a variety of lowstand,

transgressive, and highstand systems tracts (Van Wagoner and others, 1990), and are not related in any simple fashion to fluctuations in relative sea level.

SUMMARY AND CONCLUSIONS

Our analysis of Permo-Carboniferous cyclic sedimentation in the southern Cordilleran region supports the Wanless-Shepard model for glacio-eustatic control of cyclothems. Basin-wide correlation (intrabasinal) of individual shoaling-upward cycles,

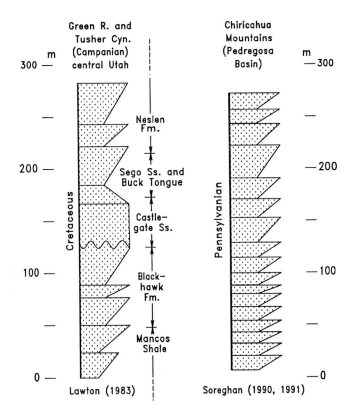

FIG. 9.—Comparison of style of stratigraphic cyclicity in Upper Cretaceous (Campanian) strata of the Sevier retroarc foreland basin (Dickinson, 1976) near the Green River in central Utah (left) to Permo-Carboniferous cyclothems of the Pedregosa basin (right) in the Ouachita-Marathon foreland. Columns show strata of comparable thickness (scale in meters) deposited in comparable time spans (6-8 my). Widths of columns schematically depict decrease in water depth (and/or increased emergence of coastal plain) from left to right (wavy line marks widespread subaerial disconformity). Sequence stratigraphic analysis interprets most of Blackhawk Formation as highstand systems tract, uppermost Blackhawk Formation and Castlegate Sandstone as lowstand systems tract, and the overlying units as multiple systems tracts (Van Wagoner and others, 1990). See text for discussion of cyclicity. See Figure 1 for locations of Green River/Tusher Canyons (Sevier foreland) and Chiricahua Mountains (Pedregosa basin).

coupled with provisional regional correlation (interbasinal) of packets of analogous cycles, is difficult to reconcile with other suggested controls. Similar stratigraphic cyclicity in both foreland and non-foreland settings of Permo-Carboniferous age, coupled with lack of comparable cyclicity in foreland successions of other ages, argues against special tectonic mechanisms as the cause of Permo-Carboniferous foreland cyclothems.

Our work also suggests that glacio-eustatic Pennsylvanian (Missourian-Virgilian) cycles, modulated by the long eccentricity cycle of planet Earth, may extend through multiple sedimentary basins from the midcontinent region to the southern and western margins of the continent. This tentative conclusion raises the possibility that Permo-Carboniferous cyclothems can potentially be used for worldwide correlation at the time scale of orbital variations, which range downward from 413,000 yr.

Success with this approach would yield time precision of the order of a tenth of a per cent of total age for upper Paleozoic rocks. In view of the fact that no isotopic or paleontologic methodology is ever apt to achieve such precision, every effort should be made to test the proposition fully.

ACKNOWLEDGMENTS

We thank J. M. Dennison and F. R. Ettensohn for their encouragement to participate in the symposium on sedimentary cycle control at the 1991 joint meeting of the Northeastern and Southeastern Sections of the Geological Society of America in Baltimore where the oral version of this paper was presented. We also thank T. J. Algeo, R. A. Armin, W. M. Connolly, P. H. Heckel, G. Dev. Klein, R. A. Knepp, R. L. Langenheim, Jr., M. R. Leeder, G. A. Smith, and J. L. Wilson for helpful preprints, discussions, and correspondence, and C. A. Ross and G. L. Wilde for identifications of fusulinids that we collected in the Pedregosa and Orogrande basins of southern Arizona and New Mexico. We are grateful to S. Y. Johnson and G. Dev. Klein for reviews that improved the manuscript. Acknowledgment is made to the donors of The Petroleum Research Fund, administered by the American Chemical Society, for partial support of this research.

REFERENCES

ALGEO, T. J., AND WILSON, J. L., 1991, Lower-Middle Pennsylvanian Gobbler Formation: Eustatic and tectonic controls on carbonate shelf cyclicity (abs.): Geological Society of America Abstracts with Programs, v. 23, p. 2.

ARMIN, R. A., 1987, Sedimentology and tectonic significance of Wolfcampian (Lower Permian) conglomerates in the Pedregosa basin: Southeastern Arizona, southwestern New Mexico, and northern Mexico: Geological Society of America Bulletin, v. 99, p. 42-65.

ARMIN, R. A., 1991, Pedregosa basin in southeastern Arizona and northern Mexico: Late Paleozoic tectonic and stratigraphic evolution (abs.): Geological Society of America Abstracts with Programs, v. 23, p. 3.

BEAUMONT, C., 1981, Foreland basins: Geophysical Journal Royal Astronomical Society, v. 55, p. 471-498.

BERGER, A., 1977, Support for the astronomical theory of climatic change: Nature, v. 268, p. 44-45.

BOARDMAN, D. R., II, AND HECKEL, P. H., 1989, Glacial-eustatic sea-level curve for early Late Pennsylvanian sequence in north-central Texas and biostratigraphic correlation with curve for midcontinent North America: Geology, v. 17, p. 802-805.

CECIL, C. B., 1990, Paleoclimate controls on stratigraphic repetition of chemical and siliciclastic rocks: Geology, v. 18, p. 533-536.

CHESNUT, D. R., JR., 1991, Eustatic and tectonic control of sedimentation in the Pennsylvanian strata of the central Appalachian basin (abs.): Geological Society of America Abstracts with Programs, v. 23, p. 16.

CLOETINGH, S., 1988, Intraplate stresses: A new element in basin analysis, in Kleinspehn, K., and Paola, C., eds., New Perspectives in Basin Analysis: New York, Springer-Verlag, p. 205-230.

CLOETINGH, S., MCQUEEN, H., AND LAMBECK, K., 1985, On a tectonic mechanism for regional sea level variations: Earth and Planetary Science Letters, v. 75, p. 157-166.

COLLIER, R. E. L., LEEDER, M. R., AND MAYNARD, J. R., 1990, Transgressions and regressions: A model for the influence of tectonic subsidence, deposition and eustasy, with application to Quaternary and Carboniferous examples: Geological Magazine, v. 127, p. 117-128.

CONNOLLY, W. M., 1991, Lithostratigraphic correlation of upper Desmoinesian and lower Missourian "cyclothem" scale cycles, SE Arizona with SE Kansas (abs.): Geological Society of America Abstracts with Programs, v. 23, p. 13.

CROWELL, J. C., 1978, Gondwanan glaciation, cyclothems, continental positioning, and climate change: American Journal of Science, v. 278, p. 1345-1372.

DICKINSON, W. R., 1976, Sedimentary basins developed during evolution of Mesozoic-Cenozoic arc-trench system in western North America: Canadian Journal of Earth Science, v. 13, p. 1268-1287.

DICKINSON, W. R., HARBAUGH, D. W., SALLER, A. H., HELLER, P. L., AND SNYDER, W. S., 1983, Detrital modes of upper Paleozoic sandstones derived from Antler orogen in Nevada: Implications for nature of Antler orogeny: American Journal of Science, v. 283, p. 481-509.

FOUCH, T. D., LAWTON, T. F., NICHOLS, D. J., CASHION, W. B., AND COBBAN, W. A., 1983, Patterns and timing of synorogenic sedimentation in Upper Cretaceous rocks of central and northeast Utah, in Reynolds, M. W., and Dolly, E. D., eds., Mesozoic Paleogeography of the West-Central United States: Denver, Society of Economic Paleontologists and Mineralogists, Rocky Mountain Section, Rocky Mountain Paleogeography Symposium 2, p. 305-336.

GALLOWAY, W. E., 1989, Genetic stratigraphic sequences in basin analysis I: Architecture and genesis of flooding-surface bounded depositional units: American Association of Petroleum Geologists Bulletin, v. 73, p. 125-142.

GOEBEL, K. A., 1991, Paleogeographic setting of Late Devonian to Early Mississippian transition from passive to collisional margin, Antler foreland, eastern Nevada and western Utah, in Cooper, J. D., and Stevens, C. H., eds., Paleozoic Paleogeography of the Western United States—II: Los Angeles, Society of Economic Paleontologists and Mineralogists, Pacific Section, Book 67, p. 401-418.

HARLAND, W. B., ARMSTRONG, R. L., COX, A. V., CRAIG, L. E., SMITH, A. G., AND SMITH, D. G., 1990, A Geologic Time Scale 1989: London, Cambridge University Press, 263 p.

HARLAND, W. B., COX, A. V., LLEWELLYN, P. G., PICKTON, C. A. G., SMITH, A. G., AND WALTERS, R., 1982, A Geologic Time Scale: London, Cambridge University Press, 131 p.

HECKEL, P. H., 1986, Sea-level curve for Pennsylvanian eustatic marine transgressive-regressive depositional cycles along midcontinent outcrop belt, North America: Geology, v. 14, p. 330-334.

HECKEL, P. H., 1990, Evidence for global (glacial-eustatic) control over upper Carboniferous (Pennsylvanian) cyclothems in midcontinent North America, in Hardman, R. F. P., and Brooks, J., eds., Tectonic Events Responsible for Britain's Oil and Gas Reserves: London, Geological Society of London Special Publication 55, p. 35-47.

IMBRIE, J., and IMBRIE, J. Z., 1980, Modeling the climatic response to orbital variations: Science, v. 207, p. 943-953.

JORDAN, T. E., 1981, Thrust loads and foreland basin evolution, Cretaceous, western United States: American Association of Petroleum Geologists Bulletin, v. 65, p. 2506-2520.

KARNER, G. O., 1986, Effects of lithospheric in-plane stress on sedimentary basin stratigraphy: Tectonics, v. 5, p. 573-588.

KLEIN, G. Dev., 1990, Pennsylvanian time scales and cycle periods: Geology, v. 18, p. 455-457.

KLEIN, G. Dev., AND WILLARD, D. A., 1989, Origin of the Pennsylvanian coal-bearing cyclothems of North America: Geology, v. 17, p. 152-155.

KLUTH, C. F., AND CONEY, P. J., 1981, Plate tectonics of the Ancestral Rocky Mountains: Geology, v. 9, p. 10-15.

LANGENHEIM, R. L., JR., 1991, Comparison of Pennsylvanian carbonate cycles in the Cordilleran region with midcontinental cyclothems suggests a common eustatic origin (abs.): Geological Society of America Abstracts with Programs, v. 23, p. 56.

LAWTON, T. F., 1983, Tectonic and sedimentologic evolution of the Utah foreland basin: Unpublished Ph.D. Dissertation, University of Arizona, Tucson, 217 p.

LOOPE, D. B., SANDERSON, G. A., AND VERVILLE, G. J., 1990, Abandonment of the name "Elephant Canyon Formation" in southeastern Utah: Physical and temporal implications: Mountain Geologist, v. 27, p. 119-130.

MAYNARD, J., AND LEEDER, M., 1992, On the periodicity and magnitude of Upper Carboniferous glacio-eustatic sea level changes: Geological Society of London Journal, v. 149, p. 303-311.

MOORE, D. G., 1959, Role of deltas in the formation of some British Lower Carboniferous cyclothems: Journal of Geology, v. 67, p. 522-539.

ROSS, C. A., AND ROSS, J. R. P., 1987, Late Paleozoic sea levels and depositional sequences, in Ross, C. A., and Haman, D., eds., Timing and Depositional History of Eustatic Sequences: Constraints on Seismic Stratigraphy: Houston, Cushman Foundation for Foraminiferal Research Special Publication 24, p 137-149.

SOREGHAN, G. S., 1990, Stratigraphic delineation of eustasy and tectonism: Preliminary observations from the Pennsylvanian of the Pedregosa basin (abs.): Geological Society of America Abstracts with Programs, v. 22, p. A283.

SOREGHAN, G. S., 1991, Late Paleozoic carbonate and mixed carbonate/clastic cycles of the Pedregosa (Arizona) and Orogrande (New Mexico) basins (abs.): Geological Society of America Abstracts with Programs, v. 23, p. 96.

SWIFT, D. J. P., HUDELSON, P. M., BRENNER, R. L., AND THOMPSON, P., 1987, Shelf construction in a foreland basin: Storm beds, shelf sandbodies, and shelf-slope depositional sequences in the Upper Cretaceous Mesaverde Group, Book Cliffs, Utah: Sedimentology, v. 34, p. 423-457.

VAN WAGONER, J. C., MITCHUM, R. M., CAMPION, K. M., AND RAHMANIAN, V. D., 1990, Siliciclastic Sequence Stratigraphy in Well Logs, Cores, and Outcrops: Concepts for High-Resolution Correlation of Time and Facies: Tulsa, American Association of Petroleum Geologists Methods in Exploration Series 7, 55 p.

VEEVERS, J. J., AND POWELL, C. McA., 1987, Late Paleozoic glacial episodes in Gondwanaland reflected in transgressive-regressive depositional sequences in Euramerica: Geological Society of America Bulletin, v. 98, p. 475-487.

WANLESS, H. R., AND SHEPARD, F. P., 1936, Sea level and climatic changes related to late Paleozoic cycles: Geological Society of America Bulletin, v. 47, p. 1177-1206.

DEPTH DETERMINATION AND QUANTITATIVE DISTINCTION OF THE INFLUENCE OF TECTONIC SUBSIDENCE AND CLIMATE ON CHANGING SEA LEVEL DURING DEPOSITION OF MIDCONTINENT PENNSYLVANIAN CYCLOTHEMS

GEORGE D. KLEIN

New Jersey Marine Sciences Consortium, Building #22, Fort Hancock, NJ 07732

ABSTRACT: New sedimentological determinations of the water depth and associated sea-level change of midcontinent Pennsylvanian cyclothems shows that they accumulated in water depths ranging from as low as 32 m to as high as 160 m, depending on which model is used to establish the deepest water facies. These depth determinations also indicate that, regardless of model, depth variations existed for different cyclothems both laterally and in time. Average water-depth determinations and sea-level change for models of Heckel (1977) and Gerhard (1991) are 96.4 m and 86.0 m respectively.

Analysis of tectonic subsidence permits calculation of the magnitude of tectonic processes and associated climatic effects, which controlled changes in sea level during deposition of Pennsylvanian cyclothems. Far-field tectonic effects, in response to regional orogenic movements, partially influenced Pennsylvanian sea-level change in the midcontinent. Organization of Virgilian and Missourian midcontinent cyclothems into four- to fivefold bundles shows that sea-level changes in midcontinent platform areas were influenced both by Milankovitch orbital parameters and longer-term climate change, whereas Desmoinesian sea-level change apparently was influenced more strongly by tectonic subsidence controlled by foreland-basin tectonism.

The magnitude of tectonically-contributed change in sea level varied laterally. In the midcontinent, tectonic subsidence accounts for approximately 5 to 20% of the total sea-level change in platform areas, and perhaps as much as 20% in basin depocenters. The remaining change in sea level is controlled by both short-term glacial eustasy (Milankovitch orbital forcing; approximately 70% of sea-level change) and long-term climate change (approximately 15% of sea-level change). These findings suggest that, away from orogenic belts, climatic change is the principal driving mechanism controlling sea level change, whereas within orogenic belts, climate becomes somewhat more subordinate as a driving mechanism for Pennsylvanian sea-level change, even though indicators of climatic change itself are preserved.

Methods discussed herein permit calculation of magnitudes of both tectonic and climatic-eustatic components of sea-level change influencing Pennsylvanian cyclothem deposition, and may be applicable to other cyclic sequences.

INTRODUCTION

Controversy exists regarding both the depth of water into which midcontinent Pennsylvanian cyclothems were deposited (see Elias, 1937, 1964; McCrone, 1964; Moore, 1964; Gerhard, 1991), and whether these Pennsylvanian cyclothems are of either tectonic or glacio-eustatic origin (see Weller, 1964; Heckel, 1984; and Leeder, 1988 for reviews). Klein and Willard (1989), Klein (1990a, 1992, 1993), and Read and Forsyth (1989) suggested that *concurrent* tectonic and glacio-eustatic processes controlled the origin and development of Pennsylvanian cyclothems. A *concurrent* lateral decrease of compressional deformation associated with orogenic belts adjacent to cratonic basins and platforms, and glacio-eustatic sea-level change controlled facies changes from the nonmarine clastic-dominated cyclothems of the Appalachians to the marine, carbonate-rich, glacio-eustatic dominated cyclothems of the midcontinent. Heckel (1986, 1991) championed a model that midcontinent cyclothems were formed by glacio-eustatic sea-level fluctuations whose periodicity ranged through the entire span of Milankovitch orbital parameters, implying that the tectonic component was minor to nonexistent.

Because both tectonic and glacio-eustatic processes appear to form Pennsylvanian cyclothems (fifth-order stratigraphic units of Busch and Rollins, 1984; Miall, 1990), the crux of the problem shifts to determining the magnitude of each of these processes which control changes in sea level during deposition.

In this paper, I report new estimates using a method to make such quantitative determinations that was described by Klein and Kupperman (1992). These estimates show that tectonic processes were one of several concurrent major controls on midcontinent Pennsylvanian cyclothem deposition. These tectonic processes involved lateral transmission of compressional forces from the Alleghenian-Hercynian orogenic belt of the Appalachian and Ouachita region of North America. Cyclic fluctuations in sea level influencing cyclothem deposition were caused by a strong far-field tectonic component (Kluth and Coney, 1981; Watney, 1985) of successive collisional events which were *concurrent* with climatically-induced eustatic changes (Heckel, 1986). These far-field tectonic events are known to influence large areas of continental plates which migrated over subduction zones (Grand, 1987).

Equally crucial to these arguments is which water depth and associated magnitude of sea-level change is represented by a Pennsylvanian cyclothem. In the midcontinent of North America, different approaches have generated different estimates of water depth and associated magnitude of sea level changes. Paleoecological approaches suggested maximum water depths and associated changes in sea level ranging from 20 m (McCrone, 1964), 55 m (Elias, 1937, 1964), and 67 m (Moore, 1964). Lithofacies interpretation of black shales suggested water depths of 100 m (Heckel, 1977, 1986, 1991) where a 100-m constant elevation of the pycnocline was assumed.

The objective herein is twofold: (1) to propose a new sedimentological method to estimate the amount of water depth change and associated magnitude of sea level change for

Tectonic and Eustatic Controls on Sedimentary Cycles, SEPM Concepts in Sedimentology and Paleontology #4
Copyright © 1994 SEPM (Society for Sedimentary Geology), ISBN 1-56576-017-4, p. 35-50.

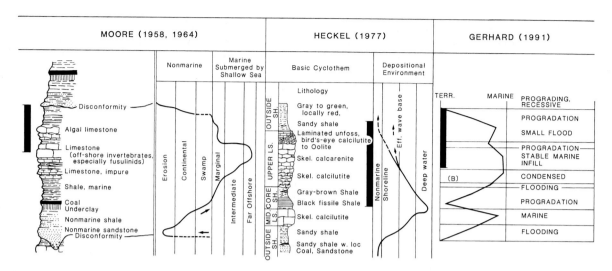

FIG. 1.—Comparison of sea-level curves for Pennsylvanian cyclothems as per Moore (1958, 1964), Heckle (1977), and Gerhard (1991). B = black shale. Vertical black bar shows interval used to sedimentologically determine magnitude of water depth and associated sea-level change in Table 1 (redrawn and modified from Gerhard, 1991).

midcontinent Pennsylvanian cyclothems, and (2) to report new calculations about the magnitude of global change in sea level produced by tectonic subsidence that forms midcontinent Pennsylvanian coal-bearing cyclothems. This approach provides a quantitative method by which to discriminate tectonically-controlled and climatically-induced eustatic influences on sea-level change during deposition of cyclic sediments.

SEDIMENTOLOGIC DETERMINATION OF WATER DEPTH

Prior estimates of water-depth changes were based on paleo-ecological (Elias, 1937, 1964; Moore, 1964; McCrone, 1964) and lithostratigraphic (Heckel, 1977, 1964, 1991) data. The paleoecological estimates were based on estimating water depths inhabited by benthonic *fusulinidae*. These water-depth estimates ranged from 20 m (McCrone, 1964), and 55 m (Elias, 1937, 1964) to 67 m (Moore, 1964, p. 367). These "deep-water" fusulinids occurred usually in open-marine limestones, and prompted Moore (1958, 1964) to suggest a sea-level curve for Pennsylvanian cyclothems (Fig. 1). Schenck (1967) and Ross (1968, 1969), however, demonstrated that the fusulinids were deposited in much shallower water, having been transported from tectonic highs. Thus, Moore's estimate of water depth was not used. Heckel's (1977, 1991) 100-m estimate of water depth and sea-level change was based on the occurrence of black shales formed under anoxic conditions below the pycnocline, which generally averages 100 m below sea level (see Fig. 1). This estimate, however, may not apply to Pennsylvanian cyclothems because of pycnocline disruption by freshwater influxes from associated nearshore deltas, such as in the Devonian sedimentary rocks of the eastern interior platform (Ettensohn and Elam, 1985; Ettensohn and others, 1988). The effect of such muddy, nutrient-rich, freshwater, turbid flows is both to limit carbonate production rate and deposition, and to result in the

shallowing of the euphotic zone (Hallock and Schlager, 1986). Thus, estimates of paleopycnocline elevations and paleodepths of the euphotic zone may be suspect. Gerhard (1991) proposed a different sea-level curve (Fig. 1) based on Holocene carbonate accumulation and production rate data and his deepest water facies appear to range from the middle of Heckel's (1977) black-shale interval to Moore's (1958, 1964) fusulinid baseline.

A sedimentological method is suggested which may resolve this problem. A prior method based on clastic barrier-island and deltaic vertical sequences (Klein, 1974) used the uncompacted thickness of such sequences to approximate water depths. A similar approach can be used by measuring the thickness from the top of the regressive facies down to the maximum transgressive elevation of a cyclothem (Klein, 1992, 1993). In Figure 1, this thickness would range from the disconformity at the top of the cyclothem to the fusulinid open-marine limestone (Moore, 1958, 1964), from the paleosol on top of the regressive limestone down to the maximum flooding of the black "core" shale (Heckel, 1977), or from the paleosol down to the lower part of the limestone representing the stable marine infill phase (Gerhard, 1991). Because the entire cyclothem records a change of sea level from nonmarine to high-stand back to nonmarine sedimentation, I assume that the marine water-depth estimate and change in sea level are all the same magnitude. This thickness of the regressive marine interval can be corrected for decompaction using empirical observations of original porosity for Holocene marine carbonates and muds. For this study, a 50% original porosity was used for limestones and an 80% original porosity for mudstones based on Deep Sea Drilling Project (DSDP) observations (e.g., Klein, and others, 1980, among others).

Sediment-accumulation rate is a second correction required to estimate water depth and associated sea-level change sedimentologically. Data concerning sediment-accumulation rates for Holocene or Mesozoic/Cenozoic counterparts of black shale, gray silty shale and limestone observed in Pennsylvanian cyclothems are sparse and variable. For limestones, Simo's (1989, his Table 1) compilation of platform carbonates was used. His accumulation rates range from 30 to 170 m/my, averaging 100 m/my. This average rate was adopted for Pennsylvanian limestones. For the gray shale, sediment accumulation rate determinations of distal portions of the delta-fed Yellow Sea by Alexander and others (1991) and Lee and Chough (1989) were used; they average 30 m/my. The Yellow Sea was selected as an appropriate counterpart because it is one of few places where shallow shelf sediment-accumulation rates are known (Alexander and others, 1991; Lee and Chough, 1989) and it is a most probable sedimentological counterpart to cratonic platforms of the past (Klein, 1982; Klein and others, 1982). Two major river deltas prograde into it, making it a relevant counterpart to Pennsylvanian conditions.

Black-shale sediment-accumulation rates are more problematic. A compilation by Ibach (1982) of black-shale sediment-accumulation rates from the first 43 legs of DSDP suggested a range from 5 to 230 m/my, averaging 35.8 m/my. These rates were challenged by Dean and others (1984a, b) and Stow and Dean (1984) as too high, based on drilling at DSDP Site 530 in the Angola basin. There, they calculated black-shale accumulation rates ranging from 3 to 14.6 m/my, which average to 8.45 m/my. This average rate was adopted herein particularly because Coveney and others (1991) suggested from geochemical analyses that Virgilian and Missourian black shales of the midcontinent were relatively deeper water counterparts to anoxic oceanic black shales.

To correct for sediment-accumulation rate, the average sediment accumulation rates for Angola basin black shales (Dean and others, 1984a, b), the Yellow Sea gray muds (Alexander and others, 1991; Lee and Chough, 1989), and global Mesozoic/Cenozoic carbonates Simo (1989) were ratioed with the black shale rate set at unity because it represents the slowest rate of sediment accumulation yielding a condensed section. This step provides a ratio of black shale: gray shale: carbonate of 1: 3.6: 11.8.

To calculate water depth (and associated sea-level change), the following formula was used:

$$\text{Water Depth} = [(T_N + T_N\Phi_o) SR_L]_L +$$
$$[(T_N + T_N\Phi_o) SR_G]_G +$$
$$[(T_N + T_N\Phi_o) SR_B]_B \quad (1)$$

where T_N is the present thickness of the limestone, gray shale, or black-shale interval in the regressive part of the cyclothem; Φ_o is original porosity; SR is the ratio value of the sediment-accumulation rate discussed previously; L is the limestone interval; G is the gray shale interval; and B the black-shale interval. Water depth and associated sea-level change estimates were calculated using the sea-level curves of Moore (1958, 1964), Heckel (1977), and Gerhard (1991) at ten outcrops of midcontinent Pennsylvanian cyclothems showing complete regressive sequences described by Heckel and others (1979) and Watney and others (1985).

Results

To estimate water depths, I used ten published outcrop stratigraphic sections for which a relative sea-level curve was provided (Heckel and others, 1979; Watney and others, 1985). Table 1 summarizes the results of these water-depth estimates and associated change in sea level and provides stratigraphic and locality information.

Data in Table 1 show considerable depth variation with respect to which model was used, as well as lateral and vertical variability for individual cyclothems. With respect to the Swope Cyclothem (Missourian), the two outcrops used for this analysis are about 2 km apart, yet they show a sharp change in water depth (Table 1). This increase in water depth is explained by local control of depositional topography (Watney, 1985; Watney and others, 1988) that was partly caused by tectonic subsidence and partly by deposition over a large carbonate bank with 60 m of subsurface relief. Similar lateral variation in the Stanton Cyclothem was caused, probably, by similar depositional topography.

The estimated water depths (Table 1) of the younger Wyandotte, Iola, and Dennis Cyclothems appear anomalously high compared to postulated water depths of Elias (1937, 1964), McCrone (1964), Moore (1964) and Heckel (1977, 1986, 1991). The increased water depth for the Dennis Cyclothem was attributed to increased tectonic subsidence (Watney, 1985; Watney and French, 1988). The Wyandotte and Iola Cyclothems probably were influenced by reactivation tectonics along the Nemaha and other fault systems (cf. Baars, in press; Carlson, 1989a, b). An unusually thick (11.5 m) upper gray ("outside") shale in the Iola Cyclothem was caused by increased deltaic influx (Heckel and others, 1979) which may have accompanied increased subsidence associated with regional tectonic reactivation. The accompanying sediment load of such a deltaic influx would cause only a 1.2-m decrease of water depth if backstripping were considered (Steckler and Watts, 1978) and is not a factor of concern.

The estimates of water depths for the ten cyclothems were averaged according to the models of Heckel (1977) and Gerhard (1991) and are 96.4 and 86.0 m, respectively (Table 1). These average values are surprisingly close with respect to the water-depth models of Heckel (1977). These average water-depth calculations are slightly less than 4% for Heckel's (1977) 100-m lithofacies approximation. Gerhard (1991) did not provide a depth estimate for his sea-level curve, however.

These findings are of significance in terms of evaluating the role of tectonics and glacially-controlled eustasy. Recent paleoclimatological modelling by Crowley and Baum (1991) suggested that Pennsylvanian glacio-eustatic change in sea level was approximately 60.0 m +/-15.0 m, providing a maximum/minimum range from 45.0 to 75.0 m. Their calculated upper values of glacially-controlled sea-level change approximates the average sea level change of 75.5 m (Table 1) using Moore's model, but is less than the sea-level change suggested by Heckel (1977) and the sedimentological determination shown in Table

GEORGE D. KLEIN

TABLE 1.—ESTIMATED WATER DEPTHS (M) AND ASSOCIATED SEA-LEVEL CHANGE, MIDCONTINENT PENNSYLVANIAN CYCLOTHEMS.

Locality	As per Moore (1958, 1964)	As per Heckel (1977)	As per Gerhard (1991)
Deer Creek Cyclothem (V) Hopper Bros. Quarry, Louisville, NB[1]	51.9	85.4	60.2
Lecompton Cyclothem (V) Kansas TPKE, Lecompton KS[1]	31.7	56.8	40.4
Oread Cyclothem (V) Kansas TPKE (Mile #199) West Lawrence, KS[1]	61.6	97.1	70.3
Stanton Cyclothem (M) Derby Stone Products Quarry Louisville, NB[1]	70.8	108.4	82.6
Stanton Cyclothem (M) Wilson CO State Lake, Chanute, KS[1]	80.4	98.3	83.3
Wyandotte Cyclothem (M) Holliday Rd, Kansas City, KS[2]	102.1	140.3	115.0
Iola Cyclothem (M) Holliday Rd, Kansas City, KS[2]	135.8	160.3	141.1
Dennis Cyclothem (M) Richfield west Quarry, Louisville, NB[1]	120.5	141.1	125.5
Swope Cyclothem (M) U.S. HWY 69, LaCygne, KS[1]	49.2	66.4	54.8
Swope Cyclothem (M) Roadcut, US-69, Jingo, KS[1]	69.6	104.0	87.3
MEAN WATER DEPTH ALL CYCLOTHEMS	75.5	96.4	86.0

[1] Localities from Heckel and others (1979; their p. 24, 28, 44, 46, 53, 54, 57)
[2] Localities from Watney and others (1985; their p. 28, 47)
(V)—Virgilian; (M)—Missourian

1 using the Heckel (1977) model. These calculations also suggest (Crowley and Baum, 1991) that, probably, the changing ice budgets during climate change of Gondwana glaciers were less than changing ice budgets of Pleistocene glaciers (cf. Fairbanks, 1989). Clearly, these findings suggest other factors contributed to Pennsylvanian changes of sea level during cyclothem deposition and both long-term climate change (Cecil, 1990) and *concurrent* tectonic influences need to be considered.

TECTONIC SUBSIDENCE AND CYCLOTHEMS

Short-Term Tectonic Change, Climate Change, and Sedimentary Cycles

Tectonic changes.—

A prevailing assumption in stratigraphic studies is that tectonic changes occur over time scales longer than estimated periodicities of cycle deposition (Heckel, 1986; Posamentier and others, 1988, among many others). This assumption ignores rapid, short-term tectonic changes in baselevel, uplift rates, fault displacements, and associated rates of sea-level change (Plafker

and Savage, 1970). These changes may be widespread. For instance, Plafker and Savage (1970) reported that, following the Chilean earthquake of May 21 and 22, 1960, an area of 200,000 km[2] subsided from 1 to 2 m along a coastal area, producing a comparable magnitude of relative sea-level change. There, historical records show a recurrence of four such major earthquake and associated subsidence and uplift events over a 400 yr period. An extrapolated rate of sea-level change would range from 1 to 2 cm/yr over this large coastal area.

Sieh (1984) and Sieh and Jahns (1984) reported that along the San Andreas fault, the recurrence interval of faulting is 145 yr, with associated sediment accumulation averaging 0.75 m per major faulting event in subsided, ponded areas. This sediment accumulation would yield an average, punctuated rate of sea-level change of 5 mm/yr in areas of coastal deposition, such as those associated with Pennsylvanian cyclothems.

Faulting in the New Madrid fault province of interior North America shows both a recurrence interval of 600 yr and subsidence of 2.0 m to 5.0 m with each faulting event (Russ, 1982). Such subsidence would generate a change in sea level of 3 to 8 mm/yr, if occurring in a cratonic coastal setting. Comparable Pennsylvanian faulting was widespread in North America (Kluth and Coney, 1981; Watney, 1985; Carlson, 1989a) and influenced sedimentation (Greb, 1989).

These extrapolated rates of neotectonic-induced rates of sea-level change from Chile, California and the New Madrid fault zone are comparable or greater than the 5 mm/yr rate of sea-level change proposed for Pennsylvanian cyclothems (Heckel, 1977, 1991).

Climate changes.—
Concurrent influences of long-term (Cecil, 1990) and short-term climatic change may control deposition of cyclothems also. According to Heckel (1986, 1991), deposition of midcontinent cyclothems was controlled by Milankovitch orbital parameters driven by glacial eustasy. His interpretation was based on calculation of cyclothem periodicity constrained by a sensitivity test of 12-, 10-, and 8-my durations for the stratigraphic intervals he studied, using the Harland and others (1982) time scale. Assuming that Holocene periodicity of eccentricity, obliquity and precession remained unchanged through geological time, Heckel (1986, 1991) concluded that such parameters influenced cyclothem deposition. Heckel (1986) suggested that this range of periodicity supported an interpretation that sea-level changes associated with cyclothem deposition were driven by Milankovitch orbital parameters that were influencing glacial eustasy.

Heckel's (1986, 1991) use of a broad range of values to infer Milankovitch orbital control on cyclothem deposition poses difficulties. First, Milankovitch orbital parameters are characterized by narrowly defined intervals (Berger, 1988; Berger and others, 1987, 1989; Fischer and others, 1990; deBoer, 1991; amongst many others). Second, Berger (1988), Berger and others (1989), and Fischer and others (1990) reported that periodicity of precession and obliquity decrease with increasing geological age. Therefore, Heckel's (1986, 1991) calculated cycle periodicities are outside the narrowly defined ranges for Pennsylvanian Milankovitch orbital parameters (Klein, 1990b; 1991a, b). Third, Klein (1990b) reported difficulties with interpreting climatic processes from such period calculations. A more accurate Pennsylvanian time scale by Lippolt and others (1984) and Hess and Lippolt (1986) demonstrated that the Pennsylvanian lasted only 19 my, whereas Harland and others (1982) reported a 34-my duration for Pennsylvanian time; this duration was reduced to 32.8 my in their 1989 time scale (Harland and others, 1990). The 45-percent reduction in time in the Lippolt and others (1984) time scale changed the cycle periods determined by Heckel (1986) significantly, showing little resemblance to the narrowly defined intervals of Pennsylvanian Milankovitch orbital parameters (Klein, 1990b). Consequently using period calculations alone to interpret the presence of Milankovitch orbital parameters is fraught with risks (Klein, 1990b, 1991b; deBoer, 1991; Kozar and others, 1990; among others).

Nevertheless, other criteria of Milankovitch orbital param-

eters are suggested for the stratigraphic record. The number of bundle elements in cyclic sediments serve as a criterion for detecting preservation of the effect of Milankovitch orbital parameters in sedimentary sequences (Herbert and others, 1986; Fischer and others, 1990; deBoer, 1991). It consists of a fivefold bundles of cycle elements comparable to the deep-sea, pelagic oxygen-isotope stratigraphic record associated with changing ice budgets driven by glacial advance and retreat and appears to be controlled by precession. DeBoer (1991) observed such fivefold bundle elements in the upper part of Heckel's (1986) midcontinent Pennsylvanian sea-level curve; thus Milankovitch orbital influences may partly control Pennsylvanian cyclothem deposition. This finding is evaluated later by examining Pennsylvanian cyclothem bundling.

The magnitude of rates of sea-level change was also considered. Heckel (1977, 1991) suggested a sea-level change for midcontinent cyclothems of 5 mm/ yr, assuming a change in sea level (and associated water depth) of 100 m. However, he did not specify a duration of time over which this change in sea level occurred. Expected changes in sea level for a rise or fall of sea level of 100 m using known Milankovitch parameters for Pennsylvanian time (Berger and others, 1989) are compared (Table 2) to expected rates of sea-level change of 120 m for the Pleistocene (Fairbanks, 1989). This table shows an expected, and obvious, time dependency on rate of sea-level change.

METHODS FOR DETERMINING TECTONIC CONTRIBUTION TO
PENNSYLVANIAN SEA-LEVEL CHANGE

Data for this study were obtained from stratigraphic cross-sections and well locations reported by McKee and Crosby (1975a, b, c). Six stratigraphic sections were selected from the midcontinent (Table 3). Lithofacies designations of McKee and Crosby (1975a, c) were adopted for decompaction procedures for a tectonic subsidence analysis. Two additional cores recovered from Kansas-type cyclothems along the northernmost flank of the Forest City basin in southwest Iowa were used (see Klein and Kupperman, 1992).

Quantitative analysis of Pennsylvanian tectonic subsidence at these localities incorporated one-dimensional backstripping methods corrected for the effect of sediment loading and compaction on basin subsidence (Steckler and Watts, 1978; Sclater and Christie, 1980; Bond and Kominz, 1984). Local isostasy was incorporated into backstripping procedures because stratigraphic studies (Burchett, 1983) showed that individual Pennsylvanian formations and members persisted laterally from the depocenter of the Salina basin onto the associated flanks without significant thickness changes. Regional tectonic analysis (Baars, in press; Carlson, 1989a, b; Nelson and Lumm, 1984) demonstrated that much of the midcontinent experienced reactivation of fault movements during Pennsylvanian time which influenced both basin subsidence and major changes in thickness. Such reactivation is amplified by far-field compressional deformation in the Forest City basin (Kluth and Coney, 1981; Watney, 1985; Klein and Hsui, 1987; Willard and Klein, 1990; Baars, in

TABLE 2.—ESTIMATED RATE OF SEA-LEVEL RISE, PENNSYLVANIAN CYCLOTHEMS, WITH
RESPECT TO PENNSYLVANIAN MILANKOVITCH ORBITAL PARAMETERS.

| | \multicolumn{5}{c}{Rates of change in Sea level (mm/year) for Sea-level Rises of:} |
	100 m[1]	121 m[2]	75.5 m[3]	96.4 m[3]	86 m[3]
17.2 ky Precession	5.8	7.0	4.4	5.6	5.0
20.5 ky Precession	4.9	5.9	3.7	4.7	4.2
33.0 ky Obliquity	3.0	3.7	2.3	2.9	2.6
40.0 ky Obliquity	2.5	3.0	1.9	2.4	2.4
100 ky Eccentricity	1.0	1.2	0.8	1.0	0.9
400 ky Eccentricity	0.3	0.3	0.2	0.2	0.2

Note: Pennsylvanian Milankovitch Orbital Parameters after Berger (1988), Berger and others (1987,1989), and Fischer and others (1990); total change in sea level after (1) Heckel (1977, 1991), (2) Fairbanks (1989), (3) Table 1 (this paper).

TABLE 3.—STRATIGRAPHIC LOCATIONS IN MIDCONTINENT, NORTH AMERICA, USED IN THIS STUDY.

Locality No. or Code.	Locality description and reference
89	East Flank, Nebraska (Salina)basin, NB. Sec 26, T7N, R10W (McKee and Crosby, 1975b, p. 130)
409	Flank up-dip from Central Nebraska (Salina) basin, NB,.Sec. 7, T16N, R 13W (McKee and Crosby, 1975b, p. 133)
1,277	Depocenter, Salina basin, KS. Sec 23, T28S, R5W (McKee and Crosby, 1975b, p. 110)
370	East Flank, Central Kansas uplift (West flank, Forest City basin), KS. Sec 23, T10S, R15E (McKee and Crosby, 1975b, p. 104)
1,085	Depocenter, Forest City basin, KS. Sec 27, T7, R13E (McKee and Crosby, 1975b, p. 109)
362	Iowa platform, IA. Sec. 35, T. 72N, R.28W (McKee and Crosby, 1975b, p.103)
Riverton Core	Fremont County, IA. Flank of Forest City basin, SE, SE, SE, SW Sec 20, T67N, R41W (Iowa Geological Survey; Recovered, 1985-1986)
Core CP-37	Clarke County, IA. Flank of Forest City basin, NE, SE, NE, Sec 2, T72N, R26W (Iowa Geological Survey; Recovered, 1976).

press; Carlson, 1989a, b).

Flexural effects were assumed to be minimal because regional stratigraphic analysis in the Salina basin (central Nebraska basin of Nebraska) demonstrated that occurrence of minor unconformities vary independently with respect to geographic location across the basin (Burchett, 1983; see comment about thickness changes above). Unconformities on the forebulge predicted from flexural modelling (Jordan, 1981; Tankard, 1986; Quinlan and Beaumont, 1984) are absent on the northeast side of the Forest City basin and on the flanks of the Salina basin.

Because of uncertainties regarding the Pennsylvanian time scale (Klein, 1990b), I used both the time scales of Harland and others (1990) and Lippolt and others (1984; Hess and Lippolt, 1986; Leeder, 1988). The numerical ages used for these tectonic

subsidence analyses for both time scales are summarized in Table 4.

To determine the amount of tectonically-driven sea-level change from a tectonic subsidence curve during individual cyclothem deposition, I used a method discussed previously by Klein and Kupperman (1992) which is summarized in Fig. 2. Key to this method is definition of the parameter ΔD as the total depth of basin subsidence during a specific interval of time. The number of cycles within that interval represented by ΔD is totalled. ΔD is then divided by the number of cycles to yield an average ΔD per cycle as per the following formula:

$$\text{Average } \Delta D/\text{cycle} = \Delta D/\text{No. of cycles} \qquad (2)$$

This method is called the *average cycle method*. The average ΔD per cycle is interpreted to represent the change in sea level

TABLE 4.—CONVERSION TABLES, PENNSYLVANIAN TIME SCALE.

	Harland and others, (1990)	Lippolt and others, (1984); Hess (1990) and Lippolt (1986)
NORTH AMERICA:		
Top, Pennsylvanian (Top of Virgilian)	290.0	300.0
Virgilian-Missourian Boundary	298.0	303.0
Missourian-Desmoinesian Boundary	303.0	306.0
Desmoinesian-Atokan Boundary	307.1	310.0
Atokan-Morrowan Boundary	311.3	315.0
Base of Pennsylvanian (Base of Morrowan)	322.8	319.0
USSR:		
Top, Carboniferous (Top of Gzhelian)	290.0	300.0
Gzhelian-Kasimovian Boundary	295.1	301.0
Kasimovian-Moscovian Boundary	303.0	306.0
Myachkovian-Podolskian Boundary	305.0	308.0
Podolskian.-Kashirian Boundary	307.1	310.0
Kashirian-Veryeian Boundary	309.2	311.5
Moscovian-Bashkirian Boundary	311.3	314.5
Base Bashkirian (Base of Pennsylvanian)	322.8	319.0
NORTHWEST EUROPE:		
Top, Carboniferous (Top of Stephanian C)	290.0	300.0
Boundary- Stephanian C and B	293.6	301.0
Boundary- Stephanian B and A	298.0	303.0
Stephanian-Westphalian Boundary	303.5	306.0
Boundary- Westphalian D and C	305.5	309.0
Boundary- Westphalian C and B	306.5	311.0
Boundary- Westphalian B and A	309.5	313.0
Westphalian-Namurian Boundary	317.0	315.0
Boundary- Namurian C and B	319.8	316.5
Boundary- Namurian B and A (Base of Pennsylvanian)	322.8	319.0

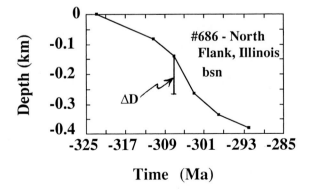

Basement Subsidence vs. Time

#686 - North Flank, Illinois bsn

ΔD

$\Delta D/Cycle = \Delta D/Number\ of\ Cycles$

FIG. 2.—Tectonic subsidence curve showing how ΔD is determined from a tectonic subsidence curve and used to calculate ΔD/cycle (after Klein and Kupperman, 1992).

associated with deposition of a cyclothem which was caused by *tectonic* subsidence.

A second method, called the *individual cycle method*, is used when better control exists on individual cycle thickness. With this method, the percentage of the thicknesses of each individual cyclothem with respect to the total thickness of all cyclothems occurring within the time represented by ΔD is computed to determine ΔD per cycle according to the following formula:

$$\Delta D_i/cycle = \Delta D\ (\%_i T_t) \qquad (3)$$

where ΔD_i is the magnitude of subsidence for an individual cycle, $\%_i$ is the thickness % of the individual cyclothem with respect to the total thickness of all cyclothems within the time interval represented by ΔD, and T_t is the total thickness of the stratigraphic section for the time during which the cyclothems of interest were deposited. ΔD per cycle calculated by the individual cycle method should approximate the change in sea level during deposition of a cyclothem that is caused by tectonic subsidence. Individual cycle method results were reported by Klein and Kupperman (1992).

TECTONIC SUBSIDENCE ASSOCIATED WITH MIDCONTINENT
CYCLOTHEMS

Figures 3 and 4 show backstripped tectonic subsidence curves used to calculate ΔD per cycle for each cyclothem. Subsidence curves for other North American sites appear in Klein and Kupperman (1992; their Fig. 3). Figures 3 and 4 demonstrate that a significant component of tectonic subsidence existed in the midcontinent during deposition of cyclothems. This deposition is also coeval with changes in tectonic regimes at the collision margins in the Appalachians and the Ouachitas. The tectonic subsidence curves (Fig. 3 and 4) suggest a strong coupling of tectonic processes at plate boundaries and interior basins and platforms, over a distance of 1,100 km from convergent plate boundaries, which fits independent interpretations for North America (Kluth and Coney, 1981; Thomas, 1983; Craddock and Van der Pluijm, 1989), and elsewhere (Ziegler, 1987; Leighton and Kolata, 1990).

The change in sea level (ΔD) per cycle caused by tectonic subsidence was determined using the average cycle method (Tables 5 and 6). Data was subdivided according to known Pennsylvanian series, or by time-stratigraphic assignment of stratigraphic groups, where known (McKee and Weir, 1975a). To determine the number of cycles for each time stratigraphic interval, I used a revised cycle chart provided by P. H. Heckel which combined earlier charts in Heckel (1986) and Boardman and Heckel (1989) for the Iowa platform, and the Salina and Forest City basins. These cyclothems are fifth-order stratigraphic units (Busch and Rollins, 1984; Miall, 1990).

In the midcontinent, changes in sea level (ΔD per cycle) attributed to tectonic subsidence ranged from 3.0 to 22.0 m in the Salina basin (Table 5), from 4.5 to 13.0 m in the Forest City basin, to 5.3 to 20.6 m on the Iowa Platform (Table 6). Of interest here is that in all three midcontinent sites, Desmoinesian-age cycles show the largest, tectonic-induced sea-level change, ranging from 13.0 to 20.6 m in basin depocenters, whereas tectonically-induced sea-level change is smaller (range from 5.3 to 13.0 m) for cyclothems of Missourian and Virgilian age, except in the Salina basin, where a 22.0-m sea-level change is attributed to subsidence during Virgilian time. This decrease in tectonically-controlled sea-level change in younger cyclothems on the Iowa platform suggests both a diminishing of tectonic forces in controlling cyclothem sea-level change and an increase in the role of glacio-eustatic influence through Middle and Upper Pennsylvanian time. The sea-level change of Desmoinesian cyclothems on the Iowa platform appears to be influenced more strongly by tectonic subsidence, approximating 20 to 30 percent of total change in sea level (depending on whether one uses Heckel's or Gerhard's model), whereas for the Missourian and Virgilian cyclothems, sea-level change created by tectonic forces is reduced. Only in the Salina basin does an increase in sea-level change due to tectonic subsidence occur again during Virgilian time that is close to the magnitudes obtained from Desmoinesian cyclothems and accounts for close to 20 to 30

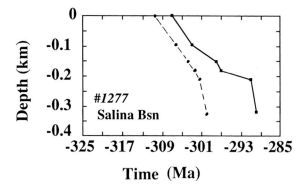

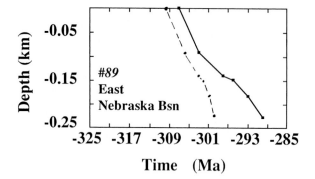

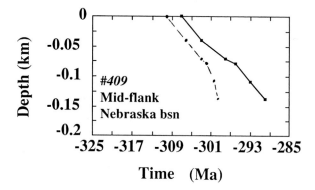

Fig. 3.—Tectonic subsidence curves for the Salina basin (Nebraska Basin of Nebraska) using time scales of Harland and others (1990, solid line) and Lippolt and others (1984, dashed line). UPPER—Salina basin depocenter, KS. MIDDLE—Eastern part, Salina basin depocenter, NB. LOWER—Mid-flank, Salina basin, NB (see Table 3 for locality description keyed to number in each panel).

Basement Subsidence vs. Time

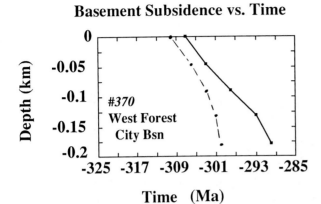

#370
**West Forest
City Bsn**

Basement Subsidence vs. Time

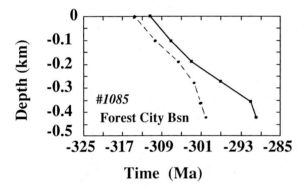

#1085
Forest City Bsn

Basement Subsidence vs. Time

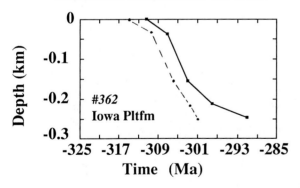

#362
Iowa Pltfm

FIG. 4.—Tectonic subsidence curves for Forest City basin and Iowa platform using time scales of Harland and others (1990, solid line) and Lippolt and others (1984, dashed line). UPPER—West flank, Forest City basin. MIDDLE—Depocenter, Forest City Basin. LOWER—Iowa platform (see Table 3 for locality description keyed to number in each panel).

percent of total sea-level change.

INTRAPLATE STRESS

In an earlier discussion (Klein and others, 1990), an attempt was made to determine the role of intraplate stress on deposition of individual cyclothems using methods developed by Cloetingh (1986, 1988) and Cloetingh and others (1989). Results obtained at three midcontinent sites are summarized in Tables 7, 8, and 9. The function ΔW/cycle in Tables 7, 8, and 9 represents the magnitude of stress-induced subsidence during deposition of a single Pennsylvanian cyclothem. It must be emphasized that DW/cycle should represent a partial contribution to the total tectonic subsidence per cycle (ΔD/cycle) inasmuch as intraplate stress is one of several different components of tectonic subsidence.

New work by Karner and others (1989; see also Sahagian and Watts, 1991) suggested that the Cloetingh method of determining intraplate stress may yield values that are too high. In part, these values may be too high because of the assumption (Cloetingh, 1986, 1988; Cloetingh and others, 1989) of lack of interface effects by the asthenosphere on the lithosphere (exemplified by the spring model; Cloetingh, 1986). Young thermal ages also appear to be a prerequisite to obtain requisite high values of uplift and subsidence that were generated by this calculation of intraplate stress (Cloetingh, 1986, 1988; Cloetingh and others, 1989). Such young thermal ages were absent in the midcontinent during Pennsylvanian time, however. Moreover, Dickinson and others (1991) reported that the inflection point associated with an inversion from extensional to inverted compressional basins does not show evidence of erosion as required by the intraplate stress basin subsidence/uplift inversion model (Cloetingh, 1988). In the Forest City and Salina basins, similar observations were reported by Burchett (1983).

Tables 7, 8, and 9 compares the ΔW/cycle for midcontinent cyclothems (as per methods of Cloetingh, 1986, 1988; Cloetingh and others, 1989) with DD/cycle. In most instances, the values for ΔW/cycle are higher than the values for ΔD/cycle. Errors differences of ΔW/ cycle and ΔD/cycle were as high as 28, 34, and 45 percent too great (Tables 7, 8, and 9). The difficulty here is that although intraplate stress can be less or equal to total tectonic subsidence, intraplate stress can never exceed total subsidence (because the part cannot exceed the whole). Consequently, it appears that intraplate stress contributed minimally, at best, to changes in Pennsylvanian sea level.

DISCUSSION

A principal goal herein was to determine how much of the magnitude of change in relative sea-level change associated with midcontinent Pennsylvanian cyclothems was controlled by tectonic subsidence. Tables 4 and 5 show that sea-level change of marine cyclothems was controlled in part by tectonic subsidence.

In the midcontinent of North America, the role of glacially-

TABLE 5.—AVERAGE SUBSIDENCE (ΔD) PER CYCLE, SALINA BASIN DEPOCENTER (#1277), MID-FLANK(#409) AND EAST-FLANK (#89).

	Depocenter Salina bsn ΔD/cycle (m) #1277	Mid-flank Salina bsn ΔD/cycle (m) #409	East-flank Salina bsn ΔD/cycle (m) #89
VIRGILIAN SERIES:			
Shawnee Group:			
(5 major cycles)	22.0	6.4	16.8
Douglas Group:			
(2 major cycles)			
MISSOURIAN SERIES:			
(10 major cycles)	8.5	3.0	4.9
DESMOINESIAN SERIES:			
(6 major cycles)	16.0	6.3	15.5

TABLE 6.—AVERAGE SUBSIDENCE (ΔD) PER CYCLE, FOREST CITY BASIN WEST FLANK (#370),AND DEPOCENTER (#l085), AND THE IOWA PLATFORM (#362).

	West-flank Forest City bsn ΔD/cycle (m) #370	Depocenter Forest City bsn ΔD/cycle (m) #1085	Iowa Platform ΔD/cycle (m) #362
VIRGILIAN SERIES:			
Shawnee Group:			
(5 major cycles)	13.0	9.6	5.3
Douglas Group:			
(2 major cycles)			
MISSOURIAN SERIES:			
(10 major cycles)	9.0	4.5	5.6
DESMOINESIAN SERIES:			
(6 major cycles)	13.0	7.8	20.6

TABLE 7.—ERROR ANALYSIS, DEPOCENTER, SALINA BASIN, KS (#1277).

	ΔW/cycle (m)[1]	ΔD/cycle (m)	Error (%)
VIRGILIAN SERIES:			
Shawnee Group:			
(5 major cycles)	5.0	22.0	/
Douglas Group:			
(2 major cycles)			
MISSOURIAN SERIES:			
(10 major cycles)	5.0	8.5	/
DESMOINESIAN SERIES			
(6 major cycles)	16.7	16.0	4.4

[1] Calculated as per Cloetingh and others (1989).

TABLE 8.—ERROR ANALYSIS, DEPOCENTER, FOREST CITY BASIN (#1085).

	ΔW/cycle[+] (m)	ΔD/cycle (m)	Error (%)
VIRGILIAN SERIES:			
Shawnee Group:			
(5 major cycles)	16.7	13.0	28.5
Douglas Group:			
(2 major cycles)			
MISSOURIAN SERIES:			
(10 major cycles)	7.9	9.0	/
DESMOINESIAN SERIES:			
(6 major cycles)	19.2	13.0	47.7

[+] Calculated using methods of Cloetingh and others (1989)

TABLE 9.—ERROR ANALYSIS, IOWA PLATFORM (#362).

	ΔW/ cycle[1] (m)	ΔD/cycle (m)	Error (%)
VIRGILIAN SERIES:			
(6 major cycles)	7.1	5.3	34.0
MISSOURIAN SERIES:			
(10 major cycles)	6.0	5.6	7.1
DESMOINESIAN SERIES:			
(6 major cycles)	20.8	20.6	1.0

[1] Calculated as per methods of Cloetingh and others (1989).

driven eustatic changes in sea level needs to be explored (Heckel, 1986; Boardman and Heckel, 1989). Reexamination of the cycle/sea-level chart of midcontinent Pennsylvanian cyclothems (Heckel, 1986; Boardman and Heckel, 1989), particularly in the Iowa platform, showed preservation of a pattern of fivefold bundles of cyclothems (deBoer, 1991; Klein, 1991b). Figure 5 summarizes a revised midcontinent Pennsylvanian cyclothem sea-level chart based on cycle charts of both Heckel (1986) and Boardman and Heckel (1989) with the number of bundles within major cycles separated by transgressive highstand black shales. All cyclothems below the middle of the Kansas City Group (Middle Missourian), including the Desmoinesian, show cycle bundling ranging from one to three bundle elements per cycle, whereas Upper Missourian and Virgilian cyclothems show a significant number of preserved cycle bundle elements ranging from four to five. Each cyclothem represents one of the elements of these cycle bundles. Comparison of average subsidence-driven sea-level change (ΔD) per cycle at the Iowa platform section (Table 6) with bundle data (Fig. 5) shows that, during Desmoinesian time, when bundle elements were three or less, 20.6 m of sea-level change could be explained by tectonic subsidence, whereas during Missourian and Virgilian time, only 5.3 to 5.6 m of sea-level change could be accounted for by tectonic subsidence. This temporal change in bundling data may also account for the presence of Desmoinesian shoreline black-

shale facies (Coveney and others, 1991), whereas the younger black shales represent a relatively deeper and more open-marine conditions of deposition. These observations and comparisons suggest that midcontinent Pennsylvanian sea-level change apparently was *caused by concurrent tectonic forces and glacio-eustatic sea-level change influenced by Milankovitch orbital forcing.*

Magnitude of Tectonic Subsidence, Glacial Eustasy, and Long-term Climate Change to Pennsylvanian Sea-Level Change

Given these findings, it is now possible to estimate the individual contribution of tectonic subsidence (herein), glacial eustasy (Heckel, 1986; Crowley and Baum, 1991) and long-term climate change (Cecil, 1990). Water-depth determinations (Table 1) provide both a range as well as an average of the magnitude of sea-level change required to deposit a midcontinent Pennsylvanian cyclothem. Paleoclimate modeling yields both an average (60 m) and a range (45 to 75 m) of glacio-eustatic sea-level change (Crowley and Baum, 1991). Tectonic subsidence analysis (Tables 5 and 6) shows that Pennsylvanian sea-level change caused by this process ranged from 3.0 to 22.0 m.

To calculate the magnitude of each process causing concurrent change in sea level during deposition of a single cyclothem, the following formula was used:

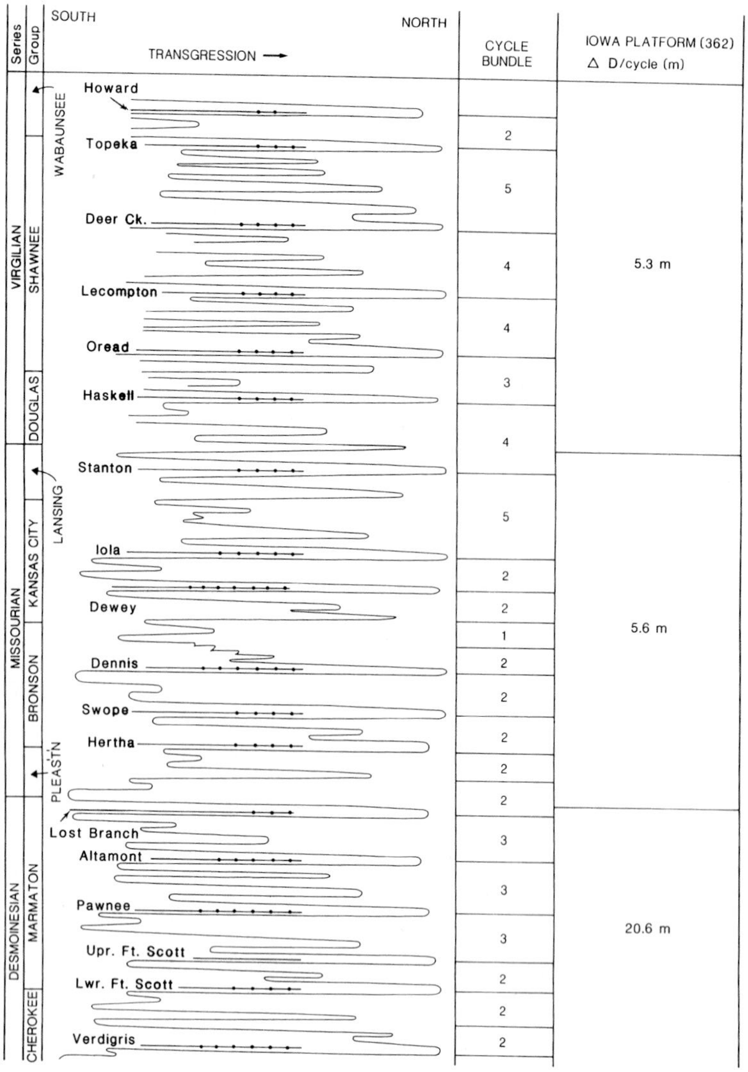

FIG. 5.—Comparison of sea-level curve for midcontinent (after Heckel, 1986; Boardman and Heckel, 1989) with number of elements of cycle bundles, and calculated average DD per cycle from the Iowa platform representing sea-level change caused by tectonic subsidence.

$$SL = T_s + G_e + C_l \qquad (4)$$

where SL is sea-level change derived from water-depth change (Table 1), T_s is the magnitude of sea-level change caused by tectonic subsidence (Tables 5 and 6), G_e is the amount of sea-level change caused by glacial eustasy (Crowley and Baum, 1991), and C_l is the amount of sea-level change caused by long-term climate change (Cecil, 1990). Because sea-level change based on water-depth estimates (Table 1), tectonic subsidence (Tables 5 and 6), and glacial eustasy (Crowley and Baum, 1991) are known, the contribution of long-term climate change is represented by the net amount of sea-level change, after subtracting the magnitudes for tectonic subsidence and glacial eustasy. The following formula was used:

$$C_l = SL - T_s - G_e \qquad (5)$$

Table 10 summarizes these numbers and a determination of the net contribution of long-term climate change. Table 10 demonstrates that neither glacial eustasy alone, nor tectonic subsidence alone, can account for the magnitude of sea-level change required to deposit a Pennsylvanian midcontinent cyclothem. The difference is interpreted to represent long-term climate change (Cecil, 1990). Table 10 shows that anywhere from a minimal value of no influence to a range from 4 to 114 m of average sea-level change is caused by long-term climate change. Given the range of values of tectonic subsidence and glacial eustasy, the contribution of long-term climate may reach as much as 63 m of sea level change in some instances. Moreover, during other periods, glacial eustasy could have been less than the 45-m minimal estimate of Crowley and Baum (1991). Thus during Pennsylvanian time, glacial eustasy caused variable magnitudes of sea level change and during certain intervals, its influence was less than calculated.

These findings pose certain implications. Since the early days of geology, stratigraphers have argued about the relative importance of intrinsic, mantle-driven, tectonic and extrinsic,

TABLE 10.—RANGE, MAGNITUDE (m) AND AVERAGE PERCENTAGE CONTRIBUTION OF
TECTONIC SUBSIDENCE, GLACIAL EUSTASY, AND LONG-TERM CLIMATE TO PENNSYLVANIAN
SEA-LEVEL CHANGE.

	Moore (1964)	Heckel (1977)	Gerhard (1991)
Range of Sea-Level Change (Table 1)	32 to 136 m	57 to 160 m	40 to 141m
LESS Range of Sea-Level by Tectonic Subsidence (Tables 5 & 6).	3 to 22 m	3 to 22 m	3 to 22 m
LESS Range, Glacial Eustasy (Crowley and Baum, 1991)	45 to 75 m	45 to 75 m	45 to 75 m
NET: Long-term Climate Change	(-16) to 39 m	(-25) to 63 m	(-8) to 44 m
Average Sea-level Change (Table 1)	76 m	96 m	86 m
LESS Range of Sea-Level by Tectonic Subsidence (Tables 5 & 6).	3 to 22 m (4—29 %)	3 to 22 m (3—23 %)	3 to 22 m (3—26 %)
LESS Average Glacial Eustasy (Crowley and Baum, 1991)	60 m (79 %)	60 m (63 %)	60 m (70 %)
NET: Long-term Climate Change	(-6) to 13 m (0—17 %)	14 to 33 m (15—34 %)	4 to 23 m (5—27 %)

solar-driven climatic processes in controlling transgressive and regressive events representing a record of absolute sea-level change. This polarized argument persisted for over 50 years with respect to Pennsylvanian cyclothems, but recently Klein and Willard (1989) and Read and Forsyth (1989), recognized that *both concurrent* tectonic processes and climatic processes (glacial advance and retreat) controlled cyclothem deposition. Results reported herein demonstrate that tectonic-subsidence processes control some of the sea-level change required to form midcontinent Pennsylvanian cyclothems, particularly during times of major regional compressional tectonic events. Both short-term glacial eustasy and longer-term climatic change (such as in the midcontinent) accounted for the remaining sea-level change. The magnitude of each of these processes can be determined by the methods discussed in this paper. Thus a quantitative, multidisciplinary approach as discussed herein can be used to discriminate the magnitude of either or both tectonic and climatic contributions which control sea-level changes during deposition of other types of cyclic sedimentary phenomena.

CONCLUSIONS

The following conclusions are drawn from this study:

1. Far-field tectonic effects influenced the tectonics of the midcontinent of North America and the deposition of associated Pennsylvanian cyclothems.

2. Because of uncertainties in the Pennsylvanian time scale,

identification of Milankovitch orbital parameters as a forcing mechanism for cyclicity from period calculations appears premature. However, preservation of a fivefold bundling of cycle elements in Upper Pennsylvanian midcontinent cyclothems shows that Milankovitch orbital parameters (particularly precession) probably played an important role in forming and preserving Pennsylvanian cyclothems during times when tectonic contributions were small and regional tectonic movements were minor. This finding confirms a model of partly concurrent tectonic and climatic controls on cyclothem deposition.

3. The contribution of intraplate stress to sea-level change during deposition of midcontinent Pennsylvanian cyclothems is minimal.

4. The tectonic contribution to Pennsylvanian sea-level change during deposition of midcontinent cyclothems accounted from five to 30 percent of sea-level change on platform areas, and perhaps as much as 30 percent of sea-level change in basin depocenters. Locally, this contribution may have been greater near areas of reactivation tectonics.

5. Water-depth estimates, using a sedimentological method, indicate that average sea-level change during deposition of cyclothems ranged from 86.0 m to 96.4 m, using models of Gerhard (1991) and Heckel (1977). Because Carboniferous glacio-eustatic sea-level change was only 60 m (+/- 15 m), both longer-term climate change and concurrent tectonic processes also contributed to the overall sea-level cycle controlling cyclothem deposition.

6. Concurrent long-term climate change may have contrib-

uted significantly to sea-level change during deposition of Pennsylvanian cyclothems (Table 10).

7. The methods used in this paper to determine the magnitude of changes in sea level caused by tectonic subsidence, glacial eustasy, and long-term climate change that are associated with deposition of Pennsylvanian cyclothems may be applicable to understanding the origin of other cyclic sedimentary sequences.

ACKNOWLEDGMENTS

I wish to acknowledge and thank the National Science Foundation (Grant EAR-90-02448) for their financial support of this research. I thank Lee C. Gerhard for providing an advance copy of his sea-level curve for Pennsylvanian cyclothems, P. H. Heckel for providing an advance copy of his revised midcontinent cycle chart used to construct Figure 5, and Donald L. Baars, Poppe L. deBoer, Thomas J. Crowley, Lee C. Gerhard, Philip H. Heckel, Albert T. Hsui, Gary D. Karner and Carol A. Stein for discussions. Finally, I wish to thank William R. Dickinson and Daniel A. Textoris for helpful and cogent comments on earlier manuscript versions of this paper.

REFERENCES

ALEXANDER, C. R., DEMASTER, D. J., AND NITTROUER, C. A., 1991, Sediment accumulation in a modern epicontinental-shelf setting: the Yellow Sea: Marine Geology, v. 98, p. 51-72.

BERGER, A., 1988, Milankovitch theory and climate: Review of Geophysics, v. 26, p. 624-657.

BERGER, A., LOUTRE, M. F., AND DEHANT, V, 1987, Influence of the variation of the lunar orbit on the astronomical frequencies of the Pre-Quaternary paleoinsolation: Louvain-la-Neuve, Scientific Report, 1987/15, Institut d'Astronomie et de Géophysique, 21 p.

BERGER, A., LOUTRE, M. F., AND DEHANT, V, 1989, Astronomical frequencies for Pre-Quaternary palaeoclimate studies: Terra Nova, v. 1, p. 474-479.

BOARDMAN, D. R., II, AND HECKEL, P. H., 1989, Glacial-eustatic sea-level curves for early Late Pennsylvanian sequence in north-central Texas and biostratigraphic correlation with curve for midcontinent North America: Geology, v. 17, p. 802-805.

DEBOER, P. L., 1991, Comment—Pennsylvanian time scales and cycle periods: Geology, v. 19, p. 408-409.

BOND, G. C., AND KOMINZ, M. A., 1984, Construction of tectonic subsidence curves for the early Paleozoic miogeocline, southern Canadian Rocky Mountains: Implications for subsidence mechanisms, age of breakup and crustal thinning: Geological Society of America Bulletin, v. 95, p. 155-173.

BURCHETT, R. R., 1983, Surface to Subsurface Correlation of Pennsylvanian and lower Permian rocks across southern Nebraska: Lincoln, Nebraska Geological Survey, Report of Investigations 8, 24 p.

BUSCH, R. M., AND ROLLINS, H. B., 1984, Correlation of Carboniferous strata using a hierarchy of transgressive-regressive units: Geology, v. 12, p. 471-474.

CARLSON, M. P., 1989a, Influence of Midcontinent rift system on occurrence of oil in Forest City basin (abs.): Bulletin of the American Association of Petroleum Geologists, v. 73, p. 340.

CARLSON, M. P., 1989b, Location of oil fields in Forest City basin as related to Precambrian tectonics (abs): Bulletin of the American Association of Petroleum Geologists, v. 73, p. 1141.

CECIL, C. B., 1990, Paleoclimate controls on stratigraphic repetition of chemical and siliciclastic rocks: Geology, v. 18, p. 533-536.

CLOETINGH, S., 1986, Intraplate stress: a new tectonic mechanism for fluctuations of relative sea level: Geology, v. 14, p. 617-620.

CLOETINGH, S., 1988, Intraplate stresses: a new element in basin analysis, in Kleinspehn, K. L., and Paola, C., eds., New Perspectives in Basin Analysis: New York, Springer-Verlag, p. 205-230.

CLOETINGH, S., KOOI, H., AND GROENEWOUD, W., 1989, Intraplate stresses and sedimentary basin evolution, in Price, R. A., ed., Origin and Evolution of Sedimentary Basins and their Energy and Mineral Resources: Washington, D, C, American Geophysical Union Monograph 48, p. 1-16.

COVENEY, R. M., JR., WATNEY, W. L., AND MAPLES, C. G., 1991, Contrasting depositional models for Pennsylvanian black shale discerned from molybdenum abundances: Geology, v. 19, p. 147-150.

CRADDOCK, J. P., AND VAN DER PLUIJM, B. A., 1989, Late Paleozoic deformation of the cratonic carbonate cover of eastern North America: Geology, v. 17, p. 416-419.

CROWLEY, T. J., AND BAUM, S. K., 1991, Estimating Carboniferous sea-level fluctuations from Gondwana ice extent: Geology, v. 19, p. 975-977.

DEAN, W. E., ARTHUR, M. A., AND STOW, D. A. V., 1984a, Origin and geochemistry of Cretaceous deep-sea black shales and multicolored claystones, with emphasis on DSDP site 530, South Angola basin, in Hay, W. W., Sibuet, J. C., and others, Initial Report of the Deep Sea Drilling Project, v. 75: Washington, D. C., United States Government Printing Office, p. 819-844.

DEAN, W. E., HAY, W. W., AND SIBUET, J. C., 1984b, Geological evolution, sedimentology and paleoenvironments of the Angola basin and adjacent Walvis ridge: synthesis of results of DSDP Leg 75, in Hay, W. W., Sibuet, J. C., and others, Initial Report of the Deep Sea Drilling Project, v. 75: Washington, D. C., United States Government Printing Office, p. 509-542.

DICKINSON, W. R., SOREGHAN, G. S., AND GOEBEL, K. A., 1991, Stratigraphic cyclicity in foreland geological systems of Permo-Carboniferous age (abs.): Geological Society of America Abstracts with Programs, v. 23, p. 21.

ELIAS, M. K., 1937, Depth of Deposition of the Big Blue (late Paleozoic) sediments in Kansas: Geological Society of America Bulletin, v. 48, p. 403-432.

ELIAS, M. K., 1964, Depth of Late Paleozoic sea in Kansas and its megacyclic Sedimentation, in Merriam, D. F., ed., Symposium on cyclic sedimentation: Lawrence, Kansas Geological Survey Bulletin 169, p. 87-104.

ETTENSOHN, F. R., AND ELAM, T. D., 1985, Defining the nature and location of a Late Devonian-Early Mississippian pycnocline in eastern Kentucky: Geological Society of America Bulletin, v. 96, p. 1313-1321.

ETTENSOHN, F. R., MILLER, M. L., DILLMAN, S. B., ELAM, T. D., GELLER, K. L., SWAGER, D. R., MRAKOWITZ, G., WOOCK, R. D., AND BARRON, L. S., 1988, Characterization and implications of the Devonian-Mississippian black shale sequence, eastern and central Kentucky, U. S.: pycnoclines, transgression, regression, and tectonism, in McMillan, N. J., Embry, A. F., and Glass, J. D., eds., Devonian of the World: Proceedings, Second International Symposium on the Devonian System: Calgary, Canadian Society of Petroleum Geologists Memoir 14, v. 2, p. 323-345.

FAIRBANKS, R. G., 1989, A 17,000-year glacio-eustatic sea level record: influence of glacial melting rates on the Younger Dryas event and deep-ocean circulation: Nature, v. 342, p. 637-642.

FISCHER, A. G., DEBOER, P. L., AND PREMOLI-SILVA, I., 1990, Cyclostratigraphy, in Ginsburg, R. N., and Beaudoin, B, eds., Cretaceous Resources, Events and Rhythms: A Program of the Global Sedimentary Geology Program: Dordrecht, Reidel, p. 139-172.

GERHARD, L. C., 1991, Stratigraphy in the modern generation: Lawrence, Kansas Geological Survey, Open File Report 91-24, 13 p.

GRAND, S. P., 1987, Tomographic inversion for shear velocity beneath the North American plate: Journal of Geophysical Research, v. 92, p. 14065-14090.

GREB, S. F., 1989, Structural controls on the formation of the sub-Absaroka unconformity in the U. S. Eastern Interior basin: Geology, v. 17, p. 889-892.

HALLOCK, P., AND SCHLAGER, W., 1986, Nutrient excess and the demise of coral reefs and carbonate platforms: Palaios, v. 1, p. 389-398.

HARLAND, W. B., COX, A. V., LLEWELLYN, P. G., PICKTON, C. A. G., SMITH, A. G., AND WALTERS, R., 1982, A Geologic Time Scale: Cambridge, Cambridge University Press, 131 p.

HARLAND, W. B., ARMSTRONG, R. L., COX, A. V., CRAIG, L. E., SMITH, A. G., AND SMITH, D. B., 1990, A Geological Time Scale 1989: Cambridge, Cambridge University Press, 262 p.

HECKEL, P. H., 1977, Origin of phosphatic black shale facies in Pennsylvanian cyclothems of Midcontinent North America: Bulletin of the American Association of Petroleum Geologists, v. 61, p. 1045-1068.

HECKEL, P. H., 1984, Changing concepts of Midcontinent Pennsylvanian cyclothems, North America, in Congrès International de Stratigraphie et de Géologie du Carbonifére, 9th, Compte Rendu, v. 3: Carbondale, Southern Illinois University Press, p. 535-533.

HECKEL, P. H., 1986, Sea level curve for Pennsylvanian eustatic marine transgressive-regressive depositional cycles along Midcontinent outcrop belt, North America: Geology, v. 14, p. 330-334.

HECKEL, P. H., 1991, Comment- Pennsylvanian time scales and cycle periods: Geology, v. 19, p. 406-407.

HECKEL, P. H., BRADY, L. L., EBANKS, W. J., JR., AND PABIAN, R. K., 1979, Pennsylvanian cyclic platform deposits of Kansas and Nebraska: Guidebook 10, Ninth International Congress of Carboniferous Stratigraphy and Geology, 79 p.

HERBERT, T. D., STALLARD, R. F., AND FISCHER, A. G., 1986, Anoxic events, productivity rhythms, and the orbital signature in a Mid-Cretaceous deep-sea sequence from central Italy: Paleoceanography, v. 1, p. 495-506.

HESS, J. C., AND LIPPOLT, H. J., 1986, $^{40}Ar/^{39}Ar$ ages of tonstein and tuff sanidines: new calibration points for the improvement of the Upper Carboniferous time scale: Isotope Geoscience, v. 59, p. 143-154.

IBACH, L. E. J., 1982, Relationship between sedimentation rate and total organic carbon content in ancient marine sediments: American Association of Petroleum Geologists Bulletin, v. 68, p. 170-188.

JORDAN, T. E., 1981, Thrust loads and foreland basin evolution, Cretaceous, Western United States: American Association of Petroleum Geologists Bulletin, v. 65, p. 2506-2520.

KARNER, G. D., CHRISTIE-BLICK, N., AND DRISCOLL, N. W., 1989, The breakup unconformity of passive continental margins: Testing the importance of inplane stress variations during the rift-drift transition on the Grand Banks, Canada (abs.): American Geophysical Union's Chapman Conference on Causes and Consequences of Long-term Sea Level Change, abstracts with program.

KLEIN, G. DEV., 1974, Estimating water depths from analysis of barrier island and deltaic sedimentary sequences: Geology, v. 2, p. 409-412.

KLEIN, G. DEV., 1982, Probable sequential arrangement of depositional systems on cratons: Geology, v. 10, p. 17-22.

KLEIN, G. DEV., 1990a, Comments on sedimentary-stratigraphic verification of some geodynamic basin models: example from a cratonic and associated foreland basin, in Cross, A. T., ed., Quantitative Dynamic Stratigraphy: Englewood Cliffs, Prentice-Hall, p. 503-518.

KLEIN, G. DEV., 1990b, Pennsylvanian time scales and cycle periods: Geology, v. 18, p. 455-457.

KLEIN, G. DEV., 1991a, Pennsylvanian time scales and cycle periods-reply: Geology, v. 19, p. 407-408.

KLEIN, G. DEV., 1991b, Pennsylvanian time scales and cycle periods-reply: Geology, v. 19, p. 409-410.

KLEIN, G. DEV., 1992, A climatic and tectonic sea-level gauge for Midcontinent Pennsylvanian cyclothems: Geology, v. 20, p. 363-366.

KLEIN, G.D., 1993, Paleoglobal change during deposition of cyclothems: calculating the contributions of tectonic subsidence, glacial eustasy and long-term climate change on Pennsylvanian sea-level change: Tectonophysics, v. 222, p. 333-360.

KLEIN, G. DEV., AND HSUI, A. T., 1987, Origin of cratonic basins: Geology, v. 15, p. 1094-1098.

KLEIN, G. DEV., AND KUPPERMAN, J. B., 1992, Pennsylvanian cyclothems: methods of distinguishing tectonically-induced changes in sea level from climatically-induced change: Geological Society of America Bulletin, v. 104, p. 166-175.

KLEIN, G. DEV., AND WILLARD, D. A., 1989, The origin of the Pennsylvanian coal-bearing cyclothems of North America: Geology, v. 17, p. 152-155.

KLEIN, G. DEV., KOBAYASHI, K., AND OTHERS, 1980, Initial Reports of the Deep Sea Drilling Project, v. 58: Washington, D. C., United States Government Printing Office, 1022 p.

KLEIN, G. DEV., CLOETINGH, S., BEEKMAN, F., AND KUPPERMAN, J. B., 1990, Magnitude of intraplate stress-induced sea-level change during deposition of Pennsylvanian cyclothems (abs.): Geological Society of America Abstracts with Programs, v. 22, p. A277.

KLEIN, G. DEV., PARK, Y. A., CHANG, J. H., AND KIM, C. S., 1982, Sedimentology of a subtidal, tide-dominated sand body in the Yellow Sea, Southwest Korea: Marine Geology, v. 50, p. 221-240.

KLUTH, C. F., AND CONEY, P. J., 1981, Plate tectonics of the Ancestral Rocky Mountains: Geology, v. 9, p. 10-15.

KOZAR, M. G., WEBER, L. J., AND WALKER, K. R., 1990, Field and modeling studies of Cambrian carbonate cycles, Virginia Appalachians- Discussion: Journal of Sedimentary Petrology, v. 60, p. 790-794.

LEE, H. J., AND CHOUGH, S. K., 1989, Sediment distribution and dispersal and budget in the Yellow Sea: Marine Geology., v. 87, p. 195-205.

LEEDER, M. R., 1988, Recent developments in Carboniferous geology: a critical review with implications for the British Isles and N. W. Europe: Geologists Association Proceedings, v. 99, p. 73-100.

LEIGHTON, M. W., AND KOLATA, D. R., 1990, Selected interior cratonic basins and their place in the scheme of global tectonics: A synthesis, in Leighton, M. W., Kolata, D. R., Oltz, D. F., and Eidel, J. J., eds., Interior Cratonic Basins: Tulsa, American Association of Petroleum Geologists Memoir 51, p. 729-798.

LIPPOLT, H. J., HESS, J. C., AND BURGER, K., 1984, Isotopische alter von pyroklastischen sanidinen aus kaolin-Kohlentonstein als korrelationsmarken fur das mitteleuropaishese Oberkarbon: Forschrift Geologische Rheinland und Westfalen, v. 32, p. 119-

150.

McCRONE, A. W., 1964, Water Depth and Midcontinent Cyclothems, *in* Merriam, D. F., ed., Symposium on Cyclic Sedimentation: Lawrence, Kansas Geological Survey Bulletin 169, p. 275-281.

McKEE, E. D., AND CROSBY, E. J., eds., 1975a, Paleotectonic investigations of the Pennsylvanian System in the United States: Washington, D. C., United States Geological Survey Professional Paper 853, Part I, 349 p.

McKEE, E. D., AND CROSBY, E. J., eds., 1975b, Paleotectonic investigations of the Pennsylvanian System in the United States: Washington, D. C., United States Geological Survey Professional Paper 853, Part 2, 192 p.

McKEE, E. D., AND CROSBY, E. J., eds., 1975c, Paleotectonic investigations of the Pennsylvanian System in the United States: Washington, D. C., United States Geological Survey Professional Paper 853, Part 3, maps and charts.

MIALL, A. D., 1990, Principles of Sedimentary Basin Analysis, 2nd ed: New York, Springer-Verlag, 668 p.

MOORE, R. C., 1958, Historical Geology, 2nd ed., New York, McGraw-Hill, 686 p.

MOORE, R. C., 1964, Paleoecological aspects of Kansas Pennsylvanian and Permian cyclothems, *in* Merriam, D. F., ed, Symposium on cyclic sedimentation: Lawrence, Kansas Geological Survey Bulletin 169, p. 287-380.

NELSON, W. J., AND LUMM, D. K., 1984, Structural geology of southeastern Illinois and vicinity: Champaign, Illinois State Geological Survey Contract/Grant Report 1984-2, p. 971-976.

PLAFKER, G., AND SAVAGE, J. C., 1970, Mechanism of the Chilean earthquake of May 21 and 22, 1960: Geological Society of America Bulletin, v. 81, p. 1001-1030.

POSAMENTIER, H. W., JERVEY, M. T., AND VAIL, P. R., 1988, Eustatic control on clastic deposition conceptual framework, *in* Wilgus, C. K., Posamentier, H. W., Ross, C. A., and Kendall, C. G. St. C., eds., Sea-level Changes: An Integrated Approach: Tulsa, Society of Economic Paleontologists and Mineralogists Special Publication 42, p. 109-125.

QUINLAN, G., AND BEAUMONT, C., 1984, Appalachian thrusting, lithospheric flexure, and the Paleozoic stratigraphy of the eastern interior of North America: Canadian Journal of Earth Sciences, v. 21, p. 973-996.

READ, W. A., AND FORSYTH, I. H., 1989, Allocycles and autocycles in the upper part of the Limestone Coal Group (Pendleian E1) in the Glasgow-Stirling region of the Midland Valley of Scotland: Geological Journal, v. 24, p. 121-137.

RUSS, D. P., 1982, Style and significance of surface deformation in the vicinity of New Madrid, Missouri, *in* McKeown, F. A., and Pakiser, L. C., eds, Investigations of the New Madrid, Missouri, earthquake region: Washington, D. C., United States Geological Survey Professional Paper 1236, p. 95-114.

SAHAGIAN, D. L., AND WATTS, A. B., 1991, Introduction of the special section on measurement, causes, and consequences of long-term sea level change: Journal of Geophysical Research, v. 96, p. 6585-6589.

SAWYER, D. S., HSUI, A. T., AND TOKSOZ, M. N., 1987, Extension, subsidence and thermal evolution of the Los Angeles Basin- a two-dimensional model: Tectonophysics, v. 133, p. 15-32.

SCLATER, J. G., AND CHRISTIE, P. A .F., 1980, Continental stretching: an explanation of the post Mid Cretaceous subsidence of the central North Sea basin: Journal of Geophysical Research, v. 85, p. 3711-3739.

SIEH, K. E., 1978, Prehistoric large earthquakes produced by slip on the San Andreas fault at Pallett Creek, California: Journal of Geophysical Research, v. 83, p. 3907-3939.

SIEH, K. E., 1984, Lateral offsets and revised dates of large prehistoric earthquakes at Pallett Creek, southern California: Journal of Geophysical Research, v. 89, p. 7641-7670.

SIEH, K. E., AND JAHNS, R. K., 1984, Holocene activity of the San Andreas fault at Wallace Creek, California: Geological Society of America Bulletin, v. 95, p. 883-896.

SIMO, A., 1989, Carbonate-sediment accumulation rates (abs.), *in* Franseen, E. K., and Watney, W. L., eds, Sedimentary Modelling: Computer Simulation of Depositional Sequences: Lawrence, Kansas Geological Survey Subsurface Series 12, p. 37-39.

STECKLER, M. S., AND WATTS, A. B., 1978, Subsidence of the Atlantic-type continental margin off New York: Earth and Planetary Science Letters, v. 41, p. 1-13.

STOW, D. A. V., AND DEAN, W. E., 1984, middle Cretaceous black shale at Site 530 off southeastern Angola basin, *in* Hay, W. W., Sibuet, J. C., and others, Initial Report of the Deep Sea Drilling Project, v. 75: Washington, D. C., United States Government Printing Office, p. 809-819.

TANKARD, A. J., 1986, Depositional response to foreland deformation in the Carboniferous of eastern Kentucky: American Association of Petroleum Geologists Bulletin, v. 70, p. 853-868.

THOMAS, W. A., 1983, Continental margins, orogenic belts, and intracratonic structures: Geology., v. 11, p. 170-272.

WATNEY, W. L., 1985, Resolving controls on epeiric sedimentation using trend surface analysis: Mathematical Geology, v. 17, p. 427-454.

WATNEY, W. L., FRENCH, J. A., AND WONG, J. C., 1989, Depositional-sequence analysis and computer simulation of Upper Pennsylvanian (Missourian) strata in the midcontinent, United States (abs.), *in* Franseen, E. K., and Watney, W. L., eds, Sedimentary Modelling: Computer Simulation of Depositional Sequences: Lawrence, Kansas Geological Survey Subsurface Geology Series 12, p. 67-70.

WATNEY, W. L., KAESLER, R. L., AND NEWELL, K. D., 1985, Recent interpretations of Late Paleozoic cyclothems: Lawrence, Midcontinent Section, Society of Economic Paleontologists and Mineralogists, 3rd Annual Meeting and Guidebook Proceedings, 279 p.

WATTS, A. B., KARNER, G. D., AND STECKLER, M. S., 1982, Lithospheric flexure and the evolution of sedimentary basins: Philosophical Transactions of the Royal Society of London, Series A, v. 305, p. 249-281.

WELLER, J. M., 1964, Development of the concept and interpretation of cyclic sedimentation, *in* Merriam, D. F., ed., Symposium on Cyclic Sedimentation: Kansas Geological Survey Bulletin 169, p. 607-621.

WILLARD, D. A., AND KLEIN, G. DEV., 1990, Tectonic subsidence analysis of the central Appalachian basin: Southeastern Geology, v. 30, p. 217-239.

ZIEGLER, P. A., 1987, Late Cretaceous and Cenozoic intraplate compressional deformations in the Alpine foreland- a geodynamic model: Tectonophysics, v. 137, p. 389-420.

EUSTATIC AND TECTONIC CONTROL OF DEPOSITION OF THE LOWER AND MIDDLE PENNSYLVANIAN STRATA OF THE CENTRAL APPALACHIAN BASIN

DONALD R. CHESNUT, JR.

Kentucky Geological Survey, 228 Mining and Mineral Resources Building, University of Kentucky, Lexington, Kentucky 40506-0107

ABSTRACT: Stratigraphic analysis of the Lower and Middle Pennsylvanian rocks of part of the central Appalachian basin reveals two orders of cycles and one overall trend in the vertical sequence of coal-bearing rocks. The smallest order cycle, the coal-clastic cycle, begins at the top of a major-resource coal bed and is composed of a sequence of shale, siltstone, sandstone, seat rock, and overlying coal bed which, in turn, is overlain by the next coal-clastic sequence.

The major marine-transgression cycle is composed of five to seven coal-clastic cycles and is distinguished by the occurrence of widespread, relatively thick (generally greater than 5 m) marine strata at its base. The Breathitt coarsening-upward trend describes the general upward coarsening of the Middle Pennsylvanian part of the Breathitt Group and includes at least five major marine-transgression cycles.

Chronologic analysis, based on averaging relative age dates determined in previous investigations, provides a duration of 20 my for the deposition of Lower and Middle Pennsylvanian strata of the central Appalachian basin. The eight major marine-transgression cycles that occurred in this interval are calculated to represent an average of 2.5 my each. The average duration of the coal-clastic cycle, in contrast, is calculated to be only about 0.4 my.

The average duration of coal-clastic cycles is of the same order of magnitude (10^5 yr) as the Milankovitch orbital-eccentricity cycles and matches the 0.4 my second-order eccentricity cycle (Long Earth-Eccentricity cycle). These orbital periodicities are known to modulate glacial stages and glacio-eustatic levels. The calculated periodicities of the coal-clastic cycles can be used to support glacio-eustatic control of the coal-bearing rocks of the Appalachian basin. The 2.5-my periodicity of the major marine-transgression cycle does not match any known orbital or tectonic cycle. The cause of this cycle is unknown, but might represent episodic thrusting in the orogen, propagation of intra-plate stresses, or an unidentified orbital cycle. The Breathitt coarsening-upward trend represents the increasing intensity and proximity of the Alleghanian orogeny.

INTRODUCTION

The Pennsylvanian System in the central Appalachian foreland basin (Fig. 1) formed by tectonic processes (Tankard, 1986; Quinlan and Beaumont, 1984; Chesnut, 1991a). Hence, tectonic processes have been interpreted by some (e.g., Tankard, 1986; Willard and Klein, 1990) as the dominant control on sedimentation in that basin. Moreover, vertical variability of the heterogenous Pennsylvanian coal-bearing strata has been interpreted as having resulted largely from tectonic processes, such as episodic thrusting (Tankard, 1986) and autocyclic processes such as delta switching (Ferm, 1979). In contrast, the subtle effects of eustasy are generally inferred by these workers to be insignificant.

However, this was not always the view. Wanless and Shepard (1936) suggested that the cyclicity of coal and overlying shales and sandstones in the Appalachian region (e.g., Ashley, 1931; Reger, 1931) was controlled by glacial eustasy. Were Wanless and Shepard right? Does evidence exist for glacio-eustatic control in the coal-bearing rocks of the central Appalachian basin? In the following discussion, the Lower and Middle Pennsylvanian rocks of the central Appalachian basin will be examined for evidence concerning the possible roles of allocyclic and autocyclic processes on the deposition of Pennsylvanian coal-bearing strata in the Appalachian region.

STRATIGRAPHIC ANALYSIS

Stratigraphic nomenclature used here has recently been accepted by the Kentucky Stratigraphic Nomenclature Committee and formalized in Kentucky by Chesnut (1992). The Lower

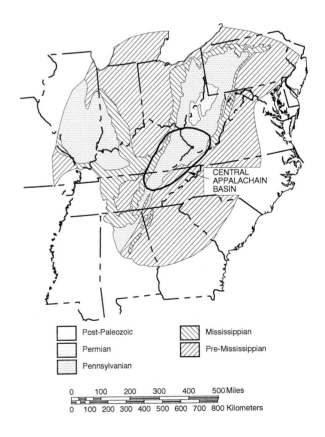

FIG. 1.—Location of central Appalachian basin.

Tectonic and Eustatic Controls on Sedimentary Cycles, SEPM Concepts in Sedimentology and Paleontology #4
Copyright © 1994 SEPM (Society for Sedimentary Geology), ISBN 1-56576-017-4, p. 51-64.

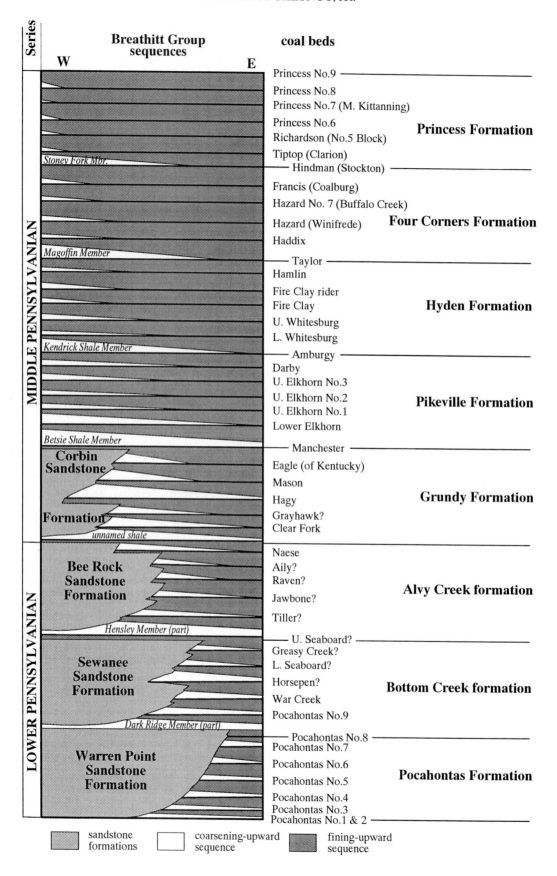

FIG. 2.—Relationship of the coal-clastic cycles, the major-marine-transgression cycle, and the Breathitt coarsening-upward trend.

and Middle Pennsylvanian rocks of the central Appalachian basin (Fig. 2) are composed of shales, siltstones, sandstones, and coal beds of the Breathitt Group (and lateral equivalents in West Virginia, Virginia, and Tennessee). The Breathitt Group is divided into eight formations; the base of the formations are defined at the base of extensive, thick marine members shown in Figure 2.

The lower part of the Breathitt Group contains massive quartzose sandstones, formerly known as the Lee Formation (Fig. 2), herein collectively and informally called Lee sandstones. Lee sandstones occur in two forms: (1) as unconformity channel fills, and (2) as large sandstone belts. Several deep channels within the mid-Carboniferous regional unconformity surface, filled with Lee-type sandstones, have been described by Rice (1984) and Beuthin (this volume). Mapping of these channels indicates flow largely to the south. The largest volume of Lee-type sandstones, however, occurs as four large, belt-shaped lenses identified, in ascending order, as the Warren Point, Sewanee, Bee Rock, and Corbin Sandstone Formations (Fig. 2). These belts trend southwest and average about 80 km (50 mi) in width. The belts are as thick as 150 m (487 ft) and are composed of a number of sandstone beds, most of which are 20 to 30 m (66-98 ft) thick. The sandstone beds are separated by thin beds of coal, rooted underclay, shale, and siltstone. Thicker sandstone beds are known but probably represent coalesced sandstone bodies. The dominant crossbed direction for Lee sandstones is to the southwest, parallel to the trend of the sandstone belts. The Lee sandstones and formations of the lower part of the Breathitt Group onlap the mid-Carboniferous unconformity (described below) surface to the northwest. Sandstone formations appear to be largely confined in a topographic low between the dipping unconformity surface to the northwest and the Breathitt Formation to the southeast.

Pennsylvanian rocks unconformably overlie Mississippian rocks throughout most of the central Appalachian basin. The unconformity is recognized from both surface and subsurface analysis. The contact at the surface is marked by such features as scours, channels, pedogenic flint clay, iron ore, and paleokarst. In the sub-surface, regional cross sections reveal sequential truncation of lowermost Pennsylvanian (Pocahontas Formation) and Mississippian (Pennington Group and underlying) rocks from the center of the basin toward the northwestern margin (Fig. 3). In the deepest parts of the basin (parts of Virginia and southeastern West Virginia), the contact is reported to be conformable (Englund, 1979). Where the lowermost Pennsylvanian rocks are removed by erosion (Fig. 3), the unconformity is an apparent "Mississippian-Pennsylvanian" unconformity. Along the western edge of the basin, strata as young as Middle Pennsylvanian directly overlie rocks as old as Early Mississippian.

Extent of Coal Beds

Cursory examination of outcrops of coal-bearing rocks in the Appalachians readily suggests that coal beds are variable in

geometry and of modest extent. Although variable, the major coal beds are much more extensive than they first appear. These beds are typically 0.3 to 2.0 m in thickness, extend in excess of 100 km, and are mined in several coal districts. The extent of the major coal beds was not evident until completion in 1978 of the joint Kentucky Geological Survey-U. S. Geological Survey geologic mapping program. Detailed geologic maps showed that major coal beds (about 50 in number) extend across eastern Kentucky (see Rice and Smith, 1980) and, based on information from other states, continued into adjoining states. The accuracy of the mapping program was tested by the Kentucky Geological Survey's coal-resource program in which more than 25,000 coal thickness and location points were collected throughout eastern Kentucky. Results of the resource study verified the occurrence of coal beds as indicated by the geologic quadrangle maps.

This is not to say that all coal beds occur without interruption. Isopach maps of the major coal beds generated by the coal resource project illustrate that, although the coal beds are extensive, they are locally absent, especially along curvilinear trends (e.g., Fire Clay coal in Brant and others, 1983). Most of these trends are occupied by scour-based channel-fill and overbank facies.

Several tonsteins (altered volcanic ash) are known from coal beds in the basin (Slusher, pers. commun., 1982; Chesnut, 1985; Burger and Damberger, 1985, p. 441-442; Outerbridge and others, 1989; Triplehorn and Finkleman, 1989; Triplehorn and others, 1989). Because several of these tonsteins can be traced within coal beds across the basin, the coals are interpreted to represent essentially isochronous units. For example, the tonstein in the Fire Clay coal bed is generally found in the lower half of the equivalent coal bed in Tennessee, Virginia, Kentucky, and West Virginia, demonstrating that the ash fell upon an extensive peat deposit (Chesnut, 1985). The widespread nature of the major coal beds demonstrates that many of the Pennsylvanian peat swamps extended beyond the present size of the central Appalachian basin.

Occurrence of Marine Strata

Coal-bearing rocks of the Appalachian Basin had been considered to represent sediments deposited during largely terrestrial conditions; coal beds are abundant, and plant fossils are common throughout the section. For many years, only a few marine beds were identified in eastern Kentucky (e.g., Morse, 1931). These included the regionally widespread Kendrick and Magoffin members which some have interpreted as uncommon marine incursions caused by sporadic tectonic pulses (e.g., Tankard, 1986).

However, geologic-map data and paleontological reports (Chesnut, 1981, 1991b), as well as new field studies (e.g., Cobb and others, 1981; Keiser and others, 1988; Martino, 1988, 1989) reveal that marine strata are abundant and occur in numerous zones. Marine strata appear to overlie all major coal beds in Lower and Middle Pennsylvanian rocks in eastern Kentucky across at least part of their occurrence. More than 40 separate

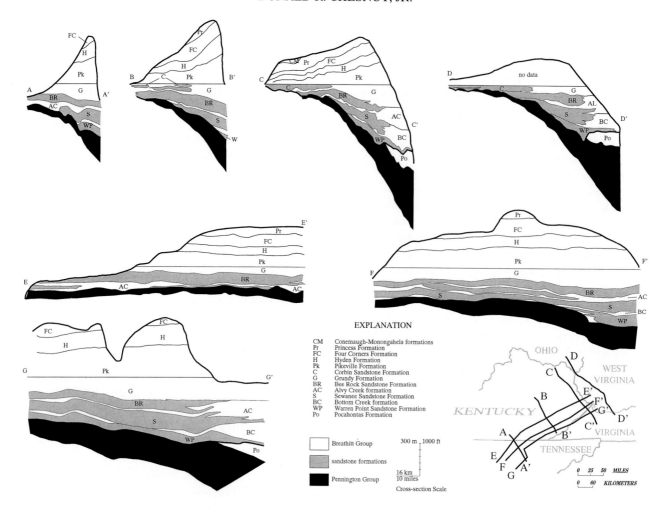

FIG. 3.—Cross sections through the central Appalachian basin. Datum is on the base of the Pikeville Fm. Units are, in descending order: CM—Conemaugh and Monongahela Fms.; Pr—Princess Fm; FC—Four Corners Fm.; H—Hyden Fm.; Pk—Pikeville Fm.; C—Corbin Sandstone Fm.; G—Grundy Fm.; BR—Bee Rock Sandstone Fm.; AC—Alvy Creek Fm. (upper part of New River Fm.); S—Sewanee Sandstone Fm.; BC—Bottom Creek Fm. (lower part of New River Fm.); WP—Warren Point Sandstone Fm.; and Po—Pocahontas Fm.

marine members ("zones") have been identified in eastern Kentucky (Chesnut, 1991b). Keiser and others (1988) also found several previously unrecognized marine units in West Virginia.

The typical marine member in the basin is a coarsening-upward sequence of shale, silty shale, siltstone, and sandstone (bayfill of Baganz and others, 1975; marine zone of Outerbridge, 1976). Body fossils occur locally, especially near the base, but are generally rare. However, trace fossils and zooturbation are generally abundant. Large calcareous concretions are common within the sequence, but thin, fossiliferous, argillaceous limestone beds are rare. Where limestones occur, they are near the base of the coarsening-upward unit.

Coarsening-upward sequences can be recognized in geophysical well logs and extend across the basin (Chesnut, 1988, 1992). Subsurface and surface stratigraphic analyses (Chesnut, 1988) indicate that many marine strata are extremely widespread and, together with coal beds, extend for more than 200 km. Detailed cross sections, shown in greatly simplified form in Figure 3, suggest that many of the marine zones can be traced

through Tennessee, Kentucky, Virginia, and West Virginia (Chesnut, 1988, 1992).

Cycles and Trends in the Coal-Bearing Rocks

The lateral extent of major coal beds, coupled with the coextensive marine members, suggests that regional transgression and regression were important controls on deposition, and that autocyclic processes such as delta switching (e.g., Ferm, 1979) were operative within and not among cycles. If relative sea-level variation controlled cyclicity, was glacial eustasy or foreland-basin tectonics the driving mechanism? Analysis of the cycles and trends within the coal-bearing rocks of the basin provides clues to this question.

To begin this analysis, seven detailed cross sections of the Lower and Middle Pennsylvanian rocks of the central Appalachian basin, shown in simplified version in Figure 3, were constructed from several thousand subsurface records (Chesnut, 1988, 1992). Examination of the original cross sections reveals two types of cycles and one overall trend in the deposition of these rocks (Chesnut, 1988, 1989, 1991c, 1991d; Chesnut and Cobb, 1989). In order of increasing magnitude, these are the coal-clastic cycle, the major-marine-transgression cycle, and the Breathitt coarsening-upward trend.

Coal-clastic cycle.—

The coal-clastic cycle begins at the top of a major-resource coal bed and is composed of a coarsening-upward sequence of shales, siltstones, and sandstones. They are in turn overlain by a fining-upward sequence of sandstones and shales, all of which are capped by a seat rock (e.g., underclay) and overlying coal bed (Fig. 4). The coarsening-upward sequence represents sediment deposited in marine or brackish-water environments (Baganz and others, 1975; Chesnut, 1981). This sequence is commonly truncated erosionally and overlain by a fining-upward sequence. The fining-upward sequence is dominated by crossbedded sandstones that contain channel-fill features such as a basal scour, basal intraformational conglomerate, epsilon crossbedding, thinning-upward bedding, and a sequence of sedimentary structures indicating waning flow. The fining-upward sequence is generally interpreted as representing fluvial channels (e.g. Gardner, 1983; Cobb and others, 1981).

The sandstone formations (Corbin, Bee Rock, Sewanee, and Warren Point) known collectively as "Lee sandstones" outcrop along the western margin of the Appalachian basin where subsidence was reduced. Several sandstone beds and numerous bounding surfaces are observed in each sandstone formation. The sandstone beds may represent coal-clastic cycles. Bounding surfaces within sandstone beds, however, (1) may reflect scouring of one sandstone body (from an earlier cycle) and deposition of a subsequent sandstone along the basin margin, or (2) may represent autocyclic processes such as channel-switching and migration. Ongoing research on the Lee sandstones will help to delineate autocyclic versus allocyclic processes affecting their deposition. Nonetheless, the sandstone formations, as well as the coal-bearing Breathitt formations, appear to have been controlled by the next higher order of cycle, the major-transgression cycle.

Major-transgression cycle.—

The next order cycle is the major-transgression cycle which corresponds to each of the eight formations shown on Figure 2. The major-transgression cycle is composed of several coal-clastic cycles. However, the basal coal-clastic sequence of this cycle contains an unusually well-developed marine or brackish-water, coarsening-upward component. The marine strata at the base of the cycles are generally thick (more than 5 m), very extensive, and are commonly fossiliferous. These major marine strata (Fig. 2), many of which are used as key stratigraphic units in geologic mapping, are, in ascending order: the Dark Ridge Member (part of), the Hensley Member (part of), an unnamed shale member (informally, the Dave Branch shale), the Betsie Shale Member, the Kendrick Shale Member, the Magoffin Member, and the Stoney Fork Member (Chesnut, 1991b).

Generalized Appalachian Cyclothem (Coal-Clastic Cycle)

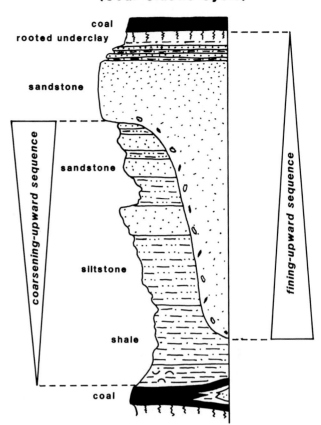

FIG. 4.—Simplified coal-clastic cycle begins at the top of a major coal bed and is composed of a marine or brackish-water, coarsening-upward sequence that is erosionally truncated and overlain by a fining-upward sequence, ending with a seatrock and overlying coal bed.

Breathitt coarsening-upward uend.—

The Breathitt coarsening-upward trend represents the general coarsening-upward grain-size of the Middle Pennsylvanian part of the Breathitt Group, exclusive of the "Lee" sandstones (Fig. 2) which have a different provenance.

Coal-clastic cycles within the Breathitt show lateral and vertical variation that reflects this trend. Thus, in the western part of the basin, the coal-clastic cycles are dominated by coarsening-upward shales. To the east, the equivalent coal-clastic cycles are dominated by fluvial fining-upward sandstones. A similar trend occurs vertically; shale-dominated coarsening-upward sequences are common in the lower part of the Breathitt Group, whereas sandstone-dominated, fining-upward sequences are prevalent toward the upper part of the Breathitt Group.

PRELIMINARY CHRONOLOGIC ANALYSIS

Some of these stratigraphic cycles in the Breathitt Group could represent glacially controlled eustatic cycles. The close association of many coal beds and marine zones in the Middle

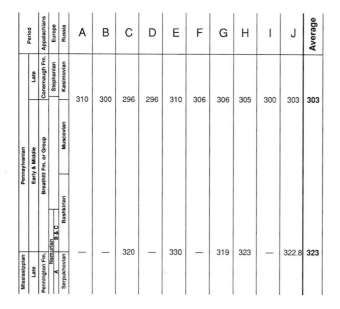

Fig. 5.—Chronologic age dates projected to the Lower and Middle Pennsylvanian strata of the central Appalachian basin. The average age in millions of years, based on these projections, for the top of the Breathitt and the base of the Pennsylvanian is shown in the right-hand column. Data attributed to A, Shell (1980); B, Odin (1982a, 1982b); C, Harland and others (1982); D, Palmer (1983); E, Salvador (1985); F, Odin (1986); G, Lippolt and others (1984), Hess and Lippolt (1986), Hess and others (1988); H, Menning (1989); I, Cowie and Bassett (1989); J, Harland and others (1990).

Pennsylvanian rocks of the Appalachian basin indicates that transgression commonly followed peat deposition. The coals were deposited as peat in extensive coastal-plain peat swamps similar to those in modern Indonesia (Grady and others, 1989; Cobb and others, 1989). Carboniferous continental glaciation coeval with Pennsylvanian coal-bearing rocks of the Appalachians is well documented (i.e., López Gamundí, 1983, 1987; Caputo and Crowell, 1985; Veevers and Powell, 1987). Pleistocene continental glaciation caused sea-level changes as much as 100-150 m around the world (Frazier, 1974; Donovon and Jones, 1979). Eustatic sea-level changes as small as a few tens of meters would have drastically affected the Carboniferous lowland environments, even those with thick raised peat domes. The discovery of stratigraphic cycles with similar periodicity as orbital cycles known to modulate glacial stages (e.g., Imbrie, 1985) would support the interpretation of eustatic control.

What is the average interval of time that the Breathitt rock cycles and trends represent? At present only one Pennsylvanian bed in the basin, the Fire Clay coal tonstein, has been radiometrically dated (311 my, Kurt Burger, pers. commun.). There are not enough data at present to establish a chronologic framework based upon radiometric dates from the Appalachian Basin. Therefore, another method had to be employed to approximate cycle duration.

A biostratigraphic framework for the Pennsylvanian rocks of the central Appalachian basin, based upon a variety of plant, spore, invertebrate, and microfossil schemes developed by other workers, has been compiled by Chesnut (1991b). According to the biostratigraphic framework, the top of the Breathitt Group (equivalent to the Middle Pennsylvanian top as used in the Appalachian basin) is near the Muscovian-Kasimovian (approx. Westphalian-Stephanian) boundary (Fig. 5). The base of the Pennsylvanian System (as used in the Appalachian basin) is near the Serpukhovian-Bashkirian (Namurian A-Namurian B) boundary.

The numerous chronologic schemes published in the last decade each have some merit, but reflect different ages (Fig. 5). The average age thus calculated for the top and bottom of the Lower and Middle Pennsylvanian rocks in the central Appalachian basin is 303 my and 323 my, respectively, suggesting an average duration of 20 my. Using extreme ages, the duration ranges from 9 to 34 my. The Lower and Middle Pennsylvanian section shown in Figure 2 is divided into eight major-transgression cycles, which are of roughly equal thickness (Chesnut, 1988, 1992). Therefore, the average duration for the major-transgression cycle is 2.5 my (with a range, based on extreme values, of 1.1 to 4.3 my). The major-transgression cycles contain five to seven major coal beds with an average of six coal-clastic cycles (Fig. 2). Coal-clastic cycles therefore average 0.4 my (with a range of 0.2 to 0.7 my).

IMPLICATIONS FOR GLACIAL EUSTASY AND TECTONICS

The Breathitt coarsening-upward trend probably represents increasing intensity of the Alleghanian orogeny. The suggested 2.5 my periodicity of the major-transgression cycle does not correspond to any known orbital periodicity; this cycle is probably controlled by either some periodic tectonic mechanism or an unknown orbital periodicity. However, the 10^5-yr order of magnitude of the coal-clastic periodicity is equivalent to eccentricity orbital cycles, and the calculated 0.4-my periodicity corresponds with the approximately 0.4-my Second Order Long Earth Eccentricity cycle (Imbrie, 1985). Although this correspondence does not, of itself, prove eustasy, the close correspondence of the duration of coal-clastic cycles with this eccentricity cycle suggests a glacio-eustatic control over Pennsylvanian sedimentation in the central Appalachian basin.

RECONSTRUCTION OF EVENTS

A possible modern analog for the low-sulfur, low-ash Appalachian coals (i.e., Lower and Middle Pennsylvanian) is provided by the coastal, domed, ombrogenous (rain-fed) peats of Indonesia (Grady and others, 1989; Cobb and others, 1989). However, present sea level is near highstand; as sea level drops, this modern peat deposit may not be preserved. On the other hand, radiocarbon dating of three buried peats in Indonesia shows a close correspondence to the last three glacial lowstands (Cobb and others, 1989). In addition, these buried peats are covered by marine sediments. Coastal peat deposition appears to occur at either highstand or lowstand when there is a relative stasis in eustatic conditions and sediment flux (Cobb and others, 1989). However, lowstand peat deposits are more likely to be extensively covered and preserved than highstand peats which would tend to be eroded.

Based on the previous discussion, many of the major coal beds of the Appalachians were probably deposited as coastal peats, perhaps at lowstand. As sea level rose, marine waters transgressed the peats, and fine-grained muds were deposited. As highstand approached, the coastline prograded and coarser clastic sediments were deposited upon the muds, creating a coarsening-upward sequence. At highstand, coastal peats may have been deposited, but as sea level dropped, high-level deposits were eroded, reworked, and carried in channels out to sea. Clastic material was transported across the basin providing a substrate for the next lowstand peat swamp.

LOWSTAND AND HIGHSTAND PALEOGEOGRAPHY

Little is known of the Pennsylvanian paleogeography of eastern North America at lowstand and highstand. Inferences can be made based on the distribution of strata within the basins and their position relative to the mid-Carboniferous unconformity (e.g., Nelson and others, 1991; Chesnut, 1992) and the distribution of biofacies within the marine strata (Chesnut, 1991c). Preliminary Pennsylvanian paleogeography maps based on these inferences are discussed here.

An Early Pennsylvanian (post mid-Carboniferous unconformity) extreme lowstand is shown in Figure 6. During Early Pennsylvanian deposition, strata progressively onlapped the newly formed Appalachian and Illinois basins, but had not covered the forebulge between the two basins. Current indicators show that immature clastics (represented by the Breathitt Group) derived from the Appalachian Mountains were transported to the west (Fig. 6) and formed a northwestward dipping clastic wedge. At lowstand, the Breathitt rivers flowed into a large trunk transport system (represented by the "Lee" sandbelts) flowing to the southwest. The Lee sandbelts were situated in a low trough between the forebulge and the Breathitt clastic wedges and may ultimately have dumped sediment into the Ouachita trough to the west via the Black Warrior-southern Appalachian basins. In the central Appalachian basin, the mechanism of transport of the Lee sands was largely fluvial at this stage. On the western side of the forebulge, the similar Caseyville sandstone belt carried sand to the northeastern part of the Ouachita trough.

An Early Pennsylvanian highstand is shown in Figure 7. Based on distribution of biofacies in eastern Kentucky (Chesnut, 1991c), marine waters entered the basin from the south. Marine waters are inferred to have transgressed from the Ouachita trough (at extreme lowstand), through the Black Warrior-southern Appalachian basins, into the central Appalachian basin (Fig. 7). Tidal and marine reworking of Lee sands may have occurred at this stage. In addition, subsequent tectonic and compactional subsidence may have formed a broad, low trough along the axis of the basin, causing the initial Lee transgression belt to widen

to the southeast. In the Illinois basin, marine waters also transgressed northeast across the Caseyville sandbelt; however, waters do not appear to have connected across the forebulge separating the basins (Fig. 7).

During most of Middle Pennsylvanian deposition (Fig. 8), trunk transport systems like the Lee and Caseyville sandbelts did not form at lowstands. Rather, Breathitt rivers apparently carried clastics through saddles formed in the forebulge (Fig. 8). The ultimate destination of the clastics may have been the Ouachita trough and areas west of the trough. After cresting of the forebulge with the Breathitt clastic wedge, there was no low trough in which the Lee sand belts could form, thus ending Lee-type deposition in the central Appalachian basin.

Preliminary biofacies distribution in marine strata in the central Appalachians indicates increasing marine conditions to the west and south during most of Middle Pennsylvanian time (Chesnut, 1991c). During rising sea level, marine waters apparently transgressed not only through the southern route to reach the central Appalachian basin, but also across saddles in the forebulge from the western basin (Fig. 9). Because there was no well-developed low trough between the Breathitt clastic wedge and the forebulge, as there had been during Early Pennsylanian sedimentation, the nature of marine transgressions may have been different. Tectonic and compactional subsidence probably formed a broad, low conduit for transgression from the south through the Black Warrior-Appalachian basins. Major transgressions may also have entered across saddles, entering the basin first through estuaries (Greb and Chesnut, 1992) and eventually covering large areas of the western-dipping clastic wedge.

During the Late Pennsylvanian lowstands, transport direction of clastics was probably similar to that of Middle Pennsylvanian transport direction (Fig. 8). However, the Ouachita trough was destroyed by late Middle and Late Pennsylvanian collisional tectonics (Houseknecht, 1986) and could not have been the final destination of the Appalachian clastics. Clastic transport must have proceeded to the west.

Late Pennsylvanian rising sea levels (Fig. 10) could not transgress from the Ouachita trough through the Black Warrior-southern Appalachian basins because the trough was destroyed, and the southern route may have been blocked by collisional tectonism. Transgressions must have proceeded through the Illinois basin and entered the southern part of the northern Appalachian basin across a much diminished forebulge north of the Jessamine dome (Fig. 10).

CONCLUSIONS

The general coarsening-upward grain size of the Lower and Middle Pennsylvanian rocks of the central Appalachian basin (Breathitt coarsening-upward trend) probably represents increasing intensity and proximity of the Alleghanian orogeny. However, the occurrence of major marine strata deposited at apparent periodic intervals (major-transgression cycle) of 2.5 my is enigmatic; no known periodicity matches this interval.

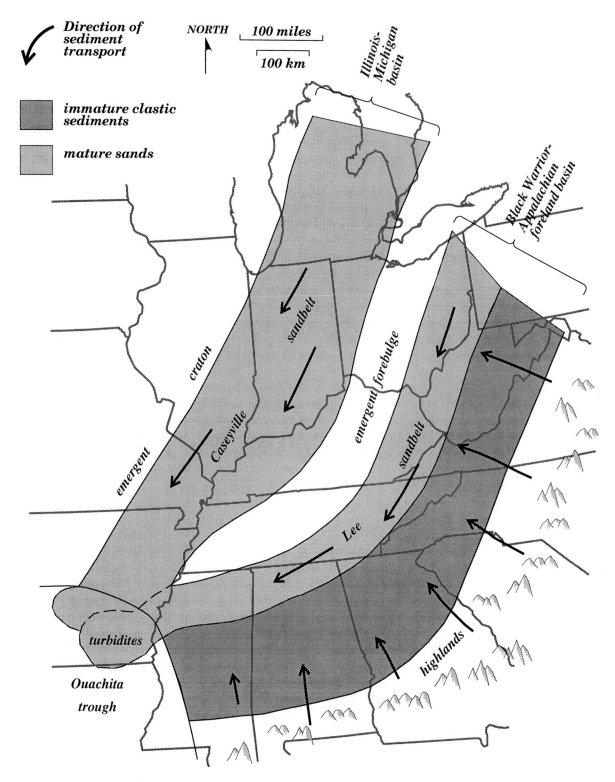

Early Pennsylvanian lowstand

FIG. 6.—Paleogeography during the Early Pennsylanian lowstand.

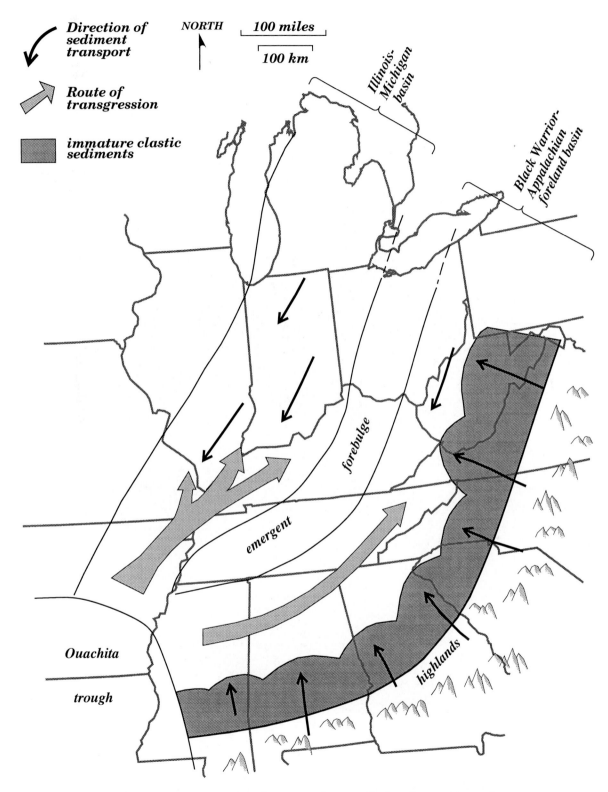

Early Pennsylvanian highstand

FIG. 7.—Paleogeography during the Early Pennsylanian highstand.

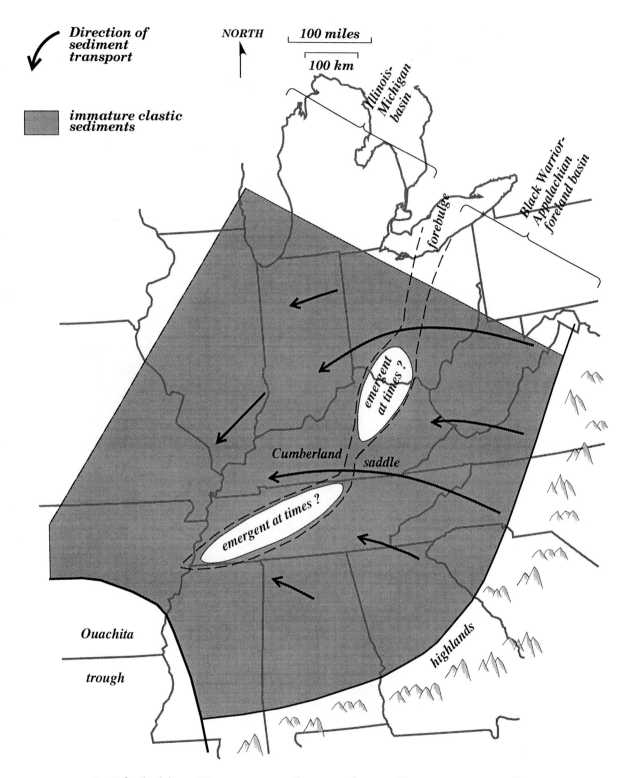

Middle Pennsylvanian lowstand

FIG. 8.—Paleogeography during the Middle Pennsylanian lowstand.

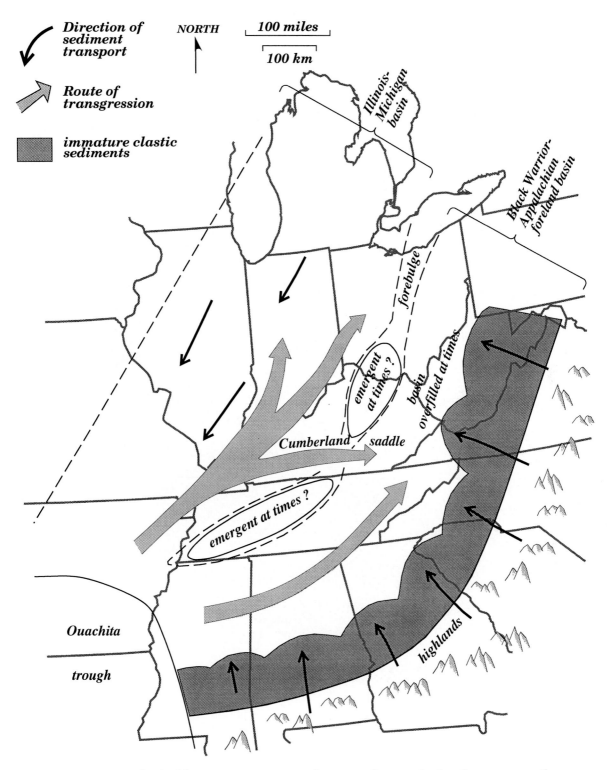

Middle Pennsylvanian highstand

FIG. 9.—Paleogeography during Middle Pennsylanian highstand.

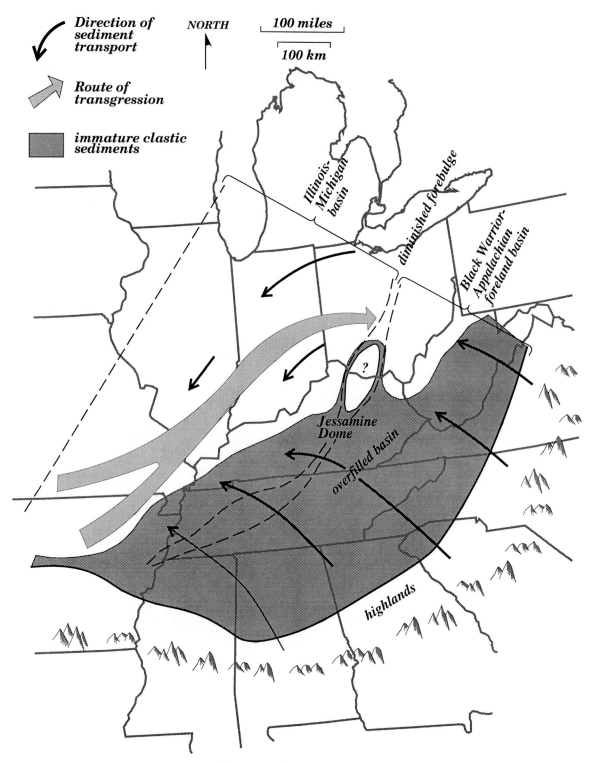

Late Pennsylvanian highstand

FIG. 10.—Paleogeography during the Late Pennsylanian highstand.

This stratigraphic cycle may be controlled either by an unidentified periodic tectonic mechanism or by an unknown orbital variation. The 0.4-my periodicity of coal-clastic cycles is interpreted to represent glacio-eustatic control which is, in turn, modulated by orbital periodicities. Autocyclic processes such as delta switching and river avulsion occurred in the central Appalachian basin, but are represented by lithologic variation found within coal-clastic cycles.

The route of marine transgression into the central Appalachian basin during the Early Pennsylvanian was northward through the Black Warrior and southern Appalachian basins. During Middle Pennsylvanian deposition, marine waters transgressed through the same route as during Early Pennsylvanian deposition and also across saddles in the forebulge. During the Late Pennsylvanian time, transgression into the Appalachian basin occurred only through the Illinois basin and north of the Jessamine dome. Regressive sediment transport during the Early, Middle, and Late Pennsylvanian was the reverse of transgressive routes.

ACKNOWLEDGMENTS

I wish to thank the following colleagues at the Kentucky Geological Survey: James C. Cobb and Stephen F. Greb (for discussions and comments), Donald W. Hutcheson (for editorial assistance), Robert Holladay and Michael Murphy (for drafting).

I would also like to thank the editors of this volume, Frank R. Ettensohn and John M. Dennison, for their considerations and comments, and the reviewers, Jack C. Pashin and Charles L. Rice, who offered many useful suggestions.

I must also thank Velbert M. Nubbins and Kumquat May for assistance in entering subsurface data into the computer data base.

REFERENCES

ASHLEY, G. H., 1931, Pennsylvanian cycles in Pennsylvania: Illinois Geological Survey Bulletin 60, p. 241-245.

BAGANZ, B. P., HORNE, J. C., AND FERM, J. C., 1975, Carboniferous and Recent Mississippi lower delta plains- A comparison: Gulf Coast Association of Geological Societies, v. 37, p. 556-591.

BRANT, R. A., CHESNUT, D. R., FRANKIE, W. T., AND PORTIG, E. R., 1983, Coal resources of the Hazard District, Kentucky: Lexington, University of Kentucky Institute for Mining and Minerals Research, Energy Resource Series, 49 p.

BURGER, K., AND DAMBERGER, H. H., 1985, Tonsteins in the coalfields of western Europe and North America, in Ninth International Congress of Carboniferous Stratigraphy and Geology, May 17-26, 1979: Washington, D.C., Compte Rendu, v. 4, p. 433-448.

CAPUTO, M. V., AND CROWELL, J. C., 1985, Migration of glacial centers across Gondwana during Paleozoic Era: Geological Society of America Bulletin, v. 96, p. 1020-1036.

CHESNUT, D. R., JR., 1981, Marine zones of the (upper) Carboniferous of eastern Kentucky, in Cobb, J. C., Chesnut, D. R., Hester, N. C., and Hower, J. C., eds., Coal and Coal-bearing Rocks of Eastern Kentucky, Geological Society of America Coal Geology Division field trip, November 1981: Lexington, Kentucky Geological Survey Series 11, p. 57-66.

CHESNUT, D. R., JR., 1985, Source of the volcanic ash deposit (flint clay) in the Fire Clay coal of the Appalachian Basin, in Dixième Congrés International de Stratigraphie et de Géologie du Carbonifère, 1983: Madrid, Compte Rendu, v. 1, p. 145-154.

CHESNUT, D. R., JR., 1988, Stratigraphic analysis of the Carboniferous rocks of the Central Appalachian Basin: Unpublished Ph. D. Dissertation, University of Kentucky, Lexington, 296 p.

CHESNUT, D. R., JR., 1989, Pennsylvanian rocks of the Eastern Kentucky Coal Field, in Cecil, C. B., and Eble, C., eds., Geology of the Carboniferous rocks of the eastern United States: Washington, D. C., 28th International Geological Congress, Field Trip No. 143, p. 57-60.

CHESNUT, D. R., JR., 1991a, Timing of Alleghanian tectonics determined by Central Appalachian foreland basin analysis: Southeastern Geology, v. 31, no. 4, p. 203-221.

CHESNUT, D. R., JR., 1991b, Paleontological survey of the Eastern Kentucky Coal Field: Invertebrates: Lexington, Kentucky Geological Survey Series 11, Information Circular 36, 71 p.

CHESNUT, D. R., JR., 1991c, Marine transgressions in the Central Appalachian Basin during the Pennsylvanian Period (abs.): Geological Society of America, Abstracts with Programs, v. 23, p. 16.

CHESNUT, D. R., JR., 1991d, Eustatic and tectonic control of sedimentation in the Pennsylvanian strata of the Central Appalachian Basin (abs.): Geological Society of America, Abstracts with Programs, v. 23, p. 16.

CHESNUT, D. R., JR., 1992, Stratigraphic and structural framework of the Carboniferous rocks of the Central Appalachian Basin: Lexington, Kentucky Geological Survey, Series 11, Bulletin 3, 42 p.

CHESNUT, D. R., JR., AND COBB, J. C., 1989, Cycles in the Pennsylvanian rocks of the Central Appalachian Basin (abs.): Geological Society of America, Abstracts with Programs, p. A52.

COBB, J. C., CHESNUT, D. R., JR., HESTER, N. C., AND HOWER, J. C., eds., 1981, Coal and Coal-bearing Rocks of Eastern Kentucky (Geological Society of America Coal Geology Division field trip, November 1981): Lexington, Kentucky Geological Survey, Series 11, 169 p.

COBB, J. C., NORRIS, J. W., AND CHESNUT, D. R., JR., 1989, Glacio-eustatic sea-level controls on the burial and preservation of modern coastal peat deposits (abs.): Geological Society of America, Abstracts with Programs, p. A26.

COWIE, J. W., AND BASSETT, M. G., 1989, 1989 Global stratigraphic chart: Episodes, v. 12, 1-page supplement.

DONOVAN, D. T., AND JONES, E. J., 1979, Causes of world-wide changes in sea level: Journal of the Geological Society of London, v. 136, p. 187-192.

ENGLUND, K. J., 1979, Mississippian System and lower series of the Pennsylvanian System in the proposed Pennsylvanian System stratotype area, in Englund, K. J., Arndt, H. H., and Henry, T. W., eds., Proposed Pennsylvanian System Stratotype, Virginia and West Virginia: The American Geological Institute, Selected Guidebook Series 1, p. 69-72.

FERM, J. C., 1979, Pennsylvanian cyclothems of the Appalachian Plateau, a retrospective view, in Ferm, J. C., and Horne, J. C., eds., Carboniferous Depositional Environments in the Appalachian Region: Columbia, Carolina Coal Group, p. 284-290.

FRAZIER, D. E., 1974, Depositional episodes: Their relationship to the Quaternary stratigraphic framework in the northwestern portion of the Gulf Basin: Austin, Texas Bureau of Economic Geology, Geological Circular 74-1, 28 p.

GARDNER, T. W., 1983, Paleohydrology and paleomorphology of a Carboniferous, meandering, fluvial sandstone: Journal of Sedimentary Petrology, v. 53, p. 991-1005.

GRADY, W. C., EBLE, C. F., AND NEUZIL, S. G., 1989, Distribution of petrographic components in a modern domed tropical Indonesian peat: A possible analog for maceral distributions in Middle Pennsylvanian coal beds of the Appalachian Basin (abs.): Geological Society of America, Abstracts with Programs, p. A25.

GREB, S. F., AND CHESNUT, D. R., JR., 1992, Transgressive channel filling in the Breathitt Formation (Upper Carboniferous), Eastern Kentucky Coal Field, USA: Sedimentary Geology, v. 75, p. 209-221.

HARLAND, W. B., ARMSTRONG, R. L., COX, A. V., CRAIG, L. E., SMITH, A. G., AND SMITH, D. G., 1990, A Geologic Time Scale 1989: Cambridge, Cambridge University Press, 263 p.

HARLAND, W. B., COX, A. V., LLEWELLYN, P. G., PICKTON, C. A. G., SMITH, A. G., AND WALTERS, R., 1982, A Geologic Time Scale: Cambridge, Cambridge University Press, 131 p.

HESS, J. C., AND LIPPOLT, H. J., 1986, 40Ar/^{39}Ar ages of tonstein and tuff sanidines: New calibration points for the improvement of the Upper Carboniferous time scale: Isotope Geoscience, v. 59, p. 143-154.

HESS, J. C., LIPPOLT, H. J., AND BURGER, K., 1988, New time-scale calibration points in the Upper Carboniferous from Kentucky, Donetz Basin, Poland and West Germany (abs.): Abstracts of Jutland Symposium FT Dating, Besaugou, unpaginated.

HOUSEKNECHT, D. W., 1986, Evolution from passive margin to foreland basin: the Atokan Formation of the Arkoma basin, south-central U.S.A.: Special Publication of the International Association of Sedimentologists, v. 8, p. 327-345.

IMBRIE, J., 1985, A theoretical framework for the Pleistocene ice ages: Journal of the Geological Society of London, v. 142, p. 417-432.

KEISER, A. F., BLAKE, B. M., JR., AND MARTINO, R. L., 1988, "Corridor G" (119) Pennsylvanian Stratigraphy: Charleston, American Association of Petroleum Geologists Eastern Section Meeting, Field Trip No. 2, 39 p.

LIPPOLT, H. J., HESS, J. C., AND BURGER, K., 1984, Isotopische alter von pyroklastischen sanidinen aus kaolin-Kohlentonstein als Korrelationsmarken fur das mitteleuroaische Oberkarbon: Forschfirst Geologisches, Rheinland und Westfalen, v. 32, p. 119-150.

LÓPEZ GAMUNDÍ, O. R., 1983, Modelo de sedimentación glacimarina para la Formación Hoyada Verde: Paleozoico superior de la provincia de San Juan, Asociación Geológica Argentina, Revista, v. 38, p. 60-72.

LÓPEZ GAMUNDÍ, O. R., 1987, Depositional models for the glaciomarine sequences of Andean Late Paleozoic basins of Argentina: Sedimentary Geology, v. 52, p. 109-126.

MARTINO, R. L., 1988, The Campbells Creek marine zone: its extent, component facies, and relation to coals of the Kanawha Formation in southern Kanawha County, West Virginia: American Association of Petroleum Geologists Bulletin, v. 72, p. 967.

MARTINO, R. L., 1989, Trace fossils from marginal marine facies of the Kanawha Formation (Middle Pennsylvanian), West Virginia: Journal of Paleontology, v. 63, p. 389-403.

MENNING, M., 1989, A synopsis of numerical time scales, 1917-1986: Episodes, v. 12, p. 3-5.

MORSE, W. C., 1931, The Pennsylvanian invertebrate fauna of Kentucky, in Jillson, W. R., The Paleontology of Kentucky: Lexington, Kentucky Geological Survey, p. 295-349.

NELSON, W. J., TRASK, C. B., JACOBSON, R. J., DAMBERGER, H. H., WILLIAMSON, A. D., AND WILLIAMS, D. A., 1991, Absaroka Sequence, Pennsylvanian and Permian Systems, in Leighton, M. W., ed.,

Interior Cratonic Sag Basins: Tulsa, American Association of Petroleum Geologists, World Basin Series, p. 143-164.

ODIN, G. S., 1982a, Numerical Dating in Stratigraphy: Chichester, Wiley, p. 1-1040.

ODIN, G. S., 1982b, The Phanerozoic time scale revisited: Episodes, v. 1982, p. 3-9.

ODIN, G. S., 1986, Recent advances in Phanerozoic time-scale calibration, in Odin, G. S., Calibration of the Phanerozoic time scale: Chemical Geology, Isotope Geoscience Section, v. 59, p. 103-110.

OUTERBRIDGE, W. F., 1976, The Magoffin Member of the Breathitt Formation: United States Geological Survey Bulletin 1422-A, p. A64-A65.

OUTERBRIDGE, W. F., TRIPLEHORN, D. M., LYONS, P. C., AND CONNOR, C. W., 1989, Altered volcanic ash below the Princess No. 6 coal zone (Middle Pennsylvanian), West Virginia and Kentucky, Central Appalachian Basin (abs.): Geological Society of America, Abstracts with Programs, v. 21, p. 134.

PALMER, A. R., 1983, The Decade of North American Geology 1983 geologic time scale: Geology, v. 11, p. 503-504.

QUINLAN, G. M., AND BEAUMONT, C., 1984, Appalachian thrusting, lithospheric flexure, and the Paleozoic stratigraphy of the eastern interior of North America: Canadian Journal of Earth Sciences, v. 21, p. 973-996.

REGER, D. B., 1931, Pennsylvanian cycles in West Virginia: Illinois State Geological Survey Bulletin 60, p. 217-239.

RICE, C. L., 1984, Sandstone units of the Lee Formation and related strata in eastern Kentucky: Washington, D.C., United States Geological Survey Professional Paper 1151-G, 53 p.

RICE, C. L., AND SMITH, J. H., 1980, Correlation of coal beds, coal zones, and key stratigraphic units in the Pennsylvanian rocks of eastern Kentucky: United States Geological Survey Miscellaneous Field Studies Map MF-1188.

SALVADOR, A., 1985, Chronostratigraphic and geochronometric scales in COSUNA stratigraphic correlation charts of the United States: American Association of Petroleum Geologists Bulletin, v. 69, p. 181-189.

SHELL, 1980, Standard legend, time stratigraphic table: Shell International Petroleum Maatschappij, The Hague.

TANKARD, A. J., 1986, Depositional response to foreland deformation in the Carboniferous of eastern Kentucky: American Association of Petroleum Geologists Bulletin, v. 70, p. 853-868.

TRIPLEHORN, D. M., AND FINKLEMAN, R. B., 1989, Replacement of glass shards by aluminum phosphates in a Middle Pennsylanian tonstein from eastern Kentucky (abs.): Geological Society of America, Abstracts with Programs, v. 21, p. 52.

TRIPLEHORN, D. M., OUTERBRIDGE, W. F., AND LYONS, P. C., 1989, Six new altered volcanic ash beds (tonsteins) in the Middle Pennsylvanian of the Appalachian Basin, Virginia, West Virginia, Kentucky, and Ohio (abs.): Geological Society of America, Abstracts with Programs, v. 21, p. 134.

VEEVERS, J. J., AND POWELL, C. M., 1987, Late Paleozoic glacial episodes in Gondwanaland reflected in transgressive-regressive depositional sequences in Euramerica: Geological Society of America Bulletin, v. 98, p. 475-487.

WANLESS, J. R., AND SHEPARD, F. P., 1936, Sea level and climatic changes related to Late Paleozoic cycles: Geological Society of America Bulletin, v. 47, p. 1177-1206.

WILLARD, D. A., AND KLEIN, G. D., 1990, Tectonic subsidence history of the Central Appalachian Basin and its influence on Pennsylvanian coal deposition: Southeastern Geology, v. 30, p. 217-239.

EVALUATION OF EVIDENCE FOR GLACIO-EUSTATIC CONTROL OVER MARINE PENNSYLVANIAN CYCLOTHEMS IN NORTH AMERICA AND CONSIDERATION OF POSSIBLE TECTONIC EFFECTS

PHILIP H. HECKEL

Department of Geology, University of Iowa, Iowa City, Iowa 52242

ABSTRACT: Pennsylvanian major marine cyclothems in midcontinent North America comprise the sequence: thin transgressive limestone, thin offshore gray to black phosphatic shale, and thick regressive limestone. Typically, the cyclothems are separated from one another by well-developed paleosols across an area of perhaps 500,000 km^2 from southern Kansas to Iowa and Nebraska. Southward, the offshore shales extend into the foreland basin of Oklahoma, and the regressive limestones and paleosols grade into deltaic to fluvial clastics derived from the Ouachita detrital source. Texas, Illinois, and Appalachian marine cyclothems are detrital-rich like those in Oklahoma, but appear to be separated by paleosols like those in the northern midcontinent. Because the black phosphatic offshore shales of the midcontinent record sediment-starved, condensed deposition below a thermocline in about 100 m of water, sea-level rise and fall of at least that amount is required over the entire northern midcontinent region to account for the widespread dark shale-paleosol cyclicity. Sparsity to absence of deltas between most cycles on the northern shelf rules out delta shifting as a control over major cyclothem formation there. Continuity of all major cyclothems across both the Forest City basin and adjacent Nemaha uplift rules out local differential tectonics on the northern shelf as a major control. Confinement of all reasonable estimates of cyclothem periods within the Milankovitch band of orbital parameters (20 ky to 400 ky), which controlled Pleistocene glacial fluctuation, points to glacial eustasy as the major control over midcontinent cyclicity. Moreover, only the documented late Quaternary post-glacial rates of sea-level rise significantly greater than 3 mm/yr are sufficient to exceed carbonate accumulation consistently and produce the characteristic thin transgressive limestone overlain by the widespread thin subthermocline black shale in each major cyclothem. Although tectonic subsidence helped provide space for sediment accumulation, *tectonic control over cyclothem deposition would require both subsidence and uplift of the midcontinent (or an equally large region) at Milankovitch band frequencies.* However, currently developed cyclic tectonic mechanisms that can achieve the required depths repeatedly in a cratonic area act at periods at the very least 5 times greater (2 my+) and at rates of sea-level rise at least 30 times too slowly (maximum of 0.1 mm/yr). Firm biostratigraphic correlation of major midcontinent Upper Pennsylvanian cyclothems with similar depositional cycles in Texas and Illinois allows a strong glacio-eustatic signal to be identified in those regions, with little evidence so far of temporally differential tectonism among them. The lithic differences between carbonate-rich midcontinent cyclothems and detrital-rich cyclothems in Texas, Illinois, and the Appalachians are attributable directly to the greater detrital influx in the latter areas, which could relate as much to appropriate climate and accessibility to detrital provenance as to tectonic activity. Preliminary correlations showing that only certain bundles of major marine transgressions extended into the Appalachian basin suggest that the absence of the others may reflect tectonic uplift there, and that with more definite correlation, a longer-term tectonic signal can be isolated in that area.

INTRODUCTION

In the years since Weller (1930) first suggested "diastrophic" (i.e., tectonic) control, and Wanless and Shepard (1936) proposed glacio-eustatic control over the formation of Pennsylvanian cyclothems (see summary in Heckel, 1984a), much progress has been made in our understanding of the processes underlying tectonics, glaciation, and sedimentation. Better understanding of sedimentary processes gave rise to a third model, delta-switching, developed by Moore (1959) and applied by Ferm (1970) to Pennsylvanian strata in the Appalachian basin, and by Shabica (1979) to the Illinois basin. Since then, Heckel (1977, 1984b, 1986, 1990) has developed several lines of evidence to support glacio-eustatic control for midcontinent cyclothems and showed (Heckel, 1980) how this model can apply to more distant areas such as the Appalachians and can accommodate delta shifting as a subsidiary effect at the shoreline. In the meantime, Quinlan and Beaumont (1984) and Tankard (1986) used flexural tectonic models to explain certain scales of cyclicity in the Appalachian basin. Recently, Klein and Willard (1989) suggested that such a model combined progressively with glacial eustasy to influence cyclothems in Illinois and to dominate

cyclothem formation in the Appalachian basin. Klein and Cloetingh (1989) further claimed that tectonic flexure was the major influence on cyclothem deposition across the entire region from the Appalachians to the midcontinent, thus reviving the old debate over tectonics versus glacial eustasy as the primary control over formation of Pennsylvanian cyclothems.

This paper summarizes and updates the evidence for glacio-eustatic control and shows how several tectonic models (at least at their present stage of development) are inadequate to account for this evidence. First, some definitions are required to set the semantic background for the discussion. *Tectonic processes* involve movement of the lithosphere, whereas *glacial processes* involve phase changes in the hydrosphere; either can have local effects on the lithosphere (e.g., local faulting and basin formation; glacial loading) or distant (eustatic) effects on sea level. Hence eustasy (or worldwide fluctuations in sea level) can be due to changing the capacity of the ocean basins (*tectonoeustasy*) or to changing the volume of ocean water sequestered in continental ice (*glacial eustasy*). More importantly for the current debate, *control over cyclothem formation means causing the cyclicity of deposition.* That is, the mechanism that controls cyclothem formation must account for both the transgression

Tectonic and Eustatic Controls on Sedimentary Cycles, SEPM Concepts in Sedimentology and Paleontology #4 and regression of the sea that together make up the cyclic signal. A mechanism, tectonic or otherwise, that simply provides a certain amount of accommodation space is inadequate.

Vertical Sequence

The upper Middle through mid-Upper Pennsylvanian succession on the northern midcontinent shelf (Iowa to southern Kansas) consists generally of an alternation of marine limestone formations and nearshore-to-terrestrial shale formations (Fig. 1A). Detailed analysis of most of this succession by myself and University of Iowa graduate students (listed in Heckel, 1989a), combined with data gleaned from other sources (long cores, literature) has allowed construction of a sea-level curve (Fig. 1B) for the region from the shelf to the Arkoma-Anadarko foreland basin of central Oklahoma. This sea-level curve records cycles of marine transgression extending from the Oklahoma basinal region various distances across the shelf (often far beyond the Iowa outcrop belt), followed by regression of the shoreline various distances toward (and commonly back into) Oklahoma.

Cycles of three informal levels of magnitude are recognized: *Major cycles* (which correspond to most of the named limestone formations in Fig. 1A) are characterized by a widespread, conodont-rich, gray- to black-shale unit across the entire shelf and into the basin, generally sandwiched between transgressive and regressive limestone members on the shelf (Fig. 2A). *Intermediate cycles* (which generally correspond to limestone units of lesser rank) are characterized by a gray, conodont-rich shale or limestone across much of the shelf, but either with limited extent on the preserved part of the shelf (e.g., Norfleet on Fig. 1B) or with only shallow-water facies at the northern limit of outcrop in Iowa and Nebraska (e.g., Exline, Iatan). *Minor cycles* typically extend as marine units only a short distance into Kansas or Missouri from the Oklahoma basinal region, or represent minor reversals within more major cycles. (Some have not been named as separate units.)

The major cycles (and several intermediate cycles) are characterized throughout all or most of their outcrop extent on the shelf by the distinctive vertical sequence of lithic members (Fig. 2A), termed a "Kansas" or "northern midcontinent" cyclothem. Each lithic member represents a particular phase of deposition within the transgressive-regressive sequence. (Most are sufficiently widespread to be recognized as named members of the limestone formations.)

Transgressive limestone.—
This member is typically a thin (0.3 to 1.0 m) marine limestone that overlies a variety of rock types ranging from mudstone paleosols to terrestrial and/or deltaic sandstone and shale that represent sea-level lowstand at or near the top of the underlying ("outside") shale formation (Fig. 2A). It locally contains shallow-water calcarenites at the base, but most of it is skeletal calcilutite with a diverse biota deposited below effective

wave base. It represents both a marine flooding unit and a deepening-upward sequence. Transgressive limestones are typically dark, nonpelleted calcilutites with neomorphosed aragonite grains and overpacked calcarenites that generally lack obvious evidence of early marine cementation or meteoric leaching or cementation. They apparently remained in the marine phreatic environment of deposition until buried by overlying marine strata, and became chemically stabilized in rock-dominated diagenetic environments. They underwent slow compaction before cementation, under decreasingly oxygenated conditions (Heckel, 1983) such that much fine-grained organic matter was preserved. Transgressive limestones that overlie coals, as is common in Middle Pennsylvanian strata, are generally little more than a thin layer of shells that often are pyritized.

Offshore ("core") shale.—
This is typically a thin (0.1 to 1.0 m) nonsandy, marine, gray to black, phosphatic shale deposited as a condensed section under conditions of near sediment starvation in water deep enough to inhibit algal production and/or preservation of carbonate mud. Those shales that are gray generally carry a diverse benthic fauna of many invertebrate phyla attesting to open-marine conditions. Carbonate mud that may have formed from breakdown of invertebrate material apparently was largely dissolved; this is suggested by corrosion of coarse skeletal debris in places in the shale (Malinky, 1984) and in overlying invertebrate calcarenites at the base of the regressive limestone (Heckel, 1983). Most offshore shales include a distinctive black, commonly fissile, facies typically underlain and overlain by the gray facies (Fig. 2A). The black facies contains from 3 to 30% organic matter (Stanton and others, 1983; Hatch and others, 1984); conspicuous peloids, laminae and nodules of nonskeletal phosphorite; and a fauna consisting mainly of conodonts, fish debris, and, in places, ammonoids and radiolarians preserved in early diagenetic calcitic nodules (bullions) and in phosphorite nodules (Kidder, 1985).

The black-shale facies was deposited in water deep enough

Fig. 1.—Late Middle through mid-Upper Pennsylvanian succession in midcontinent North America. (A) Nomenclature of groups and limestone formations (UPPER CASE), with a few members (lower case). Formations and members shown represent major (*) and intermediate-scale (×) marine cyclothems on northern midcontinent shelf; intervening shale formations are left unlabeled. (B) Sea-level curve updated from Heckel (1986, 1989b) and Boardman and Heckel (1989) showing all scales of marine transgressive-regressive cycles of deposition, from major cyclothems (largest letters on left side of curve) through intermediate (midsize letters) to minor cycles (smallest letters on left side of curve), extending from foreland basin of central Oklahoma various distances onto shelf. Offshore shales (names in parentheses) are shown by lines (solid = black, dashed = gray). Bold dots indicate nonskeletal phosphorite. Small dots indicate terrigenous detrital deposits. Oblique lines on right side of curve indicate subaerial exposure, commonly with paleosol formation.

(Continued on p. 69)

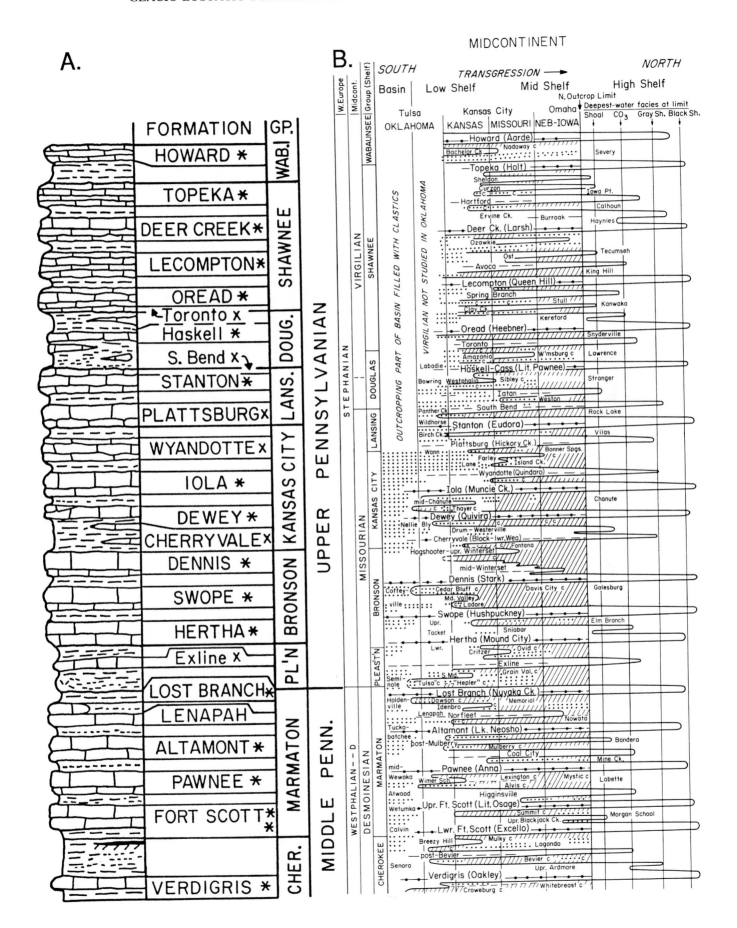

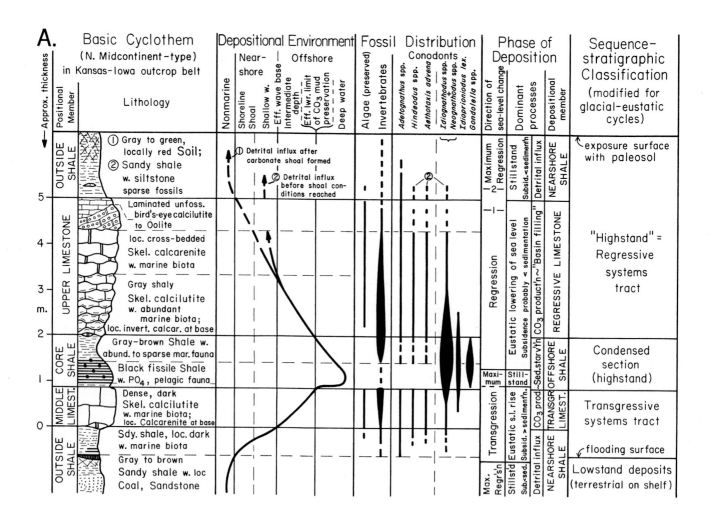

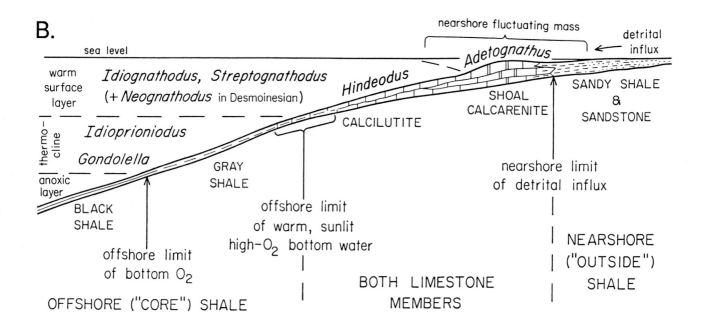

to develop a pycnocline that inhibited vertical circulation and prevented bottom oxygenation long enough to eliminate benthic organisms and preserve large amounts of organic matter that accumulated on the bottom over a long period of time. Where its distribution is patchy within the gray facies, the black facies occurs in topographic lows (Heckel and others, 1979). That this pycnocline was a thermocline is strongly suggested by the preservation of ammonoids only in early diagenetic calcitic nodules in the black facies, and by the preservation of other originally aragonitic molluscs (along with ammonoids) only in thicker portions of laterally equivalent and overlying offshore, probably distal prodeltaic, gray shales in southern Kansas and Oklahoma (Figs. 3A, 4). Apparently, the less stable aragonitic shells were dissolved in the colder water below the thermocline, unless they were preserved by early matrix mineralization in the bullion nodules or by rapid burial and replacement by siderite, phosphorite or pyrite in the distal extents of prodeltaic sediments at the periphery of the black facies. Perturbation of the thermocline by wind stress in the tropical trade wind belt led to quasi-estuarine circulation and episodic upwelling, which eventually resulted in deposition of nonskeletal phosphorite; this is happening today in low-oxygen sediment on the offshore shelf along the coast of Peru (Kidder, 1985). The offshore shale is also called the "core" shale because of its position at the "turnaround point" in the cyclothem between the transgressive and regressive limestone members, which are essentially mirror images of one another in general facies development. The core shale is not only the condensed section in sequence stratigraphic terminology, but also represents the highstand deposits of the cyclothem on the preserved mid to low part of the northern shelf.

Regressive limestone.—

This member is typically a thick (1.5 to 10 m), classic shallowing-upward, carbonate sequence consisting of skeletal calcilutite at the base grading upward into skeletal calcarenite with abraded grains, algae, and local oolite beds; this calcarenite is often cross bedded, and commonly capped with peritidal deposits and/or an exposure surface. Because this member was deposited entirely during a fall in sea level, I prefer the term "regressive" rather than "highstand" for the systems tract it represents, because only the "core" shale represents truly highstand deposits across the mid to low part of the shelf where the cyclothems are presently preserved (see French and Heckel,

FIG. 2.—(A) Basic northern midcontinent cyclothem on mid to low shelf representing one complete marine inundation and withdrawal across the entire northern shelf, modified from Heckel (1990) by addition of modified sequence-stratigraphic terminology on right side. (B) Depositional model for cyclic rock types on shelf in relation to shoreline and water depth, showing inferred living positions of major conodont genera in water masses developed at sea-level highstand, derived from generic distribution pattern shown in A, based on data from Heckel and Baesemann (1975), Swade (1985), and various student theses and dissertations listed in Heckel (1989a). (Modified from Heckel, 1990).

1994). Subaerial exposure resulted in oxidizing, undersaturated, meteoric water infiltrating the regressive limestone before much compaction took place. This water oxidized most of the original organic matter in the sediment, leached aragonitic grains, and eventually became saturated enough to precipitate blocky calcite in both intergranular and moldic voids. This preserved the original peloidal fabric, the depositional looser packing of grains, and also the porosity where cementation was incomplete. Thus, the lighter-colored, more porous and more conspicuously sparry upper part of the regressive limestone stands in contrast with the darker, denser (overcompacted), transgressive limestone and also with the lower, more offshore facies of the regressive limestone (Heckel, 1983).

In some cyclothems (e.g., Iola around Kansas City; Stanton in southeastern Kansas), a distinctly different calcarenite at the base of the regressive limestone contains only overcompacted calcitic invertebrates (echinoderms, brachiopods, bryozoans and encrusting foraminifers) with a minimal ferroan carbonate cement. It commonly shows evidence of grain corrosion, but lacks evidence of algae, grain abrasion, or cross-bedding. It therefore must have formed below effective wave base and probably below the effective photic base for algae in this sea, or at least below the level at which carbonate mud and aragonitic shells were readily dissolved. This calcarenite appears to represent proliferation of invertebrates in deep water as the pycnocline disappeared and sufficient oxygen again circulated to the bottom. Its overcompacted nature resulted from relatively deep burial before cementation as with the transgressive limestone (Heckel, 1983).

Nearshore/terrestrial ("outside") shale.—

This member is an extremely variable shale- and locally sandstone-dominated unit, which overlies the regressive limestone and forms most or all of the shale formations that separate the marine limestone formations (Fig. 1A). It consists largely of prodeltaic to deltaic, paralic, and fluvial clastic facies in many units in Missouri, Kansas and southward. Generally at the top of this member (Fig. 2A), and often forming the entire unit northward (Fig. 3A) are gray to mottled reddish, blocky mudstones (0.2 to 2.0 m thick) which, upon detailed study, have been identified as paleosols (Schutter and Heckel, 1985; Goebel and others, 1989; Joeckel, 1989), some extending for hundreds of miles along outcrop. Thus, the nearshore/terrestrial shale unit represents sea-level lowstand when the sea had withdrawn from most or all of the shelf for a substantial period of time. The blocky mudstones often are overlain by a thin shale with marine fossils, which represents the base of the transgressive systems tract resting upon the marine flooding surface (and sequence boundary) at the top of the mudstone (Fig. 2A). At several horizons, particularly in Middle Pennsylvanian rocks when the midcontinent climate generally was more humid (Schutter and Heckel, 1985), the mudstones are overlain by the most widespread and thickest coal beds in the midcontinent. These coal beds apparently formed in response to the early stages of sea-level rise of the succeeding transgression, which ponded fresh-

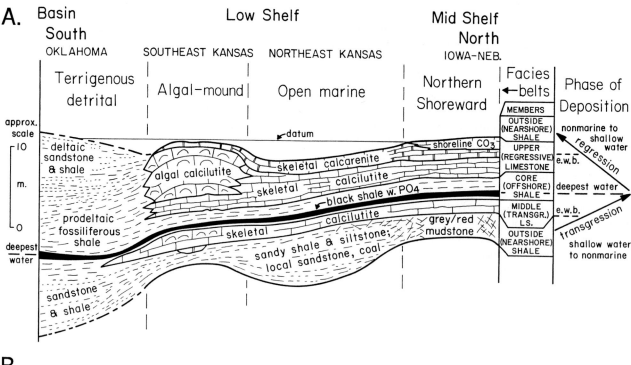

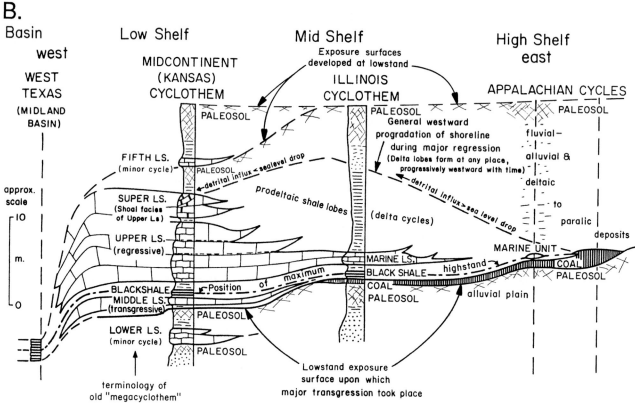

FIG. 3.—Lateral extent of Upper Pennsylvanian cyclothems in central and eastern North America. (A) Lateral facies developed in members of basic cyclothem (Fig. 2A) along midcontinent outcrop belt where only low- to mid-shelf portion is presently preserved; datum shows approximate sea level at time when detrital influx terminated deposition of regressive limestone member; "e.w.b." marks position that the effective wave base was crossed during transgression and regression in north (modified from Heckel, 1991c). (B) Inferred lateral extent of eustatic Pennsylvanian marine cyclothem bounded by exposure surfaces with paleosols from midcontinent to Appalachian basin near highstand detrital shoreline, and into deep basins of west Texas where marine deposition was continuous; cyclothems on north Texas shelf are similar to those in Illinois and Appalachian basins (modified from Heckel, 1980).

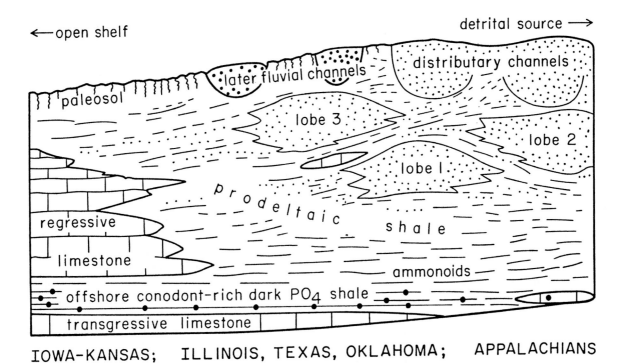

IOWA-KANSAS;   ILLINOIS, TEXAS, OKLAHOMA;   APPALACHIANS

Fig. 4.—Depositional model for eustatic marine Pennsylvanian cyclothem approaching a humid shoreline dominated by detrital influx, and incorporating delta shifting (lobes 1, 2, and 3 in succession) as a control over formation of minor local detrital cycles near shoreline during early part of eustatic phase of regression (sea-level fall) (modified from Boardman and Heckel, 1989; partly derived from Moussavi-Harami and Brenner, 1984). Bold dots indicate nonskeletal phosphorite, which reflects slow sedimentation in condensed interval during sea-level highstand. Note that widespread paleosols with incised fluvial channels mark a later phase of eustatic lowstand that terminates the eustatic transgressive-regressive cycle of deposition. State names denote general development of major Pennsylvanian marine cyclothem types common in those areas on this idealized transect.

water runoff to form broad peat swamps on the surface of low relief. These swamps then migrated up-shelf ahead of the transgression and became preserved as coal beds (Fig. 3B).

Conodont Information

Detailed work on both outcrop and core successions by Heckel and Baesemann (1975), Swade (1985), and in various theses and unpublished data, has established a distinctive vertical succession of common conodont genera characterizing all the major cyclothems on the northern midcontinent shelf (Fig. 2A). Both gray and black facies of the slowly deposited offshore ("core") shales are characterized by great abundance of conodonts—hundreds to thousands per kilogram. These faunas are strongly dominated by *Idiognathodus* (includes related *Streptognathodus*), with common *Neognathodus* (in the Desmoinesian Stage), *Idioprioniodus*, and *Gondolella* (which is generally confined to the middle of the shale). In contrast, the more rapidly deposited nearshore-marine portions of the nearshore/terrestrial ("outside") shales generally contain few conodonts, up to only 20 or so per kilogram. These faunas are typically dominated by *Adetognathus*, locally subequally with *Idiognathodus* and, less commonly, *Hindeodus* (*Anchignathodus* of previous work), but with *Idioprioniodus* and *Gondolella* conspicuously absent, and *Neognathodus* (Desmoinesian only)

generally rare. The two limestone members contain faunas that are gradational and intermediate between those of the adjacent shale members. These limestone faunas tend to be mirror images of one another, symmetrical about the offshore ("core") shale that separates them.

The distinctive differences in conodont faunas between offshore and nearshore parts of the cyclothem appear related to the nature of the different water masses that covered the shelf at different sea-level stands (Klapper and Barrick, 1978; Swade, 1985). *Idiognathodus* and *Streptognathodus* (with *Neognathodus* during Desmoinesian time) apparently dominated the normal open-marine, warm, surface-water mass (Fig. 2B) that covered most of the sea (away from strong freshwater influx) during all sea-level stands, which explains their dominance throughout most of the cyclothem. *Idioprioniodus* probably occupied the deeper, slightly cooler water mass in the top of the thermocline, which explains its occurrence mainly in the offshore shale and adjacent deeper-water parts of the limestone members. *Gondolella* apparently lived in deeper, even cooler and possibly somewhat dysoxic water, lower in the thermocline, which explains its even greater confinement within the most offshore facies, primarily the middle of the black shale. These five genera probably were pelagic, as all are generally abundant in the anoxic black-shale facies, which lacks any definitely benthic fossils even in early diagenetic nodules. At the other extreme,

Adetognathus apparently inhabited and dominated the variable nearshore, shallow-water mass, where it tolerated fluctuations in salinity and other conditions that inhibited the other genera. *Hindeodus* probably occupied a slightly more stable environment associated with carbonate sediment, as it is most commonly found in the limestone members. Both of these latter genera may have been benthic, as they are absent in the anoxic black shales.

Because these distinctive patterns in abundance and dominance of conodont genera are reasonably related to the water mass, and hence to the depositional environment and the phase of deposition within the cyclothem, they can be used to identify the phase of deposition within the overall transgressive-regressive sequence where the lithic composition is less distinctive or ambiguous. In this way, several of the intermediate cycles have been identified in the vertical succession (Fig. 1), and distinguished from the minor cycles, which are generally characterized by lower abundance of conodonts, commonly with a dominance of nearshore genera.

Just as significantly, changes in species composition, particularly of the dominant genus *Idiognathodus* (and related *Streptognathodus*) in the offshore shales in successive cyclothems (Swade, 1985; Heckel, 1989b; Barrick and Boardman, 1989; Boardman and others, 1990; Heckel, 1991a; Barrick and others, 1992) allow biostratigraphic discrimination of successive cyclothems. This further allows potential correlation of the midcontinent succession with successions in different areas, a potential that is being realized in Texas (Boardman and Heckel, 1989), Illinois (Heckel and Weibel, 1991), and the Appalachian basin (current work with J. E. Barrick).

Lateral Extent

Tracing the limestone and shale formations (Fig. 1A) along the midcontinent outcrop belt and into the near subsurface in long cores has shown that the major cyclothems vary little northward (Fig. 3A), where the main change is the increasing dominance of peritidal facies in the top of the regressive limestone. The transgressive limestones and offshore shales change little from Kansas to the present outcrop limit in Iowa and Nebraska, which is only a mid-shelf position; the cratonic shoreline was probably a significant distance farther northward at sea-level highstand. Southward from Kansas, the conodont-rich gray and black shales can be traced far beyond the limits of the two limestone members into the foreland basin region of Oklahoma (Fig. 3A). The regressive limestone typically disappears first, as it was often overwhelmed by prograding detrital influx from the southerly orogenic source as sea level dropped. The transgressive limestone commonly extends farther, as it was protected from detrital dilution by sea-level rise, when detrital influx was trapped in estuaries receding toward the orogenic detrital source. An important point to make here is that the conodont generic succession illustrated in Figure 2A maintains its lateral integrity, both northward (as might be expected) and southward as well.

Farther southward in the foreland basin of Oklahoma, where the more sandstone- and shale-dominated cyclic sequence was modeled by Bennison (1984, 1985), the offshore shale lightens in color and thickens due to increase in fine (silty) detrital influx, even during highstand, from the nearby orogenic source. However, this unit generally remains darker and finer-grained than overlying and underlying coarser beds, and its conodont fauna, although reduced in abundance because of detrital dilution, exhibits the same generic dominance and species composition as the thinner black shale facies on the northern shelf.

Other basins.—

Lateral extent of individual cyclothems beyond the midcontinent outcrop belt and subsurface can be interpreted only by comparing the midcontinent succession of marine transgressive-regressive "stratigraphic sequences" separated by exposure surfaces, terrestrial deposits, and paleosols (Figs. 1B, 2A) with similarly cyclic coeval successions in other basins. Heckel (1977, 1980) suggested an equivalency between marine units in the midcontinent, Illinois, and Appalachian basin successions as shown in Figure 3B.

In the Illinois basin, phosphatic gray to black shales that contain abundant conodont faunas of the same genera and species as their counterparts in the midcontinent similarly represent offshore shales deposited at sea-level highstand. They are generally overlain by regressive limestones that are thinner than those in the midcontinent. These limestones commonly represent only the lower subtidal carbonates, as they apparently became overwhelmed by deltaic clastics at an earlier phase of regression nearer the easterly detrital source in the direction of the Appalachian highlands. Transgressive limestones are rare, perhaps because coal beds are more common at the tops of the underlying lowstand deposits. The peat swamps that the coal beds represent probably inhibited algal carbonate production (and preservation of other fine carbonate) during the ensuing sea-level rise. The lowstand deposits intervening between the marine units are dominated by greater proportions of deltaic and terrestrial shales and sandstones, reflecting greater proximity to the easterly detrital source. They also commonly contain paleosols at the top, indicating marine withdrawal of temporal significance, and many of these are overlain by coal beds.

In the Appalachian basin, the succession is strongly dominated by detrital clastics (Fig. 3B), in which Ferm (1970) and coworkers have described conspicuous deltaic deposits. These detrital intervals also commonly contain paleosols at their tops, generally overlain by coal beds (Busch and Rollins, 1984; Busch, 1984). The marine units overlying the coal beds or paleosols are typically thin and lithologically heterogeneous. They range from argillaceous/silty/sandy skeletal limestones (wackestones to packstones) to fossiliferous, dark to light gray shales, and calcareous siltstones and sandstones. The entire range of lithotopes may be locally phosphatic, glauconitic, or sideritic, and all show various degrees of lateral facies transition (Slucher, 1988). There is no definite development of transgressive or regressive limestones or "core" shales. This is not unexpected in a region so close to such a major detrital source as the Appalachian highlands, where water depths never were very great and deltas could penetrate marine intervals even during sea-level highstands. Rather, the skeletal-dominated limestones appear to represent condensed sections developed away from or between deltaic influxes, as they commonly contain abundant conodonts as well as phosphorite and glaucony (Fahrer and Heckel, 1992).

It is worth emphasizing that even though individual marine lithotopes are not commonly traceable very far along outcrop, recent field mapping and subsurface tracing in well cores (e.g., Busch, 1984; Caudill, 1990; Chesnut, 1991; Skema and others, 1991) suggest that the overall marine intervals in which they occur are laterally persistent, except where locally cut out by channeling from higher units. The widespread phosphatic black shales typifying the offshore part of the sequence in the Illinois and midcontinent basins are lacking in the Appalachian basin, apparently because even at highstand, the water was too shallow for establishment of a long-term pycnocline. The conodont-rich, phosphatic and glauconitic limestones appear to occupy the offshore position in the Appalachian sequence (Fig. 4).

In the Midland basin of west Texas, the entire late Middle to early Upper Pennsylvanian succession is basinal sediment-starved black shale (Fig. 3B). East of the Midland basin on the north Texas shelf, however, the outcropping Pennsylvanian succession is similar to that in Illinois and to a lesser degree the Appalachians. On the north Texas shelf, conspicuous deltaic/terrestrial deposits, typically with well-developed paleosols at the top, alternate with widespread marine intervals. In these intervals, transgressive limestones, offshore shales and regressive limestones are locally recognizable and laterally persistent for short distances on outcrop in areas away from strong detrital influx (Boardman and Heckel, 1989).

As first suggested by Wanless and Shepard (1936), most facies variations in Pennsylvanian cyclothems can be related to the relative availability of detrital influx. The depositional model proposed by Boardman and Heckel (1989), for incorporating strong detrital influx (with concomitant delta shifting and fluvial channeling) into the eustatic model for Texas and Oklahoma, can be applied to the Illinois and Appalachian basins as well (Fig. 4). The potential for detrital influx varies with eustatic phase of deposition and generally increases progressively with regression. Detrital influx also is strongly affected by local climatic control over agents of transport (Cecil, 1990), and because of this, the amount of detrital influx may be less directly related to the amount of nearby tectonic activity.

Considerations regarding contemporaneity of the marine units.—
One of the recent key discoveries in Pennsylvanian stratigraphy of central and eastern North America is the recognition of paleosols and other intervals of subaerial exposure. These are as laterally continuous as the marine units they separate in the northern midcontinent. Paleosol/exposure intervals appear to be more laterally persistent than the marine intervals in the Appalachian basin (Busch, 1984), which are cut out in places by exposure surfaces (Caudill, 1990). Although just beginning to undergo detailed study (e.g., Schutter and Heckel, 1985; Goebel and others, 1989; Joeckel, 1989, 1994), many paleosols appear to be horizonated and seem to reflect long-term subaerial exposure on stable landscapes with little erosion or terrigenous input for long periods of time. This situation requires relatively long-term withdrawal of the sea and confinement of most available detrital sediment to incised local fluvial channels that are younger than (and probably distinct from) the delta distributaries involved in delta-shifting (Fig. 4). In short, the well developed, widespread underclays, flint clay horizons, redbeds, and mottled zones, commonly on the tops of the deltaic intervals in the Appalachian succession (Fig. 3B), lend a distinctly eustatic appearance of long-term sea-level lowstands to the succession, which is not readily accommodated by a delta-shifting model that assumes constant sea level and only autocyclic processes.

This cyclic signal of alternating high and low stands of sea level, each widespread across the individual basins of central and eastern North America, raises the question as to whether or not the successive highstands and lowstands are contemporaneous between these basins. The simplest, most parsimonious interpretation (Fig. 3B) would be to assume that they are, and there is no compelling evidence that they are not. This hypothesis can be tested, however, via the biostratigraphic composition of the major marine intervals. The relatively well constrained biostratigraphic correlation of the midcontinent cyclothems with the north Texas cycles by Boardman and Heckel (1989), in the face of potentially confounding deltaic activity and presumed nearby Ouachita tectonic influence to the east, provides strong support for contemporaneity of the marine horizons. The data involve ammonoids and fusulinids as well as conodonts. In addition to species evolving within lineages, they include single appearances of otherwise absent genera or species in a single marine cycle in both areas. For example, the fusulinid *Eowaeringella ultimata* and the conodont *Gondolella denuda* occur only in the Swope and upper Salesville cycles in the midcontinent and Texas, respectively, an unlikely situation if the marine units were not contemporaneous.

Preliminary correlation of the midcontinent and Illinois basin successions (Heckel and Weibel, 1991, utilizing the long Charleston, Illinois core) involves a close, cycle-by-cycle, species-level match of conodont faunas of the offshore black and gray phosphatic shales. This strongly supports the idea that the sea was closely connected between these basins during highstand (Fig. 5), as Heckel (1980) hypothetically suggested in developing the general cyclothem model (Fig. 3B).

Preliminary conodont data from marine units in the Conemaugh Group of the Appalachian basin (Brush Creek through Ames near Athens, Cambridge, and Steubenville, Ohio; and Pittsburgh, Pennsylvania) suggest a fairly close match at the species level of certain lineages with conodont faunas known from the succession of midcontinent black phosphatic shales (Fig. 1) from the Hushpuckney (Swope cycle) through the Heebner (Oread cycle) (J. E. Barrick and P. H. Heckel, in prep.).

UPPER PENNSYLVANIAN PALEOGEOGRAPHY, CENTRAL U.S.

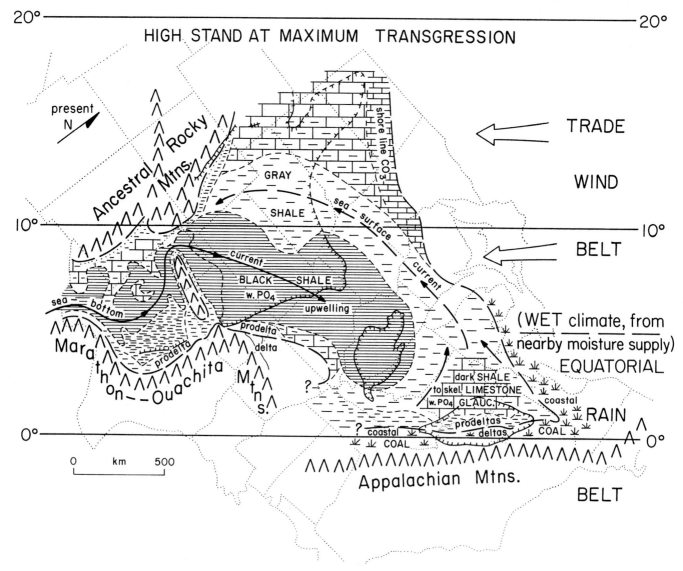

FIG. 5.—Inferred facies distribution in midcontinent sea during eustatic phase of sea-level highstand when black phosphatic offshore shale facies was deposited beneath a thermocline established over deeper area; black facies probably developed in somewhat shallower water in Illinois basin in the direction of a humid Appalachian shoreline, where a thermocline was augmented, and then replaced, by a halocline formed by strong freshwater outflow (Heckel, 1991c). Gray facies of offshore shale extended various distances outward from black facies, and covered the entire area of black facies on the shelf, both before and after black-facies deposition. Phosphatic and glauconitic skeletal limestones represent this phase of deposition in parts of Appalachian basin away from detrital influx (Fahrer and Heckel, 1992). (Modified from Heckel, 1980).

This preliminary correlation suggests that the Appalachian marine horizons may be coeval with the midcontinent marine horizons and were also probably connected at sea-level highstands. It also suggests that the more readily interpreted offshore facies in the Appalachian basin (conodont-rich, phosphatic, glauconitic limestone) is time-equivalent to the phosphatic black shale (Fig. 4), the interpreted deepest water facies in the midcontinent, as Heckel (1980) illustrated (Figs. 3B, 5), and Fahrer and Heckel (1992) have reemphasized.

POSSIBLE CONTROLLING PROCESSES

Each of the three general controls proposed for explaining the formation of Pennsylvanian cyclothems is formulated around a major process, namely delta-shifting, tectonism, or glacial eustasy. Although the proponents of each process tend to emphasize the contribution of that process (and some have recognized the probable contributions of the other processes), it is obvious that all three processes must have played some role in

Pennsylvanian sedimentation in general. Delta-shifting and tectonism are ongoing processes, and Gondwanan continental glaciation is well-documented for the Pennsylvanian (Veevers and Powell, 1987). Therefore the question becomes which process was the dominant control, or were two or all three subequal in effect? For acceptance of a process as the dominant control over Pennsylvanian marine cyclothem deposition, it must be able to account for all the previously described characteristics of these cyclothems. At the very fundamental level, *it must have controlled the cyclicity of deposition*, that is, it must have been responsible for the repetition of the widespread marine transgressions and regressions that are evident in the stratigraphic record of this period. Processes that cause only changes in thickness or facies or other characteristics within the cycles of marine deposition are merely subordinate factors.

Delta Shifting

Delta shifting is a local process that requires delta development throughout the vertical and lateral extent of the cyclic sequence and results in cycles of stratigraphic units with limited lateral extent. Delta shifting can be readily ruled out on several counts as the basic control over the major midcontinent cyclothems and therefore, by extension, as the basic control over their correlatives elsewhere. First of all, the extremely widespread extent of each of these laterally continuous major marine transgressive-regressive units in the midcontinent alone covers a presently remaining outcrop and subsurface area of roughly half of the states of Iowa, Nebraska, Missouri, Oklahoma, and almost all of Kansas, totalling perhaps 500,000 km². This compares with a generous estimate of perhaps 5000 km² for the larger individual delta lobes (Gould, 1970, p. 9) of the Holocene Mississippi River, one of the largest and most sediment-rich of modern delta systems. Second, deltas are simply lacking over most of the shelf north of Kansas City (Fig. 6), where exposure surfaces and paleosols up to 2 m thick separate nearly all the Missourian cyclothems over perhaps 200,000 km². Third, in order for well-developed (horizonated) paleosols to form, definite long-term withdrawal of the sea is required, on the order of thousands of years (Birkeland, 1984). Soils on modern active delta lobes are typically immature, and the subsidence of the lobes beneath the sea that is critical to the delta-shifting model would inhibit much further soil development after abandonment. It is clear that the few possible deltas apparent locally between the cyclothems shown on Figure 6 are bounded above and below by laterally continuous marine horizons. Thus, rather than being responsible for the cyclicity, these deltas simply resulted from availability of local sediment during a climatic regime favorable for its erosion and transport, when a particular lower sea-level stand placed the shoreline in that area for a while. By extension, the deltas that are so conspicuous in Texas, Illinois and the Appalachian basin were the products of closer, more prolific detrital sources, which, because of their proximity, affected deposition throughout more of each individual cycle of sea-level rise and fall. Even these deltas, however, were at the

mercy of the eustatic position of shoreline at any particular time.

Local Tectonism

Local tectonism, involving differing rates of subsidence of a basin coupled to a faulted uplift, could cause periodic transgression in the basin followed by apparent regression as the basin filled with detritus. The Nemaha uplift in southeastern Nebraska and the adjacent Forest City basin (Fig. 6, inset) provide an example that formed early in Pennsylvanian time and strongly affected local sedimentation through about mid-Desmoinesian time. Minor differential effects on late Desmoinesian sedimentation are shown by thinning of this part of the succession over the uplift, and by the lateral passage from black to gray color of each of the phosphatic shales in each of the five cyclothems detected in a well core from the uplift (Heckel, 1990, Fig. 6). However, the presence of each of these cyclothems (even though thinned) upon the uplift indicates that each of these marine transgressions inundated the uplift as well as the basin. This shows that the marine inundations were controlled by forces other than those that differentially controlled the uplift/basin couplet. Furthermore, the overlying Missourian cyclothems show no detectable lateral changes as they pass over the uplift (Fig. 6, core NAC), which must have been dormant at this time. Thus, from the perspective of local midcontinent tectonic features, the cyclothems are eustatic in that they have a broader or more distant control. This more distant control, however, could be either tectonic or glacial.

Broader/Distant Tectonic Controls

Broader or more distant tectonic controls include a number of models based on modern plate-tectonic and rheologic theory (Watts and others, 1982), which have been developed more recently than the original suggestions of Weller (1930, 1956). These models include the eustatic effects of large-scale but long-term cyclic tectonic processes such as orogeny, changes in ocean-ridge volume and hotspot activity summarized by Pitman and Golovchenko (1983). These processes potentially can raise and lower sea level from 100 to 500 m, but at periods between 10 and 100 million years, yielding rates of sea-level change ranging from 0.002 to 0.1 mm/yr for complete cycles of rise and fall.

More specific considerations that continental collision could cause repeated foreland flexural subsidence in response to periodic loading led to tectonic models for Lower Pennsylvanian cyclothem formation in the Appalachian region (Quinlan and Beaumont, 1984; Tankard, 1986). In these models, marine transgressions would occur in the foreland basin when it was more rapidly subsiding, followed by regressions when it was rebounding and/or being filled by detrital sediment from increased erosion of the uplifted orogen. Cloetingh (1986, 1988) developed a rheological model of similar type involving interaction between intraplate flexural stresses and deflections of the lithosphere caused by sediment loading, which could produce

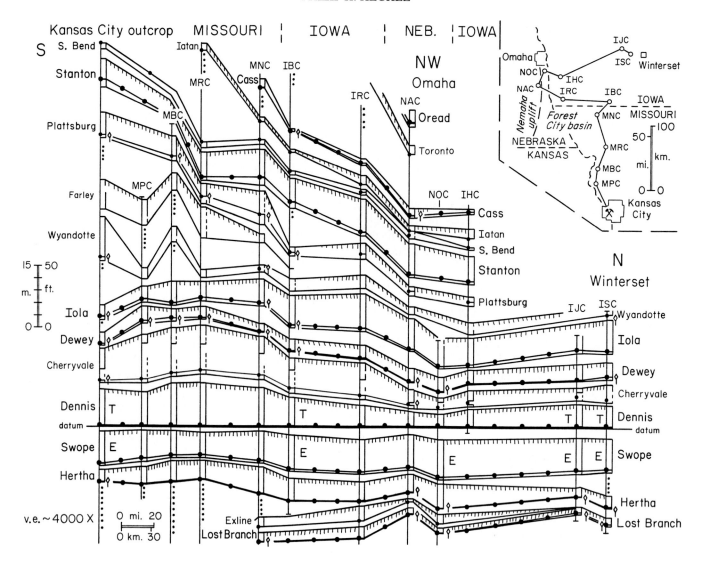

Fig. 6.—Correlation cross section of lower Upper Pennsylvanian succession (Missourian Stage = Exline through Iatan) in mid-shelf position on northern midcontinent shelf, based on long cores held by respective state geological surveys. Named units with largest letters are major marine cyclothems, with correlated black lines with spots at or above base marking black to dark gray phosphatic shales, and hachures on top marking exposure surfaces usually overlain by paleosols. Lack of hachures indicates nearshore detrital deposits culminating in deltaic or other sandstones (dots). Southward thickening of succession reflects increase in tectonic subsidence in that direction. Tailed diamond symbols for conodont faunas and letters for fusulinids (E = *Eowaeringella ultimata*; T = first *Triticites*) show biostratigraphic control for correlation (modified from Heckel, 1990).

apparent sea-level changes of more than 100 m within a few million years to yield sea-level changes of about 0.01 to 0.1 mm/yr. Klein and Willard (1989) suggested that this type of model could be combined progressively with glacial eustasy to explain the facies changes in the cyclothems from the midcontinent through Illinois to the Appalachian basin, adjacent to the orogenic belt. Klein and Cloetingh (1989) combined evidence for substantial subsidence in the midcontinent during cyclothem deposition with the probability that this subsidence was caused by the effects of far-field tectonic stresses of Appalachian-Ouachita collisional tectonics. From this they concluded both that tectonically controlled flexural stresses provided most of the necessary accommodation space in which to accumulate the

cyclothems, and further that even midcontinent cyclothem deposition was tectonically driven, with only minor additional change provided by glacial eustasy.

It is worth noting that loading mechanisms could have had quite different effects in distant areas. For example, loading could cause transgression if the downwarping were transmitted that far or, conversely, it could cause regression if the foreland downwarp diverted enough marine water into the foreland basin from distant cratonic areas. These alternatives can be tested biostratigraphically.

More recently, Cathles and Hallam (1991) proposed a different rheological model involving stress-induced changes in plate density by both collision and rifting. Although collision could cause changes in plate elevation up to 200 m, its potential period for cyclicity is greater than 10 million years. Sudden rifting of a large plate might produce enough elastic "snapback" to raise sea level on that plate on the order of 50 m, and to lower sea level on other plates by perhaps 10 m. With lithospheric readjustment over periods between 10 and 100 thousand years, these changes would be followed by similar magnitudes of sea-level change in the opposite direction.

Glacial Eustasy

First suggested for explaining Pennsylvanian cyclothems by Wanless and Shepard (1936), the glacial eustatic model has since been refined by much recent work on Pleistocene deposits (Wright, 1989). The periodicities of glacial advance and retreat, particularly as reflected in fluctuating oxygen isotope ratios in the deep marine sediment record, are statistically shown to conform closely to periodicities of the Earth's orbital parameters of roughly 20 thousand years (precession), 40 thousand years (obliquity), 100 thousand and 400 thousand years (eccentricity)—a range termed the Milankovitch band. This correspondence strongly suggests that the midlatitude fluctuations in solar insolation resulting from variations in these parameters played an important role in the waxing and waning of ice sheets (Imbrie, 1985), although exact linkages remain elusive. Finally, as Carboniferous stratigraphy of the southern continents has become better known (Crowell, 1978; Veevers and Powell, 1987), the Gondwanan glacial record can now be applied as a viable model directly to the Pennsylvanian, when cyclothems and other cyclic successions were well developed on a worldwide scale (Wells, 1960), including the Cordilleran region of western North America (Wilson, 1967; Yose and Heller, 1989; Langenheim, 1991; Dickinson and others, 1991; also this volume).

Distinguishing Broader Tectonic from Glacial-Eustatic Controls over Pennsylvanian Cyclothems

Constraints on the possible controls over the formation of marine Pennsylvanian cyclothems involve not only the more straightforward empirical data on vertical sequence and lateral extent developed previously, but also considerations of their periodicity, and the water depths and rates of transgression that are inferred from interpretation of the lithic data.

Periodicity.—
The probable durations of each of the midcontinent marine cycles of transgression and regression, based on sets of assumptions explained elsewhere (Heckel, 1986), range from 235,000 to 400,000 years for major cyclothems, 120,000 to 220,000 years for intermediate cycles, and 44,000 to 120,000 years for minor cycles. Even if the lower ends of these ranges are halved to accommodate the shorter duration of the analyzed sequence suggested by more recent radiometric dating (see Klein, 1990), the estimated ranges for all types of cycles still fall within the 20,000 to 400,000 year range (Fig. 7) of all periods of the earth's orbital cycles that are involved in the Milankovitch insolation

theory of control over Pleistocene ice ages. Klein's (1990) concern that halving the upper end of the range yields no known orbital parameter is not relevant, because Heckel (1986) simply estimated the order of magnitude of the range for all the types of cycles within a set of geologically reasonable limits (Heckel, 1991b), a point unacknowledged by Klein (1991). In fact, Klein's (1990) reduced ranges for the cycles would equally constrain any cyclic tectonic model proposed to account for the alternating rise and fall of sea level producing the cyclothems. More recent mathematical analyses of cyclothemic transgressive events in Lower to Upper Pennsylvanian successions in England, Kentucky, and Iowa show that all can be assigned to a nonrandom cause with a periodicity within the range of orbitally forced glacial eustasy and that, at this scale, tectonic mechanisms are not significant (Maynard and Leeder, 1992).

The 10-my and longer periods of the long-term tectonic models summarized by Pitman and Golovchenko (1983) are too long to account for the cyclothems (Fig. 7), as they explicitly recognized. Calculations based on the 1 to 10 cm/1000 yr (=.01 to 0.1 mm/yr) rates of sea-level change indicated by Cloetingh (1988) for the flexural stress model (invoked by Klein and Cloetingh, 1989, to account for Pennsylvanian cyclothems) would attain maximum water depths at highstand (halfway through a 400,000-year cycle) of only 2 to 20 m for a short time. As developed later, this is insufficient to account for the succession of widespread phosphatic black-shale facies overlying transgressive limestones. The time required to attain 100 m of water depth (and then reverse) would be 2 to 20 million years for each cycle. Furthermore, the flexural model of Cloetingh (1988, Fig. 2) applies to a band of finite width (<200 km) with maximum depth change along the middle, decreasing to zero along the edges, and then reversing. This model cannot be reconciled with the lateral continuity and relatively uniform thickness of individual cyclothems in the succession covering a plate-like area of 500,000 km² on the northern midcontinent shelf. Although Klein (pers. commun., 1991) has since indicated that problems in the calculations underpinning Cloetingh's model effectively slow the possible rates of sea-level change, the model failed to account for the basic characteristics of Pennsylvanian cyclothems even under its original more favorable constraints.

The whole-plate density stress model of Cathles and Hallam (1991) involves a sea-level drop of perhaps 40 to 50 m resulting from rifting of a large oceanic plate, followed by an equal rise in an equivalent time frame to produce on other plates a regressive-transgressive cycle about 60,000 years long. This is well within the periodicity of the Pennsylvanian cyclothems, and the magnitude of change is closer to that required for the major cyclothems. However, the repeatability of sea-level changes of that magnitude is probably insufficient because repeated new rifting at short periods (~400,000 years) is unlikely (Fig. 7). Therefore, although this model could be responsible for a few regressive-transgressive Pennsylvanian cycles of more minor scale, the authors emphasize that it is most appropriately applied to eustatic cycles of this scale and period in non-glacial times.

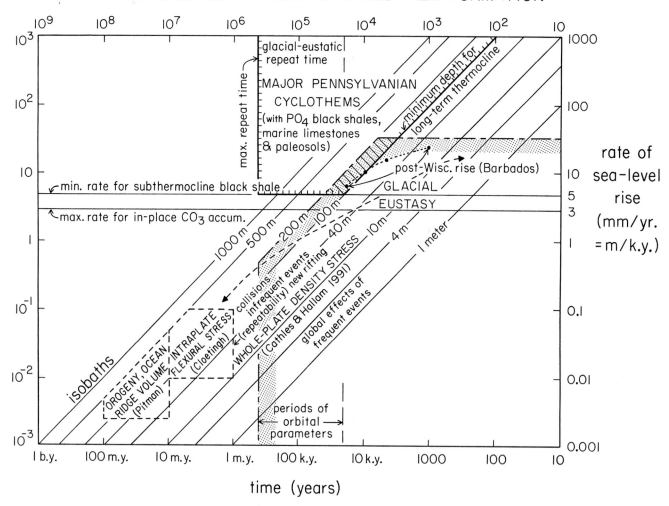

FIG. 7.—Time, transgressive rate, water depth, and repeatability constraints on mechanisms that have been considered as possible controls over the formation of midcontinent Pennsylvanian major (phosphatic black-shale-bearing) cyclothems. The field for these cyclothems (delineated by dark line with hachures) is constrained by water depth sufficient to maintain a widespread thermocline long enough to produce a widespread black shale (~100 m based on several modern analogs), a rate of sea-level rise (~5 mm/yr) sufficient to outstrip the estimated maximum rate of in-place shallow-water carbonate accumulation (~3 mm/yr) in order to consistently form this type of black shale above a thin transgressive limestone, and a repeatability of ~400 ky or less. The field for glacial-eustatic control (long-short dashed line, shaded on inside) is based on possible extremes for all parameters, but is documented by measured segments of Fairbanks' (1989) post-Wisconsinan sea-level curve for Barbados (dots connected by points), and is the only one that overlaps (oblique lining) the field for major Pennsylvanian cyclothems. Two fields for tectonic models (short dashed lines) are orders of magnitude too slow in time and rate parameters (orogeny and ocean-ridge volume; intraplate flexural stress). The third (whole-plate density stress, short dashes with arrows) lies within the proper range for time and rate of transgression, but would require new rifting or other extreme and generally infrequent tectonic events to achieve even 50 m of sea-level change, which are highly unlikely to repeat within the time frame of the cyclothems (~400 ky or less). "Jostling" of tectonic plates, which is more likely to repeat within the time frame of cyclothems, would produce sea-level changes of only a few meters; collision events that could produce a 200-m sea-level change are repeatable only on much longer time scales.

Water depth at sea-level highstand.—

Relative water depths can be rather straightforwardly interpreted in vertical sequences by noting the succession of lithic evidence for peritidal deposits, effective fair-weather wave base, and storm-wave base. This sequence can be extended further by more debatable evidence for such effects as carbonate mud generation and preservation, shell dissolution, and pycnoclines that limit vertical circulation, hence oxygen replenishment to the bottom. Determination of *absolute* water depths below the peritidal range, however, is much less certain, because of great variation in the depths of these features and in the processes responsible in different oceans and marginal seas. These variations are a function of climate and circulation patterns; the latter are dependent on irregularities in the bottom and shape of the seas, coupled with seasonal and long-term variations in climate.

Even with these reservations, it is worth considering the salient characteristics of the widespread offshore shales in major midcontinent cyclothems. Those that are gray and fossiliferous, and thus represent oxygenated bottoms, nevertheless lack algae. They also lack other aragonitic organisms such as molluscs, except in thicker, more rapidly deposited upward transitions to prodeltaic facies. Some also show evidence of corrosion of calcitic fossils. Although these shales may have been deposited below the effective photic zone for the particular algae that lived in this epeiric sea, it is possible that any indigenous calcareous algae were dissolved along with the aragonitic molluscs (the high-Mg calcite of red algae is about as soluble as aragonite). Because these shales represent condensed intervals in view of the abundance of calcitic skeletal debris, conodonts, and locally, nonskeletal phosphorite, it appears that large-scale dissolution of unstable and fine-grained carbonate was occurring over a large area for a relatively long period of time during their deposition.

Because these shales are not simply the result of a rapid detrital pulse, then non-production and/or non-preservation of carbonate mud must be responsible for the distinct identity of the offshore shale in an otherwise dominantly skeletal calcilutite sequence. Although other factors can affect carbonate mud production and preservation, a thermocline that formed in the tropical midcontinent sea at sea-level highstand and bathed the deeper areas with cooler water long enough to dissolve carbonate mud and aragonitic shells (and corrode calcitic shells) is the most plausible interpretation that accounts for the nature of the offshore shale. The fact that the tropical carbonate compensation depth is much deeper today is a result of the immense planktonic carbonate input since late Jurassic time. Bosellini and Winterer (1975) documented its depression with the rise of planktonic coccoliths and foraminifers during the Jurassic Period, and it is not unreasonable to suggest that it was much shallower in an epeiric sea 150 million years earlier during Pennsylvanian time.

The black phosphatic shale facies that the gray facies commonly surrounds attests to a pycnocline strong enough to prevent long-term vertical circulation of oxygenated surface waters to the bottom and allow preservation of a large amount of fine organic matter of both marine and distant terrestrial origin in a condensed section. It is recognized that the presence of sufficient organic matter to color sediment black depends upon the amount of input as well as preservation (Pedersen and Calvert, 1990), and that high organic productivity ("overloads") combined with rapid deposition can preserve much organic matter in shallow, more oxic water. Nevertheless, in the case of the phosphatic black-shale facies, the combination of: (1) its extremely widespread extent, (2) its offshore position as a condensed section in the cyclothem, (3) the evidence for slow deposition (the abundance of conodonts and nonskeletal phosphorite), and (4) its occurrence in topographic lows in places where its distribution is patchy, argues strongly that the organic matter derived from transitory periods of high production or

influx must have been deposited below a protective long-term pycnocline. This was probably a thermocline in view of the evidence for carbonate dissolution in the surrounding gray shale facies and the remoteness of the midcontinent from potential freshwater influx from the humid shoreline toward the Appalachian highlands.

Based on oceanographic data available at the time, Heckel (1977, p. 1057-8) suggested water depths of 100 m for the black shale facies (double that of the minimum depths of 50 m for the top of the thermocline and oxygen-minimum zone in the eastern tropical Pacific) in order to accommodate a quasi-estuarine circulation cell that promoted phosphorite deposition. This thermocline must have been relatively stable over a large area for a long period of time to protect the organic material in the condensed deposit from oxidation. The top of the thermocline in the modern tropical Atlantic Ocean fluctuates from about 50 to 80 m seasonally from 0° to 10°N, from 50 to 100 m seasonally from 10° to 20° N, and to greater depths poleward (Yentsch and Garside, 1986). Accepting the 50 to 80 m fluctuation range for the reconstructed 3° to 11°N latitude of the midcontinent sea (Fig. 5), it is worth considering that the Atlantic Ocean is a large, relatively stable water body whereas the smaller, much shallower midcontinent sea may have been subject to more extreme seasonal fluctuations and thus might have required even deeper water to establish a sufficiently permanent thermocline to protect the organic-rich black mud. Among modern epeiric seas, the central Baltic Sea maintains a pycnocline between about 60 and 90 m (Manheim, 1961) with black sapropelic muds accumulating below 120 to 140 m, but this pycnocline is largely caused by a halocline in a high-latitude sea.

It may be significant that in modern tropical, partly enclosed epeiric seas (Java Sea, 3° to 7°S; Arafura Sea, 5° to 12°S, and its arm, the Gulf of Carpentaria, to 17°S), probably the best climatic analogs for the midcontinent sea, there are no known areas of low-oxygen black-mud deposition. Water depths across much of the Java Sea average 50 to 70 m, and maximum water depths are roughly 50 m across much of the Arafura Sea, increasing to 60 to 70 m in the more enclosed Gulf of Carpentaria (Edgar and others, 1992; C. B. Cecil, pers. commun.). Although physiography, circulation patterns, climate, and amount of organic productivity also play a role, it is tempting to speculate that water depths in these modern tropical epeiric seas are too shallow to establish a thermocline permanent enough to protect widespread black-mud accumulation. In any case, I see no convincing evidence to change my original estimate of 100 m (Fig. 7) as the proper order of magnitude for water depths in the main part of the midcontinent sea (mid- to low-shelf position) during black phosphatic shale deposition (Figs. 3, 5). Toward the humid Appalachian margin of the midcontinent sea, however, widespread black-shale deposition may have become established in shallower water as the thermocline became augmented and ultimately replaced by a halocline resulting from strong freshwater outflow from the Appalachian region (Heckel, 1991c). Water depths for gray offshore shale deposition would have been somewhat shallower, extending up to the depth at which carbonate mud of any origin was consistently preserved.

Water-depth changes of 100 m can be attained by either tectonic models or glacial eustasy (Fig. 7). The greatest water-depth changes (200 to 500 m) can be attained tectonically, but only by the long-period, slow-acting processes summarized by Pitman and Golovchenko (1983) and Cloetingh (1988). Transgressive-regressive cycles with amplitudes of 100 m at the stated rates of sea-level change (0.01 to 0.1 mm/yr) would take 2 to 20 million years, assuming constant rates and instant reversal. This is 5 to 50 times too long to account for the Pennsylvanian cyclothems, but it does bracket the 3- to 5-million-year cycles estimated by Tankard (1986) for earlier Pennsylvanian foreland downwarp in the Appalachian basin. The short-period processes of elastic snapback described by Cathles and Hallam (1991) are within the 400,000-year range of the cyclothems, but these can attain even 50-m water-depth changes only under exceptional circumstances of new rifting of large plates. "Plate-jostling" could cause plate density changes and sea-level variations at the frequency of the cyclothems over single plates or groups of plates, but the sea level changes involved would be only a few meters (Fig. 7).

Water-depth changes from recent glacial eustasy are calculated to be 120 m for the post-Wisconsinan sea-level rise based on data from Barbados (Fairbanks, 1989). Potential further rise from melting the Greenland and Antarctic ice sheets is estimated to be about 50 m, for a total of 170 m of potential change from complete melting of Wisconsinan ice cover. Recent calculations of Gondwanan ice volume during Middle Pennsylvanian time by Crowley and Baum (1991) yielded a range of isostatically adjusted potential sea-level changes from 60 ± 15 m for the minimum estimated ice cover, to 170 ± 20 m for the maximum estimated ice cover. The maximum estimate is equivalent to the Pleistocene potential. Their intermediate estimate is 105 ± 15 m, which brackets the value I estimate for major midcontinent Pennsylvanian cyclothems. Clearly Gondwanan glacial eustasy can readily account for the water-depth changes required by midcontinent Pennsylvanian cyclothems that alternated between paleosols and subthermocline black shales (Figs. 6, 7).

Rate of transgression.—

A common characteristic of major midcontinent cyclothems relating to rates of sea-level rise is the sequence of a thin transgressive limestone overlain by a widespread offshore gray shale to subthermocline black shale. The transgressive limestone is almost entirely subtidal skeletal calcilutite with a diverse open marine biota of unabraded invertebrates and red algae, deposited below effective wave base. Thus, it contains remains of one of the photic carbonate producers and consists largely of lime mud, which is unlikely to have been generated by the preserved fossils because they show no abrasion. The presence of epibiotic encrusting forams without visible objects of attachment attests to the presence of a now disintegrated host, which could have been soft invertebrates such as sponges or easily disintegrated carbonate-producing forms such as green algae.

Lime mud today is known to be generated in great amounts in modern photic environments by green algae, the most dominant of modern photic carbonate producers (Neumann and Land, 1975). Some invertebrates can disintegrate completely to produce lime mud but, considering the trophic position of algae as the primary food producers and thus likely to be far more abundant, it is reasonable to assume that formation of the transgressive limestone depended to a great extent on algal carbonate production. The thinness and consistent position of this limestone below the widespread gray offshore shale (which apparently owed its existence to the demise of the algae and dissolution of any other lime mud) or the subthermocline black shale (which required water depths on the order of 100 m) indicates that sea-level rise was fast enough that water depths remained in the optimal photic zone for these algae only long enough to produce sufficient carbonate sediment to form a transgressive limestone averaging 0.3 to 1.0 m thick.

In short, the rate of sea-level rise apparently was consistently great enough at many different times to exceed the rate of algal carbonate production on a broad shelf. Even where certain transgressive limestones are quite thick (25 m for the Lost City Limestone of the Dennis cycle at Tulsa; 15 m for the Captain Creek Limestone of the Stanton cycle at Independence, Kansas), reflecting growth on a ramp or during a somewhat slower phase of transgression, they still are overlain by the offshore shale, typically the gray facies. This indicates that even where transgressive limestones are thick, the rate of sea-level rise eventually exceeded the ability of the algae to produce enough sediment to keep the sea bottom in the photic zone.

Modern rates of carbonate accumulation range from 1 to 3 mm/yr (Wilson, 1975) and up to as much as 15 mm/yr (references in Schlager, 1981) for reef tracts where the high growth rates of modern scleractinian corals dominate the total. The rates of non-reef subtidal carbonate accumulation reported by Wilson (1975) average close to 1 mm/yr, but these are all shallow-water situations where erosion and transport are reducing the potential accumulation that could occur during sea-level rise. I use the estimates of Neumann and Land (1975) that perhaps twice as much algal mud is transported out of shallow subtidal environments as remains behind (in 7 m of water depth in this case) to suggest 3 mm/yr as the average potential rate of algal carbonate accumulation alone during sea-level rise (Fig. 7). Therefore, even though other factors can influence widespread carbonate production (Schlager, 1981), in order for sea-level rise consistently to exceed carbonate accumulation over the entire shallow shelf in every cyclothem containing widespread offshore shales and to attain stable pycnocline depths of about 100 m in the major cyclothems, I estimate that a sustained rate of sea-level rise of at least 5 mm/yr is necessary (Fig. 7).

Rates of sea-level rise derived from broad cyclic tectonic models (Pitman and Golovchenko, 1983; Cloetingh, 1988) are 0.1 mm/yr or less. Those modeled by Cathles and Hallam (1991) could range up to perhaps 15 mm/year, but in circumstances where such rises are unlikely to repeat at the proper frequency. More rapid tectonic subsidence of as much as 5 mm/yr has been modeled for small (100-km) extensional "pull-apart" basins by Pitman and Andrews (1985), but this does not apply to the broad midcontinent sea extending hundreds of km in all directions (after the coupled Nemaha uplift/Forest City basin structure became dormant).

Estimates of average net subsidence of the northern midcontinent shelf required to provide accommodation space can be based on the thickness of preserved sedimentary rock. Thickness of the Verdigris-Howard interval (Fig. 1) ranges from 250 m in Iowa to 480 m near Wichita, Kansas. Taking the 10-my estimate of duration for this interval and assuming average sediment compaction to one-fourth original volume after deposition of the entire succession, yields 0.1 mm/yr required subsidence in Iowa and 0.2 mm/yr in Kansas; halving the duration estimate yields 0.2 to 0.4 mm/yr respectively for the case of a shorter Pennsylvanian Period, still an order of magnitude short of the rapid rise required. However, assuming that most compaction of the lower sediments was probably well underway while the upper sediments were being deposited substantially reduces the net subsidence rate required, which may approach the 0.025- to 0.05-mm/yr rate that assumes instantaneous compaction and a 10-my duration. This is within the 0.002- to 0.1-mm/yr range of rates estimated from the broadly acting tectonic models mentioned above.

Klein's (1991) citation of neotectonic subsidence rates of 3 to 8 mm/yr are in local tectonically active areas (San Andreas and New Madrid fault zones), which are inappropriate analogs for the post-active-Nemaha late Middle and Upper Pennsylvanian succession of the northern midcontinent. Klein and Kupperman's (1992) citation of recent rapid vertical movements over an area of 200,000 km[2] in southern Chile is also an inappropriate analog because the displacement is not only within an active orogenic belt, but also consists of 100-km wide parallel tracts of adjacent depression and elevation (Plafker and Savage, 1970), which is incompatible with the observed lateral continuity of the cyclothems over a cratonic shelf.

The post-Wisconsinan (Flandrian) glacio-eustatic sea-level rise, on the other hand, has been estimated to have ranged between 5 and 35 mm/yr at various times from 10,000 to 7,000 years BP (Cronin, 1983). More recent estimates of both rates and amounts of sea-level rise can be measured from the subsidence-corrected Flandrian sea-level curve for Barbados in Fairbanks (1989, Fig. 2). The entire rise of 120 m over the past 18,000 years was at an average rate of 6.7 mm/yr (and this includes only a 10-m rise over the past 6,000 years with a rate of 1.7 mm/yr). The 100-m rise from 15,000 to 6,000 years BP averaged 11 mm/yr, and the 24-m rise from 10,000 to 9,000 years BP averaged 24 mm/yr. Clearly glacio-eustatic sea-level rise could consistently exceed any reasonable rate of carbonate accumulation in the Pennsylvanian midcontinent sea.

Summary.—
Only glacial eustasy can raise sea level fast enough to a depth great enough to account for the unique transgressive sequence of midcontinent Pennsylvanian cyclothems (thin transgressive limestone overlain by thin offshore, commonly subthermocline shale) and to repeat these cycles within the frequencies required by the empirical data (Fig. 7). All currently proposed tectonic models fail to change sea level fast enough, to depths great enough, across an area broad enough and to repeat consistently within periods of time short enough to account for the cyclothems. Tectonic models do, however, provide bottom subsidence at rates sufficient to account for the net accommodation space required for Pennsylvanian cyclic deposits across the northern midcontinent shelf.

Detecting Possible Tectonic Signals

Combining the above lines of reasoning with the well established record of periodic glaciation during Pennsylvanian and early Permian time on the Gondwanan continents (Veevers and Powell, 1987) and the appearance of cyclicity of this scale elsewhere in North America during this time and, in fact, only during this time (Dickinson and others, 1991, also this volume), it is reasonable to conclude that the widespread transgressions and regressions of the sea responsible for the midcontinent Pennsylvanian cyclothems were primarily driven by glacial eustasy. Because this control is global, other marine cycles of similar scale elsewhere are probably also glacio-eustatic in origin. Therefore, biostratigraphic correlation of the individual marine units (already underway) across larger areas of the world can provide a data base for further analysis of the succession of marine cyclothems in different areas in order to detect possible tectonic influences on sedimentation and stratigraphic patterns. Several preliminary examples are considered.

Relative subsidence.—
The thickness of the Verdigris-Howard interval (Fig. 1) increases from 250 m in southwestern Iowa (cores IRC, ILC north of Omaha, Fig. 6) through 320 m in northeastern Kansas (sec 31-T2S-R15E, Brown Co.) and 480 m near Wichita (sec 27-T26S-R6E, Butler Co.) to 710 m in the Hugoton embayment in southwestern Kansas (sec 20-T34S-R38W, Stevens Co.; Kansas data courtesy of J. A. French). This nearly threefold increase in thickness is attributed to increase in tectonic subsidence toward the Anadarko foreland basin in Oklahoma (of which the Hugoton is an arm). Because this increase in thickness takes place without increase in the number of major cyclothems that are present in Iowa, the effect of increasing subsidence on the major cyclothems was mainly to thicken some of the regressive limestones and to insert between them several minor cycles of sea-level fluctuation southward (Watney and others, 1989), as expected on the lower shelf (see Fig. 3B). The typical northern midcontinent cyclothem sequence of transgressive limestone—offshore shale—regressive limestone—exposure surface/paleosol, however, remains readily detectable into this area of greatly increased tectonic downwarping. This challenges the strong implication in Klein and Willard (1989) that increased tectonic activity should result in the formation of more clastic-dominated Appalachian-type cyclothems.

Moreover, the approximate Appalachian correlative of the midcontinent Altamont-Oread interval (Fig. 1), the lower Conemaugh Group (Glenshaw Formation) from the Upper Freeport Coal to the Ames Limestone, maintains a 60- to 110-m thickness across Ohio, southwestern Pennsylvania, western Maryland, and northern West Virginia (Sturgeon and others, 1958; Edmunds and others, 1979; Arkle and others, 1979; Busch, 1984), thickening slightly to 126 m eastward in south-central Pennsylvania (V.W. Skema, pers. commun.). This range is somewhat less than the thickness of the equivalent midcontinent cratonic sequence in Iowa cores (130 to 150 m). This fact appears to be inconsistent with the conclusion of Klein and Willard (1989) that "Appalachian-type cyclothems were deposited in response to tectonic processes of collisional margins which caused flexural deformation and flexural basin subsidence" and with the assertion of Klein (1989) that they "appear to be formed in direct response to foreland basin flexural subsidence because they accumulated adjacent to the collision zone, thus masking glacial eustasy." If this were true, then the entire Appalachian cyclothem succession should be significantly thicker because of tectonic subsidence than the equivalent midcontinent succession. It is mainly the thick Lower to lower Middle Pennsylvanian succession in Kentucky (Lee, Breathitt Formations) and West Virginia (Pocahontas, New River, Kanawha Formations) (Rice and others, 1979; Arkle and others, 1979) that reflects foreland-basin downwarping in the Appalachian region.

The critical factor that made all Appalachian cyclothems different from northern midcontinent cyclothems was readily available detrital influx (Figs. 3B, 4). This was apparently unrelated to the amount of net subsidence, which decreased significantly in the preserved portion of the Appalachian basin from Early and Middle to Late Pennsylvanian time. It is certainly true that the readily available detrital influx in the Appalachian basin was related to nearby orogenic highlands, hence ultimately tectonic in origin, but abundant detrital influx, which is the critical factor, could also be of cratonic origin under the right climatic conditions for erosion and transport of detrital sediment (Cecil, 1990). The substantial detrital influx from the orogenic south side of the Oklahoma foreland basin was trapped within the basin and, therefore, did not affect the cyclothems forming on the north flank of the tectonically active basin, which are midcontinent in character (even though thicker than on the shelf). The major effect of differential tectonic activity on Pennsylvanian marine cyclothems appears to have been provision of different amounts of subsidence, hence accommodation space (and ultimately thickness; see Merrill, 1988), and in directing the influx of detritus into certain parts of the space provided (see Wise and others, 1991).

Possible tectonic cycles.—

Preliminary results of biostratigraphic correlation of the major marine midcontinent cyclothems with marine units in the Illinois and Appalachian successions suggest the possible detection of another type of tectonic signal in the Appalachian basin (Fig. 8). Correlation of marine cyclothems is established between the midcontinent and Illinois primarily with the Charleston core, in which all major marine units can be matched by their conodont faunas with the major midcontinent marine cyclothems in the proper order (Heckel and Weibel, 1991). Preliminary correlations with the Appalachian area, based largely on recent conodont collections, and the palynological coal correlations of R. A. Peppers (pers. commun.), suggest an interesting pattern, shown in Figure 8. The Appalachian marine units appear to be bundled in the mid-late Desmoinesian (Columbiana—Dorr Run), the mid-Missourian (lower Brush Creek—Portersville), and early Virgilian (Ames—Skelley) intervals. This means that the Appalachian basin was low enough to receive widespread glacio-eustatic marine inundations at these times. These marine bundles are separated by predominantly terrestrial intervals containing only local minor marine units that are commonly represented by a brackish-water or shoreline facies in a succession that typically displays more complex paleosol development (Joeckel, 1992). These terrestrial-dominated intervals in the Appalachian basin lack in most places the widespread marine units that occur within the coeval intervals in the midcontinent and Illinois (Altamont/post-Danville, Lost Branch/W.Franklin-Lonsdale, Hertha/Cramer units of the late Desmoinesian—early Missourian interval, and Stanton/Little Vermilion, Cass-Haskell/Omega units of the late Missourian—early Virgilian interval). Because these marine units appear to be nearly as widespread in the midcontinent and Illinois as the intervening bundles of marine units that do have widespread correlatives in the Appalachians, it appears that the Appalachian area was too elevated during these times to accommodate more than the margins of some of these major marine incursions.

The line on Figure 8 that outlines the relative extents of marine incursions in the Appalachian basin delineates a curve that shows nearly three complete longer-period cycles of presence and absence of bundles of widespread marine units in the Appalachian area. Using the estimate of 10 million years for the duration of the entire succession, these longer cycles have a period of about 3 million years, which is close to the cycle length estimated by Tankard (1986) for his tectonic loading model for Appalachian cyclicity. Although this correlation of glacio-eustatic marine cycles into the Appalachian basin is only preliminary, and some suggested correlations may require revision under more intense scrutiny, it suggests that with a firm biostratigraphic framework for the glacio-eustatic cycles, we should eventually be able to detect other, more subtle signals in the stratigraphic succession. In this preliminary case, a signal appears that is conceivably tectonic in terms of both behavior and rate. The Appalachian basin may have been slowly cycling between a condition depressed enough to accept major glacio-eustatic marine inundations and one elevated enough to inhibit extension of these incursions into this region. This cyclicity has a 3-million-year period that is more compatible with those estimated for current tectonic models.

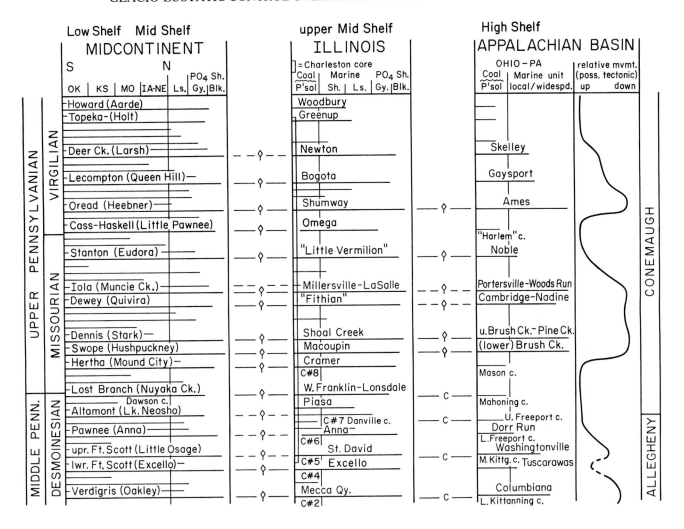

FIG. 8.—Preliminary correlation of major midcontinent eustatic cyclothems (longest horizontal lines, denoted above line by cyclothem name followed by core shale name in parentheses) summarized from Fig. 1B, with those recognized in Illinois basin, based largely on succession of conodont faunas (tailed diamond symbols) in Charleston core supplemented by outcrop collections of named units (Heckel and Weibel, 1991), and with marine units in Appalachian basin (denoted by both Ohio and Pennsylvania names); Appalachian correlations are based on coal bed correlations (c) of R. A. Peppers (pers. commun., 1991) (coal bed names/numbers are given below or opposite the horizontal lines), preliminary evaluation of recent conodont collections (J. E. Barrick and P. H. Heckel, in prep.) supplemented by illustrations in Merrill (1964, 1974), and reports on other faunal groups (Boardman and others, 1990; Boardman, pers. commun., 1991). Dashed lines represent less definite correlations. This correlation framework suggests a longer-term signal of cyclic, perhaps tectonic, movement in the Appalachian basin of roughly 3-million-year periods of depression to accommodate marine inundations followed by enough elevation to prevent much marine incursion and to cause more extensive erosion and/or paleosol formation in this region. "Harlem" coal is second coal below Ames Limestone at Ames type section reported by Sturgeon (1958), which may not be correlative with type Harlem coal 90 miles to north.

CONCLUSIONS AND IMPLICATIONS

It appears that only glacio-eustatic control provided by waxing and waning of Gondwanan ice sheets can account for all the critical characteristics of midcontinent Pennsylvanian marine cyclothems: (1) transgressive-regressive stratigraphic sequences culminating in paleosols, (2) extremely widespread lateral extent of these repeated sequences, (3) periodicity of repetition within the Milankovitch band of the earth's orbital parameters, and (4) rapid rate of transgression (>5mm/yr) to water depths of about 100 m required for the succession of thin transgressive limestones overlain by subthermocline black shales. Because this control was global, all areas of the world must have been affected and, therefore, nearly all widespread marine cycles of this periodicity during this time must have been primarily glacio-eustatic in origin. Those who would challenge this conclusion should offer an alternative explanation as to what the Gondwanan glacio-eustatic signal would have been in tropical areas near sea level during this time.

Delta shifting cannot account for the cyclothems in the northern midcontinent (where deltas essentially are lacking) or for the widespread extent of paleosols overlain by marine intervals within areas of greater detrital influx. Nevertheless, delta shifting readily explains local cycles of detrital facies within and above the major marine horizons in detrital-rich areas.

All cyclic tectonic models so far developed fail to account for several of these characteristics, although the whole-plate density stress model of Cathles and Hallam (1991) could have been responsible for some minor cycles of regression followed by transgression. The only obvious evidence of post-Nemaha tectonic activity in the midcontinent is the southward-increasing thickness of the intervals in Figures 1 and 6, reflecting increased accommodation space toward the downwarping foreland basin in Oklahoma. A longer-term cyclic tectonic model may explain subtle patterns of longer-term cyclic vertical movement suggested by preliminary late Middle-Upper Pennsylvanian correlations in the Appalachian basin.

Climate, of course, was the ultimate control over Pennsylvanian marine cyclothems by its control of periodic glaciation and deglaciation. Although glacial-interglacial shifts in climate can directly affect the tropical belt also, local cycles of climatic control over sedimentation may well have been influenced by the eustatic fluctuations in sea level, which would have controlled local sources of rainfall and temperature modulation. The climatic control shown by Cecil (1990) over the well delineated nonmarine cycles in the younger Upper Pennsylvanian (Monongahela) succession in the Appalachian basin may have been strongly related to varying proximity to the fluctuating marine shoreline toward and away from the Appalachian basin, which, at this time, remained either slightly too elevated or otherwise physiographically unconnected with the sea that still fluctuated in the midcontinent and Illinois basins.

Tectonic highlands certainly influenced the local manifestation of shifting climatic patterns, but changes in the gross physiography of this land probably occurred much more slowly than the glacial-interglacial shifts in climate. Tectonic subsidence certainly exerted control over the accommodation space for deposition of the glacial-eustatic cyclothems, and regional variations in the amount of this subsidence directed the dispersal of available detritus. However, Klein and Kupperman's (1992) assertion that "both tectonic and glacio-eustatic processes appear to form Pennsylvanian cyclothems" seems to inflate the role of tectonic processes because all they seemingly demonstrate is tectonic provision of accommodation space for cyclic deposits, not tectonic control over the cyclicity itself—which is the critical point for determining which process controlled cyclothem formation. Tectonic subsidence has provided accommodation space for both cyclic and noncyclic sedimentation throughout geologic time.

It was the amount of detrital influx that controlled the lithic nature of the eustatic cyclothems, and thus explains the difference between Appalachian, Illinois, and midcontinent cyclothems as first pointed out by Wanless and Shepard (1936). Klein and Willard's (1989) assertion that increase in certain types of tectonic activity, including greater subsidence, explains the differences between Appalachian and midcontinent cyclothems,

thereby "resolving" the controversy over glacial versus tectonic origin, is misleading. Although highlands formed by orogeny obviously can generate large amounts of detrital sediment, this sediment can get trapped in the axis of a foreland basin, and therefore not affect cyclothems developed in a tectonically active area across the basin axis (as in the southern midcontinent). Moreover, a sufficiently large amount of detrital influx probably could be generated from a cratonic source, under the proper climatic conditions for large-scale erosion and transport, to form detrital-dominated (Appalachian) cyclothems on a stable craton.

ACKNOWLEDGMENTS

I gratefully thank: J. M. Dennison, F. R. Ettensohn, and R. L. Langenheim for thoroughly reviewing the manuscript; R. M. Joeckel, G. A. Ludvigson, B. J. Witzke, J. C. Ferm, and L. M. Cathles for thoroughly reviewing parts of the manuscript; the state geological surveys of Kansas, Oklahoma, Missouri, Nebraska, Iowa, Illinois, Ohio, Pennsylvania, West Virginia, and Kentucky for various types of support, services, and information; H. R. Rollins, R. L. Martino, N. Fedorko, C. B. Cecil, R. H. Mapes, P. F. Holterhoff, and M. R. Caudill for showing me localities in the Appalachian basin; R. A. Peppers, D. R. Boardman, J. E. Barrick, R. M. Busch, C. B. Cecil, J. R. Hatch, W. L. Watney, J. A. French, S. Gray, V. W. Skema, D. R. Chesnut, L. R. Follmer, E. A. Bettis, R. G. Baker, and C. S. Yentsch for providing information and references; the donors of the Petroleum Research Fund (Grant 13647-AC2) administered by the American Chemical Society, the Allan and DeLeo Bennison Stratigraphic Research Fund, and the J. W. Swade Memorial Fund for partial support of field and laboratory work; and R. Petrick and G. Greiner for typing the many drafts of this manuscript.

REFERENCES

ARKLE, T., JR., AND OTHERS, 1979, The Mississippian and Pennsylvanian (Carboniferous) Systems in the United States-West Virginia and Maryland: Washington D. C., United States Geological Survey Professional Paper 1110-D, 35 p.

BARRICK, J. E., AND BOARDMAN, D. R. II, 1989, Stratigraphic distribution of morphotypes of Idiognathodus and Streptognathodus in Missourian-lower Virgilian Strata, north-central Texas: Lubbock, Texas Tech University Studies in Geology 2, p. 167-188.

BARRICK J. E., HECKEL, P. H., AND BOARDMAN, D. R., 1992, Morphologic evolution in early Missourian (Pennsylvanian) species of Idiognathodus and the origin of Streptognathodus (conodonts) (abs.): Geological Society of America, Abstracts with Programs, v. 24, p. 4-5.

BENNISON, A. P., 1984, Shelf to trough correlations of late Desmoinesian and early Missourian carbonate banks and related strata, northeast Oklahoma, in Hyne, N. J., ed., Limestones of the Midcontinent: Tulsa, Tulsa Geological Society Special Publication 2, p. 93-126.

BENNISON, A. P., 1985, Trough-to-shelf sequence of the early Missourian Skiatook Group, Oklahoma and Kansas, in Watney, W. L., Kaesler, R. L., and Newell, K. D., eds., Recent Interpretations of Late

Paleozoic Cyclothems: Lawrence, Proceedings of Third Annual Meeting and Field Conference, Mid-Continent Section, Society of Economic Paleontologists and Mineralogists, Kansas Geological Survey, p. 219-245.

BIRKLAND, P. W., 1984, Soils and Geomorphology: New York, Oxford University Press, 372 p.

BOARDMAN, D. R., II, AND HECKEL, P. H., 1989, Glacial-eustatic sea-level curve for early Late Pennsylvanian sequence in north-central Texas and biostratigraphic correlation with curve for midcontinent North America: Geology, v. 17, p. 802-805.

BOARDMAN, D. R., II, HECKEL, P. H., BARRICK, J. E., NESTELL, M., AND PEPPERS, R. A., 1990, Middle-Upper Pennsylvanian chronostratigraphic boundary in the midcontinent region of North America: Courier Forschungsinstitut Senckenberg, v. 130, p. 319-337.

BOSELLINI, A., AND WINTERER, E. L., 1975, Pelagic limestone and radiolarite in the Tethyan Mesozoic: a general model: Geology, v. 3, p. 279-282.

BUSCH, R. M., 1984, Stratigraphic analysis of Pennsylvanian rocks using a hierarchy of transgressive-regressive units: Unpublished Ph.D. Dissertation, University of Pittsburgh, Pittsburgh, 427 p.

BUSCH, R. M., AND ROLLINS, H. B., 1984, Correlation of Carboniferous strata using a hierarchy of transgressive-regressive units: Geology, v. 12, p. 471-474.

CATHLES, L. M., AND HALLAM, A., 1991, Stress-induced changes in plate density, Vail sequences, epeirogeny, and short-lived global sea-level fluctuations: Tectonics, v. 10, p. 659-671.

CAUDILL, M. R., 1990, Lithostratigraphy of Conemaugh Group (Upper Pennsylvanian) marine zones—Steubenville, Ohio-Weirton, West Virginia: Southeastern Geology, v. 31, p. 173-182.

CECIL, C. B., 1990, Paleoclimate controls on stratigraphic repetition of chemical and siliciclastic rocks: Geology, v. 18, p. 533-536.

CHESNUT, D. R., JR, 1991, Marine transgressions in the central Appalachian basin during the Pennsylvanian Period (abs.): Geological Society of America, Abstracts with Programs, v. 23, p. 16.

CLOETINGH, S., 1986, Intraplate stresses: A new tectonic mechanism for fluctuations of relative sea-level: Geology, v. 14, p. 617-620.

CLOETINGH, S., 1988, Intraplate stresses: A tectonic cause for third-order cycles in apparent sea level?, in Wilgus, C. K., and others, eds., Sea-Level Changes—An Integrated Approach: Tulsa, Society of Economic Paleontologists and Mineralogists Special Publication 42, p. 19-29.

CRONIN, T. M., 1983, Rapid sea level and climate change: evidence from continental and island margins: Quaternary Science Reviews, v. 1, p. 177-214.

CROWELL, J. C., 1978, Gondwanan glaciation, cyclothems, continental positioning, and climate change: American Journal of Science, v. 278, p. 1345-1372.

CROWLEY, T. J., AND BAUM, S. K., 1991, Estimating Carboniferous sea-level fluctuations from Gondwanan ice extent: Geology, v. 19, p. 975-977.

DICKINSON, W. R., SOREGHAN, G. S., AND GOEBEL, K. A., 1991, Stratigraphic cyclicity in foreland geologic systems of Permo-Carboniferous age (abs.): Geological Society of America, Abstracts, v. 23, no. 1, p. 21.

EDGAR, N. T., CECIL, C. B., GRIM, M. S., JONES, M. R., AND SEARLE, D. E., 1992, Gulf of Carpentaria, a modern analog for ancient tropical epicontinental sea sedimentation (abs.): Geological Society of America Abstracts with Programs, v. 24, p. A143.

EDMUNDS, W. E., BERG, T. M., SEVON, W. D., PIOTROWSKI, R. C.,

HEYMAN, L. AND RICKARD, L. V., 1979, The Mississippian and Pennsylvanian (Carboniferous) systems in the United States—Pennsylvania and New York: Washington D. C., United States Geological Survey Professional Paper 1110-B, 33 p.

FAHRER, T. R., AND HECKEL, P. H., 1992, Petrology and depositional significance of Conemaugh marine units in the Appalachian basin (abs.): Geological Society of America, Abstracts with Programs, v. 24, p. A321.

FAIRBANKS, R. G., 1989, A 17,000-year glacio-eustatic sea level record: influence of glacial melting rates on the Younger Dryas event and deep-ocean circulation: Nature, v. 342, p. 637-642.

FERM, J. C., 1970, Allegheny deltaic deposits, in Morgan, J. P., ed., Deltaic Sedimentation Modern and Ancient: Tulsa, Society of Economic Paleontologists and Mineralogists Special Publication 15, p. 246- 255.

FRENCH, J. A., AND HECKEL, P. H., 1994, Probable high-amplitude glacial-eustatic origin for Pennsylvanian cyclothems requires modification of applied sequence-stratigraphic classification and terminology (abs.): American Association of Petroleum Geologists Annual Meeting, Official Program, v. 3, p. 152.

GOEBEL, K. A., BETTIS, E. A., III, AND HECKEL, P. H., 1989, Upper Pennsylvanian paleosol in Stranger Shale and underlying Iatan Limestone, southwestern Iowa: Journal of Sedimentary Petrology, v. 59, p. 224-232.

GOULD, H. R., 1970, The Mississippi delta complex, in Morgan, J. P., ed., Deltaic Sedimentation Modern and Ancient: Tulsa, Society of Economic Paleontologists and Mineralogists Special Publication 15, p. 3-30.

HATCH, J. R., DAWS, T. A., LUBECK, S. C. M., PAWLEWICZ, M. J., THRELKELD, C. N., AND VULETICH, A. K., 1984, Organic geochemical analyses for 247 organic-rich rock and oil samples from the Middle Pennsylvanian Cherokee and Marmaton groups, southeastern Iowa, Missouri, southeastern Kansas, and northeastern Oklahoma: Washington, D. C., United States Geological Survey Open-file Report 84-160, 38 p.

HECKEL, P. H., 1977, Origin of phosphatic black shale facies in Pennsylvanian cyclothems of midcontinent North America: American Association of Petroleum Geologists Bulletin, v. 61, p. 1045-1068.

HECKEL, P. H., 1980, Paleogeography of eustatic model for deposition of midcontinent Upper Pennsylvanian cyclothems, in Fouch, T. D., and Magathan, E. R., eds., Paleozoic Paleogeography of West-Central United States: Denver, Rocky Mountain Section, Society of Economic Paleontologists and Mineralogists, West-Central United States Paleogeography Symposium I, p. 197-215.

HECKEL, P. H., 1983, Diagenetic model for carbonate rocks in midcontinent Pennsylvanian eustatic cyclothems: Journal of Sedimentary Petrology, v. 53, p. 733-759.

HECKEL, P. H., 1984a, Changing concepts of midcontinent Pennsylvanian cyclothems, North America: Champaign-Urbana, Neuvième Congrès International de Stratigraphie et de Géologie du Carbonifère, Compte Rendu, v. 3, p. 535-553.

HECKEL, P. H., 1984b, Factors in midcontinent Pennsylvanian limestone deposition, in Hyne, N. J., ed., Limestones of the Midcontinent: Tulsa, Tulsa Geological Society Special Publication 2, p. 25-50.

HECKEL, P. H., 1986, Sea-level curve for Pennsylvanian eustatic marine transgressive-regressive depositional cycles along midcontinent outcrop belt, North America: Geology, v. 14, p. 330-334.

HECKEL, P. H., 1989a, Current view of midcontinent Pennsylvanian cyclothems: Lubbock, Texas Tech University Studies in Geology 2, p. 17-34.

HECKEL, P. H., 1989b, Updated Middle-Upper Pennsylvanian eustatic sea-level curve for midcontinent North America and preliminary biostratigraphic characterization: Beijing, Onzième Congrès International de Stratigraphie et de Géologie du Carbonifère, Nanjing University Press, Compte Rendu, v. 4, p. 160-185.

HECKEL, P. H., 1990, Evidence for global (glacial-eustatic) control over upper Carboniferous (Pennsylvanian) cyclothems in midcontinent North America, in Hardman, R. F. P., and Brooks, J., eds., Tectonic Events Responsible for Britain's Oil and Gas Reserves: London, Geological Society Special Publication 55, p. 35-47.

HECKEL, P. H., 1991a, Lost Branch Formation and Revision of Upper Desmoinesian Stratigraphy along Midcontinent Pennsylvanian Outcrop Belt: Lawrence, Kansas Geological Survey Geology Series 4, 67 p.

HECKEL, P. H., 1991b, Comment on "Pennsylvanian time scales and cycle periods" by G. Dev. Klein: Geology, v. 19, p. 406-407.

HECKEL, P. H., 1991c, Thin widespread Pennsylvanian black shales of midcontinent North America: a record of a cyclic succession of widespread pycnoclines in a fluctuating epeiric sea, in Tyson, R. V., and Pearson, T. H., eds., Modern and Ancient Continental Shelf Anoxia: London, Geological Society Special Publication 58, p. 259-273.

HECKEL, P. H., AND BAESEMANN, J. F., 1975, Environmental interpretation of conodont distribution in Upper Pennsylvanian (Missourian) megacyclothems in eastern Kansas: American Association Petroleum Geologists Bulletin, v. 59, p. 486-509.

HECKEL, P. H., BRADY, L. L., EBANKS, W. J., AND PABIAN, R. K., 1979, Field Guide to Pennsylvanian Cyclic Deposits in Kansas and Nebraska: Lawrence, Kansas Geological Survey Guidebook Series 4, p. 4-60.

HECKEL, P. H., AND WEIBEL, C. P., 1991, Current status of conodont-based biostratigraphic correlation of Upper Pennsylvanian succession between Illinois and midcontinent, in Weibel, C. P., ed., Sequence Stratigraphy in Mixed Clastic-Carbonate Strata, Upper Pennsylvanian, East-central Illinois: Champaign, Great Lakes Section, Society of Economic Paleontologists and Mineralogists, 21st Annual Field Conference, Illinois State Geological Survey, p. 60-69.

IMBRIE, J., 1985, A theoretical framework for the Pleistocene ice ages: Journal of the Geological Society (London), v. 142, p. 417-432.

JOECKEL, R. M., 1989, Geomorphology of a Pennsylvanian land surface: pedogenesis in the Rock Lake Shale Member, Southeastern Nebraska: Journal Of Sedimentary Petrology, v. 59, p. 469-481.

JOECKEL, R. M., 1992, Contemporaneous Late Pennsylvanian paleosols in two different basins: comparison of the sub-Ames paleosol and the Lawrence Formation and Snyderville Member paleosols (abs.): Geological Society of America, Abstracts with Programs, v. 24, p. A286-287.

JOECKEL, R. M., 1994, Virgilian (Upper Pennsylvanian) paleosols in the upper Lawrence Formation (Douglas Group) and in the Synderville Shale Member (Oread Formation, Shawnee Group) of the northern midcontinent, U. S. A.: pedologic contrasts in a cyclothem sequence: Journal of Sedimentary Research, v. A64, p.

KIDDER, D. L., 1985, Petrology and origin of phosphate nodules from the midcontinent Pennsylvanian epicontinental sea: Journal of Sedimentary Petrology, v. 55, p. 809-816.

KLAPPER, G., AND BARRICK, J. E., 1978, Conodont paleoecology: pelagic versus benthic: Lethaia, v. 11, p. 15-23.

KLEIN, G. DEV., 1989, Comments on sedimentary-stratigraphic verification of some geodynamic basin models: example from a cratonic and associated foreland basin, in Cross, T. A., ed.,

Quantitative Dynamic Stratigraphy: Englewood Cliffs, Prentice Hall, p. 503-517.

KLEIN, G. DEV., 1990, Pennsylvanian time scales and cycle periods: Geology, v. 18, p. 455-457.

KLEIN, G. DEV., 1991, Reply to comment by Heckel on "Pennsylvanian time scales and cycle periods": Geology, v. 19, p. 407-408.

KLEIN, G. DEV. AND CLOETINGH, S., 1989, Tectonic subsidence during deposition of Pennsylvanian cyclothems (abs.): Washington, D.C., 28th International Geological Congress, Abstracts, v. 2 , p. 198-199.

KLEIN, G. DEV., AND KUPPERMAN, J. B., 1992, Pennsylvanian cyclothems: Methods of distinguishing tectonically induced changes in sea level from climatically induced changes: Geological Society of America Bulletin, v. 104, p. 166-175.

KLEIN, G. DEV., AND WILLARD, D. A., 1989, Origin of the Pennsylvanian coal-bearing cyclothems of North America: Geology, v. 17, p. 152-155.

LANGENHEIM, R. L., 1991, Comparison of Pennsylvanian carbonate cycles in the Cordilleran region with midcontinental cyclothems suggests a common eustatic origin (abs.): Geological Society of America Abstracts with Programs, v. 23, no. 1, p. 56.

MALINKY, J. M., 1984, Paleontology and paleoenvironment of "core" shales (Middle and Upper Pennsylvanian) midcontinent North America: Unpublished Ph. D. Dissertation, University of Iowa, Iowa City, 327 p.

MAYNARD, J. R., AND LEEDER, M. R., 1992, On the periodicity and magnitude of Late Carboniferous glacio-eustatic sea-level changes: Journal of the Geological Society (London), v. 149, p. 303-311.

MANHEIM, F. T., 1961, A geochemical profile in the Baltic Sea: Geochimica et Cosmochimica Acta, v. 25, p. 52-70.

MERRILL, G. K., 1964, Zonation of platform conodont genera in Conemaugh strata of Ohio and vicinity: Unpublished M. A. Thesis, University of Texas, Austin, 169 p.

MERRILL, G. K., 1974, Pennsylvanian conodont localities in northeastern Ohio: Columbus, Ohio Division of Geological Survey, Guidebook No. 3, 29 p.

MERRILL, G. K., 1988, Marine transgression and syndepositional tectonics; Ames Member (Glenshaw Formation, Conemaugh Group, Upper Carboniferous) near Huntington, West Virginia: Southeastern Geology, v. 28, p. 153-166.

MOORE, D., 1959, Role of deltas in the formation of some British Lower Carboniferous cyclothems: Journal of Geology, v. 67, p. 522-539.

MOUSSAVI-HARAMI, R., AND BRENNER, R. L., 1984, Deltaic sedimentation on a carbonate shelf: Stanton Formation (Upper Pennsylvanian), southeastern Kansas: American Association of Petroleum Geologists Bulletin, v. 68, p. 150-163.

NEUMANN, A. C., AND LAND, L. S., 1975, Lime mud deposition and calcareous algae in the Bight of Abaco, Bahamas: a budget: Journal of Sedimentary Petrology, v. 45, p. 763-786.

PEDERSON, T. F., AND CALVERT, S. E., 1990, Anoxia vs. productivity: what controls the formation of organic-carbon-rich sediments and sedimentary rocks?: American Association of Petrology Geologists Bulletin, v. 74, p. 454-466.

PITMAN, W. C., III, AND ANDREWS, J. A., 1985, Subsidence and thermal history of small pull-apart basins, in Biddle, K. T., and Christie-Blick, N., eds., Strike-slip Deformation, Basin Formation and Sedimentation: Tulsa, Society of Economic Paleontologists and Mineralogists Special Publication 37, p. 45-49.

PITMAN, W. C., III, AND GOLOVCHENKO, X., 1983, The effect of sea level change on the shelf edge and slope of passive margins, in Stanley, D. J., and others, eds., The Shelfbreak: Critical Interface on

Continental Margins: Tulsa, Society of Economic Paleontologists and Mineralogists Special Publication 33, p. 41-58.

PLAFKER, G., AND SAVAGE, J. C., 1970, Mechanism of the Chilean earthquakes of May 21 and 22, 1960: Geological Society of America Bulletin, v. 81, p. 1001-1030.

QUINLAN, G. M., AND BEAUMONT, C., 1984, Appalachian thrusting, lithospheric flexure, and the Paleozoic stratigraphy of the eastern interior of North America: Canadian Journal of Earth Sciences, v. 21, p. 973-996.

RICE, C. L., SABLE, E. G., DEVER, G. R., JR., AND KEHN, T. M., 1979, The Mississippian and Pennsylvanian (Carboniferous) Systems in the United States—Kentucky: Washington D. C., United States Geological Survey Professional Paper 1110-F, 32 p

SCHLAGER, W., 1981, The paradox of drowned reefs and carbonate platforms: Geological Society of America Bulletin, pt. I, v. 92, p. 197-211.

SCHUTTER, S. R., AND HECKEL, P. H., 1985, Missourian (early Late Pennsylvanian) climate in midcontinent North America: International Journal of Coal Geology, v. 5, p. 111-140.

SHABICA, C. W., 1979, Pennsylvanian sedimentation in northern Illinois: examination of delta models, in Nitecki, M. H., ed., Mazon Creek Fossils: New York, Academic Press, p. 13-40.

SKEMA, V. W., DODGE, C. H., AND SHAULIS, J. R., 1991, Lithologic character and correlation of marine units in the Conemaugh Group (Upper Pennsylvanian), western Pennsylvania (abs.): Geological Society of America, Abstracts with Programs, v. 23, no. 1, p. 128.

SLUCHER, E. R., 1988, Middle and Upper Pennsylvanian marine units of northeast Ohio, a reevaluation (abs.): Geological Society of America, Abstracts with Programs, v. 20, p. 389.

STANTON, M. R., LEVENTHAL, J. S., AND HATCH, J. R., 1983, Short range vertical variation in organic carbon, carbonate carbon, total sulfur contents, and Munsell color values in a core from the Upper Pennsylvanian Stark Shale Member of the Dennis Limestone, Wabaunsee County, Kansas: Washington, D.C., United States Geological Survey Open File Report 83-315, 8 p.

STURGEON, M. T., 1958, The geology and mineral resources of Athens County, Ohio: Ohio Geological Survey Bulletin 57, 600 p.

SWADE, J. W., 1985, Conodont distribution, paleoecology, and preliminary biostratigraphy of the upper Cherokee and Marmaton Groups (upper Desmoinesian, Middle Pennsylvanian) from two cores in South-central Iowa: Iowa City, Iowa Geological Survey Technical Information Series 14, 71 p.

TANKARD, A. J., 1986, Depositional response to foreland deformation in the Carboniferous of eastern Kentucky: American Association of

Petroleum Geologists Bulletin, v. 70, p. 853-868.

VEEVERS, J. J., AND POWELL, C. M., 1987, Late Paleozoic glacial episodes in Gondwanaland reflected in transgressive-regressive depositional sequences in Euramerica: Geological Society of America Bulletin, v. 98, p. 475-487.

WANLESS, H. R., AND SHEPARD, F. P., 1936, Sea level and climatic changes related to Late Paleozoic cycles: Geological Society of America Bulletin, v. 47, p. 1177-1206.

WATNEY, W. L., FRENCH, J. A., AND FRANSEEN, E. K., 1989, Introduction to field trip, in Watney, W. L., French, J. A., and Franseen, E. K., eds., Sequence Stratigraphic Interpretations and Modeling of Cyclothems in the Upper Pennsylvanian (Missourian) Lansing and Kansas City Groups in Eastern Kansas: Lawrence, Guidebook for Kansas Geological Society 41st Annual Field Trip, Kansas Geological Survey, p. 1-68.

WATTS, A. B., KARNER, G. D., AND STECKLER, M. S., 1982, Lithospheric flexure and the evolution of sedimentary basins: Philosophical Transactions of the Royal Society of London, v. A305, p. 249-281.

WELLER, J. M., 1930, Cyclical sedimentation of the Pennsylvanian Period and its significance: Journal of Geology, v. 38, p. 97-135.

WELLER, J. M., 1956, Argument for diastrophic control of late Paleozoic cycles: American Association of Petroleum Geologists Bulletin, v. 40, p. 17-50.

WELLS, A. J., 1960, Cyclic sedimentation: a review: Geological Magazine, v. 97, p. 389-403.

WILSON, J. L., 1967, Cyclic and reciprocal sedimentation in Virgilian strata of southern New Mexico: Geological Society of America Bulletin, v. 78, p. 805-817.

WILSON, J. L., 1975, Carbonate Facies in Geologic History: New York, Springer Verlag, 471 p.

WISE, D. U., BELT, E. S., AND LYONS, P. C., 1991, Clastic diversion by fold salients and blind thrust ridges in coal-swamp development: Geology, v. 19, p. 514-517.

WRIGHT, H. E., JR., 1989, The Quaternary: Chapter 17, in Bally, A. W., and Palmer, A. R., eds., The Geology of North America, Vol. A, An Overview: Boulder, Geological Society of America, p. 513-536.

YENTSCH, C. S., AND GARSIDE, J. C., 1986, Patterns of phytoplankton abundance and biogeography: UNESCO Technical Papers in Marine Science 49, p. 278-284.

YOSE, L. A., AND HELLER, P. L., 1989, Sea-level control of mixed-carbonate-siliciclastic, gravity-flow deposition: Lower part of the Keeler Canyon Formation (Pennsylvanian), southeastern California: Geological Society of America Bulletin, v. 101, p. 427-439.

FLEXURALLY INFLUENCED EUSTATIC CYCLES IN THE POTTSVILLE FORMATION (LOWER PENNSYLVANIAN), BLACK WARRIOR BASIN, ALABAMA

JACK C. PASHIN

Geological Survey of Alabama, P.O. Box O, Tuscaloosa, Alabama 35486-9780

ABSTRACT: The Lower Pennsylvanian Pottsville Formation in the Black Warrior foreland basin of Alabama contains abundant coal and coalbed-methane resources, and because of numerous data from geophysical well logs, offers one of the best opportunities to evaluate causes of cyclicity in Carboniferous coal-bearing strata through subsurface mapping and facies analysis. Twelve regionally mappable transgressive-regressive cycles are present in the Pottsville Formation, which is of Morrowan age. Individual cycles accumulated in an average of 0.2 to 0.5 my and thus represent high-frequency fluctuations of relative sea level.

The Black Warrior basin underwent a rapid tectonic evolution related to progressive deformational loading of the Alabama promontory as the Appalachian-Ouachita orogen developed. Subsidence rate averaged approximately 15.3 cm/1,000 yr (0.5 ft/ 1,000 yr) in the structurally deepest part of the basin, and disregarding sediment influx or sea-level change, could account for an increase of water depth of more than 75 m (250 ft) during deposition of some cycles. This rapid subsidence evidently imparted pronounced asymmetry to relative sea-level variation by amplifying marine transgression and suppressing marine regression. Whether or not extremely rapid subsidence in response to the introduction of new load elements onto the continental promontory caused regional transgression is unclear, but episodes of enhanced loading probably resulted in at least local inundation as some cycles were deposited.

Although rapid flexural subsidence may have amplified marine transgression, no tectonic causes of regional marine regression were identified that operated at the time scale of deposition of a single Pottsville cycle. For this reason, glacial eustasy is considered to have been the dominant cause of cyclicity in the study interval. Consistent distribution of fluvial-deltaic sandstone and coal in each cycle mapped indicates that, despite rapidly changing subsidence patterns, a northwest- to west-dipping coastal plain and a single sediment-dispersal system persisted in Alabama. Hence, tectonism and eustasy operated faster than sediment could be dispersed from evolving sources in the advancing orogenic belt, and the resulting paleogeography was much more sensitive to eustatic sea-level variation than to flexural changes of basin geometry.

INTRODUCTION

Cyclicity is the salient characteristic of the Pennsylvanian System in North America (Weller, 1930), yet the origin of this cyclicity remains controversial. Many geologists agree that cycles in Pennsylvanian rocks are largely the products of allogenic mechanisms, especially relative sea-level variation driven by glacial eustasy (Wanless and Shepard, 1936; Ross and Ross, 1988) and tectonism associated with supercontinent assembly (Weller, 1956; Klein and Willard, 1989). However, the relative importance of tectonism and eustasy in forming those cycles is a matter of continuing debate.

Recently, investigators have analyzed the periodicity of Pennsylvanian cycles and the subsidence regime of the sedimentary basins in which those cycles accumulated (e.g., Heckel, 1986; Klein and Kupperman, 1992). Although this approach has made important strides toward identifying the roles of tectonism and eustasy, regional subsurface mapping has largely been overlooked as a way to evaluate the effects of these factors. This report characterizes and interprets cycles in the Lower Pennsylvanian Pottsville Formation of Alabama on the basis of regional subsurface maps of cycle thickness, sandstone thickness, and coal distribution. The study focuses on the Black Creek-Cobb interval, the principal coalbed-methane target in the region, which contains most of the mineable coal beds in the Black Warrior basin and also contains conventional hydrocarbon reservoirs in sandstone.

Cycles in the Pottsville Formation of Alabama can be characterized as coarsening- and coaling-upward sequences containing (1) basal marine mudstone that coarsens upward into (2) open- to marginal-marine sandstone, which is, in turn, overlain by (3) the marginal-marine and terrestrial sandstone, mudstone, and coal that make up a coal group (Pashin and Sarnecki, 1990; Pashin, 1991a) (Fig. 1). Cyclicity in the Pottsville was originally attributed to autogenic factors like delta-lobe switching (Ferm and others, 1967; Horsey, 1981). In the past decade, however, regional subsurface maps and cross sections (Cleaves, 1981; Sestak, 1984; Pashin, 1991a) have demonstrated that the cycles extend throughout the basin and thus represent regional marine-nonmarine depositional continua. Therefore, Pottsville cycles are best explained by allogenic cyclicity related to dynamically interwoven tectonic and climatic variables. Autogenic processes were probably most effective within each cycle. This paper is devoted to characterizing the regionally extensive, allogenic cycles of the Pottsville Formation in Alabama.

METHODS

Well logs of the Black Creek-Cobb interval are numerous and provide a robust data base for regional subsurface mapping and testing the causes of cyclicity in Pennsylvanian coal-bearing strata (Fig. 2). Four rock types were distinguished using density logs that were calibrated by comparison with cores, cuttings, and drillers logs on the basis of variation in the gamma-ray, bulk-density, density-porosity, and neutron-porosity signatures. The rock types are (1) coal, (2) mudstone, (3) tight sandstone, and (4) porous sandstone. Porous reservoir sandstone is distinguished from tight sandstone by extremely low gamma count and by

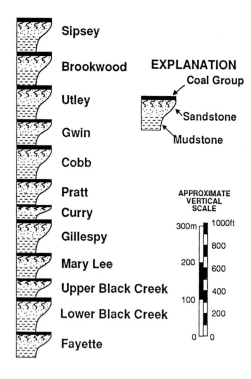

FIG. 1.—Regionally extensive coarsening- and coaling-upward cycles in the Pennsylvanian Pottsville Formation, Black Warrior basin, Alabama (cycles are named for the coal group or coal bed that caps each cycle).

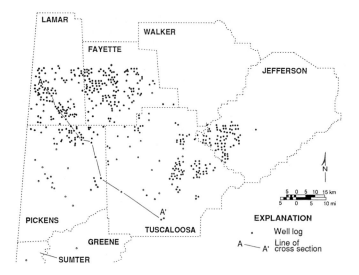

FIG. 2.—Map showing location of wells and cross section used in this study.

crossing of the neutron- and density-porosity curves.

To demonstrate stratigraphic relationships in the Black Creek-Cobb interval, regional cross sections were made using the top of the Pratt coal group as a datum. Cycles were defined on the basis of a thick mudstone unit at the base of each cycle and the presence of a coal or sandstone bed at the top. Finally, the following subsurface maps were made for each cycle: (1) cycle isopach, (2) tight-sandstone isolith, (3) porous-sandstone isolith, and (4) coal isopleth. Isopach and coal-isopleth maps also were made for the combined Black Creek-Cobb interval. Selected maps and a cross section are presented in this paper. The complete set of maps and cross sections is in Pashin and others (1990).

REGIONAL SETTING

The Black Warrior basin (Fig. 3) encompasses a triangular area in Alabama and Mississippi that is bound on the southeast by the Appalachian orogen, on the southwest by the Ouachita orogen, and on the north by the Nashville dome (Mellen, 1947; Thomas, 1988). The Black Warrior basin is separated from the Appalachian basin in the northeast by a southeast-plunging nose of the Nashville dome and is separated from the Arkoma basin in the west by the Mississippi Valley graben (Thomas, 1988, 1991). Tectonically, the Black Warrior is a late Paleozoic foreland basin that formed flexurally in response to converging thrust and sediment loads that were emplaced on the Alabama

promontory (Thomas, 1977) during Appalachian-Ouachita orogenesis (Beaumont and others, 1987, 1988; Hines, 1988).

Stratigraphy

The Pottsville Formation crops out only in the eastern part of the Black Warrior basin. Approximately two thirds of the basin is buried beneath Cretaceous and younger overburden of the Mississippi Embayment and the Gulf Coastal Plain (Fig. 3). Cretaceous strata of the Tuscaloosa Group overlie the Pottsville Formation disconformably, and adjacent to the deeply buried Ouachita orogen in Mississippi, Mesozoic and Cenozoic rocks are thicker than 1,800 m (6,000 ft).

In the Black Warrior basin, coal is in stratigraphic bundles called coal groups (McCalley, 1900), which have formed the basis of most stratigraphic subdivisions of the Pottsville Formation (McCalley, 1900; Butts, 1910, 1926; Culbertson, 1964; Metzger, 1965). Coal groups cap regressive, coarsening- and coaling-upward sequences, or cycles (Fig. 1). The cycles range in thickness from 11 m (35 ft) along the northern basin margin to more than 200 m (700 ft) in the deep subsurface and contain a basal marine mudstone that typically coarsens upward into sandstone (Pashin, 1991a). At the top of each cycle is a coal group comprising as much as 125 m (400 ft) of interbedded mudstone, sandstone, underclay, and coal.

Twelve regionally mappable coarsening- and coaling-upward cycles (Fayette through Sipsey) have thus far been identified in the Pottsville Formation of the Black Warrior basin (Fig. 1), and at least 6 younger cycles are preserved in the deepest part of the basin in Mississippi (Henderson and Gazzier, 1989). Biostratigraphic control in the Alabama Pottsville is limited, and stage boundaries have not been located precisely. However, palynomorphs indicate that the lower Black Creek through Brookwood cycles are of Westphalian A (late Morrowan) age

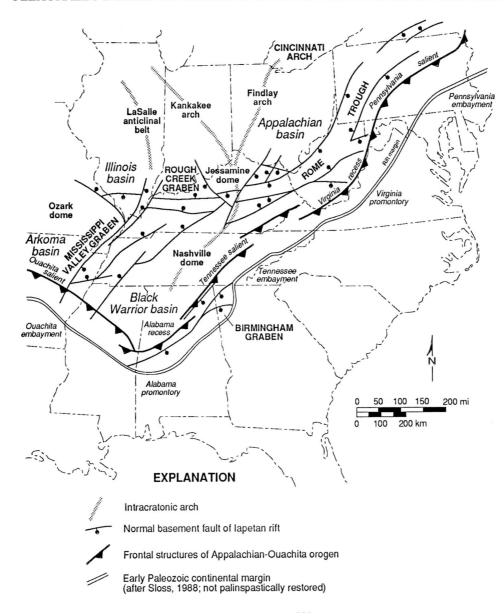

FIG. 3.—Geologic setting of the Black Warrior basin (modified from Thomas, 1988).

(Eble and Gillespie, 1989). Accounting for marked error in geochronologic evaluation of the Pennsylvanian System (Klein, 1990), as well as biostratigraphic evaluation of the Pottsville Formation in Alabama, the 10 to 12 Westphalian A cycles accumulated in 2 to 5 million years. Hence, the average span of time represented by these cycles is approximately 0.2 to 0.5 my.

Structure

The structural contour map of the top of the Mary Lee cycle (Fig. 4) shows that strata in the Black Warrior basin of Alabama are homoclinal and dip southwest. Appalachian folds and thrust faults that formed during the Allegheny orogeny are present along the southeastern margin of the Black Warrior basin. In the easternmost part of the basin, the Blue Creek anticline, Sequatchie

anticline, and Coalburg syncline strike at an approximate azimuth of 40°. These folds are detached structures that overlie decollements in Cambrian shale which ramp locally into Carboniferous strata (Rodgers, 1950; Thomas, 1985). Appalachian structures extend westward below the Gulf Coastal Plain where they override folds and thrust faults of the Ouachita orogen (Thomas, 1973, 1989) (Fig. 3).

Normal faults are numerous in the Black Warrior basin and define a series of parallel, linear to arcuate horst-and-graben systems that strike northwest and turn westward near Mississippi (Fig. 4). The faults with greatest displacement define a series of narrow grabens in Lamar and Pickens counties that extend for tens of miles and locally have throws in excess of 300 m (1,000 ft). Most faults in the Black Warrior basin parallel the Ouachita orogenic belt, and fault displacement increases toward

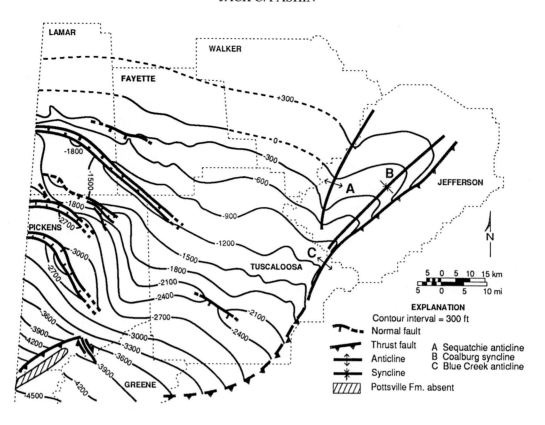

Fig. 4.—Structural contour map of the top of the Mary Lee cycle.

the orogenic front (Thomas, 1988). Synsedimentary movement of normal faults and Appalachian folds has been recognized in the Pottsville Formation (Weisenfluh and Ferm, 1984; Ferm and Weisenfluh, 1989; Pashin, 1991a). Therefore, some of these faults formed extensionally in response to deformational loading in the Ouachita orogen and concomitant flexural subsidence of the foreland basin (Hines, 1988). However, some faults in the easternmost part of the basin formed as pull-apart tear structures related to propagation of Appalachian folds and thrust faults (Pashin, 1991 b).

RESULTS: BASIN EVOLUTION AND PALEOGEOGRAPHY

Previous studies provided evidence that Ouachita tectonism caused subsidence and provided sediment sources in the western part of the Alabama promontory starting in Mississippian time and continuing well into Pennsylvanian time (Thomas, 1974, 1988; Thomas and Womack, 1983). Investigators further suggested that dispersal of sediment into the basin from Appalachian sources in the eastern part of the promontory began during Pottsville deposition but were unable to characterize paleogeography and sediment dispersal in detail because of sparse well control in the easternmost part of the basin (e.g., Horsey, 1981; Sestak, 1984). In the past five years, however, geophysical data from coalbed-methane drilling have made such characterization possible, and the following discussion focuses on the deposi-

tional systems, paleogeography, and tectonics as interpreted from well logs of the Black Creek-Cobb interval in the Alabama part of the Black Warrior basin.

Cycle Thickness

The Black Creek-Cobb interval thickens from less than 370 m (1,200 ft) in Lamar and Fayette counties toward the south and attains a thickness greater than 740 m (2,400 ft) in Sumter County (Fig. 5). Thickening is most pronounced south of the 1,800-ft (550-m) contour, which defines an arcuate area of thick sedimentary rock that extends from Tuscaloosa County into Mississippi. Cross section A-A' (Fig. 6) shows distinctive facies changes within the cycles. Tight sandstone is present in the upper part of all cycles, but porous sandstone occurs mainly in the northwest part of the cross section and is restricted to the Black Creek, Mary Lee, and Cobb cycles. Coal beds are most numerous where the cycles are thick and lack porous sandstone.

Although the Black Creek-Cobb isopach map depicts an arcuate area of thick sedimentary rock (Fig. 5), individual maps establish that cycle-thickness patterns varied systematically. For example, the combined upper and lower Black Creek cycles are more than 215 m (700 ft) thick in a depocenter in southeastern Tuscaloosa County (Fig. 7). The Gillespy and Curry cycles, however, are characterized by a sublinear depoaxis (area southeast of the 400-ft contour, Fig. 8) that traverses the southeastern

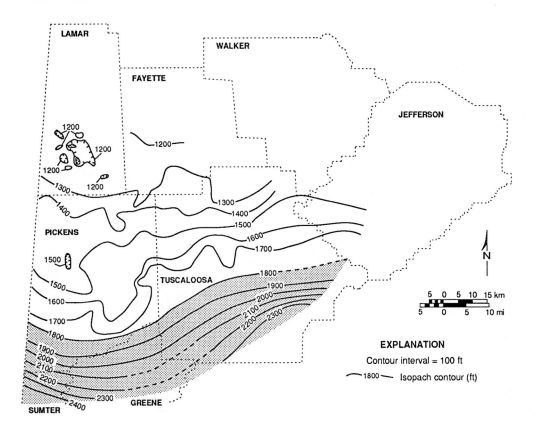

Fig. 5.—Isopach map of Black Creek-Cobb interval.

part of the study area. The depoaxis of the Cobb cycle (area south of the 350-ft contour) was arcuate and therefore resembles that of the composite Black Creek-Cobb interval (Figs. 5, 9).

Variation of the cycle-isopach pattern with time (Figs. 7-9) provides a record of subsidence history and foreland-basin evolution in the study area. Contours on the Black Creek-Cobb isopach map (Fig. 5) are oblique and are in places perpendicular to those on the structural contour map (Fig. 4), demonstrating that the modern basin differs structurally from that which existed during Pottsville deposition. The Black Creek depocenter indicates that tectonic subsidence was initially most rapid in the southeastern part of the continental promontory. Expansion of the depocenter into an arcuate depoaxis, however, reflects progressive deformational loading and merging of the relatively young Appalachian flexural moat in the eastern part of the promontory with the older Ouachita moat in the western part.

Tight Sandstone

In cores and cuttings, tight sandstone is fine to coarse grained, light gray to medium dark gray, and contains abundant rock fragments and interstitial clay that give the rock a salt-and-pepper texture. Petrographic analyses indicate that, in the Pottsville Formation, this type of sandstone is commonly litharenite and that low-grade metamorphic grains make up most of the lithic fraction (Davis and Ehrlich, 1974; Mack and others,

1983; Raymond, 1990). In well logs, tight sandstone typically has irregular, fining- and coarsening-upward log signatures (Fig. 6). Coarsening-upward signatures predominate below coal groups, whereas both types of signature are common within coal groups.

Comparing the Mary Lee, Gillespy-Curry, and Pratt sandstone-isolith maps (Figs. 10-12) establishes that tight sandstone thickens consistently toward southeastern Tuscaloosa County; the Mary Lee map also shows a thick, localized body of tight sandstone in southwestern Lamar County. Each map shows two major lobate to elongate sandstone axes; one is in eastern Tuscaloosa County, and the other is in southwestern Tuscaloosa County. In each cycle, the southwestern depositional axis is south of some faults (Figs. 10-12). In the Gillespy and Curry cycles (Fig. 11), parts of the elongate, bifurcated trend outlined by the 50-ft (15-m) contour follow faults in Tuscaloosa and Pickens counties, and the southwestern part of the bifurcated trend terminates at a fault in northern Pickens County. A similar relationship of sandstone distribution to structure is apparent in the Pratt cycle (Fig. 12) and, to a lesser extent, in the Mary Lee cycle (Fig. 10).

Tight sandstone is interpreted to represent diverse fluvial and marginal-marine depositional systems. In outcrop, similar sandstone has been interpreted to represent fluvial channels and associated flood basins (Pashin and Sarnecki, 1990), deltaic systems (Ferm and others, 1967; Horne and others, 1976;

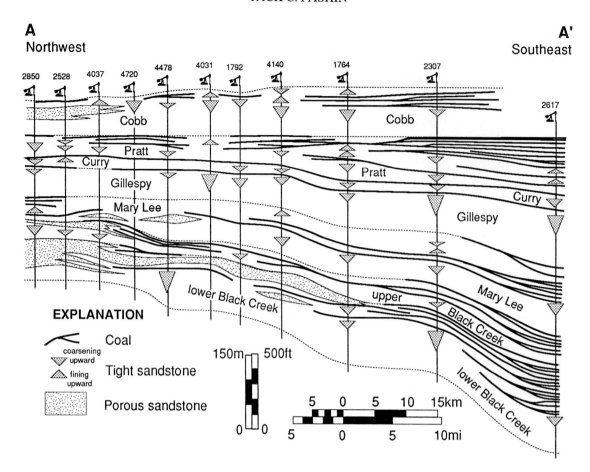

FIG. 6.—Cross section A-A' of the Black Creek-Cobb interval (location shown in Fig. 2).

Benson, 1982), and estuaries (Gastaldo and others, 1990). This diversity of depositional environment is apparent in variable well-log and isolith-contour patterns, as well as differences in stratigraphic position within cycles. For example, coarsening-upward log patterns below coal groups apparently represent delta-front deposits. In comparison, coarsening-upward patterns within coal groups probably represent crevasse-splay deposits, whereas fining-upward patterns apparently represent channel and tidal-flat deposits.

Some lobate and elongate contour forms in the sandstone-isolith maps apparently represent fluvial-deltaic systems (Fig. 13). In the Gillespy-Curry map, for example, the elongate, bifurcated trend in Pickens County (Fig. 11) may represent a constructive deltaic lobe that is in some measure analogous to the modern bird-foot lobe of the Mississippi Delta (Gould, 1970). A similar elongate trend in the Pratt cycle in this same area (Fig. 12) suggests that the trunk channel was reactivated. In the Mary Lee and Pratt cycles, however, coal beds interbedded with tight sandstone extend far beyond the lobate forms, suggesting that much of the sand was deposited in areas protected from marine influence. Sandstone lobes of this type are interpreted to represent differentially subsiding, transitive fluvial axes. Maps of individual sandstone bodies in the Mary Lee cycle

of the Coalburg syncline (Pashin, 1991a, b), moreover, indicate that tributary channels in the northeastern sandstone lobe were directed west, suggesting a more westwardly paleoslope than is apparent in the regional isolith maps.

Although cycle thickness varied in time and space (Figs. 7-9), tight-sandstone distribution varied little among cycles on a regional basis (Figs. 10-12). This relationship suggests that sand supply and dispersal were effectively independent of the changing subsidence pattern in Alabama. Persistently thick, lobate isolith patterns in southeastern Tuscaloosa County, moreover, provide evidence that a major proximal sand source lay in the Appalachian orogen (Fig. 13). Thick tight sandstone in the Mary Lee cycle of southwest Lamar County, however, has been interpreted to extend westward into deltaic deposits adjacent to the Ouachita orogen in Mississippi (Sestak, 1984).

Porous Sandstone

Porous sandstone is fine to medium grained, very light gray to yellowish gray, and is much richer in quartz and poorer in rock fragments and clay than tight sandstone. Petrographic studies indicate that light-colored, quartzose sandstone in the Pottsville Formation includes quartzarenite and sublitharenite (Davis and

(Continued on p. 98)

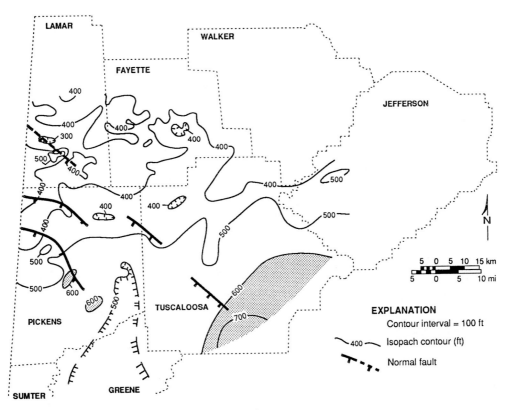

FIG. 7.—Isopach map of the combined lower and upper Black Creek cycles.

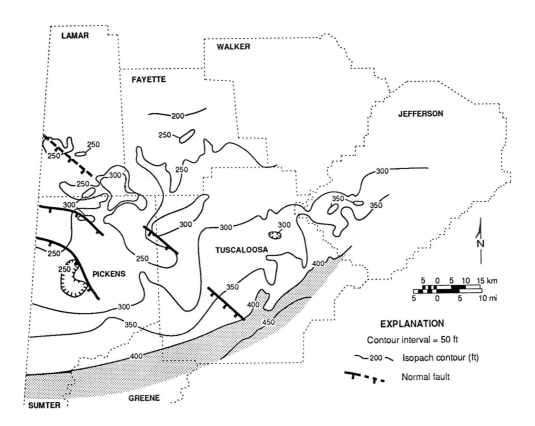

FIG. 8.—Isopach map of the combined Gillespy and Curry cycles.

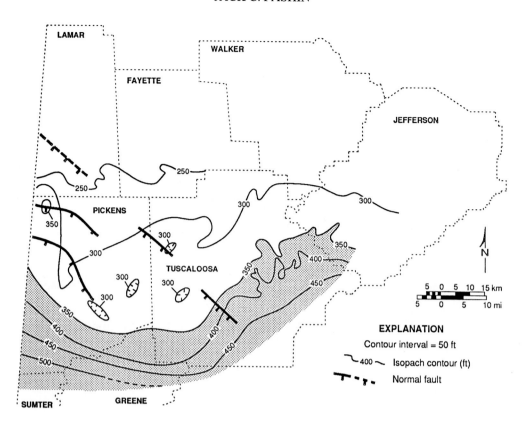

FIG. 9.—Isopach map of the Cobb cycle.

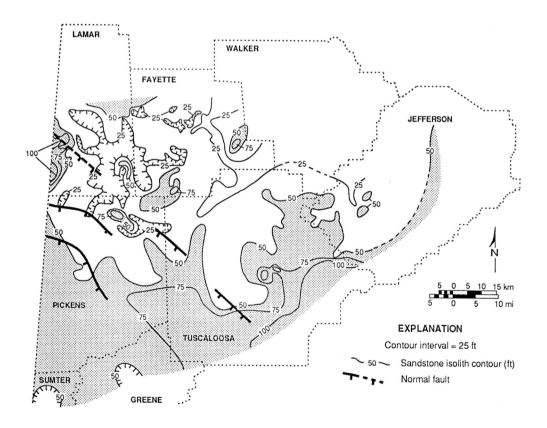

FIG. 10.—Tight-sandstone isolith map of the Mary Lee cycle.

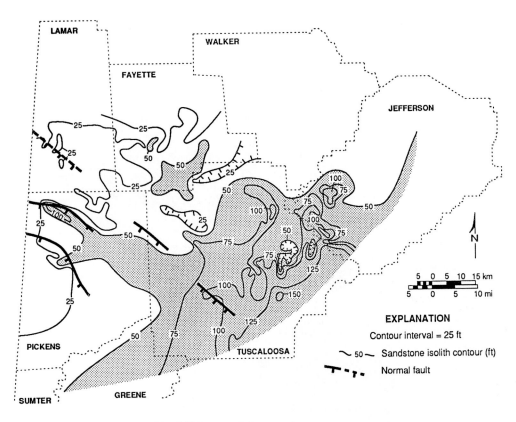

FIG. 11.—Tight-sandstone isolith map of the combined Gillespy and Curry cycles.

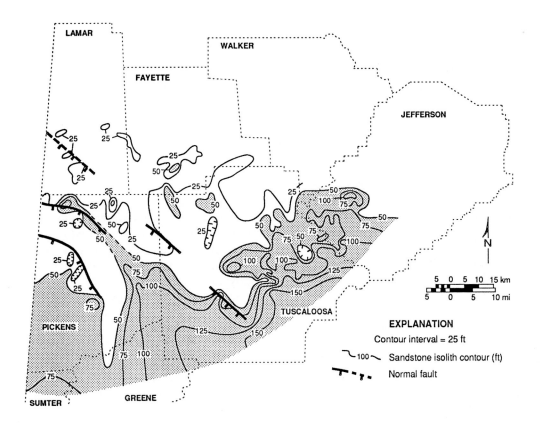

FIG. 12.—Tight-sandstone isolith map of the Pratt cycle.

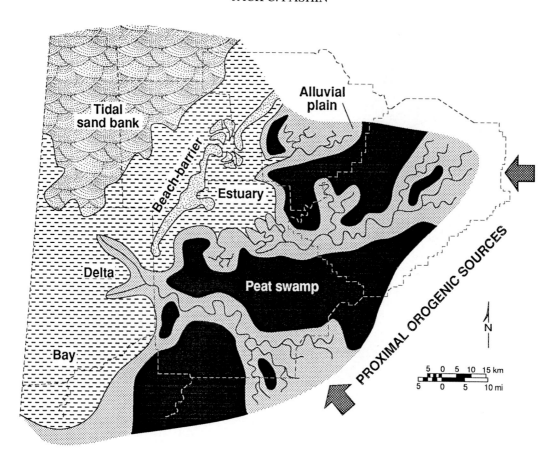

Fig. 13.—Paleogeographic model of the cyclic Black Creek-Cobb interval of the Pottsville Formation in the Black Warrior basin, Alabama.

Ehrlich, 1974; Mack and others, 1983; Raymond, 1990), but further analysis is needed to quantify the relationship between rock composition and geophysical well-log response. Porous sandstone contrasts markedly with tight sandstone in geophysical well logs not only because of low radioactivity and crossing of the porosity curves, but because porous sandstone characteristically has a blocky signature.

In the combined upper and lower Black Creek cycles, a belt of porous sandstone defined by the 50-ft (15-m) contour extends northeast from Sumter County to eastern Fayette County (Fig. 14). Locally, sandstone in this belt is thicker than 30 m (100 ft). Porous sandstone is absent in places northwest of the belt, but an irregular sandstone body that is locally thicker than 60 m (200 ft) is in Lamar County. This body terminates abruptly along a fault zone in western Lamar County, along which Mary Lee tight sandstone is thicker than 30 m (100 ft) (Fig. 10), and all or most of the Pottsville porous-sandstone units below the Black Creek-Cobb interval are absent (Engman, 1985).

Porous sandstone of the Mary Lee cycle is in a series of distinct, geographically restricted bodies (Fig. 15). Continuity of this sandstone is greatest in the northern part of the study area and decreases toward the south. Thick Black Creek and Mary Lee porous sandstone coincide in central Pickens County and in eastern Fayette County. By contrast, Mary Lee porous sand-

stone in southwest Fayette County is thicker than 15 m (50 ft) where Black Creek porous sandstone is thin or absent. However, a prominent northwest-oriented zone of thinning cuts across the thick Mary Lee sandstone of Fayette County.

Blocky geophysical log patterns in porous sandstone contrast sharply with the variable patterns in tight sandstone and signify a different set of depositional processes and environments. In outcrop, light-colored, quartzose sandstone in the Pottsville Formation has long been interpreted to represent tidally influenced beach-barrier systems (e.g., Ferm and others, 1967; Hobday, 1974; Shadroui, 1986). Large-scale, compound crossbeds characteristic of open-shelf sand waves (e.g., Allen, 1980) have also been recognized in quartzose sandstone of the Mary Lee cycle (Demko, 1990; Pashin and others, 1991). Hence, porous quartzose sandstone is interpreted to be a product of shoaling tidal currents and waves in open- to marginal-marine environments (Fig. 13).

The sandstone belt of the Black Creek cycle defines the eastern limit of porous sandstone in this area (Fig. 14) and is therefore interpreted to represent a shore-zone system, possibly with beaches. The location, thickness and irregular geometry of the sandstone in Lamar County (Fig. 14) is suggestive of a structurally influenced tidal sand bank (Fig. 13). The thickness and stratigraphic position of this sandstone further suggest that

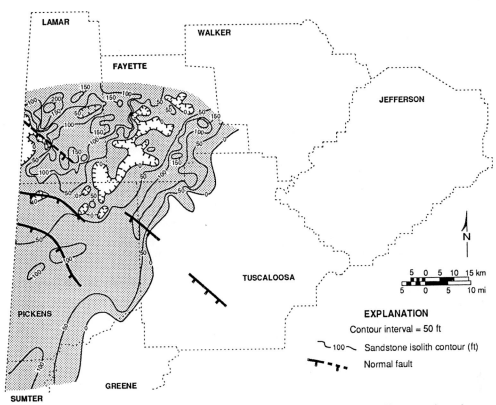

FIG. 14.—Porous-sandstone isolith map of the combined lower and upper Black Creek cycles (lower Nason sandstone).

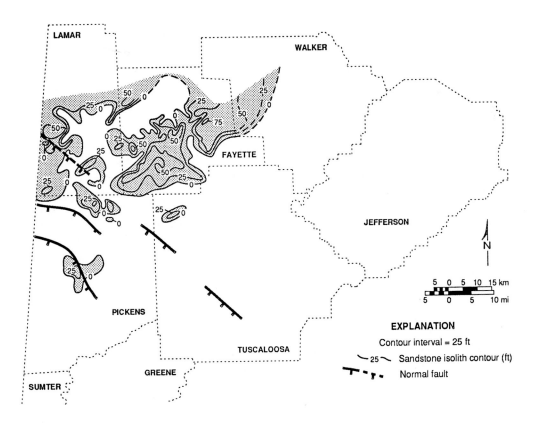

FIG. 15.—Porous-sandstone isolith map of the Mary Lee cycle (upper Nason sandstone).

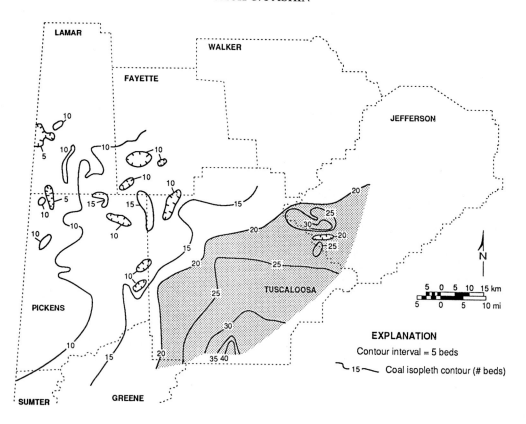

FIG.16.—Coal isopleth map of the Black Creek-Cobb interval, Black Warrior basin, Alabama.

Lamar County was a persistent site of shoaling, and extension of the sandstone above coal-bearing strata in the lower Black Creek cycle (Fig. 6) suggests that the shoal area expanded southeastward in response to marine transgression.

Offset of Mary Lee and Black Creek porous sandstone bodies (Figs. 14, 15) demonstrates the importance of relict topography and differential compaction on the distribution of porous sandstone. Coincidence of thick Mary Lee and Black Creek porous-sandstone bodies in central Pickens County and eastern Fayette County is interpreted to have resulted from shoaling on inherited topographic highs, whereas offset sandstone bodies, like those in southern Fayette County, are interpreted to have resulted from tidal currents that were directed into topographic lows. The zone of thinning that cuts across the axial sandstone in Fayette County was interpreted by Pashin and others (1991) as a mud-filled channel that formed at maximum regression; channeling may be an important cause of sandstone-body discontinuity in Lamar and Pickens counties.

Identifying source areas for porous, quartzose sandstone is difficult because it accumulated in high-energy environments where labile grains are readily removed by reworking (Davis and Ehrlich, 1974; Mack and others, 1983). Indeed, much of the sand may simply be well-washed, polycyclic sediment derived from proximal sources in the Appalachian orogen. However, Mary Lee tight sandstone contains abundant chert grains, whereas Mary Lee porous sandstone generally lacks chert (Raymond,

1990). Therefore, porous sandstone may have been derived from multiple sources, including the nearby Ouachita orogen in the west and distant parts of the Appalachian orogen in the northeast.

Coal

Coal beds in the Black Creek-Cobb interval are most numerous in southeast Tuscaloosa County where more than 40 beds are present (Fig. 16). Fewer than 10 coal beds are present in much of the northern and western parts of the study area, and fewer than 5 beds are present in parts of northwest Pickens and southwest Lamar counties. Coal isopleth maps of the Black Creek, Mary Lee, and Pratt cycles (Figs. 17-19) reflect the regional trend (Fig. 16) and show little change through time. The Black Creek and Mary Lee cycles lack coal in parts of Lamar and Pickens counties, whereas the Pratt cycle characteristically contains 2 or more coal beds throughout the study area.

Pottsville cycles represent basinwide marine-terrestrial transitions, so coal in the Black Warrior basin probably formed in a range of depositional settings (Fig. 13). For example, localized marginal-marine peat domes, which may be analogs for some localized coal bodies in the Pottsville, are forming on the Rajang delta of Borneo (Anderson, 1964) and the Klang-Langat delta of Malaysia (Coleman and others, 1970). Widespread peat bodies, which may be analogs for the thickest and most continuous

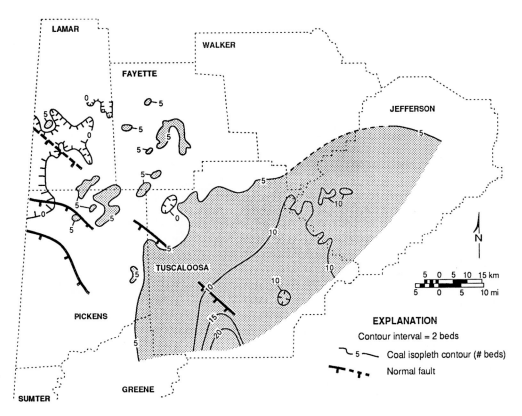

FIG. 17.—Coal isopleth map of the combined lower and upper Black Creek cycles, Black Warrior basin, Alabama.

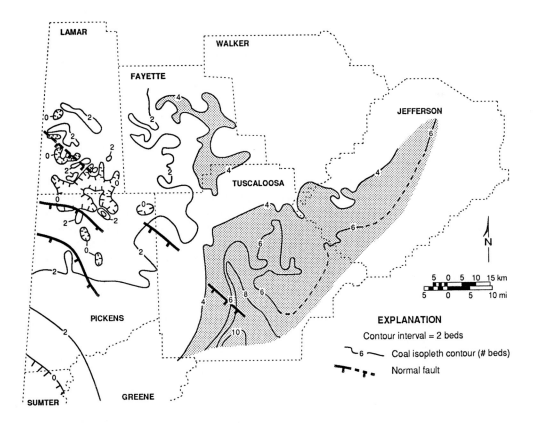

FIG. 18.—Coal isopleth map of the Mary Lee cycle, Black Warrior basin, Alabama.

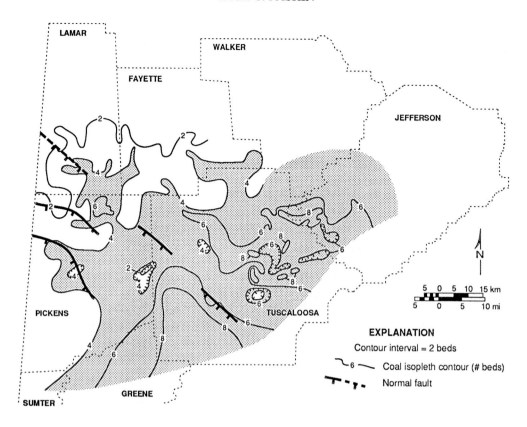

Fig. 19.—Coal isopleth map of the Pratt cycle, Black Warrior basin, Alabama.

Pottsville coal beds, are forming on alluvial plains in Indonesia (Anderson, 1983). Low-lying, planar peat accumulations are also forming as much as 70 miles inland between exposed Pleistocene beach ridges in Georgia (Cohen, 1974) and may thus be partial analogs for coal bodies associated with porous sandstone.

Indeed, styles of coal distribution in the Black Warrior basin reflect those varied settings. For example, Pashin (1991a, b) identified alluvial valley-fill and fault-bound coal bodies in the Mary Lee cycle that formed in response to synsedimentary movement of structures ancestral to the Sequatchie anticline and Coalburg syncline. In contrast, Gastaldo and others (1990) and Demko (1990) demonstrated that depositional topography inherited from porous shelf and beach sand bodies had a marked effect on the thickness and quality of marginal-marine coal bodies in the Mary Lee cycle of Walker County.

Why are coal beds most numerous in southeastern Tuscaloosa and western Jefferson Counties? Restriction of the thickest fluvial-deltaic sandstone to the southeast part of the study area reflects proximity to a sediment source and dominance of fluvial sedimentation (Fig. 13). Therefore, high sediment input and a northwestward to westward paleoslope helped maintain fluvial-deltaic platforms where peat accumulated. High subsidence rate in this area, moreover, favored accumulation of thick peat and development of numerous bed splits. The remaining parts of the study area were dominated by mud and porous-sand deposition

and were thus far from a sediment source and subject to marine influence. Consequently, peat deposition only occurred in these distal areas late in each cycle when most or all of the study area was emergent.

DISCUSSION: ORIGIN OF THE CYCLES

Having established average cycle duration and a tectonic-paleogeographic model for the study area, the causes of Pottsville cyclicity can be evaluated. An average duration of 0.2-0.5 my (Fig. 1) indicates that the cycles represent high-frequency fluctuations of relative sea level. This time span is of the same order of magnitude as the ~0.1- and 0.4-my orbital eccentricity cycles (Imbrie and Imbrie, 1980) which may have driven glacial eustasy in the midcontinent during Pennsylvanian time (Heckel, 1986). However, proximity to an active mobile belt increases the influence of tectonic subsidence on sedimentation (Klein and Willard, 1989).

By backstripping, Thomas and others (1991) calculated an average subsidence rate of approximately 15.3 cm/1,000 yr (0.5 ft/1,000 yr) for the deepest part of the Black Warrior basin in Mississippi during Pottsville deposition. In the absence of sediment influx or eustatic sea-level change, this rate of subsidence can increase water depth by more than 75 m (250 ft) in the time represented by some Pottsville cycles. Tectonic subsidence adds accommodation space and thus causes asymmetry in relative sea-level curves by amplifying the magnitude and duration of transgression and by suppressing regression (Posamentier and others, 1988); rapid subsidence may have caused pronounced asymmetry in the Black Warrior basin, especially during stratigraphic condensation related to marine transgression.

Cycle isopach maps (Figs. 7-9) demonstrate that major changes in basin geometry occurred as deformational loading of the Alabama promontory proceeded. Hence, significant spatial and temporal variations of subsidence rate occurred in the same time frame as deposition of each coarsening- and coaling-upward cycle. Loading of the continental promontory is related to diachronous docking and deformation of the diverse tectonic elements that make up the Appalachian-Ouachita orogen (Thomas, 1989; Horton and others, 1989). Thus, flexure of the lithosphere below the Black Warrior basin represents multiple events that occurred as elements of different tectonic terranes collided with and were thrust onto the Alabama promontory.

After a load element is in place, subsidence rate decreases exponentially as the lithosphere accommodates the load and relaxes stress (Watts and others, 1982). Hence, the 15.3 cm/1,000 yr subsidence rate represents an average of the extremely rapid subsidence during initial loading and the subsequent slowing of subsidence as the stress relaxed. Was the initial subsidence that occurred as each new load element was applied on part of the continental promontory rapid enough to outstrip sedimentation and cause regional transgression? The answer is debatable, but it seems that, at the very least, loading events affected the timing of some transgressions.

Although flexural subsidence is a plausible mechanism for marine transgression, identifying tectonic mechanisms that could have caused regional regression is difficult. Flexural relaxation may cause widespread marine regression in the late stages of orogeny (Beaumont and others, 1987, 1988; Ettensohn, 1991) but was probably not effective in the short time frame of Pottsville cycles or in the active loading regime that dominated the Alabama promontory during Pottsville sedimentation. Uplift associated with blind-thrust propagation apparently caused incision of valleys in the easternmost reaches of the Black Warrior basin (Pashin, 1991a, b), but such tectonism was localized and thus insignificant as an agent of regional baselevel change. Variation of horizontal intraplate stress and associated changes in plate density may cause 3rd-order (1-15 my) cyclicity (Cloetingh, 1986; Cathles and Hallam, 1991), but the capability of that stress to cause high-frequency events is questionable. Considering rapid load-related subsidence of the Black Warrior basin during Pottsville deposition, tectonism is interpreted to have had limited significance compared to eustasy as a forcing mechanism for regional marine regression and, thus, cyclicity in general.

Extreme variation of thickness among cycles (Figs. 1, 6) further supports an eustatic origin. For example, the Curry cycle, which is the thinnest regionally extensive cycle in the Pottsville Formation, almost certainly represents a shorter interval of time than the other cycles. If the Curry cycle represents

the same amount of time as the subjacent Gillespy cycle, the basin subsidence rate would have to decrease by a factor of 4 in less than 0.5 my. Indeed, it is doubtful that the lithosphere can relax stress so rapidly within the limits of crustal rigidity (Watts and others, 1982).

Sandstone and coal distribution provide insight into the effect of tectonism and eustasy on the regional paleogeographic framework. Consistency of tight-sandstone and coal distribution (Figs. 10-12, 17-19) indicates that a single sediment-dispersal system persisted in the Black Warrior basin of Alabama during Black Creek-Cobb deposition. Apparently, the most variable aspect of sediment supply in Alabama was the availability of porous sand, which may have been derived from multiple source areas (Figs. 6, 13-15). Hence, Pottsville tectonism and cyclicity operated faster than sediment could be dispersed from rapidly encroaching thrust loads. Rather, sediment was apparently supplied from relatively mature tectonic source areas throughout Black Creek-Cobb deposition. Indeed, the most striking changes in paleogeography occurred in response to the progression of depositional systems within each cycle (Fig. 6). For this reason, coastal-plain evolution is interpreted to have been much more sensitive to eustatic sea-level variation than to the tectonic changes that affected basin geometry.

ACKNOWLEDGMENTS

This report was reviewed critically by D. R. Chesnut, J. M. Dennison, F. R. Ettensohn, and W. A. Thomas; early versions were reviewed by W. B. Ayers, Jr., and W. R. Kaiser. Their insight improved the manuscript and is greatly appreciated. E. G. Rogers provided computer-drafting and technical support. Part of this research was funded by the Gas Research Institute through the Texas Bureau of Economic Geology under contract GRI 5087-214-1544.

REFERENCES

ALLEN, J. R. L., 1980, Sand waves, a model of origin and internal structure: Marine Geology, v. 26, p. 281-328.

ANDERSON, J. A. R., 1964, The structure and development of the peat swamps of Sarawak and Brunei: Journal of Tropical Geography, v. 18, p. 7-16.

ANDERSON, J. A. R., 1983, The tropical peat swamps of western Malaysia, in Gore, A. J. P., ed., Ecosystems of the World, 4B, Mires: Swamp, Bog, Fen, and Moor: Amsterdam, Elsevier, p. 188-199.

BEAUMONT, C., QUINLAN, G. M., AND HAMILTON, J., 1987, The Alleghanian orogeny and its relationship to the evolution of the eastern interior, North America: Calgary, Canadian Society of Petroleum Geologists Memoir 12, p. 425-445.

BEAUMONT, C., QUINLAN, G. M., AND HAMILTON, J., 1988, Orogeny and stratigraphy: numerical models of the Paleozoic in the eastern interior of North America: Tectonics, v. 7, p. 389-416.

BENSON, D. J., 1982, Depositional environments of coal-bearing strata in the Warrior Basin, in Rheams, L. J., and Benson, D. J., eds., Depositional Setting of the Pottsville Formation in the Black

Warrior basin: Tuscaloosa, Alabama Geological Society Guidebook, 19th Annual Field Trip, p. 15-26.

BUTTS, C., 1910, Description of the Birmingham quadrangle, Alabama: Washington D. C., United States Geological Survey Atlas, Folio 175, 24 p.

BUTTS, C., 1926, The Paleozoic rocks, in Adams, G. I., Butts, C., Stephenson, L. W., and Cooke, C. W., Geology of Alabama: Tuscaloosa, Alabama Geological Survey Special Report 14, p. 41-230.

CATHLES, L. M., AND HALLAM, A., 1991, Stress-induced changes in plate density, Vail sequences, epeirogeny, and short-lived global sea level fluctuations: Tectonics, v. 10, p. 659-671.

CLEAVES, A. W., 1981, Resources of Lower Pennsylvanian (Pottsville) depositional systems of the western Warrior coal field, Alabama and Mississippi: Oxford, Mississippi Mineral Resources Institute Technical Report 81-1, 125 p.

CLOETINGH, S., 1986, Intraplate stresses: A new tectonic mechanism for fluctuations of relative sea level: Geology, v. 14, p. 617-621.

COHEN, A. D., 1974, Petrography and paleoecology of some Holocene peats from the Okefenokee swamp-marsh complex of southern Georgia: Journal of Sedimentary Petrology, v. 44, p. 716-726.

COLEMAN, J. M., GAGLIANO, S. M., AND SMITH, W. G., 1970, Sedimentation in a Malaysian high tide tropical delta, in Morgan, J. P., and Shaver, R. H., eds., Deltaic Sedimentation: Modern and Ancient: Tulsa, Society of Economic Paleontologists and Mineralogists Special Publication 15, p. 185-197.

CULBERTSON, W. C., 1964, Geology and coal resources of the coal-bearing rocks of Alabama: Washington D. C., United States Geological Survey Bulletin 1182-B, 79 p.

DAVIS, M. W., AND EHRLICH, R., 1974, Late Paleozoic crustal composition and dynamics in the southeastern United States: Boulder, Geological Society of America Special Paper 148, p. 171-185.

DEMKO, T. M., 1990, Paleogeography and depositional environments of the lower Mary Lee coal zone, Pottsville Formation, Warrior basin, northwest Alabama: Unpublished Master's Thesis, Auburn University, Auburn, 195 p.

EBLE, C. F., AND GILLESPIE, W. H., 1989, Palynology of selected coal beds from the central and southern Appalachian basin: correlation and stratigraphic implications, in Characteristics of the Mid-Carboniferous Boundary and Associated Coal-bearing Rocks in the Appalachian Basin: Washington, 28th International Geological Congress Guidebook T352, p. 61-66.

ENGMAN, M. A., 1985, Depositional systems in the lower part of the Pottsville Formation, Black Warrior basin, Alabama: Unpublished M. S. Thesis, University of Alabama, Tuscaloosa, 250 p.

ETTENSOHN, F. R., 1991, Flexural interpretation of relationships between Ordovician tectonism and stratigraphic sequences, central and southern Appalachians, U.S.A., in Barnes, C. R., and Williams, S. H., eds., Advances in Ordovician Geology: Ottawa, Geological Survey of Canada Paper 90-9, p. 213-224.

FERM, J. C., EHRLICH, R., AND NEATHERY, T. L., 1967, A field guide to Carboniferous detrital rocks in northern Alabama: Tuscaloosa, 1967 Geological Society of America Coal Division Field Trip Guidebook, 101 p.

FERM, J. C., AND WEISENFLUH, G. A., 1989, Evolution of some depositional models in Late Carboniferous rocks of the Appalachian coal fields: International Journal of Coal Geology, v. 12, 259-292.

GASTALDO, R. A., DEMKO, T. M., AND LIU, Y., eds., 1990, Carboniferous coastal environments and paleocommunities of the Mary Lee coal zone, Marion and Walker Counties, Alabama: Tuscaloosa,

Geological Society of America Southeastern Section Guidebook, Field Trip 6, Alabama Geological Survey Guidebook Series 3-6, p. 41-54.

GOULD, H. R., 1970, The Mississippi Delta complex, in Morgan, J. P. and Shaver, R. H., eds., Deltaic Sedimentation Modern and Ancient: Tulsa, Society of Economic Paleontologists and Mineralogists Special Publication 15, p. 3-30.

HECKEL, P. H., 1986, Sea-level curve for Pennsylvanian eustatic transgressive-regressive depositional cycles along midcontinent outcrop belt, North America: Geology, v. 14, p. 330-334.

HENDERSON, K. S., AND GAZZIER, C. A., 1989, Preliminary evaluation of coal and coalbed gas resource potential of western Clay County, Mississippi: Jackson, Mississippi Bureau of Geology Report of Investigations 1, 31 p.

HINES, R. A., JR., 1988, Carboniferous evolution of the Black Warrior foreland basin, Alabama and Mississippi: Unpublished Ph. D. Dissertation, University of Alabama, Tuscaloosa, 231 p.

HOBDAY, D. K., 1974, Beach and barrier island facies in the Upper Carboniferous of northern Alabama: Boulder, Geological Society of America Special Paper 148, p. 209-224.

HORNE, J. C., FERM, J. C., HOBDAY, D. K., AND SAXENA, R. S., 1976, A field guide to Carboniferous littoral deposits in the Warrior basin: New Orleans, New Orleans Geological Society Guidebook, 80 p.

HORSEY, C. A., 1981, Depositional environments of the Pennsylvanian Pottsville Formation in the Black Warrior basin of Alabama: Journal of Sedimentary Petrology, v. 51, p. 799-806.

HORTON, J. W., JR., DRAKE, A. A., JR., AND RANKIN, D. W., 1989, Tectonostratigraphic terranes and their Paleozoic boundaries in the central and southern Appalachians: Boulder, Geological Society of America Special Paper 230, p. 213-245.

IMBRIE, J., AND IMBRIE, Z. L., 1980, Modeling the climatic response to orbital variations: Science, v. 207, p. 943-952.

KLEIN, G. DEV, 1990, Pennsylvanian time scales and cycle periods: Geology, v. 18, p. 455-457.

KLEIN, G. DEV., AND WILLARD, D. A., 1989, Origin of the Pennsylvanian coal-bearing cyclothems of North America: Geology, v. 17, p. 152-155.

KLEIN, G. DEV., AND KUPPERMAN, J. B., 1992, Pennsylvanian cyclothems: methods of distinguishing tectonically induced changes in sea level from climatically induced changes: Geological Society of America Bulletin, v. 104, p. 166-175.

MACK, G. H., THOMAS, W. A., AND HORSEY, C. A., 1983, Composition of Carboniferous sandstones and tectonic framework of southern Appalachian-Ouachita orogen: Journal of Sedimentary Petrology, v. 54, p. 1444-1456.

McCALLEY, H., 1900, Report on the Warrior coal basin: Tuscaloosa, Alabama Geological Survey Special Report 10, 327 p.

MELLEN, F. F., 1947, Black Warrior basin, Alabama and Mississippi: American Association of Petroleum Geologists Bulletin, v. 31, p. 1801-1816.

METZGER, W. J., 1965, Pennsylvanian stratigraphy of the Warrior basin, Alabama: Tuscaloosa, Alabama Geological Survey Circular 30, 80 p.

PASHIN, J. C., 1991a, Regional analysis of the Black Creek-Cobb coalbed-methane target interval, Black Warrior basin, Alabama: Tuscaloosa, Alabama Geological Survey Bulletin 145, 127 p.

PASHIN, J. C., 1991b, Subsurface models of coal occurrence, Oak Grove field, Black Warrior basin, Alabama: Tuscaloosa, The University of Alabama, 1991 Coalbed Methane Symposium Proceedings, p. 275-291.

PASHIN, J. C., OSBORNE, W. E., AND RINDSBERG, A. K., 1991, Outcrop characterization of sandstone heterogeneity in Carboniferous reservoirs, Black Warrior basin, Alabama: Bartlesville, U.S. Department of Energy Fossil Energy Report DOE/BC/14448-6, 126 p.

PASHIN, J. C., AND SARNECKI, J. C., 1990, Coal-bearing strata near Oak Grove and Brookwood coalbed-methane fields, Black Warrior basin, Alabama: Tuscaloosa, Geological Society of America Southeastern Section Guidebook, Field Trip 6, Alabama Geological Survey Guidebook Series 3-6, 37 p.

PASHIN, J. C., WARD, W. E., II, WINSTON, R. B., CHANDLER, R. V., BOLIN, D. E., HAMILTON, R. P., AND MINK, R. M., 1990, Geologic evaluation of critical production parameters for coalbed methane resources, part II, Black Warrior basin: Chicago, Gas Research Institute, Annual Report GRI-90/0014.2, Contract 5087-214-1544, 177 p.

POSAMENTIER, H. W., JERVEY, M. T., AND VAIL, P. R., 1988, Eustatic controls on clastic deposition I—conceptual framework, in Wilgus, C. K., Hastings, B. S., Kendall, C. G. St. C., Posamentier, H. W., Ross, C. A., and Van Wagoner, J. C., eds., Sea-level Changes: An Integrated Approach: Tulsa, Society of Economic Paleontologists and Mineralogists Special Publication 42, p. 109-124.

RAYMOND, D. E., 1990, Petrography of sandstones of the Pottsville Formation in the Jasper Quadrangle, Black Warrior basin, Alabama: Tuscaloosa, Alabama Geological Survey Circular 144, 48 p.

RODGERS, J., 1950, Mechanics of Appalachian folding as illustrated by the Sequatchie anticline, Tennessee and Alabama: American Association of Petroleum Geologists Bulletin, v. 34, p. 672-681.

ROSS, C. A., AND ROSS, J. R. P., 1988, Late Paleozoic transgressive-regressive deposition, in Wilgus, C. K., Hastings, B. S., Kendall, C. G. St. C., Posamentier, H. W., Ross, C. A., and Van Wagoner, J. C., eds., Sea-level Changes: an Integrated Approach: Tulsa, Society of Economic Paleontologists and Mineralogists Special Publication 42, p. 227-247.

SESTAK, H. M., 1984, Stratigraphy and depositional environments of the Pennsylvanian Pottsville Formation in the Black Warrior basin: Alabama and Mississippi: Unpublished M. S. Thesis, University of Alabama, Tuscaloosa, 184 p.

SHADROUI, J. M., 1986, Depositional environments of the Pennsylvanian Bremen Sandstone Member and associated strata, Pottsville Formation, north-central Alabama: Unpublished M.S. Thesis, University of Alabama, Tuscaloosa, 172 p.

THOMAS, W. A., 1973, Southwestern Appalachian structural system beneath the Gulf Coastal Plain: American Journal of Science, v. 273-A, p. 372-390.

THOMAS, W. A., 1974, Converging clastic wedges in the Mississippian of Alabama: Boulder, Geological Society of America Special Paper 148, p. 187-207.

THOMAS, W. A., 1977, Evolution of Appalachian-Ouachita salients and recesses from reentrants and promontories in the continental margin: American Journal of Science, v. 277, p. 1233-1278.

THOMAS, W. A., 1985, Northern Alabama sections, in Woodward, N. B., ed., Valley and Ridge thrust belt: balanced structural sections, Pennsylvania to Alabama: Knoxville, University of Tennessee Department of Geological Sciences Studies in Geology 12, p. 54-60.

THOMAS, W. A., 1988, The Black Warrior basin, in Sloss, L. L., ed., Sedimentary Cover- North American Craton: Boulder, Geological Society of America, The Geology of North America, v. D-2, p. 471-492.

THOMAS, W. A., 1989, The Appalachian-Ouachita orogen beneath the Gulf Coastal Plain between the outcrops of the Appalachian and Ouachita Mountains, in Hatcher, R. D., Jr., Thomas, W. A., and Viele, G. W., eds., The Appalachian-Ouachita Orogen in the United States: Boulder, Geological Society of America, The Geology of North America, v. F-2, p. 537-553.

THOMAS, W. A., 1991, The Appalachian-Ouachita rifted margin of southeastern North America: Geological Society of America Bulletin, v. 103, p. 415-431.

THOMAS, W. A., FERRILL, B. A., ALLEN, J. L., OSBORNE, W. E., AND LEVERETT, D. E., 1991, Synorogenic clastic-wedge stratigraphy and subsidence history of the Cahaba synclinorium and the Black Warrior foreland basin, in Thomas, W. A., and Osborne, W. E., eds., Mississippian-Pennsylvanian tectonic history of the Cahaba synclinorium: Tuscaloosa, Alabama Geological Society Guidebook, 28th Annual Field Trip, p. 37-39.

THOMAS, W. A., AND WOMACK, S. H., 1983, Coal stratigraphy of the deeper part of the Black Warrior basin in Alabama: Gulf Coast Association of Geological Societies Transactions, v. 33, p. 439-446.

WANLESS, H. R., AND SHEPARD, F. P., 1936, Sea level and climatic changes related to late Paleozoic cycles: Geological Society of America Bulletin, v. 47, p. 1177-1206.

WATTS, A. B., KARNER, G. D., AND STECKLER, M. S., 1982, Lithospheric flexure and the evolution of sedimentary basins: Philosophical Transactions of the Royal Society of London, v. A-305, p. 249-281.

WEISENFLUH, G. A., AND FERM, J. C., 1984, Geologic controls on deposition of the Pratt seam, Black Warrior basin, Alabama, U.S.A., in Rahmani, R. A., and Flores, R. M., eds., Sedimentology of Coal and Coal-bearing Sequences: Oxford, International Association of Sedimentologists Special Publication 7, p. 317-330.

WELLER, S., 1930, Cyclic sedimentation of the Pennsylvanian Period and its significance: Journal of Geology, v. 38, p. 97-135.

WELLER, S., 1956, Argument for diastrophic control of late Paleozoic cyclothems: American Association of Petroleum Geologists Bulletin, v. 41, p. 195-207.

A SUB-PENNSYLVANIAN PALEOVALLEY SYSTEM IN THE CENTRAL APPALACHIAN BASIN AND ITS IMPLICATIONS FOR TECTONIC AND EUSTATIC CONTROLS ON THE ORIGIN OF THE REGIONAL MISSISSIPPIAN-PENNSYLVANIAN UNCONFORMITY

JACK D. BEUTHIN

Department of Geology and Planetary Sciences, University of Pittsburgh-Johnstown, Johnstown, Pennsylvania 15904

ABSTRACT: Paleodrainage mapping of the Mississippian-Pennsylvanian unconformity in northwestern West Virginia verifies the existence of an incised, sub-Pennsylvanian paleovalley system there that extends for over 130 km. The paleovalleys are filled mostly with quartzose sandstones of the New River Formation. This paleovalley system was carved by a southwest-draining network of rivers that was rejuvenated during the mid-Carboniferous. Bedload-dominated streams that occupied the paleovalleys deposited most of the valley-fill sediment.

Regional paleodrainage data indicate that the sub-Pennsylvanian paleovalleys in northwestern West Virginia form the middle reach of a major paleoriver system which includes the Middlesboro paleovalley in eastern Kentucky, the Sharon paleovalley in eastern Ohio, and the Perry paleovalley in southeastern Ohio. This regional paleodrainage network (herein named the Middlesboro-Sharon-Perry paleovalley system) transported sediment from the craton north of the central Appalachian basin to the mid-Carboniferous depocenter in southwestern Virginia. Although it has been previously suggested that maximum erosional development of the Mississippian-Pennsylvanian unconformity occurred during the Early Pennsylvanian, paleoslope considerations rule against an Early Pennsylvanian age for carving of the Middlesboro-Sharon-Perry paleovalley system. Existing biostratigraphic data from the mid-Carboniferous depocenter in southern West Virginia support the existence of an Upper Mississippian (Chokierian-Alportian stages of the Namurian Series) hiatus there, suggesting that incision of the Middlesboro-Sharon-Perry paleovalley system was dominantly a Late Mississippian (Chokierian-Alportian) event.

Uplift of the craton north of the central Appalachian basin combined with subsidence within the basin that increased in rate toward the mid-Carboniferous depocenter in southern West Virginia created the generally south-dipping paleoslope which the Middlesboro-Sharon-Perry paleoriver system drained. The regional paleodrainage picture rules against tectonic uplift of the Cincinnati arch as a key factor in driving the incision of the Middlesboro-Sharon-Perry paleovalley system. Carving of this paleovalley system apparently was driven by a previously documented, Late Mississippian (Chokierian-Alportian) eustatic sea-level drop. Regional tectonic uplift during the Early Pennsylvanian may have influenced erosional development of the Mississippian-Pennsylvanian after the Late Mississippian incision of the Middlesboro-Sharon-Perry paleodrainage system.

INTRODUCTION

Mid-Carboniferous sections throughout the world are punctuated by an erosional hiatus that represents a global lowstand of sea level during which cratons were exposed and denuded (Vail and others, 1977; Saunders and Ramsbottom, 1986). On the North American craton, this event is represented by an unconformity between the Mississippian and Pennsylvanian Systems. This unconformity also delineates the boundary between the Kaskaskia and Absaroka sequences of Sloss (1963).

Historically, the contact between the Mississippian and Pennsylvanian systems in the central Appalachian basin (base of the Pottsville Group) has been interpreted to be disconformable (White, 1891; White, 1904; Wanless, 1939; Branson, 1964). Recent regional studies have shown evidence of sequential truncation of Mississippian formations beneath Pennsylvanian rocks, thus confirming a widespread sub-Pennsylvanian unconformity in the central Appalachians (Arkle, 1974; Wanless, 1975; de Witt and McGrew, 1979; Englund, 1981; Dennison, 1983; Englund and Henry, 1984; Chesnut, 1988; Englund and Thomas, 1990). Englund and others (1977) and Englund (1979) also have reported a regional unconformity at the base of the Pennsylvanian System in their stratigraphic and sedimentologic studies aimed at establishing a global reference stratotype for the Pennsylvanian System in West Virginia and Virginia.

Although the regional Mississippian-Pennsylvanian unconformity is recognized as major sequence boundary in the Paleozoic-fill of the central Appalachian basin (Colton, 1970; Dennison, 1983, 1989), there has been much debate over the relative importance of eustasy and regional tectonism as controls on the development of this unconformity. Saunders and Ramsbottom (1986) and Dennison (1989) have attributed the sub-Pennsylvanian unconformity to a eustatic sea-level fall, whereas Quinlan and Beaumont (1984), Tankard (1986), Chesnut (1988), Willard and Klein (1990) and F. Ettensohn and D. Chesnut, Jr. (pers. commun., 1991) have inferred regional tectonic uplift as the primary control on development of the sub-Pennsylvanian unconformity

Paleodrainage on an erosional unconformity can be a sensitive indicator of eustatic and tectonic controls, however this aspect of the sub-Pennsylvanian surface in the central Appalachian basin generally has been overlooked in the debate over the origin of the unconformity. The purpose of the present paper is (1) to report and discuss the results of my study of paleodrainage on the Mississippian-Pennsylvanian unconformity in northwestern West Virginia (Fig. 1), (2) to reconstruct regional sub-Pennsylvanian paleodrainage in the central Appalachian basin utilizing data from my study and from other studies, and (3) to evaluate eustasy and tectonic activity as controls on the Mississippian-Pennsylvanian unconformity based on analysis of my

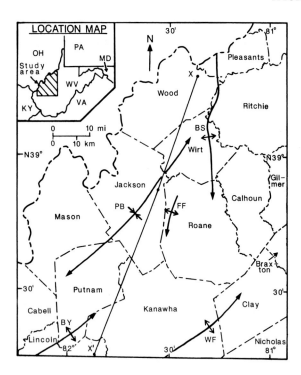

FIG. 1.—Map of study area with county names and major fold axes. Cross-section X-X' is shown in Figure 4. BS— Burning Springs anticline. BY— Byrnside anticline. FF— Flat Fork anticline. WF— Warfield anticline. PB— Parkersburg syncline. Fold axes based on Krebs (1911), Krebs (1913), Krebs (1914), Cross and Schemel (1956), Filer (1985), Sweeney (1986), and Beuthin (1989).

regional paleodrainage reconstruction.

PREVIOUS WORK IN NORTHWESTERN WEST VIRGINIA

Martens (1945) first documented the termination of progressively older Mississippian units beneath the Pennsylvanian System in northwestern West Virginia. He attributed this termination to erosional truncation along the Mississippian-Pennsylvanian unconformity.

Flowers (1956) mapped the thickness of the Mississippian Greenbrier Limestone in West Virginia. His map shows that, beyond the termination of the overlying Mississippian Mauch Chunk Group, the Greenbrier is drastically thinned along a north-trending zone in northwestern West Virginia (Fig. 2). Flowers interpreted this feature as a system of incised sub-Pennsylvanian paleovalleys. Noting that the main paleovalley split northward into two smaller "troughs" along which the Greenbrier was terminated, Flowers proposed that these two features either indicated successive lower courses of a north-flowing river, or expressed "the presence of a bend in a large river channel with a smaller river flowing into it from the southeast." As further evidence of an incised paleovalley, Flowers cited an increase in the thickness of Pottsville (basal Pennsylvanian) sandstones and conglomerates along the Greenbrier "troughs," relative to adjacent areas where the Greenbrier

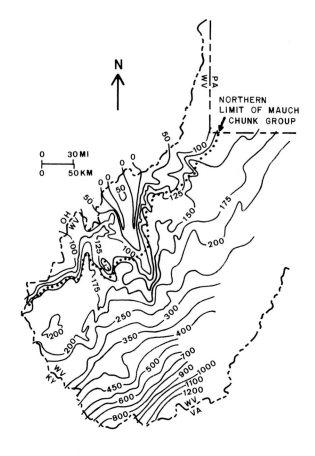

FIG. 2.—Isopachous map of the Greenbrier Limestone in West Virginia with northern limit of Mauch Chunk Group (dotted line). Thickness in feet (After Flowers, 1956, p. 2).

is not anomalously thin.

Uttley (1974) reconstructed sub-Pennsylvanian paleodrainage in eastern Ohio and northwestern West Virginia. On the basis of regional considerations, he argued that the sub-Pennsylvanian paleovalleys in northwestern West Virginia delineated by Flowers (1956) were carved by south- to southeast-flowing rivers. Uttley also traced the easternmost of the two paleovalleys in West Virginia northward into Ohio for more than 100 km, and inferred that its upper course extended much further northward. Uttley named this paleovalley the "Sharon River" for the quartzose sandstones of the Sharon Formation that fill it.

According to Wanless (1975), the Greenbrier Limestone and equivalent Mississippian carbonate units were more resistant than other strata exposed on the sub-Pennsylvanian surface, and supported a low cuesta with a northwest-facing escarpment. Wanless mapped this escarpment as extending across southeastern Ohio, just north of the present study area. He further proposed that the Greenbrier cuesta formed the divide between a northern Appalachian depositional basin and a southern Appalachian depositional basin until early Middle Pennsylvanian time.

Rice (1984) and Rice and Schwietering (1988) proposed a model of Early Pennsylvanian fluvial deposition in the central

SYSTEM			LITHOSTRATIGRAPHIC UNIT		THICKNESS
C A R B O N I F E R O U S	PENNSYLVANIAN		Pennsylvanian coal-bearing sequence, undivided		~600 m
			New River Formation		0-242 m
	MISSISSIPPIAN		upper Mauch Chunk Group		0-136 m
		lower Mauch Chunk Group	Reynolds Limestone		0-30 m
			Lillydale Shale / Webster Springs Sandstone		
			Greenbrier Limestone		0-67 m
		Pocono Group	undivided		76-167 m
			Sunbury Shale		
			Berea Sandstone		

Fig. 3.—Scheme of Carboniferous stratigraphic nomenclature for northwestern West Virginia.

Appalachians. Their model indicates a southwest-draining fluvial system and supports the conclusion of Uttley (1974) that the sub-Pennsylvanian paleovalleys in northwestern West Virginia were incised by south-draining streams. According to Rice (1984) and Rice and Schwietering (1988), the paleovalleys in northwestern West Virginia represent the site of mid-Carboniferous drainage capture through the Greenbrier cuesta of Wanless (1975). Furthermore, they hypothesized that this capture diverted the upper reaches of a cratonic drainage system into the mid-Carboniferous depocenter in southern West Virginia and southwestern Virginia.

Filer (1985) used gamma-ray logs to construct stratigraphic cross sections across the sub-Pennsylvanian paleovalleys in northwestern West Virginia. His cross sections corroborate Flowers' (1956) results, showing the extreme thinning of the Mississippian section to be a result of erosional truncation. Filer (1985) also showed that the paleovalleys are filled with quartzose basal Pennsylvanian sandstones.

Krissek and others (1986) studied the Mississippian-Pennsylvanian unconformity and the basal Pennsylvanian Sharon Sandstone in an area of southeastern Ohio that borders my study area. They mapped a northwest-trending paleovalley and inferred a northwest drainage there. These workers also concurred

with Flowers' (1956) interpretation of a north-flowing sub-Pennsylvanian drainage in West Virginia and projected the upper reaches of their paleovalley into southern West Virginia.

Beuthin (1988, 1989, 1991) and Beuthin and Neal (1990) include preliminary discussions of some results of this present report.

GEOLOGIC SETTING
Structure

This study was conducted in a 14,000 km² area of northwestern West Virginia (Fig. 1). The area lies entirely within the Appalachian Plateau physiographic province where Carboniferous beds are, for the most part, only mildly deformed. Most of the folds that involve Carboniferous strata are broad, low-dipping (<1°), northeast-trending structures that are basement-rooted. Structural deformation is generally thought to have occurred during the Late Pennsylvanian-Permian climax of the Allegheny orogeny. It is possible, however, that some growth of Allegheny fold structures occurred during the mid-Carboniferous.

Stratigraphy

The Carboniferous stratigraphy of northwestern West Virginia has been set forth in several publications of the West Virginia Geological and Economic Survey (Cross and Schemel, 1956; Haught, 1960; Overby, 1961; Haught and Overby, 1964; Cardwell, 1981; Filer, 1985; Sweeney, 1986). The Carboniferous stratigraphic scheme used in the present study (see Fig. 3) conforms to standard usage, with the exception of some operational modifications. Strata of both Mississippian and Pennsylvanian ages are present throughout the area, but the Pennsylvanian New River Formation and the Mississippian System are confined to the subsurface.

The Mississippian System comprises (in ascending order) the Pocono Group, the Greenbrier Group, and the Mauch Chunk Group.

The Pocono Group is a sequence of interbedded, marine fossiliferous sandstones, siltstones, and mostly green and gray shales. The Sunbury Shale of the Pocono Group is a thin (3-9 m), black shale at or near the base of the Mississippian System (the underlying Berea Sandstone is locally absent). The Sunbury has long been recognized as one of the most distinctive key beds in the region, and is used as such in the present study.

The Greenbrier Group is a marine limestone typically having dolomitic basal beds, oolitic middle beds, and shaley upper beds.

The Mauch Chunk Group consists of red and green shale interbedded with grey shales, carbonate, siltstone, and lenticular, argillaceous sandstones. The Mauch Chunk Group in northern West Virginia is not formally subdivided into formations; however, an informal division is utilized in this study recognizing an upper Mauch Chunk Group and a lower Mauch Chunk Group. The boundary between these two informal units is the top of the Reynolds Limestone, a relatively thin (0-12 m),

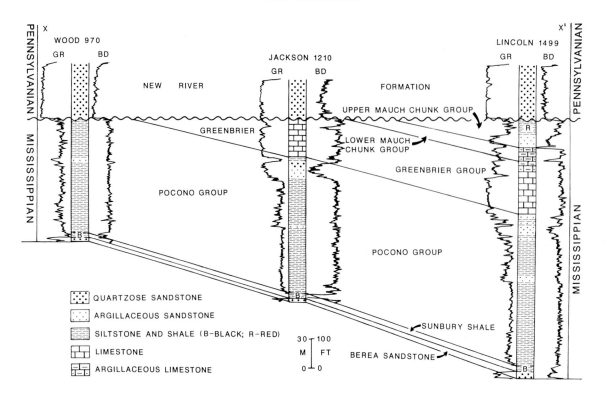

Fig. 4.—Stratigraphic cross section based on wireline logs illustrating erosional truncation of progressively older Mississippian units beneath the basal Pennsylvanian New River Formation in northwestern West Virginia. Line of section shown on Figure 1 (GR— gamma-ray log. BD— bulk density log). Lithologic interpretations based on comparison with lithologic logs from nearby wells. Angularity of sub-Pennsylvanian unconformity is exaggerated due to extreme vertical exaggeration of section.

dark gray, argillaceous marine limestone. The portion of the lower Mauch Chunk Group beneath the Reynolds Limestone consists of the Lillydale Shale (a thin, dark gray shale from 0-6 m thick) and the Webster Springs Sandstone (0-9 m of moderately argillaceous sandstone—an eastern facies equivalent of the Lillydale Shale). Whereas the lower Mauch Chunk is virtually devoid of red beds, the upper Mauch Chunk is characterized by abundant red shales.

The New River Formation is the oldest Pennsylvanian unit in the study area. Throughout most of the area, this unit consists of massive quartzose sandstone interbedded with some conglomerate and gray shale and siltstone. In the extreme southeastern part of the study area (Clay and Nicholas counties), the New River Formation undergoes a facies change; shale and siltstone dominate the unit and quartzose sandstone is the subordinate lithology. The sequence of Pennsylvanian strata that overlies the New River Formation consists of cyclically interbedded argillaceous sandstone, siltstone, shale, coal, and some limestone. In the subsurface of northwestern West Virginia, this sequence is difficult to subdivide but, in outcrop, it comprises the Kanawha Formation, the Allegheny Group, the Conemaugh Group, and the Monongahela Group. In southern West Virginia, the New River Formation is underlain by the Pennsylvanian Pocahontas Formation. The Pocahontas, New River and Kanawha formations form the Pottsville Group in West Virginia. The top

of the Pottsville Group (top of the Kanawha Formation) is difficult to delineate in the subsurface of the study area and the Pocahontas Formation is not present there; thus, I have not used the term "Pottsville" in the present paper.

Although Ferm (1974a, b) inferred that the New River Formation is partially contemporaneous with the underlying Mauch Chunk and Greenbrier Groups throughout West Virginia, neither gradation nor intertonguing between New River sandstones and underlying Mississippian strata has ever been reported in northwestern West Virginia, nor was it observed in the present study. Rather, the New River-Mississippian contact is sharp and the New River Formation overlies progressively older Mississippian units to the north because of the existence of the erosional unconformity at the base of the Pennsylvanian System throughout the study area (Fig. 4). In the northern part of the study area, sandstones and conglomerates of the New River Formation are locally absent. At those localities younger Pennsylvanian strata rest unconformably on the lower Mauch Chunk or older Mississippian units.

Mid-Carboniferous Setting

The Late Mississippian-Early Pennsylvanian tectonic setting of the central Appalachian basin is depicted in Figure 5. The Cincinnati arch and its northward extension, the Findlay arch,

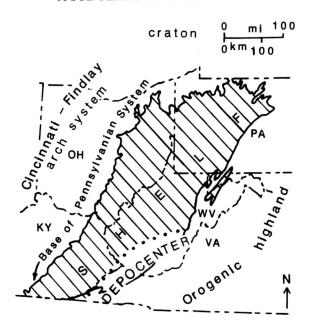

FIG. 5.—Mid-Carboniferous paleotectonic map of the central Appalachian basin. Hachures indicate known areal extent of sub-Pennsylvanian unconformity.

bounded the western margin of the basin. To the north, the Appalachian basin was bordered by the craton. The eastern margin of the basin was bordered by a tectonic highland created by deformation and uplift during the waning stages of the Late Devonian-Mississippian Acadian orogeny and the early stages of the Allegheny orogeny. A subsiding depocenter (the Pocahontas basin) existed in southeastern West Virginia and southwest Virginia. The broad, western flank of the central Appalachian basin that spans across western Pennsylvania, eastern Ohio, northern West Virginia, and eastern Kentucky was a shelf region undergoing relatively slow subsidence. According to Donaldson and Shumaker (1981) and Quinlan and Beaumont (1984), subsidence was driven largely by the emplacement of thrust loads along the eastern, orogenic margin of the basin. The present study area was situated on this shelf, just northwest of the depocenter in southeastern West Virginia. Although much of the Upper Mississippian section is absent from the shelf area, Mauch Chunk strata in southern and central West Virginia record a south to southwest Late Mississippian paleoslope (Donaldson and Shumaker, 1981; Wrightstone, 1985). Early Pennsylvanian transport direction on the Appalachian shelf and the craton was dominantly to the southwest (Siever and Potter, 1956).

Following the maximum transgression of the Late Mississippian sea in the central Appalachian basin during deposition of the Greenbrier and lower Mauch Chunk sediments, progradation of upper Mauch Chunk clastics into the basin caused a westward to southwestward regression of the Late Mississippian sea (de Witt and McGrew, 1979; Donaldson and Shumaker, 1981;

Presley, 1981). During latest Mississippian-earliest Pennsylvanian time, a major drop in base level induced emergence and subaerial erosion of the craton, resulting in the development of a regional sub-Pennsylvanian unconformity on the relatively stable, western flank of the central Appalachian basin (Fig. 5). According to Englund (1969, 1979, 1981) and Miller (1974), sedimentation in the subsiding depocenter of southern West Virginia and southwestern Virginia (Pocahontas basin) was not interrupted during the Mississippian-Pennsylvanian transition. However, Englund (1969, 1979, 1981) has reported an unconformity at the base of the Pineville Sandstone Member of the New River Formation (Lower Pennsylvanian) that progressively truncates basal New River strata and the Pennsylvanian Pocahontas Formation to the north and northwest. Chesnut (1983, 1988) also has supported the existence of an erosional unconformity between the Pocahontas and New River Formations and has inferred that the erosional development of the regional Mississippian-Pennsylvanian unconformity was an Early Pennsylvanian event. Alternatively, Rice (1985) has inferred that the Mississippian-Pennsylvanian contact (base of the Pocahontas Formation) is disconformable in the Pocahontas basin. Onlap and burial of the sub-Pennsylvanian unconformity proceeded from south to north, apparently beginning in Early Pennsylvanian time and ending in early Middle Pennsylvanian time (Arkle, 1974; Wanless, 1975; Chesnut, 1988).

DATA

Approximately 370 subsurface data points were used to construct Carboniferous stratigraphic cross sections and maps in northwestern West Virginia. Data were derived from lithologic logs and wireline logs of oil and gas prospects drilled throughout the area. All of the wireline logs, several unpublished sample descriptions and a few driller's logs were obtained from the well-data repository of the West Virginia Geological and Economic Survey in Morgantown, West Virginia. Other data sources include well-sample descriptions published in Tucker (1936) and Martens (1945), and Geologs (sample strip-logs prepared by the Geological Sample Log Company of Pittsburgh, Pennsylvania).

SUB-PENNSYLVANIAN PALEOVALLEYS IN NORTHWESTERN WEST VIRGINIA

An isopachous map of the Mississippian System with the basal Mississippian Berea Sandstone omitted is shown in Figure 6. Figure 7 is a map of the lithostratigraphic units that directly underlie the Pennsylvanian System; thus, it displays the paleogeology of the sub-Pennsylvanian unconformity.

Figure 6 indicates that the Mississippian System generally thins to the north. Thinning is partially a result of depositional pinch-out of beds within the Mississippian, and indicates a northward decrease in the rate of subsidence during deposition. Mississippian beds are also progressively truncated northward along the sub-Pennsylvanian unconformity (Fig. 7); thus, the

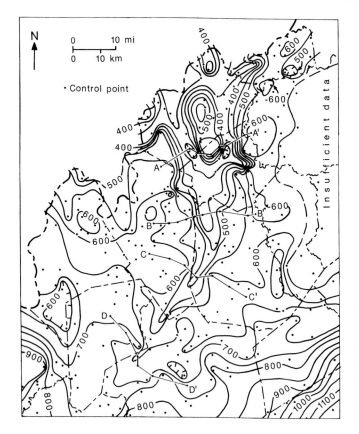

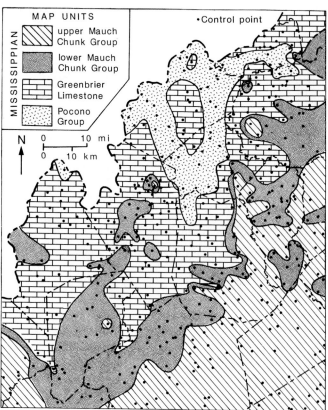

FIG. 6.—Isopachous map of the Mississippian interval from the base of the Sunbury Shale to the base of the Pennsylvanian System. Isopach interval = 15 m (50 ft) (cross-sections A-A', B-B', C-C', and D-D' are shown on Fig. 8).

FIG. 7.—Pennsylvanian subcrop map of northwestern West Virginia.

Mississippian System is also thinned due to erosional removal of beds prior to Pennsylvanian sedimentation in northwestern West Virginia. Figures 6 and 7 also show a pattern of local thinning of the Mississippian System by progressive truncation of beds beneath the Pennsylvanian System. This pattern portrays the erosional paleotopography of the sub-Pennsylvanian unconformity. The 300- through 550-foot isopachs on Figure 6 delineate a prominent system of coalesced paleovalleys. Two south-trending paleovalleys are incised into the Pocono Group in the north-central portion of the study area (compare Figs. 6, 7). These two paleovalleys merge to the south and form a single, north-south oriented paleovalley which, in turn, bends to the southwest and projects to the southwestern corner of the map area.

The two north-south oriented valleys described herein are the same as those reported by Flowers (1956), with some modification of details (compare Figs. 2, 6). However, in contrast to the results of the present study, Flowers (1956) inferred that the single paleovalley formed by the mergence of the two northern paleovalleys continued to the southeast into central West Virginia, rather than bending to the southwest. Although the isopach map of the Greenbrier Limestone prepared by Flowers (1956) does indicate that the 100-ft through 200-ft

contours "v" sharply to the southeast in central West Virginia, beds of the Upper Mississippian Mauch Chunk Group overlie the Greenbrier in that area (Fig. 2). Where the Mauch Chunk is present, changes in the thickness of the underlying Greenbrier cannot be attributed to paleotopography on the Mississippian-Pennsylvanian unconformity. Numerous workers have indicated that the base of the Greenbrier in northern West Virginia is an erosional disconformity (Martens, 1945; Youse, 1964; Arkle and others, 1979; Filer, 1985; Sweeney, 1986). It is therefore possible that anomalous changes in the thickness of the Greenbrier Limestone may be partially controlled by paleotopography on the sub-Greenbrier unconformity. Hence, Flowers' (1956) inference of a southeast-trending sub-Pennsylvanian paleovalley in the study area is not supported by data from his study, nor from mine.

Although Flowers (1956) and Krissek and others (1986) suggested that north-flowing rivers carved the sub-Pennsylvanian paleovalleys in West Virginia, this inference is not supported by the results of the present study. Quartzose sandstone of the New River Formation fills the paleovalleys (Figs. 8, 9). Inference of north-flowing paleodrainage implies that the source area for paleovalley-fill sandstones of the New River Formation was southern West Virginia or eastern Kentucky; this area, however, was near the mid-Carboniferous depocenter. Furthermore, the Mississippian System and the New River Formation

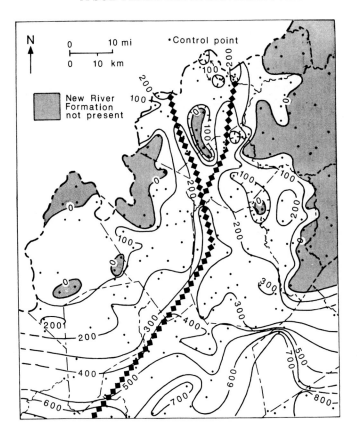

FIG. 8—Isopachous map of the New River Formation in northwestern West Virginia. Isopach interval = 60 m (100 ft) (lines of diamonds indicate axes of sub-Pennsylvanian paleovalleys shown on Fig. 10).

both increase in thickness toward the south (Figs. 6, 8, respectively), indicating that the structural slope of the basin was to the south, prior to and just after the development of the sub-Pennsylvanian unconformity. To my knowledge, there is no evidence to suggest that the structural slope of the basin was reversed during Mississippian-Early Pennsylvanian time.

Data presented herein are consistent with a southerly paleodrainage as previously hypothesized by Uttley (1974), Rice (1984) and Rice and Schwietering (1988). As stated by those workers, mid-Carboniferous drainage should have flowed south from the stable craton toward the subsiding Appalachian depocenter in southeastern West Virginia.

The paleovalleys described herein appear to be mostly a product of fluvial incision. Valley carving/stream rejuvenation occurred during a time when regional base level dropped significantly. The relatively great width of the paleovalleys (12-25 km) suggests that these features are true fluvial valleys, not simply incised channels. The rivers that drained and incised the sub-Pennsylvanian surface probably had channels that were much smaller than the paleovalleys which they ultimately eroded. The paleovalleys appear to be the net geomorphic effect of a complex set of synergistic erosional processes including channel downcutting, lateral erosion, and mass-wasting. Where paleovalley segments are carved into carbonate units (Green-

brier and Reynolds Limestones), karst processes may have also contributed to the erosional development of these features, but to a lesser degree than surficial processes.

Figure 10 portrays my reconstruction of paleodrainage on the sub-Pennsylvanian unconformity in northwestern West Virginia at the time of maximum downcutting of that surface. My reconstruction is based on interpretation of the paleotopographic and paleogeologic configurations of the sub-Pennsylvanian surface as portrayed in Figures 6 and 7. Although it is possible that the system of trunk paleovalleys was formed by superposition of palimpsest streams or drainages, I interpret this system as having been carved by a single, integrated drainage network for several reasons. First, the two "upper" trunk paleovalleys merge in the inferred downstream direction. Second, these same two paleovalleys are similar in overall dimensions and both are filled with quartzose sandstones of the New River Formation (Fig. 9). Third, a regional paleodrainage network can be linked together if a local integrated paleodrainage network is inferred. In sum, there is no physical evidence that makes it necessary to postulate that the paleovalleys were carved by more than one drainage system; parsimony supports the inference of a single, integrated paleodrainage network.

The fact that the paleovalleys in northwestern West Virginia are most deeply incised where they are cut into the Greenbrier Limestone (Figs. 6, 7) supports the inference of Wanless (1975), Rice (1984), and Rice and Schwietering (1988) that the Greenbrier Limestone supported a cuesta on the sub-Pennsylvanian surface. The Pocono and Mauch Chunk Groups appear to have been less resistant to erosion, probably because the sandstones and shales that compose these units were unconsolidated or weakly consolidated during the mid-Carboniferous.

The portion of the New River Formation which fills paleovalleys is mostly massive sandstone (Figs. 8, 9). Without sedimentological data, it is difficult to constrain the depositional environment(s) of the New River sandstones which fill the paleovalleys; however, some interpretations can be derived from the results presented herein. Given that a southerly drainage on the sub-Pennsylvanian surface has been inferred in the present study, and that Siever and Potter (1956) reconstructed a southwestern paleoslope for basal Pennsylvanian sandstones on the western flank of the Appalachian basin, the New River paleovalley-fill sandstones probably were derived from the craton, north of the Appalachian basin. Pennsylvanian subcrop maps of Arkle (1974) and Dennison (1983) show that the lower Mississippian Pocono and Waverly Groups were widely exposed on the sub-Pennsylvanian surface in western Pennsylvania and eastern Ohio. Sandstones of these Mississippian units also may have been source beds for some of the New River paleovalley-fill deposits. Many of the beds that fill the paleovalleys were probably deposited by the streams which initially carved the valleys. Rice (1984) and Rice and Schwietering (1988) have supported a general model for fluvial deposition of Lower Pennsylvanian quartzose sandstones in the central Appalachian basin. Valley alluvium consisting almost

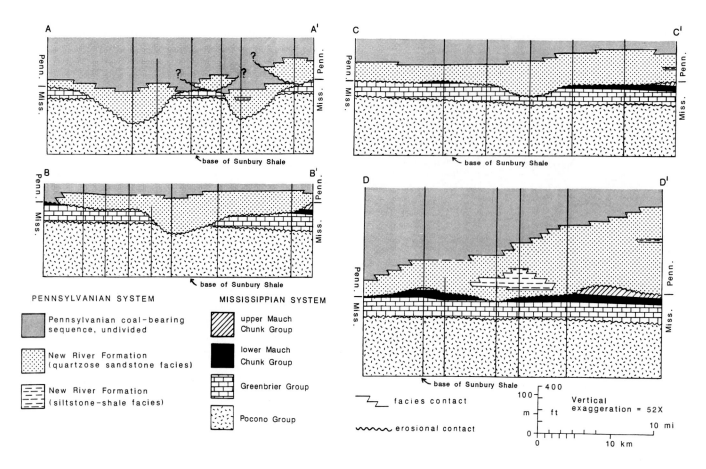

FIG. 9—Carboniferous cross sections transverse to axes of sub-Pennsylvanian paleovalleys in northwestern West Virginia (lines of cross sections shown of Fig. 6). Datum at base of each cross section is base of Sunbury Shale.

entirely of sand and gravel is a result of aggradation of bedload-dominated channels that have low sinuosity and a high width-to-depth ratio (Schumm, 1968). The general lack of fine-grained sediments in the New River paleovalley-fill beds suggests that alluviation of the paleovalleys was dominated by bedload Pennsylvanian surface flowed from the craton north and northwest of the Appalachian basin into the mid-Carboniferous depocenter of southwestern Virginia.

AGE OF PALEOVALLEY INCISION

Biostratigraphic data from strata which fill the Middlesboro, Sharon and Perry paleovalleys are generally lacking, thus only indirect evidence is available to bracket the relative age of valley incision.

Several studies have reported that the regional unconformity that forms the Mississippian-Pennsylvanian boundary throughout most of the central Appalachian basin originates in the lower part of the Lower Pennsylvanian New River Formation in southern West Virginia and southwestern Virginia (the Pocahontas basin) and progressively truncates strata of the Lower Pennsylvanian Pocahontas Formation toward the northwest (Englund, 1969, 1979; Chesnut, 1988; F. Ettensohn and D.

Chesnut, Jr., pers. commun., 1991). According to theses studies, the major erosional development of the Mississippian-Pennsylvanian unconformity occurred during the Early Pennsylvanian Period; allegedly, the widespread occurrence of Mississippian rocks below the unconformity resulted from Early Pennsylvanian truncation of basal Pennsylvanian strata.

The hypothesis that the maximum erosional development of the Mississippian-Pennsylvanian unconformity occurred during the early Pennsylvanian is difficult to reconcile with the paleodrainage data presented herein. Figure 11 indicates that the Middlesboro paleovalley, which drained to the southwest, is incised into Upper Mississippian strata just beyond the northwestern limit of the Lower Pennsylvanian Pocahontas Formation. If it is assumed that the Pocahontas-New River unconformity in the Pocahontas basin is the eastern extension of the regional Mississippian-Pennsylvanian unconformity, then it is reasonable to infer that drainage on the Pocahontas-New River surface flowed west to northwest into the Middlesboro paleovalley. However, the progressive northwestward termination of the Pocahontas Formation and Upper Mississippian units beneath the New River and Lee Formations as mapped by Englund and Henry (1984) and Chesnut (1988) suggests that the paleoslope on the Pocahontas-New River surface was to the southeast.

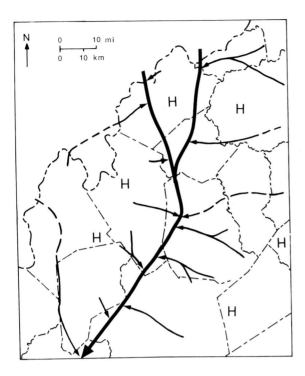

FIG. 10—Paleodrainage map of the sub-Pennsylvanian unconformity in northwestern West Virginia showing axes of trunk streams (heavy solid lines), tributary streams (thin solid lines), inferred flow direction (indicated by arrows), and areas of paleotopographic highs (indicated by "H"s).

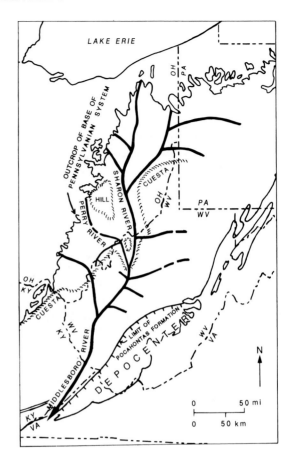

FIG. 11.—Regional paleodrainage map of the sub-Pennsylvanian unconformity in the central Appalachian basin. Paleodrainage reconstruction based on data from present study, D. W. Neal (personal data), Uttley (1974), Craig and Connor (1979), Rice (1985), and Rice and Schwietering (1988). Western and northern limit of the Lower Pennsylvanian Pocahontas Formation after Englund and Henry (1984).

Chesnut (1988), in fact, indicated that the paleoslope on unconformity in the Pocahontas basin was to the southeast. A southeast-dipping paleoslope on the Pocahontas-New River surface is feasible only if a drainage divide existed between the Middlesboro paleovalley and the Pocahontas basin. Chesnut (1988, Figs. 13, 26) and Englund and Thomas (1990, Plate I) have constructed detailed stratigraphic cross sections of Carboniferous rocks in southeastern Kentucky and southwestern Virginia. Although these northwest-southeast oriented cross sections diagram a sub-New River unconformity and indicate an incised paleotopographic low along the projected axis of the Middlesboro paleovalley, the paleotopography of the New River surface as configured on these cross sections does not indicate the existence of a paleodrainage divide between the Middlesboro paleovalley and the Pocahontas basin. The paradox presented by the apparently southeast-dipping paleoslope on the Pocahontas-New River unconformity, and the lack of evidence for a paleodrainage divide between the Middlesboro paleovalley and the Pocahontas basin, suggests that incision of the Middlesboro-Sharon-Perry paleovalley system was not coeval with the development of the Pocahontas-New River unconformity.

In contrast to the prevailing view that the Mississippian-Pennsylvanian boundary is conformable in the Pocahontas basin, Rice (1985) suggested that the systemic boundary there

(base of the Pocahontas Formation) is an erosional unconformity. Rice (1985) also correlated the sandstone-fill of the lower reach of the Middlesboro paleovalley in Virginia with the lower part of the Lower Pennsylvanian Pocahontas Formation. Shepherd and others (1986) challenged Rice's inference of a sub-Pennsylvanian unconformity in the Pocahontas basin, but they were unable to disprove it. Rice (1986) subsequently defended and reiterated his interpretations.

Although Rice (1985) did not support his postulation of a Mississippian-Pennsylvanian unconformity with biostratigraphic data, I have found that existing biostratigraphic data from Upper Mississippian-Lower Pennsylvanian strata in the Pocahontas basin indicate that the Mississippian-Pennsylvanian boundary there is disconformable (Fig. 12). On the basis of megafloral remains, Gillespie and Pfefferkorn (1977, 1979), Pfefferkorn and Gillespie (1982), and Gillespie and others (1989) demonstrated that the base of the Pennsylvanian System in the Pocahontas basin (base of Pocahontas Formation) corresponds to the base of the Namurian B Series (base of the Kinderscoutian

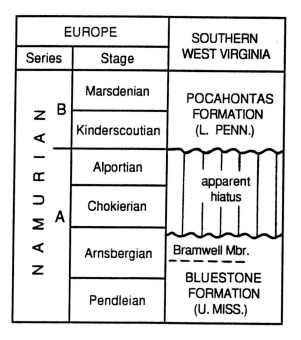

EUROPE		SOUTHERN WEST VIRGINIA
Series	**Stage**	
N A M U R I A N B	Marsdenian	POCAHONTAS FORMATION (L. PENN.)
	Kinderscoutian	
N A M U R I A N A	Alportian	apparent hiatus
	Chokierian	
	Arnsbergian	Bramwell Mbr.
	Pendleian	BLUESTONE FORMATION (U. MISS.)

FIG. 12.—Correlation of the Upper Mississippian Bluestone Formation and Lower Pennsylvanian Pocahontas Formation with the standard mid-Carboniferous sequence in western Europe illustrating an apparent Mississippian-Pennsylvanian unconformity in southern West Virginia. Correlation of the Pocahontas Formation with the Namurian B Series (Kinderscoutian and Marsdenian stages) is after Gillespie and Pfefferkorn (1977, 1979), Pfefferkorn and Gillespie (1982), and Gillespie and others (1989). Assignment of the Bramwell Member of the Bluestone Formation to the Arnsbergian Stage of the Namurian A Series is based on biostratigraphic data reported in Repetski and Henry (1983).

Stage) of Europe (Fig. 12). According to Gillespie and others (1989), the Upper Mississippian Bluestone Formation, which directly underlies the Pocahontas Formation, correlates with the Namurian A Series of Europe. The Bramwell Member of the Bluestone Formation is the youngest Mississippian unit in the Pocahontas basin (Englund, 1979). Repetski and Henry (1983) reported that a conodont faunule recovered from the Bramwell Member is no older than the *Adetognathus unicornis* conodont zone, but is possibly as young as the lower part of the *Rhacistognathus muricatus* conodont zone of Lane and Straka (1974). According to Brenckle and others (1977) and Sutherland and Manger (1984), the *A. unicornis* and *R. muricatus* conodont zones correlate with the upper Arnsbergian Stage in the Namurian A of Europe. Existing data therefore indicate that the youngest Mississippian beds in the Pocahontas basin (Bramwell Member) are no younger than the Arnsbergian Stage of the Namurian A Series. The standard Namurian A Series of Europe comprises (in ascending order) the Pendleian, the Arnsbergian, the Chokierian, and the Alportian Stages. In the Pocahontas basin, Kinderscoutian-age beds of the Namurian B Series (lower part of the Pocahontas Formation) rest directly on Arnsbergian-age beds of the Namurian A Series (Bramwell Member). Thus, on

the basis of existing biostratigraphic evidence, a Chokierian-Alportian hiatus occurs between Mississippian and Pennsylvanian beds in the Pocahontas basin (Fig. 12). I therefore agree with Rice (1985) that the Mississippian-Pennsylvanian (Bluestone-Pocahontas) contact in the Pocahontas basin is disconformable

On the basis of the foregoing discussion, I suggest that incision of the Middlesboro-Sharon-Perry paleovalley system was predominantly a late Mississippian (Chokierian-Alportian age) event. Also, although the Pocahontas-New River unconformity does appear to represent an Early Pennsylvanian episode of exposure and erosion in the central Appalachian basin, I suggest that this event postdates the maximum erosional development of the Middlesboro-Sharon-Perry paleovalley system.

TECTONIC AND EUSTATIC CONTROLS ON PALEOVALLEY INCISION AND UNCONFORMITY DEVELOPMENT

According to Siever and Potter (1956), the craton north of the Appalachian basin was uplifted during the mid-Carboniferous. During the same time, a subsiding depocenter existed in southern West Virginia and southwestern Virginia (Arkle, 1974). These factors indicate that the south- to southwest-dipping paleoslope on which the Middlesboro-Sharon-Perry paleoriver system flowed was created by a combination of epeirogenic uplift north of the Appalachian basin, and subsidence within the basin that increased in rate toward the depocenter in southern West Virginia and southwestern Virginia.

As previously demonstrated in the present study, the Middlesboro-Sharon-Perry paleovalley system was incised during Late Mississippian time (Chokierian and Alportian ages of the Namurian A). Hypothetically, the Late Mississippian episode of fluvial incision in the central Appalachian basin could have been induced by tectonic uplift within the basin, a eustatic sea-level fall, or both.

Several recent studies have modeled the tectonic evolution of the Appalachian basin, and have supported intrabasinal uplift as the major control on development of the Mississippian-Pennsylvanian unconformity. Quinlan and Beaumont (1984) and Tankard (1986) inferred that flexural uplift of a north- to northeast-trending peripheral bulge in the vicinity of the present-day Cincinnati-Findlay arch system controlled the development of sequence-bounding unconformities in the Paleozoic of eastern North America, including the Mississippian-Pennsylvanian unconformity. According to those studies, a Late Mississippian phase of tectonic quiescence in the Appalachian orogenic belt allowed viscoelastic rebound of a flexural bulge on the cratonic margin of the Appalachian basin. Isostatic uplift and eastward migration of this peripheral bulge toward the Appalachian orogenic belt allegedly drove the erosional development of the sub-Pennsylvanian unconformity in the central Appalachian basin. F. Ettensohn and D. Chesnut, Jr. (pers. commun., 1991) also attributed the Mississippian-Pennsylvanian unconformity to uplift and migration of a northeast-south-west oriented flexural bulge along the western margin of the central Appalachian basin. In contrast to the model developed by Quinlan and Beaumont (1984) and Tankard (1986), F. Ettensohn and D. Chesnut, Jr. (pers. commun., 1991) inferred that the bulge migrated northwestward, away from the orogenic belt. Furthermore, F. Ettensohn and D. Chesnut, Jr. (pers. commun., 1991) suggested that uplift and migration of the bulge occurred during the Early Pennsylvanian Period, and that this flexural event was driven by emplacement of a thrust load on the continental margin during an early phase of the Alleghany orogeny.

Regardless of the specific mechanism of flexural uplift and the direction of bulge migration, an east- to southeast-dipping regional paleoslope would have existed on the western flank of the central Appalachian basin if uplift of a northeast-southwest oriented bulge occurred as hypothesized by Quinlan and Beaumont (1984), Tankard (1986), and F. Ettensohn and D. Chesnut, Jr., (pers. commun., 1991). However, the Middlesboro-Sharon-Perry paleoriver system indicates that regional paleoslope on the western flank of the Appalachian basin was to the south and southwest during the development of the Mississippian-Pennsylvanian unconformity. The results of the present paper do not rule out the possibility that flexural uplift of a northeast-southwest trending bulge occurred along the western flank of the basin during Late Mississippian of Early Pennsylvanian time. However, paleodrainage evidence does rule against flexural uplift of the western flank of the Appalachian basin as a major control on the regional paleoslope and on the incision of the Middlesboro-Sharon-Perry paleoriver system. If flexural-bulge uplift did play a role in incision of this paleoriver system, then the uplift must have occurred along the northern periphery of the Appalachian basin in order for the south-draining paleovalley system to have developed.

Saunders and Ramsbottom (1986) summarized evidence for a global mid-Carboniferous unconformity and inferred that this unconformity represents a eustatic lowstand during the Chokierian and Alportian Ages of the Namurian Epoch of Europe (late Chesterian Epoch of the Mississippian Period of North America). Ross and Ross (1985, 1988) have also indicated that the Chokierian and Alportian stages are missing in many mid-Carboniferous shelf successions around the world. As indicated previously, the Middlesboro-Sharon-Perry paleovalley system apparently was incised during Chokierian-Alportian time. On the basis of the synchronism of the two events, I infer that incision of the Middlesboro-Sharon-Perry paleovalley system was dominantly controlled by the mid-Carboniferous eustatic lowstand documented by Saunders and Ramsbottom (1986). In a study of the sub-Dakota (mid-Cretaceous) unconformity on the southeastern Colorado Plateau, Aubrey (1989) demonstrated that eustasy played a major role in driving fluvial incision more than 320 km inland from the Cretaceous seaway of the western interior of North America. The upper reaches of the Middlesboro-Sharon-Perry paleovalley system are a comparable distance from the mid-Carboniferous depocenter in southwestern Virginia (Fig. 11). Therefore, it is plausible to infer that eustasy was a dominant control on Late Mississippian fluvial incision throughout the central Appalachian basin.

The results of the present study indicate that eustasy played a key role in the development of the regional Mississippian-Pennsylvanian unconformity in the central Appalachian basin insofar that fluvial incision driven by a Late Mississippian eustatic lowstand apparently created most of the local relief on the sub-Pennsylvanian surface. However, much of this surface was not onlapped and buried until late Early Pennsylvanian or early Middle Pennsylvanian time (Arkle, 1974; Wanless, 1975; Chesnut, 1988), indicating that subaerial exposure and denudation continued long after the Late Mississippian eustatic lowstand. It therefore seems likely that intrabasinal tectonic activity also was a major control on development of the Mississippian-Pennsylvanian unconformity, especially during the Early Pennsylvanian Period. As inferred by F. Ettensohn and D. Chesnut, Jr. (pers. commun., 1991), Early Pennsylvanian exposure and erosion in the central Appalachian basin may have resulted from migration of a northeast-southwest oriented flexural bulge away from the Appalachian orogenic belt.

In sum, although many recent studies have indicated that tectonic evolution of the central Appalachian basin dominated over eustasy in controlling the origin of the regional Mississippian-Pennsylvanian unconformity (Quinlan and Beaumont, 1984; Tankard, 1986; Willard and Klein, 1990; F. Ettensohn and D. Chesnut, Jr., pers. commun., 1991), paleodrainage considerations indicate that the eustasy did play a significant role. Future studies, therefore, should be directed at resolving the interplay between eustasy and intrabasinal tectonics in generating the Mississippian-Pennsylvanian unconformity, rather than wholly attributing this major sequence boundary to one process or the other.

ACKNOWLEDGMENTS

Significant portions of this work are based on my doctoral dissertation submitted to the University of North Carolina at Chapel Hill. Funding for my dissertation research was provided by the Department of Geology at the UNC-CH (Martin fellowship and Appalachian Basin Industrial Associates fellowship), the Graduate School at UNC-CH, Arco Oil and Gas Company, Sigma Xi Scientific Society and the Southeastern Section of the Geological Society of America. I extend my appreciation to Professor John M. Dennison for suggesting and supervising my dissertation study, and for reviewing the present manuscript. Material support for manuscript preparation was provided by the Department of Geology at East Carolina University while I was a visiting faculty member there. I thank Professor Donald W. Neal of East Carolina University for generously allowing use of his personal data in reconstructing sub-Pennsylvanian paleodrainage in southern West Virginia. Wallace de Witt, Jr. and Charles L. Rice carefully and insightfully reviewed an early version of this manuscript. Joseph F. Schwietering, Donald R. Chesnut, Jr., and Frank R. Ettensohn all reviewed the present manuscript, and their critical comments helped to improve the final version of the manuscript.

JACK D. BEUTHIN

REFERENCES

ARKLE, T., JR., 1974, Stratigraphy of the Pennsylvanian and Permian systems in the central Appalachians, in Briggs, G. ed., Carboniferous of the southeastern United States: Boulder, Geological Society of America Special Paper 148, p. 5-29.

ARKLE, T. JR., BEISSEL, D. R., LARESE, R. E., NUHFER, E. B., PATCHEN, D. G., SMOSNA, R. A., GILLESPIE, W. H., LUND, R., NORTON, W., AND PFEFFERKORN, H. W., 1979, The Mississippian and Pennsylvanian (Carboniferous) Systems in the United States- West Virginia and Maryland: Washington, D. C., United States Geological Survey Professional Paper 1110-D, p. 1-35.

AUBREY, M. W., 1989, Mid-Cretaceous alluvial-plain incision related to eustasy, southeastern Colorado Plateau: Geological Society of America Bulletin, v. 101, p. 443-449.

BEUTHIN, J. D., 1988, Subsurface evidence for the pre-Absaroka unconformity (mid-Carboniferous) in the Pennsylvanian stratotype area (West Virginia) (abs.): Geological Society of America, Abstracts with Programs, v. 20, p. A267.

BEUTHIN, J. D., 1989, Genetic character of the Mississippian-Pennsylvanian contact in the northern part of the proposed Pennsylvanian stratotype area, West Virginia: Unpublished Ph. D. Dissertation, The University of North Carolina at Chapel Hill, Chapel Hill, 83 p.

BEUTHIN, J. D., 1991, Tectonic vs. eustatic controls on the Mississippian-Pennsylvanian unconformity, central Appalachians: constraints from paleodrainage considerations (abs.): Geological Society of America, Abstracts with Programs, v. 23, p. 8,

BEUTHIN, J. D., AND NEAL, D. W., 1990, A mid-Carboniferous paleovalley system in the central Appalachians- and ancient "big river"? (abs.): Geological Society of America, Abstracts with Programs, v. 22, p. A44.

BRANSON, C. C., 1964, Pennsylvanian System in the central Appalachians, in Branson, C. C., ed., Pennsylvanian System in the United States: Tulsa, American Association of Petroleum Geologists, p. 74-96.

BRENCKLE, P., LANE, H. R., MANGER, W. L., AND SAUNDERS, W. B., 1977, The Mississippian-Pennsylvanian boundary as an intercontinental biostratigraphic datum: Newsletters in Stratigraphy, v. 6, p. 106-116

CARDWELL, D. H., 1981, Oil and gas report and map of Gilmer and Lewis Counties, West Virginia: Morgantown, West Virginia Geological and Economic Survey Bulletin B-18A, 42 p.

CECIL, C. B., AND ENGLUND K. J., 1989, Origin of coal deposits and associated rocks in the Carboniferous of the Appalachian basin, in Edmunds, W. E., Eggleston, J. R., and Englund, K. J., eds., Characteristics of the Mid-Carboniferous Boundary and Associated Coal-bearing Rocks in the Appalachian Basin (Twenty-eighth International Geological Congress Field Trip Guidebook T352): Washington, D. C, American Geophysical Union, p. 67-72.

CHESNUT, D. R., JR., 1983, A preliminary study of the Upper Mississippian and Lower Pennsylvanian rocks of eastern Kentucky and nearby areas: Lexington, Kentucky Geological Survey, Series 11, Open-File Report OF-83-02, 23 p.

CHESNUT, D. R., JR., 1988, Stratigraphic analysis of the Carboniferous rocks of the Central Appalachian basin: Unpublished Ph. D. Dissertation, University of Kentucky, Lexington, 297 p.

COLTON, G. W., 1970, The Appalachian Basin— its depositional sequences and their geologic relationships, in Fisher, G. W., Pettijohn, F. J., Reed, J. C., Jr., and Weaver, K. N., eds., Studies in Appalachian geology: New York, Wiley, p. 5-47.

COSKREN, T. D., AND RICE, C. L., 1979, Contour map of the base of the Pennsylvanian System, eastern Kentucky: Washington, D. C., United States Geological Survey Miscellaneous Field Studies Map MF-1100, scale 1:250,000.

CRAIG, L. C., AND CONNOR, L. W., eds., 1979, Paleotectonic investigations of the Mississippian System in the United States, Part I, Introduction and regional analyses of the Mississippian System: Washington, D. C., United States Geological Survey Professional Paper 1010, p. 14-48.

CROSS, A. T., AND SCHEMEL, M. P., 1956, Geology and economic resources of the Ohio River valley in West Virginia: Part I, Geology: Morgantown, West Virginia Geological and Economic Survey Volume 22, 149 p.

DENNISON, J. M., 1983, Sedimentary Tectonics of the Appalachian Basin: Dallas, Dallas Geological Society Short Course, 129 p.

DENNISON, J. M., 1989, Paleozoic Sea-level Changes in the Appalachian Basin (Twenty-eighth International Geological Congress Filed Trip Guidebook T354): Washington, D. C., American Geophysical Union, 56 p.

DE WITT, W., JR., AND MCGREW, L. W., 1979, The Appalachian Basin region, in Craig, L. C., and Connor, L. W., eds., Paleotectonic investigations of the Mississippian System in the United States, Part I, Introduction and regional analyses of the Mississippian System: Washington, D. C., United States Geological Survey Professional Paper 1010, p. 14-48.

DONALDSON, A. C., AND SHUMAKER, R. C., 1981, Late Paleozoic molasse of central Appalachians, in Miall, A. D., ed., Sedimentation and Tectonics in Alluvial basins: Toronto, Geological Association of Canada Special Paper 23, p. 99-124.

ENGLUND, K. J., 1969, Relationship of the Pocahontas Formation to the Mississippian-Pennsylvanian boundary in southwestern Virginia and southern West Virginia (abs.): Geological Society of America, Abstracts with Programs, v. 13, p. 446.

ENGLUND, K. J., 1979, Mississippian System and Lower Series of the Pennsylvanian System in the proposed Pennsylvanian System stratotype area, in Englund, K. E., Arndt, H. H., and Henry, T. W., eds., Proposed Pennsylvanian System Stratotype, Virginia and West Virginia (Ninth International Congress of Carboniferous Stratigraphy and Geology Meeting Field Trip 1): Alexandria, American Geological Institute Selected Guidebook Series 1, p. 69-72.

ENGLUND, K. J., 1981, Regional aspects of the Mississippian-Pennsylvanian boundary in the central Appalachian basin (abs.): Geological Society of America, Abstracts with Programs, v. 13, p. 446.

ENGLUND, K. J., ARNDT, H. H., GILLESPIE, W. H., HENRY, T. W., AND PFEFFERKORN, H. W., 1977, A field guide to the proposed Pennsylvanian System Stratotype, West Virginia: Washington, D. C., American Association of Petroleum Geologists Annual Meeting Guidebook, Field Trip 2, 80 p.

ENGLUND, K. J., AND HENRY, T. W., 1984, The Mississippian-Pennsylvanian boundary in the central Appalachians, in Sutherland, P. K., and Manger, W. L., eds., Ninth International Congress of Carboniferous Stratigraphy and Geology, Compte Rendu, v. 2, Biostratigraphy: Champaign-Urbana, Southern Illinois University Press, p. 330-336.

ENGLUND, K. J., AND THOMAS, R. E., 1990, Late Paleozoic depositional trends in the central Appalachian basin: Washington, D. C., United States Geological Survey Bulletin 1839-F, 19 p.

FERM, J. C., 1974a, Carboniferous environmental models in the eastern

United States and their significance, *in* Briggs, G., ed., Carboniferous of the Southeastern United States: Boulder, Geological Society of America Special Paper 148, p. 79-95.

FERM, J. C., 1974b, Carboniferous paleogeography and continental drift: Krefeld, Seventh International Congress of Carboniferous Stratigraphy and Geology, Compte Rendu, v. 3, p. 9-25.

FILER, J. K., 1985, Oil and gas report and maps of Pleasants, Wood and Wirt Counties, West Virginia: Morgantown, West Virginia Geological and Economic Survey Bulletin B-11A, 87 p.

FLOWERS, R. R., 1956, Subsurface study of the Greenbrier Limestone in West Virginia: Morgantown, West Virginia Geological and Economic Survey Report of Investigations 15, 17 p.

GILLESPIE, W. H., CRAWFORD, T. J., AND RHEAMS, L. J., 1989, Biostratigraphic significance of compression-impression plant fossils near the Mississippian-Pennsylvanian boundary in the southern Appalachians, *in* Edmunds, W. E., Eggleston, J. R., and Englund, K. J., eds., Characteristics of the Mid-Carboniferous Boundary and Associated Coal-bearing Rocks in the Appalachian Basin (Twenty-eighth International Geological Congress Field Trip Guidebook T352): Washington, D. C, American Geophysical Union, p. 55-60.

GILLESPIE, W. H., AND PFEFFERKORN, H. W., 1977, Plant fossils and biostratigraphy at the Mississippian/Pennsylvanian boundary in the Pennsylvanian System stratotype area, West Virginia/Virginia (abs.): Botanical Society of America Annual Meeting, Abstracts of Papers, p. 37.

GILLESPIE, W. H., AND PFEFFERKORN, H. W., 1979, Distribution of commonly occurring plant megafossils in the proposed Pennsylvanian System stratotype, *in* Englund, K. J., Arndt, H. H., and Henry, T. W., eds., Proposed Pennsylvanian System Stratotype, Virginia and West Virginia (Ninth International Congress of Carboniferous Stratigraphy and Geology Meeting Field Trip 1): Alexandria, American Geological Institute Selected Guidebook Series 1, p. 87-96.

HAUGHT, O. L., 1960, Oil and gas report and map of Kanawha County, West Virginia: Morgantown, West Virginia Geological and Economic Survey Bulletin B-19, 24 p.

HAUGHT, O. L., AND OVERBY, W. K., JR., 1964, Oil and gas map and report of Braxton and Clay Counties, West Virginia: Morgantown, West Virginia Geological and Economic Survey Bulletin B-29, 19 p.

HOBDAY, D. K., AND HORNE, J. C., 1977, Tidally influenced barrier island and estuarine sedimentation in the Upper Carboniferous of southern West Virginia: Sedimentary Geology, v. 18, p. 97-122.

KREBS, C. E., 1911, Jackson, Mason and Putnam Counties: Morgantown, West Virginia Geological and Economic Survey County Geologic Report 9, 387 p.

KREBS, C. E., 1913, Cabell, Wayne and Lincoln Counties: Morgantown, West Virginia Geological and Economic Survey County Geologic Report 4, 483 p.

KREBS, C. E., 1914, Kanawha County: Morgantown, West Virginia Geological and Economic Survey County Geologic Report 11, 679 p.

KRISSEK, L. A., KETRING, C. L., JR., AND KULIKOWSKI, D. L., 1986, Lower Pennsylvanian sandstones of southeastern Ohio: Implications for sediment sources and depositional environments in the north-central Appalachian basin: Morgantown, Appalachian Basin Industrial Associates Program, v. 11, p. 109-142.

LANE, H. R., AND STRAKA, J. J., II, 1974, Late Mississippian and Early Pennsylvanian conodonts, Arkansas and Oklahoma: Boulder, Geological Society of America Special Paper 152, 144 p.

MARTENS, J. H. C., 1945, Well-sample records: Morgantown, West Virginia Geological and Economic Survey, v. 17, 889 p.

MILLER, M. S., 1974, Stratigraphy and coal beds of Upper Mississippian and Lower Pennsylvanian rocks in southwestern Virginia: Charlottesville, Virginia Division of Mineral Resources Bulletin 84, 211 p.

OVERBY W. K., JR., 1961, Oil and gas report of Jackson, Mason and Putnam Counties, West Virginia: Morgantown, West Virginia Geological and Economic Survey Bulletin B-23, 26 p.

PFEFFERKORN, H. W., AND GILLESPIE, W. H., 1982, Plant megafossils near the Mississippian-Pennsylvanian boundary in the Pennsylvanian System Stratotype, West Virginia/ Virginia, *in* Ramsbottom, W. H. C., Saunders, W. B., and Owens, B., eds., Biostratigraphic Data for a Mid-Carboniferous Boundary: Leeds, Subcommission of Carboniferous Stratigraphy, p. 128-133.

PRESLEY, M. W., 1981, The Mississippian-Pennsylvanian boundary in the central Appalachian basin as a record of changes in basin geometry and clastic supply (abs.): Geological Society of America, Abstracts with Programs, v. 13, p. 532-533.

QUINLAN, G. M., AND BEAUMONT, C., 1984, Appalachian thrusting, lithospheric flexure and the Paleozoic stratigraphy of the eastern interior of North America: Canadian Journal of Earth Science, v. 21, p. 973-996.

REPETSKI, J. E., AND HENRY, T. W., 1983, A Late Mississippian conodont faunule from area of proposed Pennsylvanian System stratotype, eastern Appalachians: Fossils and Strata, no. 15, p. 169-170.

RICE, C. L., 1984, Sandstone units of the Lee Formation and related strata in northeastern Kentucky: Washington, D. C., United States Geological Survey Professional Paper 1151-G, 53 p.

RICE, C. L., 1985, Terrestrial vs. marine depositional model— A new assessment of subsurface Lower Pennsylvanian rocks of southwestern Virginia: Geology, v. 13, p. 786-789.

RICE, C. L, 1986, Reply to Comment on "Terrestrial vs. marine depositional model—A new assessment of subsurface Lower Pennsylvanian rocks of southwestern Virginia: Geology, v. 14, p. 801-802.

RICE, C. L., AND SCHWIETERING, J. F., 1988, Fluvial deposition in the central Appalachians during the Early Pennsylvanian: Washington, D. C., United States Geological Survey Bulletin 1839-B, 10 p.

ROSS, C. A., AND ROSS, J. R. P., 1985, Late Paleozoic depositional sequences are synchronous and worldwide: Geology, v. 13, p. 194-197.

ROSS, C. A., AND ROSS, J. R. P., 1988, Late Paleozoic transgressive-regressive deposition, *in* Hastings, B. S., Posamentier, H., Ross, C. A., St. C. Kendall, C. G., Van Wagoner, J., and Wilgus, C. K., eds., Sea-level Changes: An Integrated Approach: Tulsa, Society of Economic Paleontologists and Mineralogists Special Paper 42, p. 227-247.

SAUNDERS, W. B., AND RAMSBOTTOM, W. H. C., 1986, The mid-Carboniferous eustatic event: Geology, v. 14, p. 208-212.

SCHUMM, S. A., 1968, Speculations concerning paleohydrologic controls of terrestrial sedimentation: Geological Society of America Bulletin, v. 79, p. 1573-1588.

SHEPHERD, R. G., PASHIN, J. C., AND GREB, S. F., 1986, Comment on "Terrestrial vs. marine depositional model— A new assessment of subsurface Lower Pennsylvanian rocks of southwestern Virginia: Geology, v. 14, p. 800-801.

SIEVER, R., AND POTTER, P. E., 1956, Sources of basal Pennsylvanian sediments in the Eastern Interior basin; part 2, Sedimentary petrology: Journal of Geology, v. 64, p. 317-335.

SLOSS, L. L., 1963, Sequences in the cratonic interior of North America:

Geological Society of America Bulletin, v. 74, p. 93-114.

SUTHERLAND, P. K., AND MANGER, W. L., 1984, The Mississippian-Pennsylvanian boundary in North America, *in* Sutherland, P. K., and Manger, W. L., eds., Ninth International Congress of Carboniferous Stratigraphy and Geology, Compte Rendu, v. 2, Biostratigraphy: Champaign-Urbana, Southern Illinois University Press, p. 319-329.

SWEENEY, J., 1986, Oil and gas report and maps of Wirt, Roane and Calhoun Counties, West Virginia: Morgantown, West Virginia Geological and Economic Survey Bulletin B-40, 102 p.

TANKARD, A. J., 1986, Depositional response to foreland deformation in the Carboniferous of eastern Kentucky: American Association of Petroleum Geologists Bulletin, v. 70, p. 853-868.

TUCKER, R. C., 1936, Deep well records: Morgantown, West Virginia Geological and Economic Survey Volume 7, 560 p.

UTTLEY, J. S., 1974, The stratigraphy of the Maxville Group of Ohio and correlative strata in adjacent areas: Unpublished Ph. D. Dissertation, The Ohio State University, Columbus, 252 p.

VAIL, P. R., MITCHUM., R. M., JR., AND THOMPSON, S., 1977, Seismic stratigraphy and global changes of sea level, Part 4: Global cycles of relative changes of sea level, *in* Payton, C. E., ed., Seismic Stratigraphy- Applications to Hydrocarbon Exploration: Tulsa, American Association of Petroleum Geologists Memoir 26, p. 83-97.

WANLESS, H. R., 1939, Pennsylvanian correlations in the Eastern Interior and Appalachian coal fields: Washington, D. C., Geological Society of America Special Paper 17, 130 p.

WANLESS, H. R., 1975, Appalachian region, *in* McKee, E. D., and Crosby, E. J., coordinators, Paleotectonic investigations of the Pennsylvanian System in the United States: Part I. Introduction and regional analyses of the Pennsylvanian System: Washington, D. C., United States Geological Survey Professional Paper 853, p. 17-62.

WHITE, D., 1904, Deposition of the Appalachian Pottsville: Geological Society of America Bulletin, v. 15, p. 267-282.

WHITE, I. C., 1891, Stratigraphy of the bituminous coal field in Pennsylvania, Ohio and West Virginia: Washington, D. C., United States Geological Survey Bulletin 65, 212 p.

WILLARD, D. A., AND KLEIN, G. DeV., 1990, Tectonic subsidence history of the central Appalachian basin and its influence on Pennsylvanian coal deposition: Southeastern Geology, v. 30, p. 217-239.

WRIGHTSTONE, G. W., 1985, The stratigraphy and depositional environment of the Ravencliff formation in McDowell and Wyoming counties, West Virginia: Unpublished M. S. Thesis, West Virginia University, Morgantown, 99 p.

YOUSE, A. C., 1964, Gas producing zones of the Greenbrier (Mississippian) Limestone, southern West Virginia and eastern Kentucky: American Association of Petroleum Geologists Bulletin, v. 48, p. 465-486.

EVIDENCE FOR ORBITALLY-DRIVEN SEDIMENTARY CYCLES IN THE DEVONIAN CATSKILL DELTA COMPLEX

JAY VAN TASSELL

Science Department, Eastern Oregon State College, La Grande, Oregon 97850

ABSTRACT: Recent computer climate modeling has revealed possible mechanisms which allow Milankovitch orbital parameters to influence climate and sedimentation in tropical areas with monsoonal climates such as the setting of the Devonian Catskill Delta complex of the Appalachians. There is evidence that the shoreline, shelf, slope, and basin sediments of the Catskill Delta complex contain sedimentary cycles with depositional periods of approximately 1-3 ka, 20-40 ka, 100 ka, 400 ka, and, possibly, 1.3, 2.0 and 3.5 Ma, which may be related to orbital precession, obliquity and eccentricity cycles, plus harmonics of these forcing periods. Preliminary correlations suggest that at least some of these rhythms may be traceable throughout the Appalachian basin and some may be of global extent. Until more complete age-dating and improved stratigraphic correlation becomes available, the synchronism of these events will remain unproven.

INTRODUCTION

Late Quaternary glacio-eustatic and climatic fluctuations have been convincingly correlated with the Milankovitch parameters (precession, obliquity and eccentricity) of the Earth's orbit (Hays and others, 1976; Berger and others, 1984). It therefore seems reasonable to suggest that sea level may have responded to the same orbital forcing mechanisms during previous glacial episodes in the Earth's history, for example, during the Carboniferous Period (Crowell, 1978; Heckel, 1986; Collier and others, 1990). Is it also reasonable to suggest that orbital variations may have produced climate and sea-level fluctuations and influenced deposition during non-glacial periods, as proposed by Van Tassell (1987) for the Upper Devonian Catskill Delta complex (Figs. 1, 2) of the Appalachians?

Crowley and others (1992) describe records from early Pleistocene Epoch, Pliocene Epoch, Miocene Epoch, Cretaceous Period and Triassic Period which indicate that orbitally-induced 100-ka and sometimes 400-ka climate fluctuations were occurring during times when there is either little evidence for the presence of extensive ice sheets or the ice sheets were fluctuating at other dominant periods. Crowley and others' (1992) energy balance climate models indicate that these proxy records, many of which are from tropical regions, may be related to both 100-ka and 400-ka temperature increases produced by the twice yearly passage of the sun across the equator and the seasonal timing of perihelion over low-latitude land areas involved in monsoonal fluctuations. These climate changes could also change atmospheric circulation patterns, producing cyclic fluctuations in sediment yield such as those described for Miocene fan delta systems by De Boer and others (1991) and for Devonian fluvial systems in East Greenland by Olsen (1990).

These studies suggest that the presence of orbitally-driven cycles might be expected in the Devonian Catskill Delta complex, which was deposited in the Appalachian foreland basin during the middle and late Devonian Period at paleolatitudes between 15° and 20° S in a seasonally wet and dry (monsoonal)

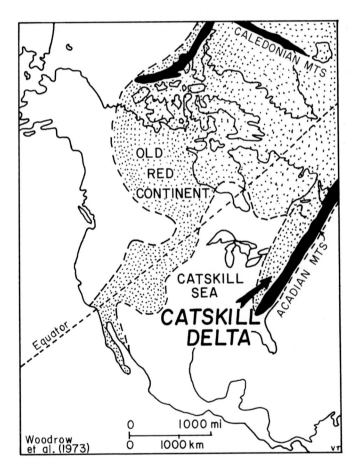

FIG. 1.—Middle to late Devonian paleogeography, based on Woodrow and others (1973), showing the location of the Catskill Delta complex in the monsoonal belt at paleolatitudes between 15° and 10°S.

climate (Heckel and Witzke, 1979; Woodrow, 1985; Kent and Opdyke, 1985; Miller and Kent, 1986a, b). Recent calculations by Berger and others, (1992) indicate that during the Devonian Period the main periods of precession (approximately 17 and 20

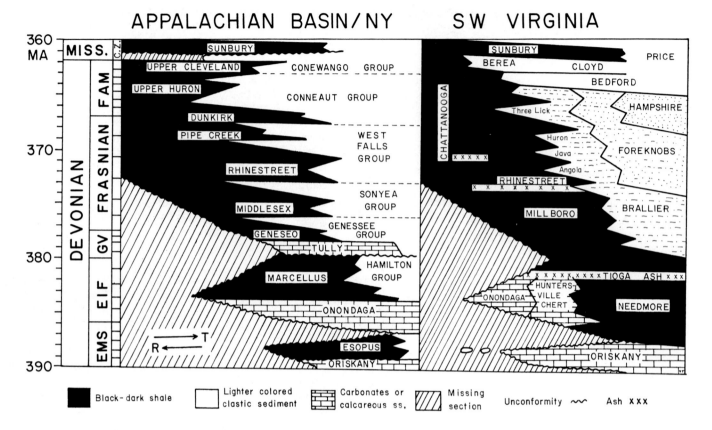

FIG. 2.—Overall stratigraphy of the Catskill Delta complex, based on Ettensohn (1984), House (1985), and Dennison and others (1992). Note the 1-2 Ma cycles in the Givetian, Frasnian and Famennian deposits of southwestern Virginia. The repetition of tongues of dark shales in the Appalachian basin has been cited as possible evidence of depositional cycles with periods of 2-4 Ma (Bayer and McGhee, 1985; Ettensohn, 1985).

ka) and obliquity (approximately 33 and 41 ka) were shorter than during the present day, while eccentricity periods (95, 123, 131 and 400 ka) may be assumed to have been more or less constant over the last half-billion years. Berger's (1976) calculations also indicate the possibility of periodicities of 1.3, 2.0 and 3.5 Ma, but these have been documented only in rare cases (Olsen, 1986).

Elrick and others (1991), attempting to explain the 0.6- to 2.85-ka periodicities of alternating Mississippian distal storm deposits and quiet-water sediments, described nonlinear climate-oscillator models which suggest that short-term (approximately 2 ka) paleoclimate periodicities recorded in Antarctic ice cores and Quaternary deep-sea sediments may have been produced by one of several harmonics of the precession and obliquity orbital forcing periods. Evidence from Permian evaporite varves and the 14C content of Holocene tree rings suggests that periodicities of less than 3 ka may also be attributed to changes in solar activity which affect global temperatures and cause climate changes.

These studies indicate that solar variations and orbitally-induced climate, rainfall and sea-level variations during the deposition of the Catskill Delta sequence may have produced sedimentary cycles with depositional periods of approximately 1-3 ka, 20-40 ka, 100 ka, and, possibly, 1.3, 2.0 and 3.5 ka.

CATSKILL DELTA DEPOSITIONAL CYCLES

Cycles of coarse and fine units in the nonmarine portion of the Catskill Delta were recognized early in this century by Barrell (1913) and Chadwick (1936) and were later interpreted as the result of stream channels shifting across flood plains by Allen and Friend (1968) and Walker and Harms (1971). Early studies of fluvial fining-upward cycles have been succeeded by detailed quantitative reconstructions of Catskill river systems. Bridge and Gordon (1985), Gordon and Bridge (1987) and Willis and Bridge (1988) have shown that Catskill river deposits include grey to reddish-grey sandstone bodies that are interpreted as the results of aggrading single-channel rivers that migrated laterally within well-defined channel belts. Each sandstone body consists of one or more stories defined by a 'major' basal erosion surface and lateral-accretion surfaces that extend from the base to near the top of the storey. The river deposits also include interbedded sandstone, siltstone, mudstone and shale organized into meter-scale fining-upward bedsets which represent the deposits of overbank floods on a floodplain (Fig. 3). The spatial distribution of sandstone bodies (channel-belt deposits) relative to sandstone-mudstone deposits (overbank deposits) can be explained by periodic channel-belt

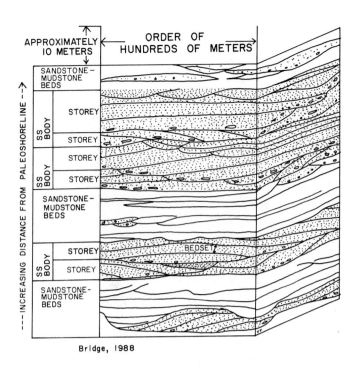

FIG. 3.—Schematic bedding geometry of sandstone bodies (stippled) and sandstone-mudstone beds in Devonian fluvial deposits of New York State, based on Bridge (1988). The spatial distribution of sandstone bodies (channel-belt deposits) relative to sandstone-mudstone deposits (overbank deposits) appears to be related to channel-belt avulsion at intervals of 10^2–10^3 years during intervals of net floodplain deposition.

avulsion during net floodplain deposition (Gordon and Bridge, 1987).

The Devonian Catskill Delta complex also contains larger-scale depositional cycles which are characterized by a regular, repeated and predictable display of lithofacies and biofacies, and a rather consistent thickness of strata per cycle. Walker and Harms (1971) and Walker (1971) recognized marine-nonmarine transition cycles ("Irish Valley motifs") in the Irish Valley Member at the base of the upper Devonian Catskill Formation in central Pennsylvania (Fig. 4). These depositional cycles record marine transgression, nondeposition, winnowing and bioturbation; a marine shoaling phase which is dominantly shale; and, finally, progradation of a quiet muddy shoreline and the development of an alluvial plain. These cycles range in thickness from 4 to 34 m.

Somewhat similar cycles occur in the shallow-marine shelf strata of the Upper Devonian West Falls Group of south central New York (Fig. 5). The shelf strata of the West Falls Group, described by Craft and Bridge (1987) include centimeter- to decimeter-thick beds of sandstone and siltstone interbedded with mudstone. These beds are typically arranged into larger-scale coarsening-upward sequences which are typically 15 to 30 m thick and contain an upward increase in the abundance of hummocky cross-stratification, plant debris, coquinites, claystone interclasts and the number of certain bivalves. These sequences

are overlain somewhat abruptly by thin bioturbated sandstones and thickening mudstones, making these sedimentary cycles markedly asymmetrical (cf. Woodrow and Isley, 1983).

Van Tassell (1987, 1988) has shown that the upper Devonian (Frasnian-Famennian) Brallier, Scherr and Foreknobs Formations of Virginia and West Virginia show a spectrum of cycle patterns ranging from fining-upward to symmetrical to coarsening-upward (Fig. 5). These cycles contain varying percentages of basin and slope shales; turbidite siltstones; reddish "slope" shales possibly deposited offshore during periods of major river flooding; lower shoreface, parallel-laminated siltstones; nearshore (upper shoreface) ripple-bedded, cross-bedded and planar-bedded sandstones with flute casts, mudchips and plant fragments; and low-angle cross-stratified sandstones which may represent beach deposits. Overall, the percentage of shallow-water facies in the cycles increases upward, recording the upward transition for the turbidite slope environments of the Brallier Formation to the shelf and shoreline environments of the Foreknobs Formation (which, in turn, is overlain by fluvial deposits with prominent fining-upward cycles).

In contrast to the cycles of clastic deposits described above, the late Eifelian to middle Givetian Hamilton Group of New York includes thin carbonate deposits (Brett and Baird, 1985, 1986). This 90- to 580-m eastward thickening sequence grades westward from alluvial and coastal deposits into marine siltstones and sandstones, and then into basinal medium to dark gray mudrock, and finally into light gray calcareous mudstones and argillaceous limestones in the west (Fig. 6). Recent work suggests that these limestone members were deposited in turbulent shallow-shelf settings. The limestones overlie unconformity surfaces and may record the initial periods of transgressions (Brett, 1992, pers. commun.). Small-scale (0.5-3 m) upward-coarsening cycles appear to be recorded in the form of thin (0.5-1.0 m) alternations of fossiliferous, concretionary carbonate beds, each about 20 to 30 cm thick, alternating with noncalcareous mudstone. Larger, distinctly cyclic packages up to tens of meters thick also form a dominant motif in the Hamilton Group. These major cycles are similar in nature to the smaller cycles but are distinctly thicker, record a much broader range of facies from base to top, and are more laterally extensive. These cycles grade eastward across facies boundaries from black clay shale-mudstone-packstone and grainstone cycles in western New York through transitional calcareous, silty mudstones into distinct upward-coarsening shale-sandstone cycles in areas closest to the delta margin. All of the cycles, in turn, are superimposed on three larger transgressive-regressive cycles recognized by Johnson and others (1985).

The discussion above illustrates that depositional cycles occur in all of the sedimentary environments of the Catskill Delta complex and, in some cases, can be traced across facies boundaries from one environment to another. Another unusual characteristic of the Catskill Delta sequence is also evident: it is composed of microrhythms often sharing microfacies elements of the larger rhythms. House (1985), describing mid-Paleozoic ammonoid evolutionary events, noted that this characteristic is

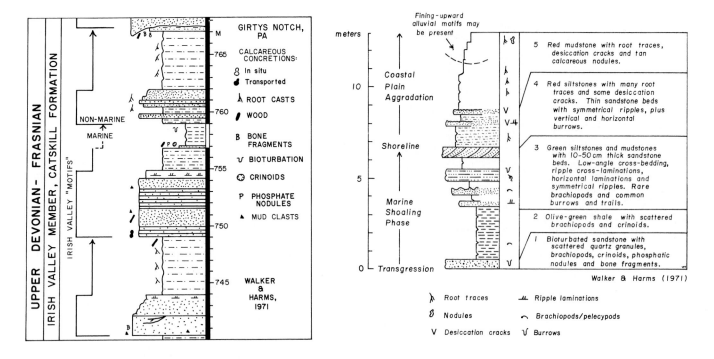

FIG. 4.—Cycles of interbedded marine and nonmarine facies in the Irish Valley Member of the Catskill Formation at Girtys Notch, Pennsylvania, and detailed interpretation of an idealized "Irish Valley motif", interpreted to represent one of the oscillations of a prograding muddy shoreline in the central Pennsylvania area (Walker and Harms, 1971). These cycles average 100 ka in duration.

shared by some of the major sedimentary rhythms of the Devonian, Carboniferous and Jurassic and pointed out that this is difficult to explain in wholly eustatic terms. House (1985) concluded that a more likely explanation is a correlation of such microrhythms with Milankovitch-type cycles and related spread of cold conditions, CO_2 variations, changes in ocean currents and precipitation patterns, and other factors.

CYCLE PERIODICITIES

Precise age-dating is critical in order to identify sedimentary cycles produced by climatic and eustatic variations which may have been induced by orbital and solar variations via atmospheric and oceanic coupling. At present, however, the 290.0±0.5 Ma Tioga Ash Bed near Lewisburg, Pennsylvania (Roden and others, 1990) is the only precisely dated unit in the Appalachian basin sequence (Dennison and others, 1992). Rough estimates of the depositional periods of Catskill Delta units and the cycles they contain can be obtained by calculations based on the stratigraphic positions and thickness of the units within the boundaries of the Devonian stages dated by Harland and others (1982, 1990), but these estimates are only approximate because they are based on an assumption that sedimentation rates remained approximately constant during the deposition of the sequence. It is important to note that the error bars on the dates for the boundaries of the Devonian stages are on the order of ±10-15 Ma, adding additional uncertainty. This is highlighted by the observation that the date for the Tioga Ash Bed is about

9-million-years older than the age indicated by the Harland and others (1990) time scale based on the stratigraphic position of the Tioga Ash Bed.

Lack of precise age dating is thus the biggest obstacle in proving that orbital variations had a major influence on the Catskill Delta sequence. The estimates of cycle periodicities presented in this paper must be viewed as preliminary until more exact age-dating becomes available.

HIERARCHY OF CATSKILL DELTA CYCLES

There appears to be a hierarchy of cycles present in the Catskill Delta complex. Several lines of evidence indicate that storm activity at intervals of 10^2-10^3 years had a major influence on deposition on the Catskill Delta margin. This hypothesis is supported by counts of turbidite beds (Lundegard and others, 1985; Van Tassell, 1987) and shelf storm beds (Duke, 1985; Craft and Bridge, 1987). Studies of fluvial deposits (Schumm, 1968; Woodrow and others, 1973; Willis and Bridge, 1988) also indicate the presence of cycles with similar periodicities which may have been produced by lateral shifting of river channels, delta lobes and other depositional features during storms. This remains to be documented, however.

Longer-period cycles are poorly documented in Catskill Delta alluvial deposits, but are common in Catskill Delta shelf, slope and basin deposits. Calculations of the periodicities of these cycles repeat the same theme: (1) the 25 prograding muddy shoreline motifs in the Irish Valley Member of the Catskill

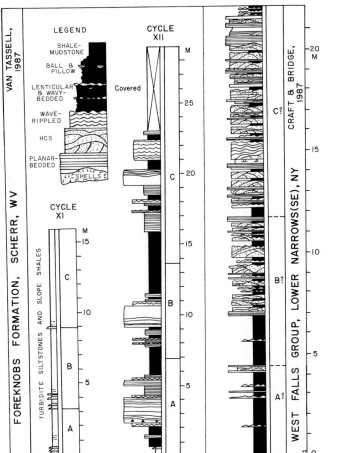

FIG. 5.—Representative depositional cycles from the Foreknobs Formation of West Virginia (Van Tassell, 1987) and the West Falls Group of New York (Craft and Bridge, 1987). Each cycle has an estimated depositional period of approximately 10^3-10^4 ka. Note the variations in cycle patterns from fining-upward to coarsening-upward and the apparent subdivision of the larger cycles into smaller packets (A, B, C).

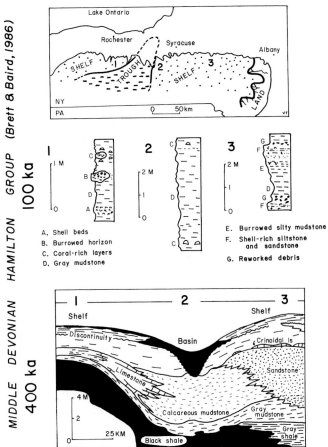

FIG. 6.—Regional stratigraphy of the Upper Hamilton Group (Brett and Baird, 1986). Note the lateral variation of the 100 ka cycles, including cycles with numerous thin limestone beds in the Genesse Valley area of New York (1), cycles dominated by mudstone in the Finger Lakes (Cayuga Valley) area of west-central New York (2), and the numerous upward regressive hemicycles dominated by siltstones in the central New York (Tully Valley) area (3). A representative Hamilton Group 400 ka Cycle, the Centerfield Limestone and equivalent Chenago Sandstone, in western, west-central, and central New York clearly shows the lateral variation of facies within cycles (Brett and Baird, 1985).

Formation of central Pennsylvania were deposited over an interval of approximately 2.0-2.5 Ma, suggesting that the average depositional period of each of the cycles was approximately 100 ka (Van Tassell, 1987), (2) the 15- to 30-m thick coarsening-upward sequences in the Upper Devonian West Falls Group of central New York, which was deposited over an interval of 2-3 Ma, have average recurrence periods of 10^4-10^5 years (Craft and Bridge, 1987), (3) the approximately 20 cycles recognized in the type section of the Foreknobs Formation in West Virginia, deposited over an interval of 2.0-2.5 Ma, have average durations on the order of 100 ka (Van Tassell, 1987), and (4) careful study of Walker and Sutton's (1967) measurements of turbidites in the Sonyea Group of central New York and Walker's (1971) measurements of turbidites in the Brallier Formation at Shamokin Dam, Pennsylvania and Woodmont, Maryland (Fig. 7) indicates that there are rhythmic patterns of deposition in these cycles with a range of 50 to 200 beds per cycle and an overall average of

approximately 100 beds per cycle. This indicates an average periodicity of approximately 10^4-10^5 years assuming the average intervals of turbidite deposition cited above.

Thus there are several lines of evidence suggesting the presence of depositional cycles with periods ranging from 10^4-10^5 years in the Catskill Delta shelf and slope deposits, with the majority of estimates clustering around 100 ka (6th-order cycles of Busch and Rollins, 1984). Van Tassell (1987) pointed out that many of these cycles appear to be divisible into 3 subcycles (Fig. 5). This may be a hint of smaller-scale subcycles with periodicities of approximately 30 ka.

Brett and Baird (1986) estimated that the small-scale (0.5-3 m) upward-coarsening cycles in the middle Devonian Hamilton Group represent 100-125 ka fluctuations, while the 15 major

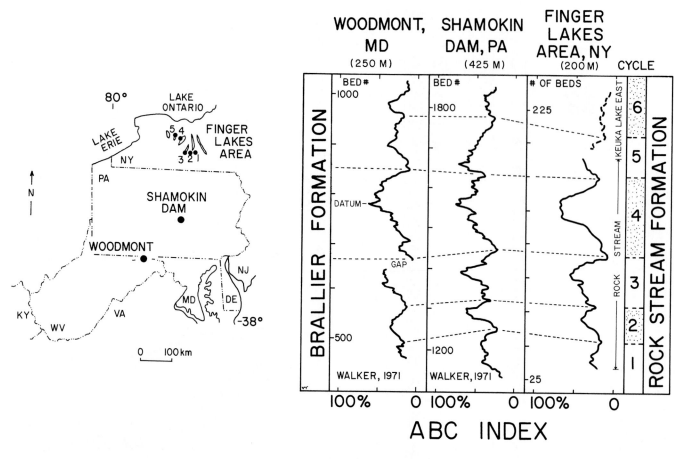

FIG. 7.—Comparison of the long-term cyclic depositional patterns in turbidites of the Rock Stream Formation of the Sonyea Group in the Finger Lakes area of New York (Walker and Sutton, 1967) with the depositional patterns evident in Walker's (1971) measurements of turbidites in the upper Devonian Brallier Formation of Maryland and Pennsylvania. Dashed lines show the boundaries of depositional patterns evident in Walker's (1971) measurements of turbidites in the upper Devonian Brallier Formation of Maryland and Pennsylvania. Dashed lines how the boundaries of depositional cycles with similar patterns; correlation of individual cycles with specific cycles in the other regions has not been established. Each cycle represents, on the average, approximately 100 ka of deposition.

transgressive-regressive (T-R) cycles with thicknesses of up to 10 m now recognized over the 6-7 Ma span of the Hamilton Group record oscillations on the order of 400-500 ka in duration. They are thus very similar to fifth-order allocycles such as the midcontinent cyclothems described by Heckel (1977, 1986). The observation that the Hamilton Group has also been previously subdivided into three fourth-order (0.8-1.5 Ma) cycles by Johnson and others (1985) provides a hint that longer period depositional rhythms may be present in the Catskill Delta succession.

The possibility that there are long-term cycles in the Catskill Delta complex is highly controversial. Bayer and McGhee (1986) hypothesized that the New York Devonian section contained depositional and biological rhythms on the order of 1 conodont zone (approximately 2.3 Ma), the lower resolution of their data, in addition to longer-period cycles. House (1985), focusing on Devonian ammonoid extinction events, described similar rhythmic fluctuations with an average period of approximately 4.3 Ma, although he speculated that the period could be somewhat high if some of the extinction events had not been detected. Following a different line of evidence, Ettensohn (1985) concluded that the Catskill Delta complex is composed of five major cycles of alternating dark shales and coarser clastics throughout much of the Appalachian basin. Four Upper Devonian cycles were completely developed, beginning, in ascending order, with the Geneseo, Middlesex, Rhinestreet and Dunkirk black shales (Fig. 2). Each cycle lasted approximately 2-3 Ma based on the chronology of Harland and others (1982) and, according to Ettensohn (1985), each cycle represents the development of a new peripheral basin within the larger Appalachian basin as a result of tectonism.

Johnson and others (1985) recognized transgressive-regressive (T-R) cycles in their Devonian sea-level curve for Euramerica, which was based in part on the Devonian sequence in New York. These cycles ranged from 1-4 conodont zones in length (approximately 0.5-2 Ma). Johnson and others (1985) attributed these 0.5-2 Ma T-R cycles to episodes of mid-plate uplift and submarine volcanism that abruptly reduced the capacity of the Devonian ocean basins.

Recent stratigraphic studies (Dennison and others, 1992)

also indicate the possible presence of cycles with periods of 1-2 Ma in the Catskill Delta sequences of southwestern Virginia (Fig. 2). Van Tassell (1987) pointed out that recognizing the 100-ka cycles in the Catskill Delta succession provides a way to subdivide the larger radiometrically-dated intervals of the complex into smaller subdivisions until better age-dating becomes available. Basically, this involves (1) piecing together a composite stratigraphic column, (2) picking a starting point (in this case, the Frasnian-Famennian boundary) which has been radiometrically dated, and (3) marking in the boundaries of 100-ka cycles, and using these boundaries as indicators of 100-ka increments of time throughout the section. In many respects, this is similar to dividing the approximately 10-Ma duration of the Frasnian sequence (Harland and others, 1990) into 100 equal thickness units because the thickness of the cycles tends to remain relatively constant. In reality, though, it is not strictly valid to assume that each of the sedimentary cycles has a period of exactly 100 ka. Because cycle boundaries are not always clearly marked, this technique involves considerable interpretation and judgement.

Despite these limitations, curves produced by plotting the changes in estimated relative depths during successive 100-ka cycles, effectively filtering out the effects of the 100-ka and shorter cycles, help illustrate the possible longer-period rhythms in the succession more readily (Fig. 8). These preliminary and highly speculative curves, based on measurements of turbidites in the Brallier Formation at Woodmont, Maryland and Shamokin Dam, Pennsylvania by Walker (1971) and measurements of the Scherr Formation at Briery Gap Run, West Virginia and in the Foreknobs Formation at Scherr, West Virginia (Van Tassell, 1987, 1988) indicate a strong rhythm with an average period of approximately 1.2 Ma in the sequence. Deviations on either side of this 1.2-Ma cycle highlight the presence of shorter-period rhythms with average periods of 400 ka. In addition, a less obvious cycle with a period of approximately 2 Ma is evident if the asymmetry of individual 1.2-Ma cycles is carefully examined. It must be emphasized, however, that these results are highly speculative until better age-dating of the sequence becomes available.

The evidence cited above suggests that there are strong sedimentary rhythms in the Catskill Delta complex with depositional periods similar to the dominant approximately 100-ka, 400-ka, 1.3- and 2-Ma eccentricity periods and weaker superimposed rhythms with periodicities similar to orbital obliquity and precession periods. There is also evidence of Catskill Delta storm activity with recurrence intervals in the range of known climatic variations triggered by variations due to the harmonics of the precession and obliquity cycles or solar activity cycles. The evidence is not sufficient, however, to prove that these sedimentary cycles were in fact triggered by orbital variations.

TENTATIVE CYCLE CORRELATIONS

Lateral tracing of Catskill Delta sedimentary-cycle patterns has just begun. Prior to the surge of interest in facies models in the late 1960's and early 1970's, it was often assumed that units in the upper Devonian Catskill Delta sequence correlated across long distances (e.g., Dyson, 1963; Frakes, 1967). Since the 1970's, however, sedimentologists (e.g., Craft and Bridge, 1987) have criticized the long-distance correlations of Catskill Delta stratigraphers because biostratigraphic correlation is still primitive and lithostratigraphic correlation is difficult because distinctive beds are lacking and the structural deformation of the area. This does not mean that lateral correlation is impossible, however. One 100-ka cycle, the Minnehaha Springs Member of the Scherr Formation, has been correlated several hundred kilometers across several lobes of the Catskill Delta (Lyke, 1986) and more than 100 km across the outcrop belt and into the Appalachian basin (Barrell and Dennison, 1986). Filer (1991) has recently correlated distal coarse-grained tongues of sediment, including parts of the Foreknobs Formation, along the Catskill Delta margin and in the Appalachian basin by tracing the signature of these sediment tongues on geophysical logs of wells penetrating the Famennian Huron (Dunkirk) black shales and the dark shales of the Frasnian Angola, Java and Rhinestreet Formations. This demonstrates the possible synchronism of small scale progradational events originating off widely separated depositional systems along the Catskill Delta margin.

Brett and Baird (1986) have been able to trace some 100- to 125-ka cycles in the Hamilton Group from one side of their study area across the basin to the other side, but other 100- to 125-ka cycles in their west and central New York study area are not so widespread. Brett and Baird (1986) suggested that the widespread cycles may be the result of allocyclic processes, either sea level or climate. In the case of the cycles of more restricted extent, autocyclic processes such as episodic subsidence, tectonic adjustments of the basin margin, or, possibly, local delta progradation cannot be ruled out. In contrast to the 100-125 ka cycles, Brett and Baird (1986) have correlated all the major 400-500-ka shale-limestone cycles in the Hamilton Group in western New York across the basin to the eastern side of the basin closer to the delta margin. At least several of these cycles can be tentatively correlated with T-R cycles in the rest of the Appalachian basin, in the Cordilleran region, Europe, Morocco and elsewhere using the conodont stratigraphy of Johnson and others (1985).

The bases of the Frasnian Middlesex, Rhinestreet and Pipe Creek shales, which mark the bases of major cycles in the Catskill Delta complex, have been tentatively correlated with major breaks in other successions throughout Euramerica on the basis of conodont biostratigraphy (Johnson and others, 1985). The major sedimentary rhythms and Frasnian extinction events in the New York sequence have been correlated with marked environmental changes and discontinuous sedimentation patterns in Belgium, France and Canada (Fig. 9), plus a rather similar pattern and an abrupt termination by drowning of the stromatoporoid reef complexes in western Australia (House, 1985). The synchronism of these events remains untested, however, since conodont and ammonoid dating are both incomplete. We seem to be tantalizingly close to being able to say that

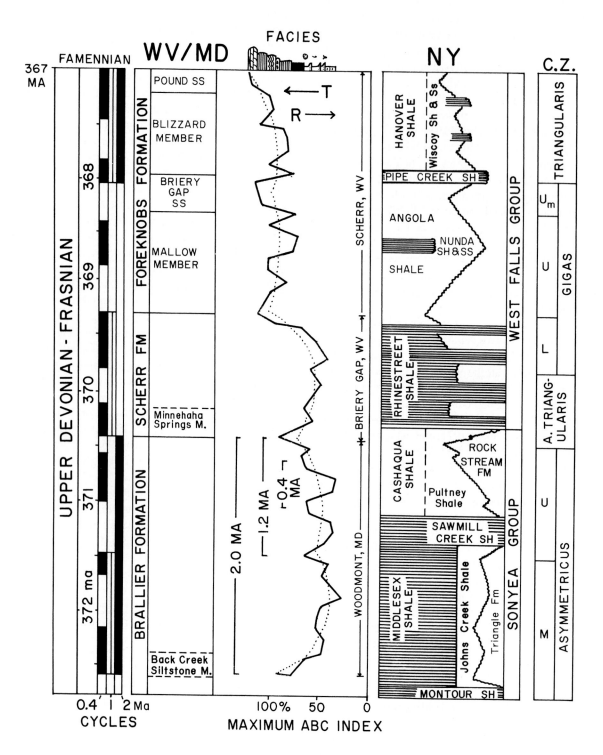

Fig. 8.—Long-term depositional cycles in the Frasnian Catskill Delta complex based on the hypothesis that 100-ka cycles are a major feature of the depositional sequence. Note the approximately 1.2-Ma and 400-ka variations. A suggestion of a 2-Ma rhythm can be seen if the asymmetry of individual 1.2-Ma cycles is carefully examined. The stratigraphy of the New York succession is modified from Rickard (1975). The age of the Frasnian-Famennian boundary is from Harland and others (1982, 1990).

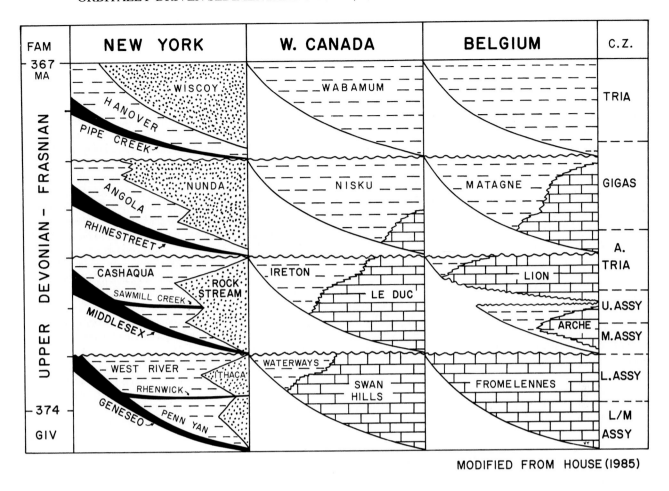

MODIFIED FROM HOUSE (1985)

FIG. 9.—Possible global correlations of New York Catskill Delta major transgressive-regressive cycles with those in sequences in Belgium anu western Canada, modified from House (1985). Note the suggestion of 1- and 2-Ma cycles and the overall progradation of clastic sediments in the New York area as carbonate sedimentation declines in the western Canada and Belgium areas. Due to the uncertainty in conodont dating techniques, the synchronism of these events remains to be proven.

at least some of the depositional cycles in the Catskill Delta complex can be traced globally and are thus related to global climate or sea level changes, but the evidence is still too tentative to confirm this.

CONCLUSIONS

Global climate modeling has provided several new mechanisms by which sedimentary cycles can be produced by orbital forcing in tropical regions during periods without glaciation. These models indicate that the Catskill Delta setting was very conducive for the formation of orbitally- induced sedimentary cycles.

There appears to be a hierarchy of cycles in the Catskill Delta complex which matches known Milankovitch periodicities, including previously unknown harmonics in the 1- to 3-ka range and longer-period cycles in the 1- to 4-Ma range, but we cannot say for sure whether or not Milankovitch variations actually produced these variations until better absolute dating becomes available.

Sedimentary cycles are especially prominent in the shelf, slope and basinal environments of the Catskill Delta complex. Quantitative models such as those presented by Koltermann and Gorelick (1992) and Collier and others (1990) may help us better understand the complexity of the signature of Milankovitch variations in Catskill Delta alluvial fan and alluvial plain environments.

At least some of the cycles in the Catskill Delta sequence appear to be traceable across the Appalachian basin area, if not globally. These correlations remain tentative, however, until the Devonian conodont and ammonoid stratigraphy is further refined and tied to radiometric dating more completely. Many Catskill Delta cycle patterns can be traced across facies boundaries, emphasizing the difference between lithostratigraphic and cyclostratigraphic correlations.

It is very likely that the interaction of eustatic sea-level changes, climate changes, and tectonism, along with local lateral facies shifting, combined to produce the complex sedimentary patterns present in the Catskill Delta complex. Because these factors are interrelated and interdependent, greatly improved age dating, improved stratigraphic correlation, and statistical analyses such as time-series analysis will be necessary in order to sort out the individual effects of each factor. After a century of research, our study of the Catskill Delta sequence has just begun.

ACKNOWLEDGMENTS

This paper has been greatly improved by the reviews of John Dennison, William Sevon, Frank Ettensohn, and Jonathan Filer. The author has benefited also from the criticism and suggestions for improvement made by Poppe de Boer, David Smith, John Bridge, Carlton Brett, Donald Woodrow, Roger Walker, Edward Cotter, Edward Clifton, Robert Suchecki, Charles Harris, Jim Smith, and George McGhee, Jr.

REFERENCES

ALLEN, J. R. L., AND FRIEND, P. F., 1968, Deposition of the Catskill facies, Appalachian regions, with notes on some other Old Red Sandstone basins, in Klein, G. Dev., ed., Late Paleozoic and Mesozoic Continental Sedimentation, Northeastern North America: Boulder, Geological Society of America Special Paper 160, p. 21-74.

BARRELL, J., 1913, The Upper Devonian delta of the Appalachian geosyncline: Part 1, The delta and its relations to the interior sea: American Journal of Science, v. 36, p. 429-472.

BARRELL, S. M., AND DENNISON, J. M., 1986, Northwest-southeast stratigraphic cross section of Devonian Catskill Delta in east-central West Virginia and adjacent Virginia: Morgantown, Appalachian Basin Industrial Associates Fall Meeting Program, v. 11, p. 7-32.

BAYER, U., AND MCGHEE, G. R. JR., 1986, Cyclic patterns in the Paleozoic and Mesozoic: Implications for time scale calibrations: Paleoceanography, v. 1, p. 383-402.

BERGER, A. L., 1976, Obliquity and precession for the last 5,000,000 years: Astronomy and Astrophysics, v. 51, p. 127-135.

BERGER, A., IMBRIE, J., HAYS, J., KUKLA, G., AND SALTZMAN, B., 1984, Milankovitch and Climate: Understanding the Response to Orbital Forcing: Boston, D. Reidel, 510 p.

BERGER, A., LOUTRE, M. F., AND LASKAR, J., 1992, Stability of the astronomical frequencies over the Earth's history for paleoclimate studies: Science, v. 255, p. 560-566.

BRETT, C. E., AND BAIRD, G. C., 1985, Carbonate-shale cycles in the middle Devonian of New York: An evaluation of models for the origin of limestones in terrigenous shelf sequences: Geology, v. 13, p. 324-327.

BRETT, C. E., AND BAIRD, G. C., 1986, Symmetrical and upward shoaling cycles in the Middle Devonian of New York State and their implications for the punctuated aggradational cycle hypothesis: Paleoceanography, v. 1, p. 431-445.

BRIDGE, J. S., 1988, Devonian fluvial deposits of the western Catskill region of New York State: Tulsa, 1988 Annual Field Trip Guidebook, SEPM Eastern Section, 23 p.

BRIDGE, J. S., AND GORDON, E. A., 1985, Quantitative interpretation of ancient river systems in the Oneonta Formation, Catskill magnafacies, in Woodrow, D. L., and Sevon, W. D., eds., The Catskill Delta: Boulder, Geological Society of America Special Paper 201, p. 163-182.

BUSCH, R. M., AND ROLLINS, H. B., 1984, Correlation of Carboniferous strata using a hierarchy of transgressive-regressive units: Geology, v. 12, p. 471-474.

CHADWICK, G. H., 1936, History and value of the name "Catskill" in geology: New York State Museum Bulletin, v. 307, 116 p.

COLLIER, R. E. L., LEEDER, M. R., AND MAYNARD, J. R., 1990, Transgressions and regressions: A model for the influence of tectonic subsidence, deposition and eustacy, with application to Quaternary and Carboniferous examples: Geological Magazine, v. 127, p. 117-128.

CRAFT, J. H., AND BRIDGE, J. S., 1987, Shallow-marine sedimentary processes in the late Devonian Catskill Sea, New York State: Geological Society of America Bulletin, v. 98, p. 338-355.

CROWELL, J. C., 1978, Gondwanan glaciation, cyclothems, continental positioning, and climate change: American Journal of Science, v. 278, p. 1345-1372.

CROWLEY, T. J., KIM, K. Y, MENGEL, J. G., AND SHORT, D. A., 1992, Modeling 100,000-year climate fluctuations in Pre-Pleistocene time series: Science, v. 255, p. 705-707.

DE BOER, P. L., PRAGT, J. S. J., AND OOST, A. P., 1991, Vertically persistent sedimentary facies bound aries along growth anticlines in the thrust-sheet-top South Pyrenean Tremp-Graus foreland basin: Basin Research, v. 3. p 63-78.

DENNISON, J. M., BAMBACH, R. K., DOROBEK, S. L., FILER, J. K., AND SHELL, J. A., 1992, Silurian and Devonian unconformities in Southwestern Virginia: Chapel Hill, University of North Carolina at Chapel Hill Geologic Guidebook 1, 233 p.

DUKE, W. L., 1985, Hummocky cross-stratification, tropical hurricanes, and intense winter storms: Sedimentology, v. 32, p. 167-194.

DYSON, J. L., 1963, Geology and Mineral Resources of the New Bloomfield Quadrangle: Harrisburg, Pennsylvania Geological Survey Atlas 137A, 63 p.

ELRICK, M., READ, J. F., AND CORUH, C., 1991, Short- term paleoclimatic fluctuations expressed in Lower Mississippian ramp-slope deposits, southwestern Montana: Geology, v. 19, p. 799-802.

ETTENSOHN, F. R., 1984, The nature and effects of water depth during deposition of the Devonian-Mississippian black shale: Lexington, University of Kentucky Institute for Mining and Minerals Research, 1984 Eastern Oil Shale Symposium Proceedings, p. 333-346.

ETTENSOHN, F. R., 1985, Controls on development of Catskill Delta complex basin-facies, in Woodrow, D. L., and Sevon, W. D., eds., The Catskill Delta: Boulder, Geological Society of America Special Paper 201, p. 65-77.

FILER, J. K., 1991, Small scale sequences with a portion of the Upper Devonian clastic wedge, Appalachian basin- origins and implications (abs.): Geological Society of America, Abstracts with Programs, v. 23, no. 1, p. 28.

FRAKES, L. A., 1967, Stratigraphy of the Devonian Trimmers Rock in Eastern Pennsylvania: Pennsylvania Geological Survey Bulletin GV51, 208 p.

GORDON, E. A., AND BRIDGE, J. S., 1987, Evolution of Catskill (Upper Devonian) river systems: Journal of Sedimentary Petrology, v. 57, p. 234-249.

HARLAND, W. B., ARMSTRONG, R. L., COX, A. V., CRAIG, L. E., SMITH, A. G., AND SMITH, D. G., 1990, A Geologic Time Scale 1989: Cambridge, Cambridge University Press, 263 p.

HARLAND, W. B., COX, A. V., LLWELLYN, R. G., PICTON, C. A. G., SMITH, A. G., AND WALTERS, R., 1982, A Geologic Time Scale: Cambridge, Cambridge University Press, 131 p.

HAYS, J. D., IMBRIE, J., AND SHACKLETON, N. J., 1976, Variations in the earth's orbit: pacemaker of the ice ages: Science, v. 194, p. 1121-1132.

HECKEL. P. H., 1977, Origin of phosphatic black shale facies in Pennsylvanian cyclothems in midcontinent North America: American Association of Petroleum Geologists Bulletin, v. 61, p. 1045-1068.

HECKEL, P. H., 1986, Sea-level curve for Pennsylvanian eustatic marine transgressive-regressive depositional cycles along midcontinent outcrop belt, North America: Geology, v. 14, p. 330-334.

HECKEL, P. H., AND WITZKE, B. J., 1979, Devonian world paleogeography determined from the distribution of carbonates and related lithic paleoclimate indicators, in House, M. R., Scrutton, C. T., and Bassett, M. G., eds., The Devonian System: Tulsa, Special Papers in Paleontology 23, p. 99-123.

HOUSE, M. R., 1985, Correlation of mid-Paleozoic ammonoid evolutionary events with global sedimentary perturbations: Nature, v. 313, p. 17-22.

JOHNSON, J. G., KLAPPER, G., AND SANDBERG, C. A., 1985, Devonian eustatic fluctuations in Euramerica: Geological Society of America Bulletin, v. 96, p. 567- 597.

KENT, D. V., AND OPDYKE, N. D., 1985, Multi- component magnetizations from the Mississippian Mauch Chunk Formation of the central Appalachians and their tectonic implications: Journal of Geophysical Research, v. 90, p. 5371-5383.

KOLTERMANN, C. E., AND GORELICK, S. M., 1992, Paleoclimate signature in terrestrial flood deposits: Science, v. 256, p. 1775-1782.

LUNDEGARD, P. D., SAMUELS, N. D., AND PRYOR, W. A., 1985, Upper Devonian turbidite sequence, central and southern Appalachian basin: contrasts with submarine fan deposits, in Woodrow, D. L., and Sevon, W. D., eds., The Catskill Delta: Boulder, Geological Society of American Special Paper 201, p. 107-121.

LYKE, W. L., 1986, The stratigraphy, paleogeography, depositional environment, faunal communities, and general petrology of the Minnehaha Member of the Scherr Formation, an upper Devonian turbidite sequence, central Appalachians: Southeastern Geology, v. 26, p. 173-192.

MILLER, J. D., AND KENT, D. V., 1986a, Paleomagnetism of the upper Devonian Catskill Formation from the southern limb of the Pennsylvania salient: Possible evidence of oroclinal rotation: Geophysical Research Letters, v. 13, p. 1173-1176.

MILLER, J. D., AND KENT, D. V., 1986b, Synfolding and prefolding magnetizations in the upper Devonian Catskill Formation of Eastern Pennsylvania: Implications for the tectonic history of Acadia: Journal of Geophysical Research, v. 91, p. 12791-12803.

OLSEN, H., 1990, Astronomical forcing of meandering river behavior: Milankovitch cycles in Devonian of East Greenland: Palaeogeography, Palaeoclimatology and Palaeoecology, v. 79, p. 99-115.

OLSEN, P. E., 1986, A 40-million-year lake record of Early Mesozoic orbital climate forcing: Science, v. 234, p. 842-848.

RICKARD, L. V., 1975, Correlation of the Silurian and Devonian rocks in New York State: New York State Museum and Science Service, Geological Survey Map and Chart Series No. 4.

RODEN, M. K., PARRISH, R. R., AND MILLER, D. S., 1990, The absolute age of the Eifelian Tioga Ash Bed, Pennsylvania: Journal of Geology, v. 98, p. 282-285.

SCHUMM, S. A., 1968, Speculations concerning paleo-hydrologic controls of terrestrial sedimentation: Geological Society of America Bulletin, v. 79, p. 1573-1578.

VAN TASSELL, J., 1987, Upper Devonian Catskill Delta margin cyclic sedimentation: Brallier, Scherr, and Foreknobs Formations of Virginia and West Virginia: Geological Society of America Bulletin, v. 99, p. 414-426.

VAN TASSELL, J., 1988, Upper Devonian Catskill Delta Milankovitch cycles, in Dennison, J. M., ed., Geologic Field Guide, Devonian Delta, East-Central West Virginia and Adjacent Virginia: Charleston, Appalachian Geological Society, p. 77-84.

WALKER, R. G., 1971, Non-deltaic depositional environments in the Catskill clastic wedge (Upper Devonian) of central Pennsylvania: Geological Society of America Bulletin, v. 82, p. 1305-1326.

WALKER, R. G., AND HARMS, J. C., 1971, The "Catskill Delta", a prograding muddy shoreline in central Pennsylvania: Journal of Geology, v. 79, p. 218-299.

WALKER, R. G., AND SUTTON, R. G., 1967, Quantitative analysis of turbidites in the Upper Devonian Sonyea Group, New York: Journal of Sedimentary Petrology, v. 37, p. 1012-1022.

WILLIS, B. J., AND BRIDGE, J. S., 1988, Evolution of Catskill river systems, New York State: Calgary, Proceedings of the 2nd International Symposium on the Devonian, p. 85-106.

WOODROW, D. L., 1985, Paleogeography, paleoclimate, and sedimentary processes of the late Devonian Catskill Delta, in Woodrow, D. L., and Sevon, W. D., eds., The Catskill Delta: Boulder, Geological Society of America Special Paper 201, p. 51-64.

WOODROW, D. L., FLETCHER, F. W., AND AHRNSBRAK, W. F., 1973, Paleogeography and paleoclimate at the deposition sites of the Devonian Catskill and Old Red facies: Geological Society of America Bulletin, v. 84, p. 3051-3063.

WOODROW, D. L., AND ISLEY, A. M., 1983, Facies, topography, and sedimentary processes in the Catskill Sea (Devonian), New York and Pennsylvania: Geological Society of America Bulletin, v. 94, p. 459-470.

HIGH FREQUENCY EUSTATIC AND SILICICLASTIC SEDIMENTATION CYCLES IN A FORELAND BASIN, UPPER DEVONIAN, APPALACHIAN BASIN

JONATHAN K. FILER

Geology Department, The University of North Carolina, Chapel Hill, North Carolina 27599

Abstract: During the Late Devonian, a thick clastic wedge, derived from a tectonically active source area to the east, was deposited in the Appalachian basin. The basinal facies of the wedge is composed of blackish-gray or dark-brown organic carbon-rich shales alternating with gray or greenish-gray non-organic shales. Five large (third-order or approximately 2 Ma or longer) black/gray shale cycles have been previously recognized in the New York outcrop belt, as well as in the subsurface throughout the basin. Eleven newly defined higher frequency basinwide organic/inorganic shale cycles occur within the relatively inorganic portion of one of the larger cycles. These higher frequency lithologic cycles represent the basinal expression of fourth-order (0.1-0.3 Ma) cycles of parasequence scale.

The stratigraphic position of these cycles can be traced into nearershore silty and sandy facies using subsurface gamma-ray logs, revealing a synchronous cyclicity. These correlations show that the eleven cycles can be subdivided into two stacked sets of basinwide progradational parasequence sets. This pattern is confirmed by correlation into the most distal portions of the basin and examination of patterns of thinning and convergence. A previously recognized major transgression occurs at the top of the interval which rapidly initiated a major period of organic-rich shale deposition (Huron/Dunkirk Shales). These stratigraphic patterns are interpreted to result from a series of minor eustatic sea-level cycles during Upper Devonian time.

INTRODUCTION

Discussions of the nature, scales, and causes of cyclic deposition of sediments have a long history in the geologic literature. Various models have been proposed to explain relatively thin (typically 1-10's m), high-frequency cycles which appear to be deposited in lateral continuity over tens to thousands of kilometers. Many of these models have invoked climate-driven eustatic sea-level changes, and include the well known cyclothems (Wanless and Shepard, 1936; Heckel, 1986). Recent models have been proposed for purer carbonate sequences (Goodwin and Anderson, 1985; Read and others, 1986; Goldhammer and others, 1987). In siliciclastic rocks, recent work has interpreted high frequency, siliciclastic cycles in the context of sequence stratigraphy (Van Wagoner and others, 1990; Mitchum and Van Wagoner, 1991). Van Tassell (1987, this volume) has applied the punctuated aggradational cycle model (Goodwin and Anderson, 1985) to siliciclastic outcrops that are, in part, coeval with the interval examined in this study.

Sea-level cycles are typically grouped into hierarchies (orders) based on interpretations of their duration. While there may be general consensus among authors as to the duration of typical cycles, a different numerical order may be applied to cycles of the same general length. The numerical order of anyone's cycle, then, depends in part on whose orders are followed. For this study, the proposed cycle orders of Mitchum and Van Wagoner (1991) have been followed (Fig. 1). Both this study and Mitchum and Van Wagoners' (1991) study deal with siliciclastics and are largely based on subsurface data. Their apparent timing of cycles is of similar duration to those discussed below as well.

In this paper, the regressive portion of a third-order Upper Devonian cycle will be examined in detail throughout much of the Appalachian basin. Results described are preliminary and are based on ongoing correlation and mapping of high frequency sedimentation cycles throughout the basin using a network of several hundred subsurface control points (Fig. 2). The data base consists primarily of gamma-ray logs from oil and gas test wells.

First, a well defined fourth-order cyclicity is described, based on organic carbon preservation within an "inner distal" facies (just basinward of the limit of significant gamma-ray log-detectable turbidite siltstone bundles). Subsequently, the time-correlative interval is examined in nearershore facies to test for the presence of similar cyclicity in the deposition of coarser clastic intervals. Finally, more distal stratigraphic patterns are examined for patterns of stratigraphic downlap and convergence. The stratigraphic patterns observed and their meaning in terms of sea-level changes are then related to published interpretations of Late Devonian global sea levels.

STRATIGRAPHIC NOMENCLATURE OF THE STUDY INTERVAL
Outcrop Stratigraphy

The study area, which includes most of the preserved Appalachian basin west of the Valley and Ridge, is shown in Figure 2. Because of the wide geographic area covered by this study, different nomenclatures relate to different portions of the study area. The stratigraphic nomenclatures pertinent to this paper are presented in Figures 3 and 4.

In the western, distal portions of the basin, stratigraphic divisions have been based on the cycles of organic carbon-rich and organic carbon-poor deposition. In total, five major cycles of alternating blackish-gray or dark-brown organic carbon-rich shales and light-gray or greenish-gray shales have been recognized in the outcrop of New York and Ohio and in the subsurface within the Appalachian basin Upper Devonian strata (Fig. 3;

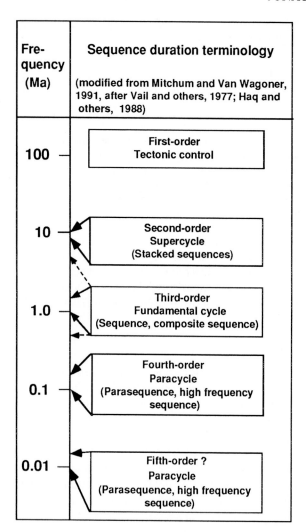

Fig. 1.—Terminology applied to and duration of various orders of cycles or sequences.

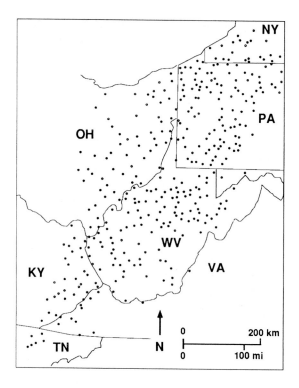

Fig. 2.—Subsurface control utilized in this study. Gamma-ray well logs were utilized from all this locations, in most cases supplemented by bulk density well logs.

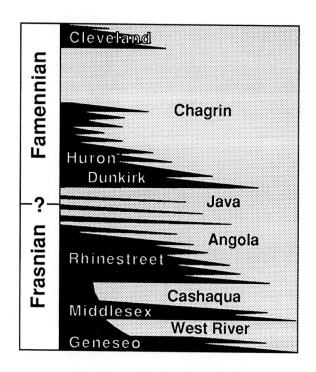

FIG. 3.—The Upper Devonian of the basinal facies of the Appalachian basin consists of five alternating third-order cycles of basal black organic-rich shale (black above) overlain by dominantly gray-green shale (gray above). This paper focuses on the upper half of the third cycle (Java Formation and Angola Member of the West Falls Group).

Rickard, 1975; Roen, 1981, 1983; Ettensohn, 1985). This study focuses on the primarily inorganic portion of the third cycle and the transition to the organic-rich portion of the fourth cycle (Fig. 3).

In New York, the study interval incorporates the upper Rhinestreet Shale, the Angola Shale, and the Java Formation of the West Falls Group, as well as the overlying Dunkirk Shale and the eastern correlatives of these units (primarily the Nunda and Wiscoy sandstones). At the northern end of the basin, the outcrop belt trending east-west across New York is roughly parallel to depositional dip. Organic-rich black shale tongues which have been traced eastward into coarser-grained facies have been considered to represent transgressions and have been used to divide the more proximal strata into formal units with chronostratigraphic significance (Fig. 4, Column 7, for details of this correlation, see Rickard, 1975).

Southward into northern and central Pennsylvania and along the eastern margin of the basin, the outcrop belts are oriented

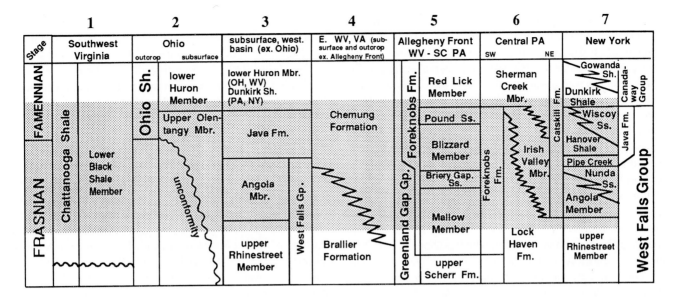

Fig. 4.—Stratigraphic nomenclature around the Frasnian-Famennian boundary from throughout the Appalachian Basin. The interval examined in this study is highlighted in gray.

more nearly parallel to depositional strike. The stratigraphic nomenclature applied in these areas reflects the overall progradational nature of deposition. Named units (Fig. 4, Column 6) reflect a series of facies passing upward from distal marine to nonmarine without time significance.

Farther south, in southern Pennsylvania, Maryland, and northern West Virginia, along the Allegheny Front (the structural boundary between the Allegheny Plateau and the Valley and Ridge structural provinces in the Appalachian basin) the progradational sequence has been subdivided into the various units of the Greenland Gap Group (Fig. 4, Column 5) on the basis of laterally continuous coarser intervals thought to reflect eustatic sea-level drops (Dennison, 1971; McGhee and Dennison, 1976; Avary and Dennison, 1980; Lyke, 1986).

Recently, the Greenland Gap Group has been extended farther north into central Pennsylvania (Warne, 1986; Warne and McGhee, 1991) and into the Valley and Ridge of Virginia (Barrell, 1986; Dennison and others, 1988; Sinclair, 1991). Brachiopod biostratigraphy has been employed in order to correlate the Greenland Gap Group with its New York equivalents (McGhee, 1977; McGhee and Dennison, 1980; Warne, 1986; and Warne and McGhee, 1991).

In the southern end of the basin, the study interval occurs within the lower black-shale member of the Chattanooga Shale of Swartz (1927) in Virginia and Tennessee (Fig. 4, Column 1).

Subsurface Stratigraphy

The subsurface stratigraphy of the western shale facies of the Appalachian basin was extensively studied by numerous workers in the 1970's as part of a series of projects funded by the U. S. Department of Energy. The purpose of that work was to understand and enhance the extraction of natural gas from those formations. A basinwide network of gamma-ray log cross sections was constructed and formal outcrop nomenclature was extended into the subsurface. Again, pertinent portions of this nomenclature are shown in Figure 4 (Columns 2 and 3). The results of this long range subsurface correlation project have been summarized by Roen (1981, 1983; specific authors are referenced therein).

In the subsurface of the eastern parts of the basin, the marine portion of the Upper Devonian is a complexly interbedded generally progradational wedge of gray shales, siltstones, and sandstones (Fig. 4, Column 4). In West Virginia and Pennsylvania, the section has been extensively drilled for oil and gas. A regionally diverse and inconsistently applied informal nomenclature has been attached to sandy intervals. In northwestern West Virginia, the stratigraphic position of many of these oil and gas "sands" relative to the formal shale stratigraphy has been illustrated in more detailed, local gamma-ray cross-sections (Filer, 1985; Sweeney, 1986). Piotrowski and Harper (1979) provided a generalized correlation between zones of drillers' terminologies and Devonian Shale formations in Pennsylvania. Farther to the east, in West Virginia, correlations between the outcropping Greenland Gap Group and oil and gas producing horizons have also been presented (Dennison, 1971; Barrell, 1986; Dennison and others, 1988; Filer, 1988; Boswell and others, 1988).

FOURTH-ORDER CYCLES OF THE BASINAL FACIES
New York Outcrop

In the terminology applied to the New York Upper Devonian shale outcrop, the section investigated includes the upper Rhinestreet Member, the Angola Member, the Java Formation (of the West Falls Group), as well as the overlying Dunkirk

Shale. The Dunkirk is equivalent to the basal part of the Huron Member of the Ohio Shale (Roen, 1981, 1983). The West Falls Group is the third of the five major black shale/gray shale cycles of the Upper Devonian (Fig. 4). The study interval includes most of this cycle, as well as the transition into the overlying black shale portion of the next cycle (Dunkirk/Huron shale). In New York, a higher frequency cyclicity of organic-carbon preservation has long been recognized. This cyclicity is well illustrated in published descriptions, stratigraphic columns, and correlation diagrams (Pepper and deWitt, 1950, 1951; Pepper and others, 1956; deWitt, 1960; Rickard, 1975).

For example, deWitt's (1960) columnar section of the type Java Formation at Java Village shows a basal black shale, the Pipe Creek Member which is about 6 m (20 ft) thick. The overlying Hanover Shale Member, which is about 27 m (90 ft) thick, is dominantly medium-gray to greenish-gray shale and silty mudstone but contains several intervals dominated by black organic-rich shale. These intervals are typically 0.5 to 1.0 m (1.6 to 3.3 ft) thick (deWitt, 1960, see his Fig. 2). Pepper and DeWitt (1950) traced the Java Formation west from the type section until the outcrop belt disappeared under Lake Erie. To the west, it thins slightly, becomes less silty, and the organic shale bands become more common. Additionally, the overlying massive black Dunkirk Shale thickens westward from less than 6 m (20 ft) to over 10 m (33 ft). To the east, the corresponding interval thickens, and the Hanover and Pipe Creek Shale Members are replaced by the more proximal Wiscoy Sandstone. A similar scale cyclicity of carbon preservation, expressed as black-shale bands, was described from outcrops of the Angola Shale in New York (Pepper and others, 1956).

Subsurface Gamma-Ray Stratigraphy

A high frequency, fourth-order cyclicity is detectable in gamma-ray log data throughout the western portions of the basin as well.

Gamma-ray logs are ideal for determining the presence of organic carbon in marine shales. Natural rock radioactivity is due almost entirely to emissions from three radioisotopes, ^{40}K, ^{238}U, and ^{232}Th (Adams and Gasparini, 1970). Carbon-rich shales are enriched in authigenic uranium which is reduced from solution in sea water and occurs in these shales as urano-organic complexes or adsorbed onto the surfaces of organic and inorganic material (Durrance, 1986). The result is an increase in total gamma-ray emissions over base level of adjacent non-organic shales. This base level is, in turn, controlled by the mineralogy of the shales (Fertl, 1983). Figure 5 is excerpted from a more detailed gamma-ray log cross section extending in the subsurface from Chautauqua County, New York to Washington County, Virginia (Fig. 2 shows complete well control). The correlation shows that along the line of this section and during the time of the upper Rhinestreet Shale through the Java Formation, eleven high frequency cycles of shale deposition, each initiated by a period of significantly higher organic carbon preservation can be observed throughout the basin. These

organic marker beds are indicated by the elevation of total gamma-ray emissions above a non-organic shale base line.

Within the upper Rhinestreet and lower Angola shales, six cycles, (cycles 1-6, Fig. 5) each with a basal radioactive shale (indicating carbon preservation) are present. As illustrated in Figure 5, these cycles are typically about six to ten meters (20 to 33 ft) thick, with about one meter (3.3 ft) of organic shale at the base. During upper Angola deposition, a significantly thicker cycle (about 30 meters (100 ft) in Fig. 5) was deposited. The organic carbon-rich basal Pipe Creek Member of the Java Formation is at the top of this cycle. Four more cycles make up the remaining Java Formation, again six to ten meters (20 to 33 ft) thick. The final cycle is capped by the Dunkirk/Huron Shale, and initiates a major period of organic-carbon accumulation. Portions of this cyclicity have previously been recognized in the subsurface. Neal (1979) and Dowse (1980) distinguished cycles eleven and seven as separate informal members of the Java and Angola Shales respectively in the subsurface of western West Virginia. The double radioactive kick that caps cycles seven and eight has been designated as the subsurface equivalent of the Pipe Creek Member of the Java Formation by numerous workers on subsurface cross sections as summarized by Roen (1981).

The line of cross section excerpted in Figure 5 was chosen to illustrate as much detail as possible of the internal stratigraphy of the interval of interest, and essentially follows closely the 120-m (400-ft) isopach of the entire package. Less detail of this fourth-order shale cyclicity is visible to the west of the line of cross-section A-A' because of depositional thinning and downlaps. To the east and toward the source of clastic input, both the loss of the organic-rich cap at the top of each cycle and an increasing content of bedded silts and fine sand within the interval obscure much of the cyclicity.

The true thickness of each of the eleven shale cycles illustrated in Figure 5 has been adjusted by stretching or shrinking, so that the vertical height of each cycle in this diagram is proportional to its mean thickness within the entire package for the four wells illustrated. The stick diagram at the bottom right of Figure 5 shows, to scale, the preadjustment true thickness of each cycle. The purpose for adjusting the vertical scales is to emphasize the detailed internal correlation of the level of gamma-ray emissions within each cycle. Excursions of gamma-ray intensity to the right of a gray shale base line mark the position of periods of significantly enriched organic carbon preservation, and were the points of correlation on which vertical scales were adjusted.

There are clearly basinwide subcycles within the eleven fourth-order cycles (Fig. 5). These subcycles are consistently visible on logs in the data base within the pathway illustrated on Figure 5. The internal subcycles may represent fifth-order cycles within individual radioactive shale capped fourth-order cycles. They are expressed in the number, shape, thickness, and strength of internal deflections in each of the eleven fourth-order cycles.

The gamma-ray log has been widely used as a subsurface tool for correlation, and its use for this purpose is well estab-

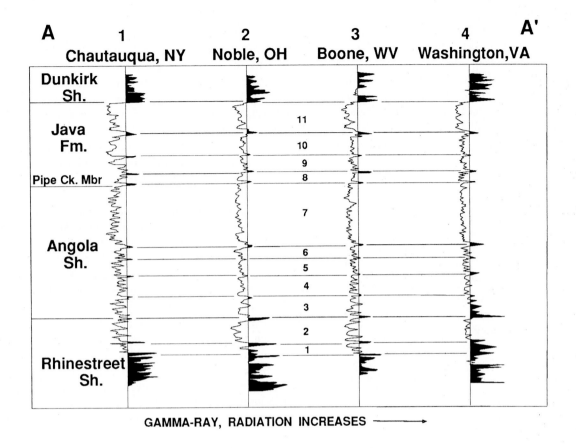

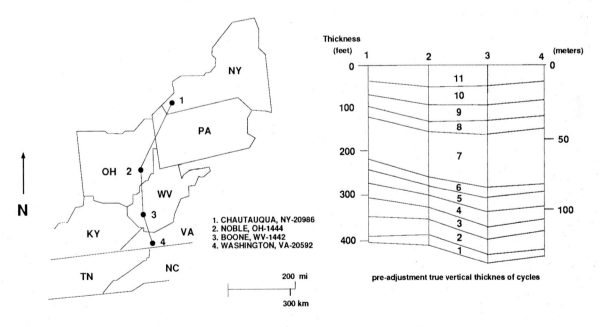

Fig. 5.—The gamma-ray log cross section at top illustrates the lateral continuity of radioactive black-shale bands within the dominantly gray shale of the study interval. The cross section extends for about 600 km along depositional strike within the basinal shale facies. Deflections to the right of the vertical line with each gamma-ray log indicate higher than normal radiation levels of organic-rich shales. The true vertical thickness of each cycle has been adjusted and is displayed with a vertical dimension proportional to its mean thickness in the four wells. This adjustment was done to highlight the consistency of the internal gamma-ray signal within individual cycles. The thicknesses are displayed to scale in the stick figure to the bottom right.

lished (Rider, 1986). The degree of correlation observable on cross-section A-A', where individual deflections down to the scale of a meter to a few meters can be traced over 600 km (375 mi), must be considered unique.

In order to more fully understand the nature and causes of the organic/non-organic cyclicity discussed above, it is necessary to examine correlations into both subsurface nearshore facies to examine for concurrent cyclicity in silt and sand deposition, and into more distal facies to examine stratigraphic patterns as the section thins.

CORRELATION INTO NEARERSHORE AND MORE DISTAL FACIES

In the preceding sections of this paper, the deposition of eleven fourth-order cycles of carbon preservation within a portion of the basinal facies of the Upper Devonian of the Appalachians was demonstrated. These cycles were illustrated in what was termed an "inner distal" facies, immediately basinal of deposition of gamma-ray log detectable siltstone bundles. In the following sections, stratigraphic patterns within coeval sediments are examined in nearershore facies and more distal facies. The case will be made that these eleven fourth-order cycles can be divided into two basinwide "parasequence sets" (in the sense of Van Wagoner and others, 1990; see discussion below), each capped by a particularly regressive cycle.

Nearshore Equivalents

Figures 6 and 7 are gamma-ray cross sections which illustrate correlation with nearershore facies (see Fig. 2 for cross-section locations). Cross-section B-B' (Fig. 6) extends from central Ohio to near the Allegheny Front in West Virginia. Cross-section C-C' (Fig. 7) extends from northcentral Ohio to near the outcrop belt in southwestern New York. Cross-section A-A' (Fig. 5) ties together B-B' and C-C'. As can be seen on B-B' and C-C', it is within cycle 7 (top of the Angola Shale) and cycle 11 (top of the Java Formation) that significant input of coarser material is first encountered in correlating eastward toward the source area in both New York and West Virginia (see Ritchie- 5455 well, Fig. 6, and compare with Cattaragus- 20751, Fig. 7). These tongues of silt and very fine sand represent the upper Nunda and upper Wiscoy Sandstones, respectively, of the New York outcrop. In the subsurface western West Virginia, cycle 7 is informally known as the "Alexander sand", and cycle 11 as the "Benson sand". The correlation of these two zones with the upper Angola Shale and the Java Formation respectively has been previously illustrated (Filer, 1985; Sweeney, 1986).

These two cross sections illustrate the synchronous occurrence, in two widely separated parts of the basin, of the most distal deposition of coarser clastics within the study interval. Additionally, it can be seen that the organic shales which mark the base of each cycle lose their radioactive signature at about the same position in the basin. In general, cyclicity farther to the east is expressed as a series of generally coarsening and cleaning upwards turbidite siltstone bundles.

Cross-section B'-B' (Fig. 6) continues across West Virginia to near the Allegheny Front. Continuing toward the source of clastic input, the section becomes increasingly silty and sandy. Within the nearershore facies, however, cycles 7 and 11 ("Alexander" and "Benson" sands of drillers) remain generally more sand-rich. Distinct widespread shales (equivalent to the Pipe Creek Member at the top of cycle 7 and the basal Lower Huron Shale of Ohio or Dunkirk Shale of New York at the top of cycle 11) top these sandy cycles. These combinations of sandy cycles capped by distinct shales serve as useful datums on which the rest of the package can be correlated and depositional patterns examined.

The two cycles capping each parasequence set (cycles 7 and 11) develop into clean, relatively massive sands in West Virginia (see the Randolph- 939 well, Fig. 6). These two subsurface sands have previously been correlated with the Pound and Briery Gap Sandstone Members of the Foreknobs Formation along the Allegheny Front outcrop belt (Lewis, 1983; Barrell, 1986). The Pound and Briery Gap Sandstones have been interpreted to represent deposition during eustatic sea-level falls (Dennison, 1971, 1985). In conjunction with the most basinward deposition of coarser clastics which also occurs within these cycles, this establishes their position as the two most regressive cycles within the package of eleven fourth-order cycles.

The study interval thickens from west to east towards the Acadian orogenic highlands and the sources of sediment input. In the eastern edge of the study area, near the Allegheny front, cycles are typically 20 to 30 m (60 to 100 ft) thick. The exception is again cycle 7 which maintains its character as a distinctly thicker interval.

Cyclic deposition of sandstones within sequences in general may not be as readily obvious as the cyclic deposition of the contemporaneous basinal facies discussed above. It is in fact not most clearly expressed in the most proximal well on section B-B' (Randolph- 939). Better developed cyclicity in seen in wells slightly basinward, represented by the Barbour-1200 well (Fig. 6), suggesting some sediment bypassing to more distal shelf locations during less regressive cycles. The correlation in the Barbour-1200 well (Fig. 6) in general shows spiky cleaning upward patterns in each parasequence. Some cycles are a composite gamma-ray pattern showing multiple bundles within them. These may be a reflection of the same fifth-order cyclicity as was observed in the basinal shale facies. Each set is topped by the more regressive cycles and a widespread prominent shale marker.

Some of the correlations shown in Figure 7 may seem subtle, and they represent a best attempt to delineate within the nearshore facies the detailed cyclicity observed within coeval basinal shales. Certainly, to some extent autocyclic depositional processes within the higher energy nearershore environments have obscured portions of this cyclicity. Ongoing more detailed correlations and subsequent mapping of these cycles by the author should lead to refined understanding of depositional controls at this time.

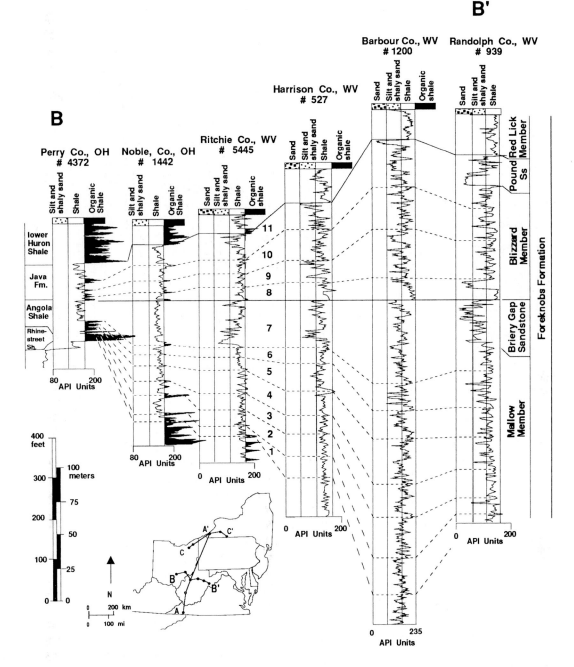

Fig. 6.—Gamma-ray cross-section B-B' illustrates the correlation from basinal shale facies of Ohio east into nearershore silty and sandy facies in northern West Virginia. The westernmost occurrence of significant siltstones is within cycles 7 and 11, as is the case with the westernmost occurrence of clean sandstones. To the west, the cross section shows thinning and convergence of cycles, but cycles 7 and 11 remain prominent.

Distal Equivalents

The western end of both section B-B' and C-C' (Figs. 7 and 8) illustrate the stratigraphic patterns which are observed within the study interval in correlating into more distal facies from the depositional strike section illustrated in section A-A' (Fig. 5). The line of A-A' was selected to illustrate the maximum internal detail of gamma-ray stratigraphy observable.

The Perry County, Ohio-4372 well (section B-B', Fig. 6) and the Geauga County, Ohio-197 well (section C-C', Fig. 7) are the most distal wells illustrated. Readily noticeable, by examination of the correlation from the east to the west to these two wells, is the progressive thinning, from the base upward, of cycles 1-6 beneath cycle 7 and cycles 8-10 beneath cycle 11.

In the Geauga, Ohio-616 well (section B-B', Fig. 6), the four cycles (numbers 8-11) of the Java Formation are still evident.

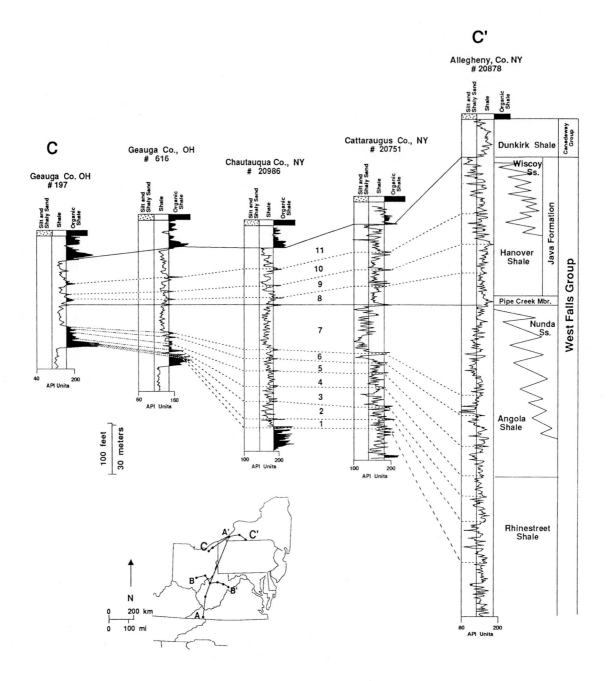

Fig. 7.—Gamma-ray cross-section C-C' illustrates the correlation from basinal shale facies in northern Ohio east into nearershore silty and sandy facies in southwestern New York. As is the case in West Virginia (see cross-section B-B', Fig. 6) the westernmost occurrence of siltstones is within cycles 7 and 11. This figure and Figure 6 also show similar patterns of thinning and convergence to the west.

Cycles 1 and 2 of the underlying Upper Rhinestreet and Angola Shales have converged by downlap and merged into the underlying radioactive portion of the Rhinestreet Shale. Cycles 3-6 are still detectable as alternations of radioactive and nonradioactive shale, but their internal character is lost due to thinning. The most regressive cycle at the top of the Angola shale, cycle 7, is still a distinct massive inorganic shale zone, although it has thinned to about 18 m (50 ft).

In more distal positions, the Perry, Ohio- 4372 (section B-B', Fig. 6) and the Geauga- 197 (section C-C', Fig. 7) wells show continued downlap and merging of basal cycles of the Angola Shale (cycles 1-6) and Java Formation (cycles 8-10) into radioactive organic shale, while cycles 7 and 11 which cap the successive formations remain as distinct inorganic shales.

In summary, moving progressively westward from the line of section A-A', a stratigraphic pattern of thinning and eventual

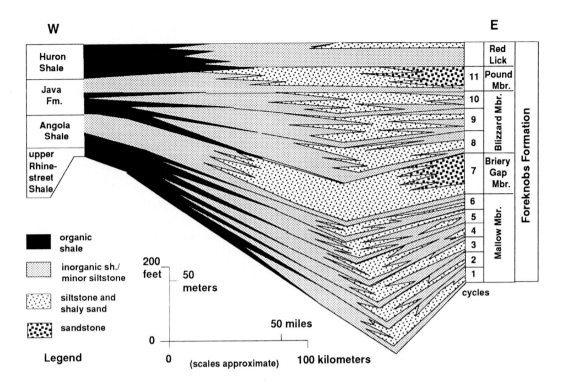

Fig. 8.—Schematic diagram summarizing lithofacies patterns observed along the line of cross-section B-B' (Fig. 7).

merging of progressively younger shale cycles is seen within two sets of cycles. Cycles 7 and 11 are the particularly regressive cap of each successive set. As discussed above, it is within these cycles that correlation into nearershore facies shows both the first appearance of distinct siltstones and very fine sandstones, and, eventually in West Virginia, the first appearance of clean, massive, nearshore sands. Thus, cycles 7 and 11 can be seen to represent the most regressive cycles capping the two successive sets of cycles in a series of facies from distal to proximal in widely separated parts of the Appalachian basin.

CYCLE STACKING PATTERNS

Eleven basinwide fourth-order cycles of siliciclastic deposition have been illustrated in previous sections of this paper. These occur within the regressive portion of a third-order cycle of sedimentation. These fourth-order cycles can be grouped into two progradational sets of cycles each capped by a particularly regressive cycle. This was demonstrated by examining the patterns of downlap and convergence of the cycles into basinal organic rich shales, where the progressive convergence from the base upwards of each set is obvious. Distal siltstone bundles were deposited most basinward in both West Virginia and New York in the cycles which cap each set. The most regressive cycles were correlated with prominent clean, nearshore sand bodies which outcrop along the Allegheny Front in West Virginia. Figure 8 schematically summarizes these observations (compare Fig. 8 with Fig. 6 and Fig. 7).

In the recently described terminology of well log, core, and outcrop based high resolution sequence stratigraphy (Van Wagoner and others, 1990; Mitchum and Van Wagoner, 1991; see Fig. 1 of this paper), the regressive portion of a third-order cycle is the lowstand system tract portion of a depositional sequence. Fourth-order cycles occur within all portions of a depositional sequence. Sediments deposited during a fourth-order cycle constitute a parasequence, and may be stacked into progradational, retrogradational, or aggradational parasequence sets, depending on the relationship between the rate of deposition versus the rate of creation of accommodation space (see Mitchum and Van Wagoner, 1991, their Fig. 5). The stratigraphic pattern illustrated above is one of two sets of progradational parasequence sets stacked on top of each other, within the lowstand wedge. The shoreward shift of facies that occurs at the top of cycle 7 (equivalent to the Pipe Creek Member of New York) suggests a somewhat larger sea-level rise than the average during this time. Subsequent cycles reflect this in the initiation of the second progradational parasequence set. Regressive cycle 11 occurs at the top of the second parasequence set, and is capped by the initiation of a third-order cycle of organic-rich shale deposition (Lower Huron/Dunkirk Shale). This deposition was initiated by a more significant transgression. In sequence stratigraphic terminology, this marks the initiation of the transgressive systems tract. The base of this tract is the prominent shale which caps cycle 11 across section B-B' and C-C' (Figs. 7 and 8).

DURATION OF FOURTH-ORDER CYCLES

Five major cycles of alternating organic-rich and non-organic shale were deposited in the distal Appalachian basin during Late Devonian time (Fig. 3). Three of these cycles were deposited during the Frasnian, beginning with deposition of the Geneseo Member of the Genesee Group (organic-rich shale) and ending with deposition of the Java Formation and Angola Member of the West Falls Group (generally non-organic shale, see Fig. 3). Two later cycles were deposited during Famennian time, beginning with the Huron/Dunkirk organic-rich shale. The time scale of Harland and others (1990) assigns a 10-Ma duration to Frasnian time. Thus, the three major cycles of the Frasnian stage have an approximate average duration of 3.3 Ma. This study has identified eleven higher frequency cycles within the non-organic portion of the final Frasnian major cycle (West Falls Group of New York, Fig. 3). If it is assumed that this non-organic, regressive portion of the cycle was deposited over one-half of the average major cycle time period, then the eleven higher frequency cycles have an average duration of approximately 150 ka. Note that this is an average duration, as variations in the thickness of cycles exist (Fig. 5), and that thickness may not be proportional to duration, especially during a period of generally regressive basin filling. With this limitation in mind, the thickness variations illustrated in Figure 5 suggest that cycles range in duration from about 100 ka to 300 ka.

Based on this simplistic analysis, the larger cycles discussed here fall within the range (although longer than the average) of the third-order cycles of Mitchum and Van Wagoner (1991), while the higher frequency cycles fit within the range of their fourth-order cycles (Fig. 1). As in any analysis of the absolute time scale of sedimentation cycles, uncertainties exist in this calculation. Uncertainties in this analysis include: (1) uncertainties in current radiometric time scales; and (2) the assumptions that the major cycles observed in the Upper Devonian are of equal duration, and that the non-organic portion of any of the third-order shale cycles equals half of the total length of that cycle.

DISCUSSION

This paper has presented some preliminary results of an ongoing study of cyclic deposition within a siliciclastic foreland basin fill in part of the Upper Devonian of the Appalachian basin. Eleven high-frequency, fourth-order cycles of alternating organic and inorganic basinal shales were deposited within the primarily inorganic, regressive portion of a third-order depositional cycle. It is also suggested that a subtle fifth-order basinwide gamma-ray cyclicity exists.

Examination of stratigraphic patterns in both the most distal and nearer shore facies shows that these cycles can be grouped into two sets of progradational parasequences within a lowstand systems tract. These underlie the marine portion of a subsequently deposited transgressive systems tract. At the top of each parasequence set, a particularly regressive fourth-order cycle

occurs. Previous workers have attributed the third-order Late Devonian lithologic cycles of organic and inorganic shales in the Appalachians to cycles of transgression and regression (Rickard, 1975; McGhee and Bayer, 1985). The Devonian eustatic sea-level curve of Johnson and others (1985) shows two significant regressive pulses just prior to the boundary between the Frasnian and Famennian stages (Fig. 9). This interpretation was based in part on the New York section in the northern end of the study area, and reflect the regressive pulses of the Nunda and Wiscoy sands, separated by the Pipe Creek Shale and overlain by the Dunkirk Shale. Dennison (1985) pointed out that these two pulses are recorded by deposition of the Pound and Briery Gap Sandstones in the central Appalachians (cycles 7 and 11).

Based on the higher order cyclicity demonstrated in this work, a preliminary modification of the upper Frasnian portion of the Johnson and others (1985) sea-level curve is suggested (Fig. 9). This modified curve reflects the basin-wide parasequence stacking patterns observed in the Appalachian basin. While it is not based on global data, it is suggested this fourth-order cyclicity, which appears to have a periodicity on the order of 0.1 to 0.3 Ma, also reflects global sea level. The basinal black-shale bands represent the transgressive phase of cycles. Concurrent silt and sand deposition in distal turbidite to shelf facies shows a matching, if less perfect, cyclicity in progradational bundles which represent the regressive phase of cycles. Thus, a process or set of processes must have acted synchronously and moderated both the preservation of organic carbon in the distal basin and the reorganization of nearershore depositional systems. While simple organic-productivity cycles might explain black-shale cycles, a change in the depositional energy regime is implied by the nearershore cycles.

The fact that the cyclicity is basinwide precludes autocyclicity of depositional systems as a causative mechanism. A tectonically controlled model for Appalachian basin black shales deposition has been proposed (Ettensohn, 1985, 1987) which incorporates the observed southward migration of the Acadian Orogenic source area with time (Dennison, 1985; Ettensohn, 1987; Ferrill and Thomas, 1988). This model, however, deals with the larger third-order shale cycles, and offers no tectonic mechanism which could explain relatively uniform widespread higher frequency cycles.

Perhaps the most compelling reason for interpreting these cycles as eustatic is their occurrence in a structure of stacked sets, each capped by a globally recognized sea-level cycle (cycles 7 and 11). The modified sea-level curve in Figure 9 shows cycles of similar magnitude, except for the apparently more regressive capping cycle of each set. The vertical scale in Figure 9 should not be considered to be proportional to absolute time. The sea-level cycles shown were fitted into the Johnson and others (1985) diagram without altering their vertical scale. Further analysis, including basinwide mapping of the lithology and thickness of each cycle, should lead to a more highly refined sea-level curve for this time.

Recent discussions of high resolution siliciclastic sequence stratigraphy have emphasized sand-rich depositional facies, and

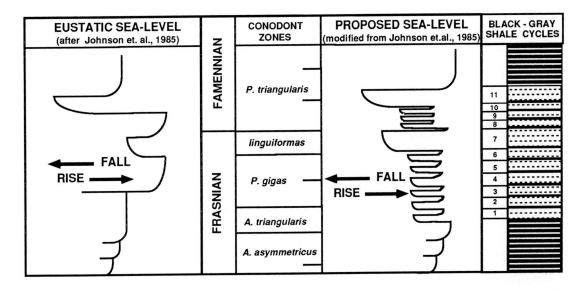

Fig. 9.—The eustatic sea-level curve of Johnson and others (1985) for the interval bracketing the Frasnian-Famennian boundary is displayed on the left. The global regression-transgression cycles they identified during the *P. linguiformis* and middle *P. triangularis* conodont zones correspond to cycles 7 and 11 of this study. It is suggested that other sedimentation cycles (1-6 and 8-10) identified in the Appalachian Basin are the expression of a continuous series of sea-level cycles at this time. These cycles appear to have been of lessor magnitude and thus have not yet been recognized globally.

have suggested that basinal deposition is relatively insensitive in recording high-frequency sequences (Van Wagoner and others, 1990; Mitchum and Van Wagoner, 1991). Apparently, during this portion of the Upper Devonian, the organic-carbon preservation capacity of the distal part of the Appalachian basin was sensitive to high frequency sea-level changes. This signal allows widespread correlation through the basinal facies and into nearershore facies and establishes time equivalence of parasequences, from widely separated, independent depositional systems (shown in Figs. 7 and 8). Similar parasequence stacking patterns can be observed in widely separated areas as well. This then confirms the geographic scale of parasequence development to be basinwide.

CONCLUSIONS

Sedimentation in the Appalachian basin during the latest Frasnian and earliest Famennian Stages took place within a framework of a series of basinwide parasequences. Eleven parasequences can be identified within an interval of about 2 Ma duration. The duration of individual parasequences ranges from about 0.1 to 0.3 Ma. Parasequences are expressed in basinal facies by alternations of basal, thin [1 to 2 m (3 to 6 ft)] organic black shales overlain by thicker [6 to 10 m (20 to 33 ft)] gray shales.

Concurrent cycles exist in more proximal facies as well. In siltstone turbidite facies, it is expressed in generally coarsening and cleaning upward bundles. Clean, nearshore shelf sandstones are developed within two parasequences. In these facies, cycles are in general 2 to 3 times thicker than in basinal facies. Two parasequences are consistently the most regressive through-

out all facies and in areas supplied by widely separated sediment inputs. These cap two basinwide sets of progradational parasequence sets, and confirm that the depositional control of the cycles acted synchronously in time and similarly in magnitude throughout the area of deposition.

Autocyclic control of parasequence development is ruled out by their geographic extent. Tectonically controlled variations of subsidence or sedimentation rates are unlikely due to both the geographic extent and relatively short duration of cycles. Regressive eustatic sea-level signals have been recognized globally at the time of the two most regressive pulses within the series of eleven cycles in the Appalachian basin. It is suggested that all eleven parasequences were deposited basinwide in response to eustatic sea-level cycles, only the most significant of which have been recognized globally to date.

ACKNOWLEDGMENTS

This work was supported, in part, by grants from the Sigma Xi Research Foundation, the Appalachian Basin Industrial Associates, the Southeastern Section of the Geological Society of America, and the Martin Fund of the University of North Carolina. John M. Dennison is thanked for his advice throughout this project, as well as for his helpful reviews of this manuscript. Additional thoughtful reviews by Donald L. Woodrow and Charles K. Paull greatly improved the manuscript.

REFERENCES

ADAMS, J. A. S., AND GASPARINI, P., 1970, Methods in Geochemistry and Geophysics: Amsterdam, Elsevier, 295 p.

AVARY, K. L., AND DENNISON, J. M., 1980, Back Creek Siltstone Member of Devonian Brallier Formation in Virginia and West Virginia: Southeastern Geology, v. 21, p. 121-153.

BARRELL, S. M, 1986, Stratigraphy and depositional environments of Upper Devonian rocks in east-central West Virginia and adjacent Virginia: Unpublished M.S. Thesis, University of North Carolina at Chapel Hill, Chapel Hill, 113 p

BOSWELL, R. M., DONALDSON, A. C., AND LEWIS, J. S., 1988, Subsurface stratigraphy of the Upper Devonian and Lower Mississippian of northern West Virginia: Southeastern Geology, v. 29, p. 105-131.

BUSCH, R. M., AND ROLLINS, H. B., 1986, Correlation of Carboniferous strata using a hierarchy of transgressive-regressive units: Geology, v. 12, p. 471-474.

DENNISON, J. M., 1970, Stratigraphic divisions of Upper Devonian Greenland Gap Group (Chemung Formation) along Allegheny Front in West Virginia, Maryland, and Highland County, Virginia: Southeastern Geology, v. 12, p. 53-82.

DENNISON, J. M., 1971, Petroleum related to Middle and Upper Devonian deltaic facies in central Appalachians: American Association of Petroleum Geologists Bulletin, v. 55, p. 1179-1193.

DENNISON, J. M., 1985, Devonian eustatic fluctuations in Euramerica: Discussion: Geological Society of America Bulletin, v. 96, p. 1595-1597.

DENNISON, J. M., BARRELL, S. M., AND WARNE, A. G., 1988, Northwest-southeast cross section of Devonian Catskill Delta in east-central West Virginia and adjacent Virginia, in Dennison, J. M., ed., Geologic Field Guide Devonian Delta in East-Central West Virginia and Adjacent Virginia: Charleston, Appalachian Geological Society, p. 12-35.

DE WITT, W, 1960, Java Formation of late Devonian age in western and central New York: Bulletin of the American Association of Petroleum Geologists, v. 44, p. 1933-1939.

DOWSE, M. E., 1980, The subsurface stratigraphy of the Middle and Upper Devonian clastic sequence in northwestern West Virginia: Unpublished Ph.D. Dissertation, West Virginia University, Morgantown, 177 p.

DURRANCE, E. M., 1986, Radioactivity in Geology: Principles and Applications: Chichester, Ellis Horwood Limited, 441 p.

ETTENSOHN, F. R., 1985, The Catskill Delta complex and the Acadian orogeny: a model, in Woodrow, D. L., and Sevon, W. D., eds., The Catskill Delta: Boulder, Geological Society of America Special Paper 201, p. 39-50.

ETTENSOHN, F. R., 1987, Rates of relative motion during the Acadian orogeny based on the spatial distribution of black shales: Journal of Geology, v. 95, p. 572-582.

FERRILL, B. A., AND THOMAS, W. A., 1988, Acadian dextral transpression and synorogenic sedimentary successions in the Appalachians: Geology, v. 16, p. 604-608.

FERTL, W. H., 1983, Gamma-ray spectral logging: A new evaluation frontier— part III, measuring source rock potential: World Oil, v. 196, p. 147-155.

FILER, J. K., 1985, Oil and gas report and maps of Pleasants, Wood, and Ritchie counties, West Virginia: West Virginia Geological and Economic Survey Bulletin B-11A, 87 p.

FILER, J. K., 1988, Chronostratigraphy and facies of the Upper Devonian clastic wedge, West Virginia, in Dennison, J. M., ed., Geologic Field Guide Devonian Delta in East-Central West Virginia and adjacent Virginia: Charleston, Appalachian Geological Society, p. 67-76.

GOLDHAMMER, R. K., DUNN, P. A., AND HARDIE, L. A., 1987, High frequency glacio-eustatic sealevel oscillations with Milankovitch characteristics recorded in Middle Triassic platform carbonates in Northern Italy: American Journal of Science, v. 287, p. 853-892.

GOODWIN, P. W., AND ANDERSON, E. J., 1985, Punctuated aggradational cycles: A general hypothesis of episodic stratigraphic accumulation: Journal of Geology, v. 93, p. 515-533.

HARLAND, W. B., ARMSTRONG, R. L., COX, A. V., CRAIG, L. E., SMITH, A. G., AND SMITH, D. G., 1989, A Geologic Time Scale: Cambridge, Cambridge University Press, 263 p.

HECKEL, P. H., 1986, Sea-level curve for Pennsylvanian eustatic marine transgressive-regressive depositional cycles along midcontinent outcrop belt, North America: Geology, v. 14, p. 330-334.

JOHNSON, J. G., KLAPPER, G., AND SANDBERG, C. A., 1985, Devonian eustatic fluctuations in Euramerica: Geological Society of America Bulletin, v. 96, p. 567-587.

LEWIS, J. S., 1983, Reservoir rocks of the Catskill delta in northern West Virginia— a stratigraphic basin analysis emphasizing depositional systems: Unpublished M.S. Thesis, West Virginia University, Morgantown, 148 p.

LYKE, W. L., 1986, The stratigraphy, paleogeography, depositional environment, faunal communities, and petrology of the Minnehaha Member of the Scherr Formation, an Upper Devonian turbidite sequence, central Appalachians: Southeastern Geology, v. 26, p. 173-192.

McGHEE, G. R., 1977, The Frasnian-Famennian (Late Devonian) boundary within the Foreknobs Formation, Maryland, and West Virginia: Geological Society of America Bulletin, v. 88, p. 806-808.

McGHEE, G. R., AND DENNISON, J. M., 1976, The Red Lick Member, a new subdivision of the Foreknobs Formation (Upper Devonian) in Virginia, West Virginia, and Maryland: Southeastern Geology, v. 18, p. 49-57.

McGHEE, G. R., AND DENNISON, J. M., 1980, Late Devonian chronostratigraphic correlations between the central Appalachian Allegheny Front and central and western New York: Southeastern Geology, v. 21, p. 279-286.

McGHEE, G. R., AND BAYER, U., 1985, The local signature of sea-level changes, in Friedman, G. M., ed., Sedimentary and Evolutionary Cycles: Berlin, Springer-Verlag, p. 98-112.

MITCHUM, R. M., AND VAN WAGONER, J. C., 1991, High-frequency sequences and their stacking patterns: Sequence-stratigraphic evidence of high-frequency eustatic cycles: Sedimentary Geology, v. 70, p. 131-160.

NEAL, D., 1979, Subsurface stratigraphy of Middle and Upper Devonian clastic sequence in southern West Virginia and its relation to gas production: Unpublished Ph.D. Dissertation, West Virginia University, Morgantown, 142 p.

PEPPER, J. F., AND DE WITT, W., 1950, Stratigraphy of the Upper Devonian Wiscoy Sandstone and the equivalent Hanover Shale in western and central New York: Washington, D. C., United States Geological Survey, Oil and Gas Investigations Preliminary Chart OC 37, 2 sheets.

PEPPER, J. F., AND DE WITT, W., 1951, Stratigraphy of the late Devonian Perrysburg Formation in western and west-central New York: Washington, D. C., United States Geological Survey, Oil and Gas Investigations Chart OC 45, 1 sheet.

PEPPER, J. F., DE WITT, W., AND COLTON, G. W., 1956, Stratigraphy of the West Falls Formation of late Devonian age in western and west-central New York: Washington, D. C., United States Geological Survey, Oil and Gas Investigations Chart OC 55, 1 sheet.

PIOTROWSKI, R. G., AND HARPER, J. A., 1979, Black shale and sandstone facies of the Devonian "Catskill" clastic wedge in the subsurface of western Pennsylvania: Morgantown, United States Department of Energy, Eastern Gas Shales Project number 13, 40 p.

READ, J. F., GROTZINGER, J. P., BOVA, J. A., AND KOERSCHNER, W. F., 1986, Models for generation of carbonate cycles: Geology, v. 14, p. 107-110.

RICKARD, L. V., 1975, Correlation of Silurian and Devonian rocks in New York State: Albany, New York State Museum and Science Services Map and Chart series number 24.

RIDER, M. H., 1986, The Geological Interpretation of Well Logs: Glasgow, Blackie and Son, Limited, 175 p.

ROEN, J. B., 1981, Regional stratigraphy of the Upper Devonian black shales in the Appalachian Basin, in Roberts, T. G., ed., Geological Society of America Cincinnati '81 Field Trip Guidebooks, V. II, Economic Geology, Structure: Falls Church, American Geological Institute, p. 324-334.

ROEN, J. B., 1983, Geology of the Devonian Black Shales of the Appalachian Basin: Organic Geochemistry, v. 5, p. 241-254.

SINCLAIR, J. F., 1991, Middle and Upper Devonian lithostratigraphy of Elliot Knob, Virginia: Unpublished M.S. Thesis, University of North Carolina at Chapel Hill, Chapel Hill, 151 p.

SWARTZ, J. H., 1927, The Chattanoogan age of the Big Stone Gap Shale: American Journal of Science, v. 14, p. 485-499.

SWEENEY, J., 1986, Oil and gas report and maps of Wirt, Roane, and Calhoun counties, West Virginia: West Virginia Geological and Economic Survey Bulletin B-40, 100 p.

VAN TASSELL, J., 1987, Upper Devonian Catskill Delta margin cyclic sedimentation: Brallier, Scherr, and Foreknobs Formations of Virginia and West Virginia: Geological Society of America Bulletin, v. 99, p. 414-426.

VAN WAGONER, J. C., MITCHUM, R. M., CAMPION, K. M., AND RAHMANIAN, V. D., 1990, Siliciclastic Sequence Stratigraphy in Well Logs, Cores, and Outcrops: Concepts for High-Resolution Correlation of Time and Facies: Tulsa, American Association of Petroleum Geologists Methods in Exploration Series 7, 55 p.

WANLESS, H. R., AND SHEPARD, F. P., 1936, Sea level and climatic changes related to late Paleozoic cycles: Geological Society of America Bulletin, v. 47, p. 1177-1206.

WARNE, A. G., 1986, Stratigraphic analysis of the Upper Devonian Greenland Gap Group and Lockhaven Formation near the Allegheny Front of central Pennsylvania: Unpublished M.S. Thesis, New Brunswick, Rutgers University, 219 P.

WARNE, A. G., AND MCGHEE, G. R., 1991, Stratigraphic subdivisions of the Upper Devonian Scherr, Foreknobs, and Lock Haven Formations near the Allegheny Front of Central Pennsylvania: Northeastern Geology, v. 13, p 96-109.

ROLES OF EUSTASY AND TECTONICS IN DEVELOPMENT OF SILURIAN STRATIGRAPHIC ARCHITECTURE OF THE APPALACHIAN FORELAND BASIN

WILLIAM M. GOODMAN[1] AND CARLTON E. BRETT

Department of Earth and Environmental Sciences, University of Rochester, Rochester, New York 14627

ABSTRACT: Refined intrabasinal correlation of medial-Silurian strata has led to recognition of discrete eustatic and tectonic controls on the stratigraphic architecture of the Appalachian foreland basin. Major unconformities and disconformities are used to define third- and fourth-order sea-level cycles (depositional sequences and sub-sequences). Although they are asymmetric, most unconformities are present along both basin margins, and their timing corresponds with sea-level lowstands in other Silurian basins. The correlation with apparent, global, sea-level lowstands suggests a eustatic component to Silurian Appalachian basin unconformities.

Silurian sequences are divisible into systems tracts that are correlatable across the basin. Transgressive systems tracts are laterally correlative, retrogradational, carbonate and sandstone successions that onlap unconformities on northwestern and southeastern margins of the basin, respectively. Highstand systems tracts are thickest in the basin center and thin laterally toward each margin. They are divisible into early highstand phases, typified by aggradational, fine-grained siliciclastic and argillaceous carbonate successions, and late highstand (or regressive) phases, that characteristically exhibit a general upward-coarsening (progradational) pattern. These regressive deposits are typically divisible into two or more sub-sequences whereas the transgressive and early highstand systems tracts comprise a single subsequence.

Smaller-scale, discontinuity-bound, stratal packages that are interpreted to record fifth- and sixth-order sea-level changes, are analogous to parasequences sets and parasequences, respectively. Many of the small-scale cycles can be mapped basin-wide or until they are truncated under marginal unconformities. High-frequency, eustatic, sea-level changes or climatic oscillations are plausible mechanisms to explain the pervasiveness of these cycles.

Documented tectonic controls on stratigraphic architecture include lateral migration of the foreland-basin axis and uplifts along both the cratonic arch and the tectonically active eastern basin margin. These tectonic signatures are imprinted on marginal unconformities and are also recorded by progressive lateral shift in the locus of thickest accumulated sediment and deepest facies. Tectonically imprinted unconformities are more fully developed on one basin margin and are distinguished from purely eustatically generated unconformities by their asymmetry.

Consequently, large-scale sedimentary cycles, that are bounded by unconformities and correlate with sea-level changes in other basins, record an interplay between foreland basin geodynamics and eustatic processes of similar rate. In contrast, smaller-scale, sedimentary allocycles, that have recurrence intervals that outpace rates of foreland-basin flexure, more clearly record high-frequency, low-amplitude, eustatic sea-level changes or climatic oscillations.

INTRODUCTION

A key goal in the study of ancient foreland basins is to discriminate the stratigraphic signatures of eustatic processes and local tectonic flexure. Interest in eustatic sea-level fluctuations as a causative agent of sedimentary patterns in foreland basins and other settings has been fostered by the recognition of widespread small scale sedimentary cycles (Einsele and Seilacher, 1982; Bush and Rollins, 1984; Aigner, 1985; Goodwin and Anderson, 1985; Read and others, 1986; Grotzinger, 1986a, b; Heckel, 1990; Einsele and others, 1991). Furthermore, the broad application of sequence stratigraphic concepts (Vail and others, 1977; Van Wagoner and others, 1988) has underscored the global and probable eustatic nature of many large scale cycles. At the same time, models of foreland basin tectonics have been developed to explain large-scale architectural patterns (Quinlan and Beaumont, 1984; Beaumont and others, 1988). These models relate changes in rates and trends of subsidence to thrust loading and subsequent relaxation of the lithosphere. Conse-

quently, both lithospheric flexure and eustatic sea-level changes have been considered possible mechanisms for cyclicity in the rock record of foreland basins. This paper attempts to assess the relative importance of eustatic processes and local tectonics in the development of Silurian cycles in the Appalachian foreland basin.

SILURIAN ALLOCYCLES OF THE APPALACHIAN BASIN

Sedimentary cycles are a pervasive feature in Silurian strata of the Appalachian basin. Correlations of major carbonate and sandstone tongues between the Ontario-New York and Pennsylvania-Maryland Silurian outcrop belts were established as early as 1934 by C. K. Swartz. Hunter (1970) provided a major contribution to the understanding of these strata by correlating key limestone and sandstone units in the New York and Pennsylvania outcrop belts through the use of a large subsurface data base. Hunter's work corroborated earlier interpretations of Swartz (1934) that Silurian carbonate and siliciclastic tongues extend basinward from opposite margins and are time-correlative at the resolution of the established biostratigraphic zonations.

The sea-level curve published by Dennison and Head (1975)

---

[1] Present address: The Sear-Brown Group, 85 Metro Park, Rochester, New York 14623

Tectonic and Eustatic Controls on Sedimentary Cycles, SEPM Concepts in Sedimentology and Paleontology #4

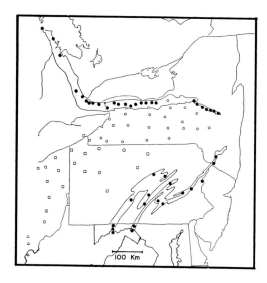

FIG. 1.—Map depicting location of key sections and well logs. Outcrops are shown as solid black circles; open circles are unpublished gamma-ray profiles from D. J. Crowley and others (unpublished data); open triangles are gamma-ray profiles from a dissertation by Lukasik (1988).

for the Siluro-Devonian succession of the Appalachian basin, together with more recent correlations, focused more attention on the apparent synchroneity of vertical facies changes in different parts of the basin. Cotter (1983, 1988, 1990), Brett (1983) and Brett and others (1990a, b) have illustrated the occurrence of large- and small-scale cyclicity at key sections in the Pennsylvania-Maryland and New York-Ontario outcrop belts.

The primary objective of the present study was to establish physical stratigraphic correlations of sufficient resolution to test prevalent models of allocyclicity and basin geodynamics. In order to make these correlations, data from over 100 outcrops in Ontario, New York and Pennsylvania were compiled (Fig. 1). Unpublished subsurface data of L. V. Rickard and D. J. Crowley, and data from a dissertation by D. M. Lukasik (1988) provide additional detail to the larger-scale basinal cross-sections of Hunter (1970).

We have focused primarily on the late Llandoverian and Wenlockian Series of the northern Appalachian basin to illustrate allocyclicity (Fig. 2). This interval is relatively thin (approximately 50 meters or less), is fairly well exposed, and is well correlated through the subsurface.

In previous papers, Brett and others (1990a, b) have divided the lower and medial Silurian succession into a series of depositional sequences and sub-sequences, most of which are laterally traceable across the basin. These papers also attempted to reconcile the genetic hierarchy model of Busch and Rollins (1984) with Exxon type sequence stratigraphy (Table 1). Third-order cycles (as defined by Busch and Rollins, 1984) can be interpreted as depositional sequences (as defined by Vail and others, 1977). Smaller-scale, fourth- through sixth-order cycles, can be interpreted as key components of sequences, such as sub-

sequences (as defined by Brett and others, 1990b; "simple sequences" of Vail and others, 1991), parasequence sets, and parasequences. The general characteristics of these different scales of cycles are discussed in the following sections.

Large Scale Sequences: Third-Order Cycles

In the first, classic treatment of Exxon-type sequence stratigraphy (Vail and others, 1977), unconformity-bound depositional sequences were interpreted to be generated by cyclic, eustatic, sea-level drops with a 1- to 10-million year recurrence interval. Recently, Vail and others (1991) redefined third-order sequences as spanning approximately 0.5 to 5.0 Ma. Similarity in unconformity patterns on various continental shelves has been the primary basis for the eustatic interpretation. Although part of this interpretation has been questioned (Pitman, 1978; Watts, 1982; Summerhayes, 1986), defense of, and refinement to, the original interpretation and resultant Phanerozoic sea-level curve has followed published critiques (Haq and others, 1988).

Brett and others (1990a, b) have divided Silurian strata of the Appalachian foreland basin into at least eight, third-order, depositional sequences (Fig. 3). Silurian sequences consist of mixed siliciclastic and carbonate sediments and are bounded by sharp, regionally angular unconformities that are particularly accentuated along basin margins. These third-order sequences represent intervals ranging from approximately 3.0 to 5.0 Ma based upon absolute dates for Silurian series and stages (McKerrow and others, 1985; Gale, 1985).

In general, sequences have a crudely lenticular geometry, being thinnest (commonly less than 5 m) and most carbonate dominated to the northwest, adjacent to the Algonquin Arch. Sequences are thickest (up to 40 m) to the southeast where a laterally migrating depocenter trended northeast-southwest (Bolton, 1957; Winder and Sanford, 1972; Brett and others, 1990b) from western to central New York and Pennsylvania. Sequences also thin as they grade into paralic facies to the southeast (proximal to siliciclastic sources).

Sequences are markedly incomplete along basin margins. Lowstand deposits are thin or absent. Transgressive to early highstand deposits comprise the bulk of the preserved strata. Late highstand deposits are commonly truncated beneath successive marginal unconformities.

Near the basin center or depocenter, sequences are not only thicker, but are more stratigraphically complete. The sequence bounding unconformities become less distinctive as lacunas decrease toward the depocenter. Lowstand deposits, however, remain poorly represented. Late highstand regressive deposits comprise most of the preserved strata.

Transgressive systems tracts.—

Transgressive systems tracts form the lower portions of sequences in all areas except depocenters. They tend to be relatively thin (centimeters to a few meters) and to contain relatively uniform facies across the basin. Along the northwest

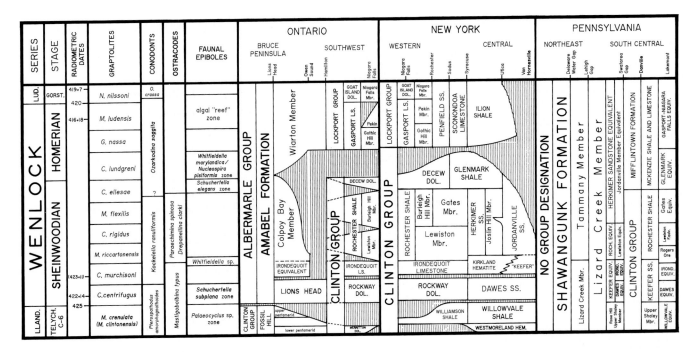

FIG. 2.—Chronostratigraphy of medial Silurian (late Llandoverian to early Ludlovian) strata in the northern Appalachian Basin and adjacent areas. Abbreviated time divisions: Lland.=Llandoverian; Telych.=Telychian; Lud.=Ludlovian; and Gorst.=Gorstian. Radiometric dates from Siveter and others (1988).

carbonate ramp (southern Ontario and western New York), transgressive deposits are represented by dolomitic carbonates, typically pelmatozoan-rich, storm-winnowed packstones and grainstones (Bolton, 1957; Zenger, 1965). These western ramp systems tracts have sharply defined bases, commonly with slight irregularities and rip-up clasts, derived from erosion at underlying sequence-bounding unconformities.

Transgressive carbonates exhibit either tabular or irregular geometries dependent upon component depositional systems. For example, the Irondequoit Limestone (Sequence V) in western New York and Ontario contains level-bottom encrinites and small bryozoan bioherms. The Irondequoit Formation displays only gradual westward thinning from about 10 to 5 m (Fig. 4). In contrast, other transgressive deposits pinch and swell and, probably reflect a series of large-scale submarine dunes or bars. For example, the lower or Gothic Hill Member of the Gasport Formation (Sequence VI) may range from 1 to 10 m in thickness over a distance of a few kilometers (Zenger, 1965; Fig. 5). Common sedimentary structures include planar and trough cross-stratification that reflect migration of bedforms.

Faunas of transgressive, skeletal limestones are dominated by pelmatozoans, especially plates of camerate crinoids and rhombiferan cystoids. Brachiopods are represented by robust, pentamerids, atrypids or athyrids (for example, *Whitfieldella*). Tabulate and rugose corals, stromatoporoids and bryozoans are typically present and may form local biostromes or small bioherms. For example, corals are prolific in the transgressive carbonates of Sequence IV (Fossil Hill Formation; see Bolton, 1957; Winder and Sanford, 1972), and Sequence VI (Gasport

and lower Goat Island Formations; see Zenger, 1965). Collectively, these faunas indicate a benthic assemblage (BA) 2 to 3 position (classification of Boucot, 1975; see Brett, in press). These fossils represent shallow ramp communities that lived in near-wave-base environments under euphotic zone conditions.

Upper surfaces of the transgressive skeletal carbonates are sharply defined and may contain firmground features. For example, they may be overlain by very thin, shell-rich, glauconite-bearing, condensed beds ("downlap" and "backlap" shell beds of Kidwell, 1991). Small bioherms may extend upward from the condensed deposits into the overlying highstand facies (see Fig. 4). For example, in Sequence V, algal-fistuliporoid mounds occur at the top of the Irondequoit Limestone and protrude up to 2 m into the overlying Rochester Shale (Hewitt and Cuffey, 1985). In Sequence VI, larger stromatoporoid and tabulate-dominated bioherms extend up to 5 m from the top of the Gothic Hill Member (Gasport Formation) into the argillaceous dolostones of the Pekin Member (Zenger, 1965; Brett and others, 1990a; see Fig. 5). These mounds apparently represent "keep-up" structures (Sarg, 1988) that grew vertically in response to sea-level rise.

In more basinward sections (for example, in west-central New York), crinoidal packstones and grainstones, comprising the transgressive systems tract, grade into fossiliferous, nodular, argillaceous pack- or wackestone. Pelmatozoans typically decrease in abundance in these more basinal, finer-grained carbonates. Corals and stromatoporoids are rare and small; diverse brachiopods indicate a BA-4 position (Brett, in press). Fossil preservation tends to be rather poor with most shells disarticu-

TABLE 1: RELATIONSHIP BETWEEN NOMENCLATURE FOR HEIRARCHY OF ALLOCYCLE AND PREVIOUS SEQUENCE STRATIGRAPHIC OR CLASSICAL TERMINOLOGY. T-R UNIT (TRANSGRESSIVE-REGRESSIVE UNIT) MODIFIED FROM BUSCH AND ROLLINS (1984). SEQUENCE TERMINOLOGY ADAPTED FROM VAIL AND OTHERS (1991).

Magnitude of T-R Unit	Sequence Equivalent	Previous Nomemnclature
1st-Order: > 50 Ma	Mega-sequence	1st-Order Depositional Sequences: Vail and others (1977)
2nd-Order: 5 to 50 Ma	Super-sequence, Super-sequence Sets	2nd-Order Depositional Sequences: Vail and others (1977, 1991) Synthems: Chang (1975), Ramsbottom (1979)
3rd-Order: 1.5 to 5 Ma	Sequences	3rd-Order Depositional Sequences: Vail (1977) 3rd-Order Depositional Sequences (in part): Vail (1991), Einsele and others (1991)
4th-Order: 0.9 to 1.5 Ma	Sub-sequences	Mesothems: Chang (1975), Ramsbottom (1979) 3rd-Order Depositional Sequences (in part): Vail and others (1991)
5th-Order: 40,000 to 450,000 years	Parasequences Sets	Cyclothems: Chang (1975), Heckel (1977), Heckel and others (1979), Ramsbottom (1979) Megacyclothems: Moore (1936), Fischer (1982) PAC Sequences: Goodwin and Anderson (1985) 4th-Order Cycles (in part): Vail and others (1991), Einsele and others (1991)
6th-Order: 50,000 to 100,000 years	Parasequence	Punctuated Aggradational Cycles (PACs): Goodwin and Anderson (1985) 5th-Order Cycles (in part): Einsele and others (1991) Parasequences: Einsele and others (1991), Vail and others (1991)
7th-Order:	Rhythms	6th-Order Cycles: Vail and others (1991) 5th-Order Cycles (in part): Einsele and others (1991)

lated and partially comminuted. The presence of glauconite and/ or phosphatic granules, bioturbated fabrics, and firmgrounds or hardgrounds indicates gradual accumulation of carbonate sediment in quiet water. Such facies are well represented in the Manitoulin Dolostone (Sequence I), Merritton Dolostone (Sequence IV), and portions of the Irondequoit Limestone (Sequence V) (Brett and others, 1990a, b).

In the deepest, basin-center facies belt, carbonate deposition failed completely; transgressive deposits are represented only by a very thin lag of quartz pebbles or granules, phosphatic or glauconitic granules, and thin coquinite. For example, the base of the Williamson Shale (Sequence IV), at basin center localities, is marked by the Second Creek phosphatic bed, a 0.5- to 10.0-cm thick lag deposit that rests on an unconformity. The Second Creek bed is overlain by black monograptid-bearing shale, indicating deep shelf or basinal, BA-5 conditions (Eckert and Brett, 1989). Evident reworking of fossil debris and mixed benthic faunas indicate that the phosphate bed is a highly condensed, transgressive, lag deposit.

Southeast of the basin center, most Silurian transgressive systems tracts are represented by distinctive beds of sandy, hematitic, crinoidal packstones, commonly containing complex mixtures of reworked fossils. These beds also contain quartz and phosphatic granules or nodules, and oolites or spastoliths (see Cotter, 1990, 1992). These condensed beds have sharp bases and may contain rip-up clasts. Thin concentrated layers of

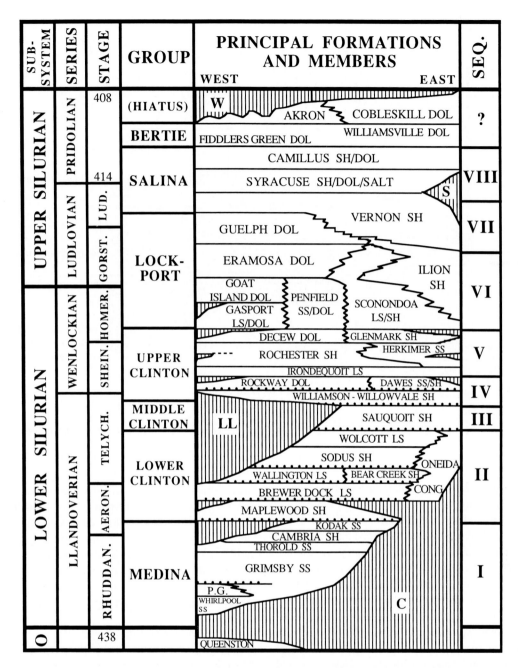

FIG. 3.—Generalized Silurian stratigraphy of New York State illustrating division into third-order depositional sequences. Vertical axis is scaled to time; horizontal is west to east distance along the New York outcrop belt from Niagara Gorge to near Utica, N.Y., approximately 300 km. Abbreviations: 0=Ordovician; Rhuddan.=Rhuddanian; Aeron.=Aeronian; Telych.=Telychian; Shein.= Sheinwoodian; Homer.=Homerian; Gorst.=Gorstian; Lud.=Ludfordian. Ages listed are estimates for beginning and end of Silurian Period and end of Ludlovian (in millions of years). For formations and members: P.G.=Power Glen Shale, DOL=dolostone, LS=limestone, SH=shale, SS=sandstone; dotted lineds indicate persistent hematitic/phosphatic beds; major unconformities are labeled with upper case letter C=Cherokee (basal Silurian); LL=Late Llandovery; S=Salinic; W=Wallbridge; SEQ.=sequences; see Brett and others (1990b), and text for explanation. Chart modified from Rickard (1975).

oolitic hematite form caps on these beds. Fossil evidence suggests a BA-3 to 4 mid-ramp position. Such complex transgressive beds include several of the well known "Clinton iron ores" of the New York outcrop belt, such as the Furnaceville, the Sterling Station, and Wolcott Furnace ores in the lower Clinton Group (Sequence II), the Westmoreland ore at the base of

Sequence IV and the Kirkland Hematite at the base of Sequence V (Gillette, 1947; Brett and others, 1990b).

Still farther to the southeast, the hematitic, arenaceous carbonates pass laterally into predominantly siliciclastic facies. These proximal transgressive deposits are divided into two facies. In some cases, the hematitic carbonates grade laterally

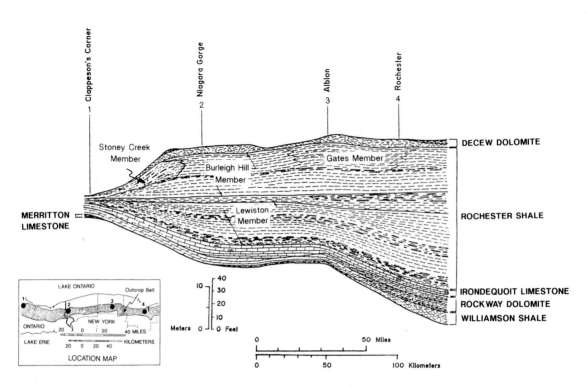

FIG. 4.—Cross section of the upper Clinton Group (Sequences IV and V) balanced on Lewiston E subunit between Hamilton, Ontario and Sodus, New York.

into green or dark gray clay shales containing a restricted fauna, including eocoeliid or lingulid brachiopods and, in some cases, eurypterids (BA-1 to 2 classification of Boucot, 1975). These shales, in turn, grade laterally toward the strand line into purple or maroon mudstones containing thin hummocky cross-stratified sandstone interbeds, as are well exposed in the Rose Hill Formation of Pennsylvania (Cotter, 1988). Such green to purple clay shales are thought to represent low energy inner ramp "lagoonal" muds that pass laterally into tidal flat facies of the Lizard Creek Member (Shawangunk Formation). These facies characterize transgressive systems tracts of Sequences II and III (for example, Sodus and Sauquoit Shales in New York and laterally equivalent Rose Hill Formation in Pennsylvania). Analogous shaly sediments are also seen in the near shore transgressive facies of the lower Lockport Group (Sequence VI). Green to black Ilion-McKenzie shales grade eastward into Vernon or Bloomsburg red claystones (Zenger, 1965; Swartz, 1935).

An alternate proximal facies spectrum for the transgressive systems tracts is seen in Sequences I and V. In these cases, transgressive, phosphatic or hematitic carbonates are replaced to the southeast by medium to thick-bedded, hummocky to trough cross-stratified, quartz arenites. Thus, for example, the Kirkland Hematite grades into the quartz arenites of the Keefer Sandstone (in Pennsylvania) or lower Jordanville Sandstone (in east central New York; Zenger, 1971; Brett and others, 1990b). Body fossils

are rare or absent in the sandstones, but *Skolithos* and other trace fossils, in conjunction with planar cross-stratification and lamination, suggest a very shallow subtidal to intertidal (BA-1 to 2) depositional environment. Thin concentrations of hematitic ooids at the tops of sandstone-rich intervals may represent condensed sections associated with surfaces of maximum starvation. These sheetlike sandstone bodies probably represent reworking of lowstand sand deposits during initial sea-level rise.

Highstand systems tracts.—
The highstand deposits of Silurian sequences are sharply separated from transgressive systems tracts by distinctive surfaces of sediment starvation. These surfaces are nearly as laterally traceable as sequence boundaries.

Early highstand deposits consist primarily of fine-grained siliciclastics or argillaceous carbonates. In northwestern sections, early highstand deposits generally consist of dolomitic mudstone with sparse skeletal debris. Thin, shell-rich storm layers may also be present, as for example, in the lower part of the Rochester Shale in Sequence V (see Fig. 4). As noted previously, small bioherms may extend upward into and interfinger with, these early highstand mud-rich deposits. However, carbonate deposition appears generally to have decreased during the early, rapid sea-level rise phase of the highstand systems tracts.

In more basinal areas, early highstand facies consist of dark-

WEST

EAST

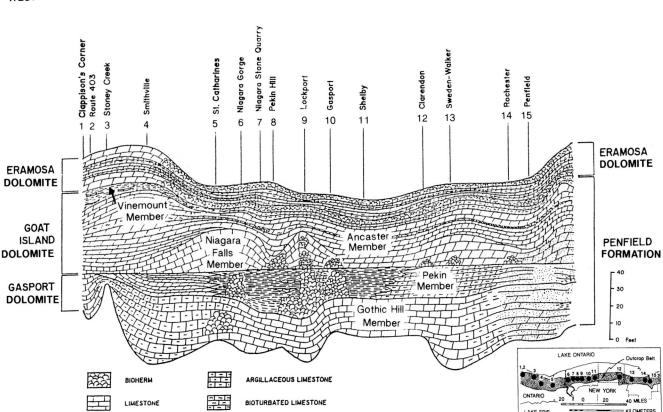

FIG. 5.—Cross section of lower Lockport Group (Sequence VI) with base on Niagara Falls Member of Goat Island Formation as datum. Note retrogradational, crinoidal grainstone to biohermal, argillaceous wackestone sub-sequences represented by the Gasport and Goat Island Formations.

gray to black shales or argillaceous carbonates. These fine-grained facies may contain thin, lenticular beds of skeletal debris or silt that represent tempestites. Such deposits occur in the lower Cabot Head Shale (Sequence I, Eckert, 1984) and in the lower Williamson and Willowvale Shale (Sequence IV, Eckert and Brett, 1989). Early highstand muds accumulated in low energy settings below normal wave base, but were occasionally affected by storm waves. Diverse brachiopod, bryozoan, and trilobite faunas indicate a BA-3 to 4 assemblage in these deposits.

On the southeastern ramp, highstand dark-gray shales grade laterally upslope into medium to greenish-gray shales with interbedded bioturbated siltstones and cross-stratified, fine sandstones. Body fossils tend to be sparse, but trace fossils, especially crawling and resting traces (*Cruziana* and *Rusophycus*) are common on the bases of thin sandy tempestites. Such sandy tempestites are common in the Joslin Hill facies of the Herkimer Sandstone of Sequence V (Zenger, 1971). A benthic assemblage-2 to 3 fauna is represented by the sparse body fossils in these beds. These early highstand facies pass laterally into thinly bedded, gray sandstones, alternating with green shale beds that may contain small *Skolithos*, *Planolites* or *Arthrophycus* trace

fossils indicative of lower intertidal depositional environments.

Late highstand or regressive facies are incompletely preserved at the basin margins. These condensed facies are typically shell-rich in the west, but pass eastward into sandy, phosphatic and/or hematitic carbonates. Where preserved, these deposits exhibit a coarsening-upward and bed-thickening upward succession, indicating progradation. In northwestern sections, this pattern is recorded in upward thickening and amalgamation of silty carbonate beds. In certain southeastern sections, proximal to the siliciclastic source area, progradation is recorded in amalgamated, hummocky cross-stratified sandstones facies. Faunal changes from BA-4 to BA-3 assemblages indicate upward shallowing within these deposits from below-wave-base to near wave-base conditions.

Progradational patterns are best displayed in the late highstand facies of Sequence V, in the Herkimer Sandstone. The Joslin Hill facies contain two progradational successions (sub-sequences, see below). Each commences with a hematitic shelly deposit representing a transgressive lag. This lag deposit is sharply overlain by greenish-gray clay shales that display an upward increase in the frequency of thin sandstone beds. Clay shales are, in turn, capped by trough cross-stratified, reddish-brown sandstones with lenses of conglomerate. Each of these coarse sandstone intervals is sharply overlain by fossiliferous sandstones and grainstones representing marine flooding intervals.

Causal mechanisms of Silurian sequences.—

Silurian depositional sequences in the Appalachian basin appear to have been the result of basin wide fluctuations in relative sea level. Transgressive, early highstand, and late highstand (regressive) systems tracts are recognizable in all cases. Except in basinal areas, sequences are sharply bounded at their bases by erosional surfaces that are very low angle, regionally angular unconformities. Surfaces of maximum starvation (maximum flooding surfaces) are also well defined across the basin profile.

Each sequence records a major cycle of sea-level rise and fall, probably representing about 50 to 100 m of water depth change (see Brett and others, 1993 for estimates of absolute depths of Silurian benthic assemblages). The sequence-bounding unconformities formed by subaerial to submarine erosion of basin margins during peak lowstands of sea level. Equivalent, relatively coarse-grained intervals in the basin center may represent lowstand sediments that have prograded basinward from erosional, shallow ramp settings.

Thin, storm-winnowed, sediment blankets of the transgressive systems tract were deposited in shallow-water, relatively sediment-starved environments during initial sea-level rise; maximum sediment starvation, recorded as condensed deposits or disconformities, occurred as relative sea level rose at a maximum rate. Coarse-grained siliciclastics were probably trapped in coastal estuarine areas.

Early highstand muds accumulated under deepest water (maximally transgressive) conditions. Finally, the late highstand reflects initial lowering of sea level, typically coupled with progradation of shallow water facies as increased volumes of coarse-grained siliciclastics were delivered to the basin.

Vail and others (1991) noted that local tectonism creates accommodation space and may enhance or subdue sequence and systems tract boundaries on continental shelves; however, tectonism alone does not *create* these surfaces. Their reason for advocating a eustatic mechanism lies in the apparent synchroneity of unconformity-generating sea-level falls recorded on passive continental margins. If Appalachian basin sequence boundaries are eustatically produced, erosional events should be associated with major sea-level falls documented elsewhere.

Johnson and others (1985) have documented four major, global sea-level highstands and intervening lowstands during the Llandoverian, and we have recognized major lowstands during the early and late Wenlockian in North America and Great Britain. The basal Silurian Cherokee Unconformity is a widespread, probably global, signal in the rock record (McKerrow, 1979). The sea-level fall associated with this unconformity has long been considered glacio-eustatic in origin and is also known to coincide with the climax of a major biological extinction event. The tectonic significance of the

Cherokee Unconformity in the Appalachian basin has only recently been emphasized (Quinlan and Beaumont, 1984; Tankard, 1986; Middleton, 1991).

The base of Sequence II is placed at the contact between the Medina and Clinton Groups. The age of this group boundary, based upon ostracod and conodont biostratigraphy is mid-Llandoverian (Aeronian B_3 to C_1). A mid-Llandoverian sea-level fall is also recorded in Poland and Wales (McKerrow, 1979) as well as many regions within North America (Johnson, 1987).

Sequence III commences at the boundary between the lower and middle portions of the Clinton Group. The timing of the sea-level fall linked to this unconformity is mid- to late Llandoverian (Telychian C_4) on the basis of ostracods and conodonts. A sea-level fall near the Aeronian/Telychian boundary is also described by Johnson (1987) and Johnson and others (1991) for many regions of North America. The global pattern, however, is less clear. Regressive trends are noted by McKerrow (1979) in Wales, Shropshire, and Girvan, but the basal Telychian sea-level lowstand is not apparent elsewhere.

The Sequence IV boundary occurs at the contact between the middle and upper portions of the Clinton Group. This unconformity was generated by a widespread late Llandoverian (Telychian C5) sea-level lowstand (Johnson, 1987).

The base of Sequence V is placed in the upper Clinton Group. The lowstand represented by unconformity is early Wenlockian (early Sheinwoodian). A sea-level fall of this age has been noted by Dennison and Head (1975) throughout the Appalachian basin and is also well-documented in the Welsh Borderland type area (Siveter and others, 1989), in Europe (McKerrow, 1979) and elsewhere in North America (Johnson, 1987).

The base of Sequence VI is the sharp contact between the Clinton and Lockport Groups and equivalent strata of the Mifflintown Formation. The sea-level fall responsible for the generation of this unconformity is late Wenlockian (Homerian) on the basis of conodonts and is also recognized in the Welsh Borderland type area (Siveter and others, 1989) and Gotland (McKerrow, 1979).

Thus, it appears that a high degree of interbasinal correlation of sea-level patterns across continents exists. Eustatic sea-level lowstands are probably the primary cause of third-order, sequence-bounding unconformities, although the driving mechanism remains poorly understood. Local tectonic processes may enhance or alter the patterns of unconformity, as noted in the last section of this paper.

Sub-sequences: Fourth-Order Cycles

Silurian depositional sequences encompass approximately 3 to 5 Ma. However, in all cases, these sequences are further divisible into cycles of similar motif that contain less erosive bounding disconformities. These "sub-sequences" (Brett and others, 1990b) are considered to reflect shorter duration, approximately 1.0- to 1.5-Ma cycles in the range referred to by Busch and Rollins (1984) as fourth-order cycles. Sub-sequences

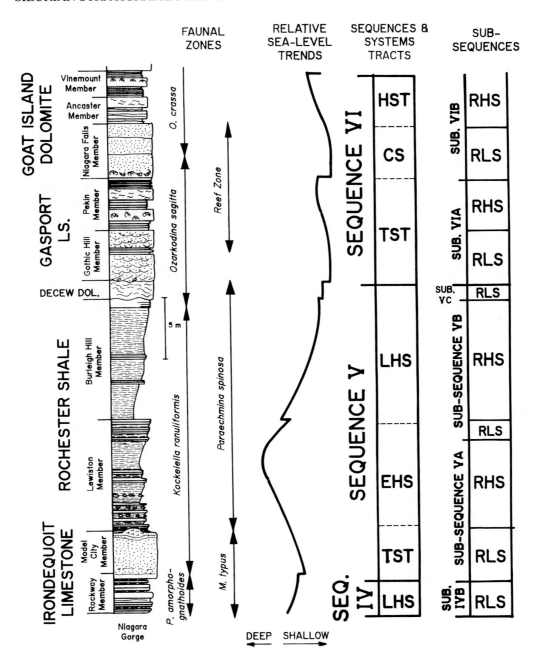

FIG. 6.—Columnar section, biostratigraphic zones, relative sea-level trends, and component system tracts of Sequence V at Niagara Gorge, New York. Abbreviations are as follows: TST=transgressive systems tract; EHS=early highstand; LHS=late highstand; CS=condensed section; RHS=relative highstand; and RLS= relative lowstand.

are also comparable to Carboniferous mesothems described by Ramsbottom (1979) that apparently record rapid regressions followed by progressive, stepwise transgressions.

Other researchers, including Vail and others (1991), have recently broadened the temporal range of third-order sequences to approximately 0.5 to 5.0 Ma. This broader definition of third-order sequences would also include sub-sequences of Brett and others (1990b). As noted, we chose to retain the Busch and Rollins (1984) classification of allocycles and to consider the smaller-scale sub-sequences as a discrete scale in a hierarchy, because: a) the bounding discontinuities (except those which

also coincide with sequence boundaries) involve less erosional truncation than do third-order sequence boundaries, particularly in basinal areas; and b) two or more sub-sequences consistently occur within each Silurian sequence, suggesting a nested hierarchy of allocycles.

The nested nature of sub-sequences within larger third-order sequences can also be reconciled with the systems tract concept of Exxon sequence stratigraphy. The first (lowest) sub-sequence within each Silurian sequence comprises the transgressive and early highstand systems tracts of that sequence. Hence, the basal, bounding surfaces of the sequence and the lowest sub-

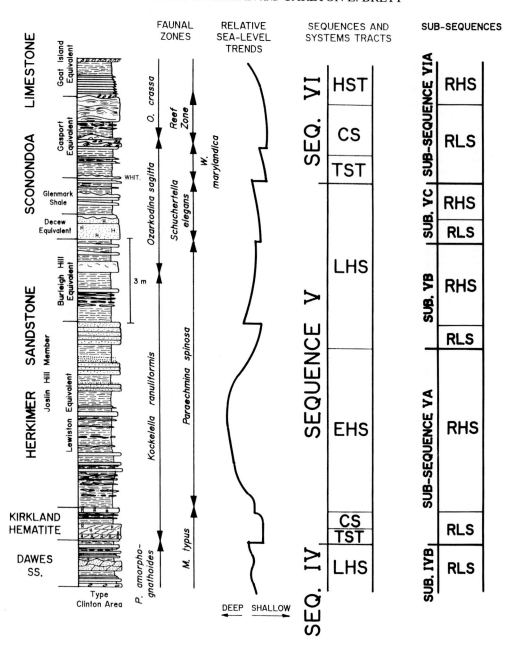

FIG. 7.—Columnar section, biostratigraphic zones, relative sea-level trends, and component systems tracts of Sequence V in type Clinton area of east-central New York. Abbreviations are as follows: TST=transgressive systems tract, EHS=early highstand, LHS=late highstand, CS=condensed section; and WHIT.=*Whitfieldella*.

sequence are coincident. The second and higher sub-sequences, together, comprise the aggradational to progradational, highstand or regressive systems tract of the third-order sequence. The highstand systems tract may contain two sub-sequences that correspond to early (retrogradational to aggradational) and late (aggradational to progradational) phases.

Each sub-sequence exhibits a sequence-like internal facies pattern, commencing with relatively shallow water sediments ("relative lowstand deposits" of Brett and others, 1990b) that are sharply overlain by deeper water facies ("relative highstand deposits"). These two phases may be separated by a thin

condensed section or sharp surface of starvation (marine flooding surface).

As in the case of sequences, sub-sequences display a roughly lenticular cross-sectional geometry across the foreland basin. Sub-sequences are thickest near the axis of maximum basin subsidence. The bounding surfaces of sub-sequences are subtle and conformable near the depocenter but become progressively sharper and more erosional toward both eastern and western basin margins.

The systems tract analogs within sub-sequences display a regular pattern of facies change across the basin. The relative lowstand deposits (analogous to transgressive systems tracts of the third-order cycles) grade from thin (0.5 m) skeletal lag deposits along the cratonic arch to bundles of silty or shell-rich beds near the basin axis and finally into sandy hematitic or phosphatic carbonates and thin (0.5-3 m) sandstone tongues along the southeastern basin margin. The relative highstand deposits of each sub-sequence tend to consist of argillaceous carbonates or calcareous shale along the Algonquin Arch, dark gray or green shales in the basin center, and siltstone or silty mudstone with sandy tempestites to southeast.

Brett and others (1990) describe some 15 sub-sequences in the Silurian of the Appalachian basin. For example, in western New York, the Wenlockian Series (Fig. 6) is divisible into at least five sub-sequences, three in sequence V and two in sequence VI; each sub-sequence is bounded by sharp, locally erosional surfaces.

Sub-sequence VA consists of the Irondequoit Limestone and the lower portion of the Lewiston Member of the Rochester Shale. The Irondequoit Limestone represents the transgressive systems tract of third-order Sequence V, and also comprises the relative lowstand interval of fourth-order Sub-sequence VA. The Irondequoit Limestone contains a retrogradational or deepening facies trend that continues through the basal beds of the Lewiston Member (informally designated Lewiston A and B submembers by Brett, 1983). In contrast, the overlying upper beds of the Lewiston Member (Lewiston C and D submembers), comprising the relative highstand of Sub-sequence VA, are arranged in a progradational facies succession.

The base of Sub-sequence VB is marked by a thin interval of bryozoan-rich packstone and grainstone beds located near the middle of the Rochester Shale (informally designated as Lewiston E submember by Brett, 1983). In addition to representing the relative lowstand interval of Sub-sequence VB, these fossiliferous beds mark the base of the progradational, late highstand systems tract of third-order Sequence V. The Lewiston E packstones are abruptly overlain by sparsely fossiliferous gray shales at a surface of sediment starvation (marine flooding surface). The overlying, sparsely fossiliferous shales (Burleigh Hill Member) exhibit an aggradational to progradational facies trend that is characteristic of the highstand deposits of the larger sequences.

Correlative sub-sequences are recognizable in east-central New York, where Sub-sequence VA is represented by the crinoidal Kirkland Hematite (relative lowstand interval) and overlying lower shaley facies (relative highstand interval) of the Herkimer Sandstone (Fig. 7). Sub-sequence VB commences with crinoidal sandy dolostone that is located near the middle of the Herkimer Sandstone. This dolostone (relative lowstand interval) is sharply overlain by a crudely progradational succession (relative highstand interval) of green shales and cross-bedded red sandstones.

Comparable facies successions representing sub-sequences VA and VB are also found in the Lizard Creek and Tammany Members of the Shawangunk Conglomerate of eastern Pennsylvania. These Shawangunk sub-sequences confirm an along-strike relationship with easternmost sections of the Herkimer Sandstone in New York and facilitate physical stratigraphic correlation. Based upon correlation of facies patterns and disconformity surfaces between the Niagara region and eastern Pennsylvania, sub-sequences appear to be circumbasinal.

Causal mechanisms of sub-sequences.—

Fourth-order boundaries resemble more widely spaced third-order surfaces located in the same section. Both types of cycle boundaries are typically sharp and erosional, and juxtapose relatively shallow-water facies over deeper water deposits. Fourth-order cycles are clearly analogous to sequences, but represent shorter-term and smaller-scale fluctuations in relative sea level, probably amounting to 20 to 50 meters of net depth change (see Brett and others, 1993). The major distinction between the two scales of bounding surfaces lies in the extent of erosion along transects perpendicular to the basin axis.

Fourth-order cycle boundaries are circumbasinal indicating that they represent allocyclic lowstands. Again, in some cases, correlation with sea-level falls in other basins suggests a strong eustatic control. For example, a refined relative sea-level curve for the medial Silurian interval can be compared with curves generated for the Wenlock Edge type area (Fig. 8) and elsewhere (see Sivieter and others, 1988; Johnson and others, 1991). Brett and others (1990b) and Kemp (1991) have pointed to the obvious similarities of stratigraphic successions in mid-Silurian shelf facies on the Laurentian (North American) and Avalonian (Welsh Borderland) sides of the Iapetus Ocean. Thus, the latest Llandoverian Williamson-Willowvale Shales of the New York outcrop belt exhibit facies similarities with the British Purple Shales. The early Wenlockian shallow ramp carbonates of the Irondequoit Limestone are analogous and correlative to the Buildwas Beds. The mid-Wenlockian Rochester Shale correlates with the Coalbrookdale Shale. The late Wenlockian Lockport carbonates correlate with the Much Wenlock Limestone.

Evidence for early and late Wenlockian lowstands and a mid-Wenlockian highstand has been found in deep water, graptolite facies of the Denbigh Trough, Lake District, and southern uplands of Great Britain; in the Federickton Trough-Metapedia Basin of Maine and New Brunswick (Kemp, 1991) and in shallow water carbonates of Gotland and Estonia (Johnson and others, 1991). Moreover, sea-level lowstands in the early and late Wenlockian, corresponding to the Irondequoit and Gasport Limestones in New York, have been identified on other paleocontinents including Siberia and Gondwanaland (Johnson and others, 1991). Large-scale, probably eustatically driven, regressive maxima both at the base and top of the Wenlock Series appear in nearly all areas and are likely to be related to 3rd order cycle boundaries. The medial Wenlockian regression occurring about midway in the Rochester Shale of the Appalachian Basin has yet to be recognized in the Coalbrookdale Beds of Wenlock Edge. However, the yype Wenlock Series is rather thick and some intervals are poorly exposed. Further work in the type area might result in more refined sea-level curves for

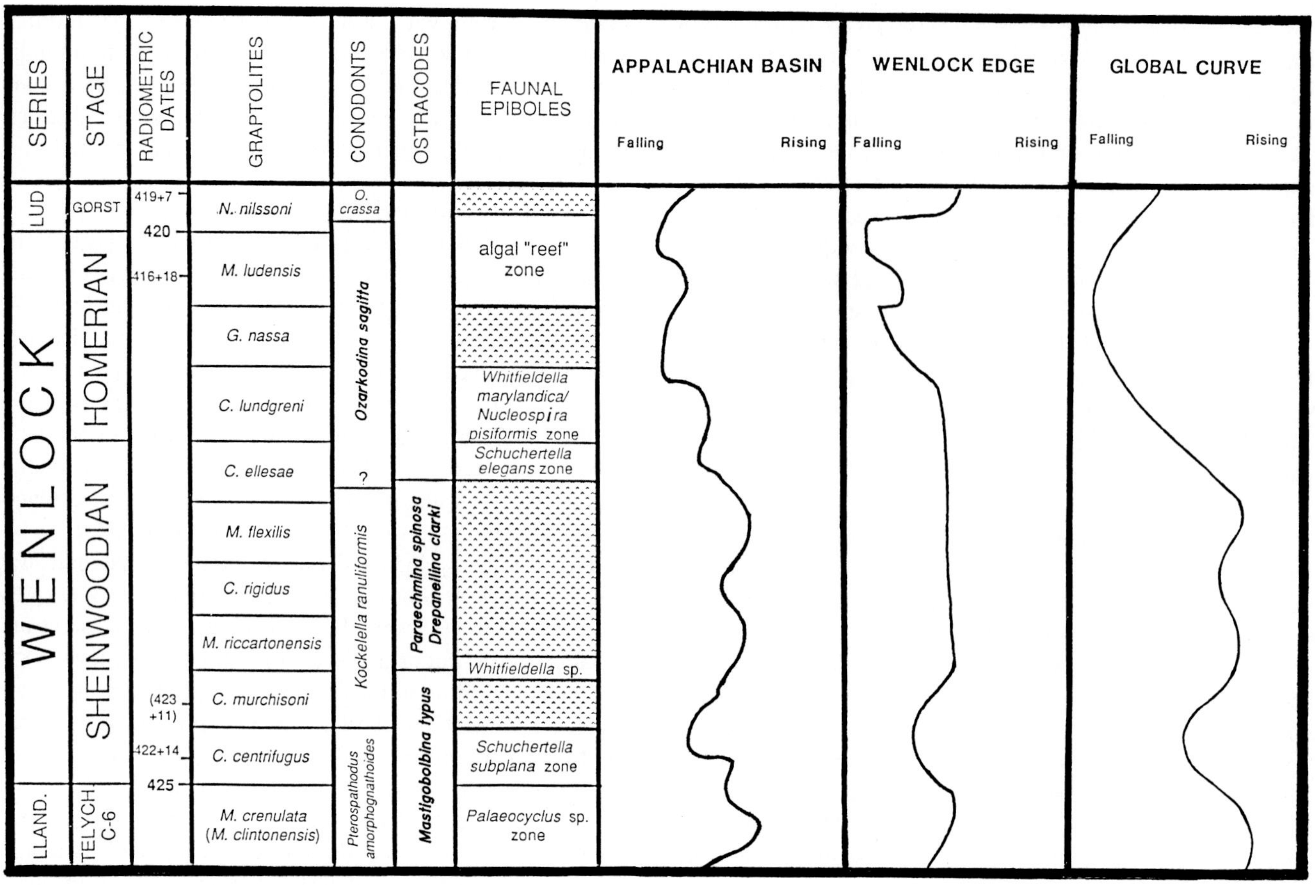

SERIES	STAGE	RADIOMETRIC DATES	GRAPTOLITES	CONODONTS	OSTRACODES	FAUNAL EPIBOLES	APPALACHIAN BASIN (Falling / Rising)	WENLOCK EDGE (Falling / Rising)	GLOBAL CURVE (Falling / Rising)
LUD.	GORST	419+7	N. nilssoni	O. crassa		algal "reef" zone			
		420	M. ludensis	Ozarkodina sagitta					
	HOMERIAN	416+18	G. nassa			Whitfieldella marylandica/ Nucleospira pisiformis zone			
WENLOCK			C. lundgreni						
			C. ellesae	?		Schuchertella elegans zone			
			M. flexilis	Kockelella ranuliformis	Paraechmina spinosa Drepanellina clarki				
	SHEINWOODIAN		C. rigidus						
			M. riccartonensis			Whitfieldella sp.			
		(423 +11)	C. murchisoni		Mastigobolbina typus				
		422+14	C. centrifugus	Pterospathodus amorphognathoides		Schuchertella subplana zone			
LLAND.	TELYCH C-6	425	M. crenulata (M. clintonensis)			Palaeocyclus sp. zone			

FIG. 8.—Relative sea-level curve (smoothed) for Appalachian Basin and type Wenlockian of Britain. These curves are compared against the global Silurian sea-level curve of Johnson and others (1991).

comparative analysis. A minor regression corresponding in time to the mid-Rochester Shale event is shown on the Silurian global sea-level curve of Johnson and others (1991). This diagram indicates that evidence for a minor shallowing event in the mid-Sheinwoodian exists in Sweden and Estonia, but the evidence is not discussed in the text.

Detailed study of the Much Wenlock Limestone demonstrates two closely spaced regressive maxima (Lower and Upper Quarried Limestones) separated by a minor transgression (Nodular Beds) (Collins, 1989). This pattern of regression-transgression-regression is reflected in the Appalachian basin by the lower grainstone (Gothic Hill Member) of the Gasport Limestone, the upper argillaceous dolostone (Pekin Hill Member) of the Gasport, and the lower grainstone (Niagara Falls Member) of the overlying Goat Island Dolostone. Such evidence suggests that at least some fourth-order sub-sequences, like third-order sequences, are also interbasinal and probably eustatic in origin.

Cyclothems or Parasequence Sets: Fifth-Order Cycles

Sub-sequences are typically divisible into at least two smaller-scale cycles that correspond to the relative lowstand and highstand

intervals. These small-scale cycles range in thickness from less than 1.0 m to a few meters. Their recurrence interval is estimated to be about half that of fourth-order sub-sequences, and, therefore, is approximately 0.4- 0.5 Ma. This time range coincides with the range typical of cyclothems or fifth-order cycles of Busch and Rollins (1984) (note that cyclothems would be considered fourth-order cycles by Vail and others, 1991).

Fifth-order cycles as defined herein display varied internal facies patterns and lateral changes in symmetry comparable to those noted by Brett and Baird (1986) and Einsele and Bayer (1991) for Devonian and Jurassic cycles, respectively. Asymmetric retrogradational (upward deepening) facies successions are typical of the basal fifth-order cycle in Silurian sequences. This basal cycle includes the lower sequence-bounding unconformity and the transgressive systems tract of each third-order sequence. The overlying fifth-order cycles may be slightly aggradational to symmetrical in the lower portion of the highstand systems tract or progradational in the late highstand systems tract.

For example, fourth-order Sub-sequence IVA contains two fifth-order cycles (Fig. 9). In central New York, this interval commences with a thin hematite and phosphate grainstone

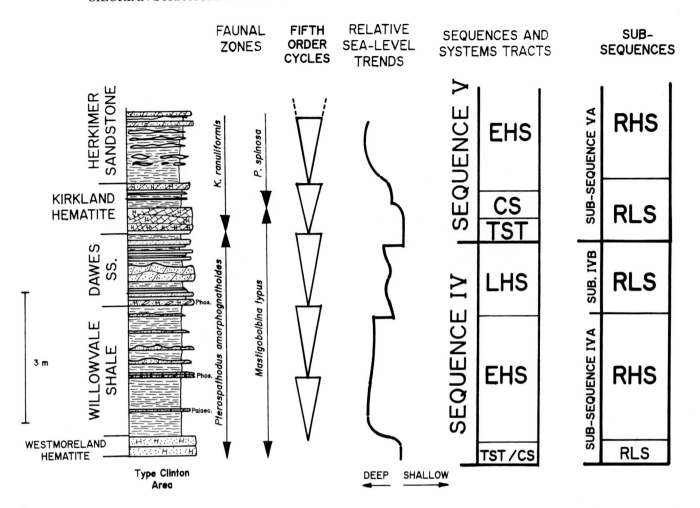

FIG. 9.—Columnar section, biostratigraphic zones, fifth-order cycles, relative sea-level trends, component systems tracts, and sub-sequences of Sequence IV in the type Clinton area of east-central New York. Abbreviations for sequences are as follows: TST=transgressive systems tract; EHS=early highstand; LHS=late highstand; CS=condensed section. Abbreviations for sub-sequences are as follows: RLS=relative lowstand deposits; and RHS=relative highstand deposits. Phos.=phosphate pebble bed; and Palaeo.=*Palaeocyclus*.

(Westmoreland Hematite) which is, itself, a thin representative of the transgressive systems tract of third-order Sequence IV. This thin, condensed unit is overlain by greenish-gray, fossiliferous shale (Willowvale Shale). A thin, phosphate pebble-bearing dolostone separates the Willowvale Shale into two subequal parts. These two aggradational fifth-order cycles of the Willowvale Shale comprise the early highstand systems tract of Sequence IV and the relative highstand interval of Sub-sequence IVA.

Subsequence VA also contains two component fifth-order cycles. The first cyclothem in Subsequence VA consists of the Irondequoit-Kirkland carbonates and the overlying 1-5 m of the Rochester Shale (Lewiston Member) or lower Herkimer Formation (see Figs. 4, 6). The second, nearly symmetrical cycle begins with brachiopod- and bryozoan-rich limestone of the Lewiston B-3 submember. These fossiliferous beds pass upward into sparsely fossiliferous shales (Lewiston C) which are, in turn, overlain by bryozoan-rich beds (Lewiston D). Total faunal change suggests that these cycles deepen from BA-3 to

deeper BA-4 biofacies close to the depocenter where the fifth-order cycles are subsymmetrical. In areas proximal to siliciclastic sources, the cycles are markedly progradational as predicted in models of Einsele and Bayer (1991). For example, the two fifth-order cycles of Sub-sequence VA are represented in the Herkimer Sandstone by two shale to sandstone successions. These cycles are capped by thin hematitic shell-rich beds that represent major flooding surfaces. The better preservation of regressive rather than transgressive parts of these cycles reflects increased input of siliciclastic into the basin during fifth-order regressions (Brett and Baird, 1986; Einsele and Bayer, 1991). Thinner, transgressive shell beds record sediment starvation associated with base-level rise.

Causal mechanisms.—

Fifth-order cycles appear to record minor fluctuations in relative sea level, ranging from a few to, perhaps, ten meters. The boundaries of fifth-order cycles in the medial-Silurian succession of the Appalachian Basin coincide with sharp, verti-

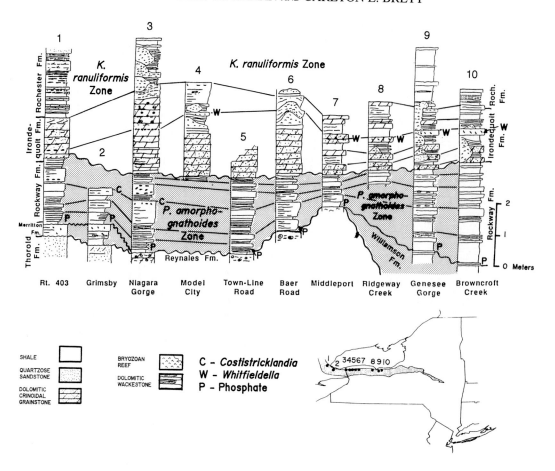

FIG. 10.—Correlated columnar sections of Rockway and Irondequoit Formations and basal beds of Lewiston Member (Rochester Shale) illustrating division into sixth-order cycles.

cal facies changes that have been used previously as boundaries between members. The bounding surfaces are commonly overlain by thin, fossiliferous condensed stratigraphic intervals. For example, in Sequence IV, the base of the Westmoreland Hematite, the Second Creek Bed, an unnamed middle Willowvale phosphatic bed and the Salmon Creek Bed (Lin and Brett, 1988) are all widely traceable units that overlie significant stratigraphic surfaces.

Correlation of cycle boundaries through successive facies belts, suggests that the control on fifth-order cycle formation is not intrinsic to any specific depositional system. Direct correlation of cycles between active and passive margins of the basin suggests a process, probably eustasy, that operated independently of basin subsidence. Unfortunately, intracontinental correlations of Silurian strata are insufficient to correlate fifth-order cycles in different basins. Cyclothems of similar temporal magnitude (400-500 Ka), however, have been correlated globally in the Carboniferous (Ross and Ross, 1988).

PACs or Parasequences: Sixth-Order Cycles

Superimposed upon the pattern of 1- to 10-meter scale fifth-order cycles are smaller bundles of beds, commonly arranged in

upward shallowing motifs, ranging between 0.5 to over 1.0 m in thickness. These bundles resemble punctuated aggradational cycles (PACs) of Goodwin and Anderson (1985). Such sixth-order cycles which are estimated to have an average recurrence interval of 100,000 years (Busch and Rollins, 1984) or less (Goodwin and Anderson, 1985), are commonly equated with parasequences of Exxon-type depositional sequences (but see Anderson and Goodwin, 1992 for counter arguments).

Sixth-order cycles have been documented in Silurian strata in both the New York-Ontario and Pennsylvania-Maryland outcrop belts (Cotter, 1988; Brett and others, 1990b). However, these cycles are not preserved or recorded uniformly in all facies. In western New York, sixth-order cyclicity is well-defined in the Rockway Dolostone, Irondequoit Limestone and Lewiston Member of the Rochester Shale (Fig. 10).

Sixth-order cycles in the Rockway Dolostone are typically 30 to 60 cm thick and consist of calcareous shale and argillaceous, dolomitic wackestone. In basinal sections, the cycles are composed primarily of greenish-gray, fossiliferous shales bearing a *Clorinda-Eoplectodonta* fauna (BA-5) with thin caps of nodular limestone. In shallower shelf settings, the argillaceous wackestones dominate the cycles and shale is subordinate. These fine-grained carbonates contain a poorly preserved fauna

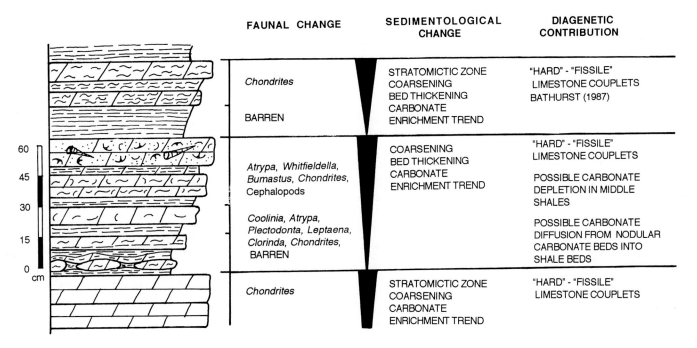

FIG. 11.—Example of Rockway sixth-order cycle in the Genesee River Gorge, Rochester, NY. Description of faunal, textural, and diagenetic components of same sixth-order cycle are provided to the right of the stratigraphic column.

consisting of *Costistricklandia*, orthoconic cephalopods, dendroid graptolites, and rare favositid corals (BA 4).

The Rockway cycles are well-developed in facies that represent depositional environments exclusively below wave base. The preservation and resolution of sixth-order cycles diminishes in wave-base environments. For example, in the Irondequoit Limestone, individual cycles can be correlated from the basin center sections, where they are readily discernible, to shallow shelf localities where these cycles become amalgamated and are less distinct.

At the Genesee Gorge section in Rochester, New York, Irondequoit cycles are typically 30 to 60 cm thick and contain calcareous shale, bryozoan mudmounds and crinoidal or brachiopod-bearing wackestones and packstones. At least four sixth-order cycles are arranged in a retrogradational succession that forms the larger-scale Irondequoit fifth-order cycle (see previous section). The basal cycle is thin and consists predominantly of a condensed, crinoidal packstone unit. The overlying cycles contain progressively better developed, maximally transgressive lower halves (Fig. 10). At the Niagara Gorge, the Irondequoit Limestone is a massive, crinoidal grainstone. Cycles at this locality are cryptic, recognizable mainly by stylolitic bounding surfaces.

Similar to cycles in the Rockway and Irondequoit Formations, cycles in the correlative Keefer Sandstone vary in motif dependent upon position along an onshore-offshore transect. Distal shelf sections of Keefer Sandstone contain shale-prone cycles with fossiliferous sandstone caps. Sections recording wave-base environments consist largely of massive, sparsely fossiliferous, well-washed quartz arenites (Cotter, 1990); cycles

are amalgamated and are poorly preserved and/or recorded. This condition is analogous to the patterns of preservation of cycles in the correlative Irondequoit Limestone at the Niagara Gorge.

Sixth-order cycles in sections of Keefer-equivalent Shawangunk Formation contain sandstone-based, upward fining cycles. These cycles culminate in barren, slightly bioturbated, ruddy brown, silty shales. These cycles are interpreted to represent muddy tidal flat parasequences.

The bounding surfaces of sixth-order cycles are commonly sharp, but are generally conformable when they do not coincide with the boundaries of larger-scale allocycles. The bounding surfaces are more sharply defined in the condensed carbonate sequences of the Rockway Dolostone and the Irondequoit Limestone. In less condensed sections of Rochester Shale, the boundaries of these cycles become more diffuse.

Smaller-scale Rhythms

The boundaries of sixth-order cycles are probably the finest-scale, widely traceable surfaces which have utility for correlation purposes. However, in some cases, sixth-order cycles are divisible into smaller-scale rhythms. This rhythmic bedding of a probably cyclic nature is recognized in some intervals of the medial-Silurian succession. For example, polished drill cores of the Williamson Shale exhibit alternating, decimeter-scale, green and black beds with sharp boundaries. Similar scale alternations of carbonates and calcareous shales occur in the Rockway Dolostone. Sixth-order cycles display 4 to 5 small-scale, shale-limestone rhythms (see Fig. 11). These rhythms are traceable laterally for 10's of kilometers. The carbonate bed of each successive rhythm becomes more distinct and thicker, with the highest form a thick cap to the meter-scale cycle.

Causal Mechanisms

At present, we have not attempted to correlate the sixth-order cycles beyond the Appalachian basin. Certainly, the fifth-order regressive peaks can be recognized throughout New York, Pennsylvania, and Ontario. Sixth-order cycles can be traced for up to 200 km in the New York-Ontario outcrop belt. The apparent nested hierarchy of cycles and their estimated recurrence intervals suggest eustatic effects, possibly related to Milankovitch periodicity. Decimeter-scale rhythms may be the result of variations in the pycnocline or in input of fine-grained, siliciclastic detritus due to climatic oscillation.

Sixth-order cycles represent very minor fluctuations in relative sea level- perhaps spanning a few meters of rise and fall. The mechanism of short-term minor sea-level fluctuation during the Silurian remains uncertain. Glacial evidence is unknown in most of the Silurian (but see Bjorlykke, 1985; Hambrey, 1985), but possible mountain glaciation may have left little detectable record. Tectono-eustasy appears to operate too slowly to produce these high frequency facies changes. Other eustatic mechanisms include snap-back effects of plate-rifting (Cathles and Hallam, 1991) and thermal expansion of water (Wigley and Raper, 1987). We cannot effectively argue for any one of these mechanisms, but merely note that cycle periodicities for the Silurian are comparable to those of widely accepted glacio-eustatic sea-level fluctuations in the Carboniferous Period (see Heckel, 1990).

The rhythms within Silurian parasequences resemble limestone-marl rhythms described primarily from the Mesozoic Era (see Fischer, 1991, for review). We interpret the Silurian cyclic bedding as the result of climatic fluctuations. The presence of four to five such rhythms within sixth-order parasequences suggests that they may represent the 26 ka precessional cycle of the Milankovitch band.

TECTONIC EFFECTS: THE ARCHITECTURAL RESPONSE TO THRUST LOADING, RELAXATION, AND UNLOADING

In the previous section we argue that most Silurian cycles were not produced by local tectonic effects. However, we do not deny that regional tectonics played a significant role in the evolution of the Appalachian foreland basin and in the architecture of its sedimentary fill during medial Silurian time. In fact, the facies belts, thickness trends, and stacking patterns of successive, nearly synchronous sequences and sub-sequences provide a detailed framework for examining tectonic influences.

Although Silurian through Early Devonian time is generally viewed as a relatively quiescent interlude in the evolution of the Appalachian Orogen, there is evidence of some compressional tectonism during the Silurian (Quinlan and Beaumont, 1984; Tankard, 1986). At least two cycles of tectonic activity are recognized: (1) an early and incomplete phase spanning the early to middle Llandoverian (S_1); and (2) a later, Salinic phase (S_2) beginning in the late Llandoverian and terminating in the Ludlovian. Each episode probably involved renewed thrust-faulting in the south-central Taconic region. These "tectonic cycles," each consisting of a thrust-loading event and ensuing phases of lithospheric relaxation, and peneplanation and isostatic rebound of the hinterland, apparently controlled lateral basin axis/forebulge migration. Basin-axis shift and forebulge development and propagation are readily interpreted from distinctive patterns and timing of regional, asymmetrically developed unconformities, and stacking patterns of unconformity-bound stratal packages.

Evidence for tectonic influence on stratigraphic architecture includes the following: (1) timing and regional patterns of angular-unconformity development; (2) migration of the depocenter and area of deepest water facies between successive, unconformity-bound units; and (3) alternations of coarse- and fine-grained siliciclastics in the foreland basin.

Sequence-Bounding Unconformities

Dennison and Head (1975) illustrate that in a roughly circular basin, sea-level falls will result in a basinward shift of high energy wave-base, paralic, and nonmarine environments circumbasinally. Consequently, regardless of whether the erosional process responsible for generating sequence-bounding unconformities is submarine or subaerial, the magnitude of the circumbasinal unconformity should be uniform, if eustatic sea-level lowering is the sole cause of erosion.

However, mapping of successive, sequence-bounding unconformities in the Silurian Appalachian foreland-basin reveals a distinctive alternation in the severity of erosion between the passive (northwestern) and tectonically active (southeastern) basin margins (Fig. 12). This pattern of alternating unconformities is not predicted by purely eustatic models. In fact, the only suggested mechanism which may produce such a pattern of alternating unconformities is foreland basin flexure. The model which best explains the observed pattern is that of Quinlan and Beaumont (1984) and Beaumont and others (1988) that invokes viscoelastic lithospheric flexure in response to episodic, thrust-load implacement. As envisioned by these workers, episodic thrust-loading in foreland basin settings generates large-scale, compressional "tectonic cycles" that consist of the following four phases: (1) initial load implacement; (2) relaxation of the lithosphere immediately under and adjacent to the load; (3) erosional unloading; and (4) isostatic uplift of the orogen and the adjacent molasse wedge followed by beveling of the hinterlandward portions of the clastic wedge.

The Silurian strata of the Appalachian foreland basin appear to exhibit two of these large-scale tectonic cycles (Fig. 12). Sequence-bounding unconformities record the progression of the basin floor through the various phases of the flexural cycle. In addition, the geometry of strata internal to Silurian sequences records lateral-basin axis migration predicted by the viscoelastic

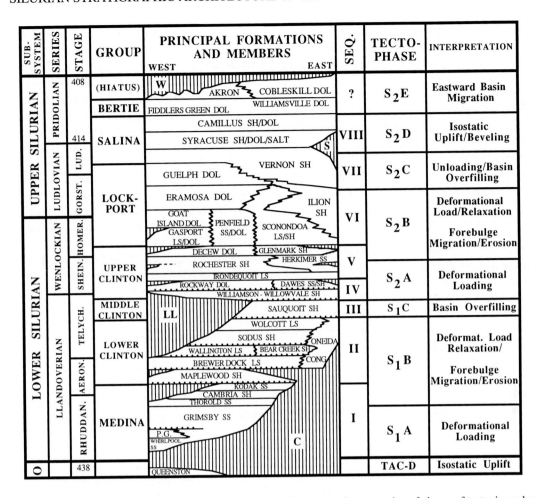

FIG. 12.—Tectophases of Sequences I-VIII across New York State. Right side presents interpretation of phases of tectonic cycles. Tectophases designated as TAC-D=Taconic late phase; S₁A-S₁C and S₂A-S₂E are inferred phases of tectonic cycles during the Silurian (see text for discussion). For other abbreviations, see Figure 3.

flexural model.

Two sequence-bounding unconformities, the basal Silurian Cherokee Unconformity (base of Sequence I) and the Salinic Unconformity (base of Sequence VII) are prominently developed along the tectonically active (eastern to southeastern) margin of the Appalachian basin. These unconformities truncate (progressively eastward) two relatively thick wedges of red, fine-grained, alluvial to marginal-marine deposits (Queenston-Juniata and Vernon-Bloomsburg Formations; Fig. 12). At their easternmost preserved limits, these unconformities also bevel strata which underlie the above-mentioned molasse. For example, near Oneida, New York, the Cherokee Unconformity juxtaposes Ordovician Frankfort Shale and Silurian Oneida Conglomerate (Rickard, 1975). The entire Queenston Shale and Medina Group are missing in this area. Similarly, east of Utica, New York, the Salinic Unconformity juxtaposes the late Llandoverian Willowvale Shale and the Pridolian Syracuse Shale. The entire Wenlockian and Ludlovian Series are missing in this area. Net thicknesses of strata removed by these unconformities are substantial (up to 300 meters) in eastern New York.

In contrast, at least three other unconformities (bases of Sequences II, IV, and VI) appear to be more prominently developed on the passive, western basin ramp (see Fig. 12). These unconformities truncate thin, typically condensed strata which presumably accumulated in shallow marine environments atop the Algonquin Arch.

Two of these unconformities the late Llandoverian basal Sequence IV boundary and the late Wenlockian basal Sequence VI boundary, display major truncation of strata. Along both of these unconformities, erosion is greatest in vicinity of the Algonquin Arch in Ontario. In both cases, the area of maximum erosion defines a NE/SW trending arch that can be traced from near Hamilton, Ontario into central Ohio (see Brett and others, 1990b). It is also noteworthy that both erosion surfaces were developed concurrently with, or immediately following, deposition of relatively thick, siliciclastic deposits in the basin. In the first case, erosive beveling occurred during or slightly after deposition of the 30- to 40-meter-thick Sauquoit Shale and laterally correlative red and black, marginal marine sandstones of the Otsquago Formation in eastern New York. The upper 100 meters of the Rose Hill Shale succession in Pennsylvania also comprise the correlative, fine-grained, basin fill. In the second case, erosion on the Algonquin Arch began after deposition of 20 to 40 meters of Rochester Shale and equivalent Herkimer Sandstone in the foreland basin.

The unconformities generated predominantly along the active margin are interpreted to have developed during late-phase isostatic uplift of the proximal clastic wedge and/or early phase thrust-load implacement (Fig. 12, phases S_1B, S_2B). Conversely, unconformities generated predominantly along the cratonic arch complex developed as a result of lithospheric relaxation and concomitant uplift and migration of a peripheral bulge.

Basin Axis and Arch Migration

We have previously documented two intervals of progressive, unidirectional (eastward) basin axis migration separated by an intervening period of westward migration during the medial Silurian (Brett and others, 1990a, b; Fig. 12, 13 herein). This pattern is consistent with other evidence suggesting two orogenic episodes during the Silurian (labeled S_1 and S_2 on Fig. 12). The first episode occurred during the early Llandoverian and resulted in deposition of the thick Tuscarora Sandstone-Conglomerate wedge in the basin depocenter (Brett and others, 1990b). During this time, a minor peripheral bulge may have separated the depocenter of the Tuscarora from that of finer-grained and more open-marine Medina Group (Duke, 1991). Duke postulates that this partition trended south-southwest from the Rochester, New York area. The Medina Group appears to extend into the Michigan basin with little or no evidence of separation by a cratonic arch in the Niagara region. However, during mid-Llandoverian time, uplift and truncation of Medina strata in western New York and Ontario suggests initial development of a minor forebulge or incipient Algonquin Arch (Fig. 12, phase S_1C). During later Medina deposition, there is also evidence for eastward migration of the eastern paleoshoreline from near Niagara Gorge to near Oswego, New York, a distance of approximately 200 kilometers (Fig. 13).

During deposition of the overlying lower and middle Clinton Group (Sequences II, III) the eastern strandline migrated an additional 150 kilometers eastward to the vicinity of Richfield Springs, New York (see Fig. 13). The depocenter (area of thickest sediment accumulation) also can be seen to migrate progressively from west to east through successive lower and middle Clinton cycles. Migration of the basin axis, depocenter and eastern strandline toward the hinterland is consistent with other evidence of tectonic quiescence and lithospheric relaxation (phase S_1B, Fig. 12).

Following deposition of the Sauquoit-Otsquago wedge (Sequence III) and development of the major, regional, angular unconformity at the base of Sequence IV, the direction of basin axis migration reversed (Fig. 12, phase S_2A; Fig. 13). This reversal suggests renewed thrust faulting and load implacement. The coarse-grained siliciclastic wedge of the Keefer Sandstone in Pennsylvania and the Herkimer Sandstone and Tammany Member of the Shawangunk Formation in New York (Prave and

others, 1989) appears to record the cannibalization and redeposition of proximal basin-margin fill in response to renewed tectonism. A shale-prone basin again developed during mid-Wenlockian (Rochester Shale) time. Shale-dominated basin fill suggests waning phases of thrust load emplacement and a transition to ensuing lithospheric relaxation. The forebulge appears to have been reactivated in the Hamilton, Ontario area and some lateral migration may have occurred during the late Wenlockian. During deposition of the lower part of the Lockport Group (Sequence VI), the forebulge may have migrated or a subsidiary uplift developed farther to the east near Rochester, New York where a NE/SW trending sandy shoal developed (Penfield Sandstone, Zenger, 1965; Crowley, unpubl. manuscript; Fig. 12, phase S_2B). Southeast of this area, fine-grained, dark gray, organic-rich muds of the Sconondoa-Ilion and McKenzie Formations accumulated in a relatively shallow trough. This sub-basin was bordered to the east by a prograding, muddy tidal flat complex and to the west by the Penfield Sandstone (Fig. 12).

During Ludlow time, the Bloomsburg-Vernon alluvial plain environment replaced earlier, higher energy strandline facies. These red-bed units prograded across the sub-basin and ultimately overfilled the foreland basin (Fig. 12, phase S_2C). Concurrently, carbonate deposition (Eramosa-Guelph) continued along the passive margin with little or no separation of strata between Appalachian and Michigan Basins.

The Ilion-McKenzie and Vernon-Bloomsburg siliciclastics appear to record erosional unloading of orogenic areas during medial Silurian time. The Salinic unconformity that truncates the eastern limits of the Vernon Shale in New York State may record final isostatic uplift (Fig. 12, phase S_2D). The overlying Salina Group displays the pattern of eastward basin axis migration ("antiperipheral bulge"; Fig. 12, phase S_2E), as predicted during late phases of unloading (Beaumont and others, 1988).

Hence, asymmetric Appalachian sequence-bounding unconformities are associated with terminal phases of foreland orogenic activity. Again, as predicted by Beaumont and others (1988), in the unloading phase, the strata overlying the main unconformity record an eastward migration of the shoreline and basin center facies. This is especially well displayed in the Salina and Bertie Groups that show progressive eastward migration of the basin center, the locus of halite deposition (Rickard, 1969).

Timing of Tectonic Phases

Patterns of unconformity and basin-axis shift during early and medial Silurian deposition appear to corroborate predictions of the viscoelastic thrust-leading model of Quinland and Beaumont (1984). Field data also provide constraints on the timing and geometry of the lithospheric response to thrust loading. Two cycles of thrusting, relaxation and at least partial unloading apparently occurred within about 25 million years between early Llandoverian and late Ludlovian time. A pulse of early Silurian tectonism followed basin overfilling and isostatic uplift of the

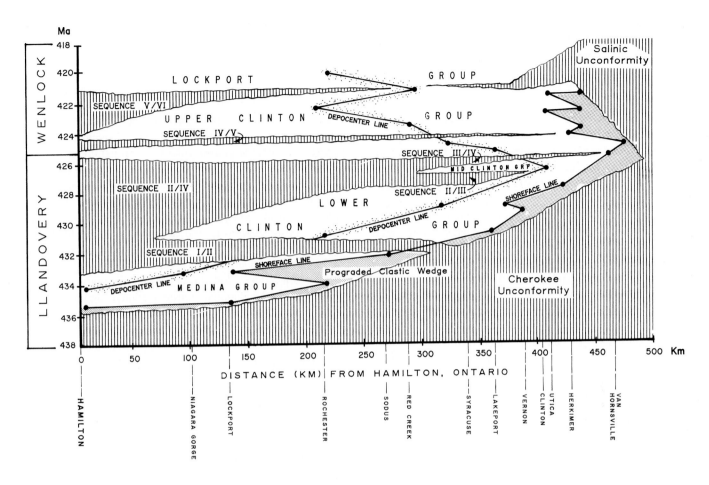

FIG. 13.—Migration of shoreface and depocenter for the Appalachian Basin during early to medial Silurian time in section along New York State outcrop belt. Major sequence-bounding unconformities are depicted by vertical ruling.

Late Ordovician unloading phase of the Taconic Orogeny (TAC-D of Fig. 12). This tectonic cycle began in earliest Silurian time and continued to the late Telychian Stage. The relaxation phase commenced during the later half of the Aeronian Stage and was particularly accentuated during late Telychian time when a major unconformity developed on the uplifted forebulge. Hence, relaxation spanned about 5 million years. During this same interval, the eastern shoreline and depocenter migrated from northwest to southeast approximately 250 kilometers (Fig. 13).

The second tectophase (S_2) apparently was initiated in latest Llandoverian time, approximately coincident with the major latest Telychian (C_6) sea-level highstand. From this time to late in the Wenlockian Epoch, the direction of the basin-axis migration reversed and the depocenter migrated approximately 230 kilometers northwestward between Oneida and Niagara County, New York (Fig. 13). A forebulge uplift was again accentuated during mid to late Wenlockian time resulting in a second, regionally angular unconformity below the Lockport Group (Sequence VI basal boundary), and less pronounced truncation surfaces within the Rochester Shale and Lockport strata. Thus, the relaxation phase commenced about 4 million years after the

appearance of coarse siliciclastic (Dawes-lower Keefer) in latest Llandoverian/earliest Wenlockian time. The relaxation phase of this tectonic cycle is brief, only about 1.0 to 1.5 million years. Consistent with this brief time frame, lateral migration of the basin axis is limited to about 150 kilometers. During latest Wenlockian to late Ludlovian time (about 8 to 10 million years), the third or unloading phase took place and culminated with overfilling of the foreland basin by red Vernon-Bloomsburg molasse followed by isostatic uplift and regional, eastward truncation of the red beds in the latest Ludlovian or earliest Pridolian time. Finally, continued unloading resulted in a renewed phase of eastward basin axis shift during the 20 million years of the Late Silurian-Early Devonian time.

DISCUSSION

Correlation of cyclic facies changes across environmental belts in a basin is one basis for discriminating between extrinsically controlled allocycles and autocycles which are local responses to intrinsic environmental dynamics. Patterns of vertical facies change that are similar and synchronous in various environmental belts of the basin cannot be explained by local, autogenic, environmental processes such as delta lobe switching; an extrinsic controlling mechanism is required. The traceability of most Silurian sedimentary cycles across environmental boundaries firmly establishes that these facies successions are not related solely to the dynamics of particular depositional systems. Several scales of cyclicity appear to be circumbasinal in extent and are represented in very disparate facies, ranging between pure carbonates and quartz arenites. However, a key issue has remained regarding apparently synchronous facies changes in the Appalachian basin: are these basin-wide relative sea-level changes driven by eustasy or tectonism?

With the advent of models attempting to link episodic thrust loading and subsequent basin response to patterns in the stratigraphic record, certain workers have suggested that some sedimentary allocycles may be tectonically generated (Blair and Bilodeau, 1988; Heller and others, 1988; Miall, 1991). Nonetheless, as in the case of the Silurian System of the Appalachian basin, cycles that are circumbasinal and cross cut patterns of basin-axis and forebulge migration, must have been generated by an extrinsic process unrelated to local thrust-loading.

Until recently, no tectonic models had been advanced that could satisfactorily explain basin-wide patterns. However, new models linking continental-scale, base-level fluctuations (yo-yo tectonics of Friedman, 1987) with high frequency, stress-induced changes in lithospheric plate density (Klein and Willard, 1989; Cathles and Hallam, 1991) provide possible alternatives to eustasy. Such fluctuations would be plate specific and, therefore, a critical test for eustasy must involve attempts at intercontinental correlation of cycles. Although much more detailed correlations are still required, general synchroneity of larger-scale Silurian cycles in the Appalachian basin, Europe, and elsewhere (Johnson and others, 1985, 1991; data presented herein) suggest a strong eustatic component to the allocycles.

On the other hand, the pattern of regional facies change *within* allocycles cannot be fully explained by sea-level fluctuations. Tectonism probably played a critical role in generation of large-scale stratigraphic architecture in the Silurian Appalachian basin. Patterns of depocenter and basin axis migration, for example, are attributable to differential subsidence related to deformational or thrust load emplacement, lithospheric relaxation and sediment deposition in the basin (Pitman, 1978; Quinlan and Beaumont, 1984).

With careful correlation of cycles, using through-going surfaces and marker beds, it has been possible to discriminate between stratigraphic patterns attributable to Silurian-age Appalachian basin flexure and cyclic patterns of a more widespread, probably non-tectonic nature. Refined correlations and interpretations have led to the present account of Silurian sea-level fluctuations and basin flexure. Our reconstructions depict the Silurian Appalachian foreland basin as fairly dynamic for a period of time generally considered to be a quiescent interlude between two major orogenic events (Taconic and Acadian). Although these results are preliminary, we are closer to defining the relative roles of eustasy and tectonics in the generation of

stratigraphic architecture.

CONCLUSIONS

1) The Silurian succession in the Appalachian Foreland Basin is subdivisible into at least eight large-scale (third-order), unconformity-bound, depositional sequences (Sequences I-VI). Silurian sequences resemble coastal margin sequences recognized by seismic stratigraphers in scale and in exhibiting internal units analogous to transgressive and highstand systems tracts. However, these sequences display no lowstand deposits and are divisible into sub-sequences (fourth-order cycles). The sub-sequences in the highstand systems tracts permit division of these intervals into early and late phases. Late highstand systems tracts may comprise one or more sub-sequences.

2) Fourth-order sub-sequences are further divisible into two or three fifth-order cyclothems or parasequence sets of variable motif (e.g., progradational, retrogradational, aggradational). Fifth-order cycles are, in turn, divisible into as many as five discrete, meter-scale upward shallowing PACs (sixth-order cycles). These small-scale cycles, which correspond to parasequences, are best developed in shallow shelf and paralic facies.

3) All orders of allocycles (third-through sixth-orders) are widespread to circumbasinal. The correlation of third- and fourth-order lowstands in the Silurian Appalachian basin with those elsewhere in North America, Baltica, and China suggests an overriding eustatic influence. Major mid-Silurian highstands occurred in latest Telychian, mid-Wenlockian, and early Ludlovian time; major lowstands occurred during the earliest, middle and late Llandoverian, and early and late Wenlockian time.

4) The effects of local tectonic thrust loading are also manifest in a) local accentuation of sequence-bounding unconformities in areas of arching or forebulge development; b) differential development and migration of the depocenter and shoreline of the foreland basin. Two tectonic cycles can be identified in the Silurian of the Appalachian basin, one beginning in the earliest Silurian and the other commencing approximately in the latest Llandoverian. During each tectonic cycle, three phases of tectonism can be recognized in the sedimentary record. Early thrust load phases are recorded by influxes of coarse siliciclastics. Relaxation phases resulted in the formation of shale-dominated basin fills and the development of an accentuated forebulge near the "Algonquin Arch." A later unloading phase, recognizable only in the second tectonic cycle, resulted in overfilling of the Appalachian basin with red molasse. Finally, isostatic uplift of the active basin margin resulted in an eastward (hinterlandward) expanding unconformity which truncated the older molasse. The uplift was followed by eastward migration of an incipient basin over the truncated sedimentary wedge.

5) Stratigraphic accumulation during the Silurian of the Appalachian basin cannot be viewed simply as the result of tectonics or eustasy, but rather must be viewed as the net result of the interplay of these mechanisms. The role of eustasy is to

produce the nested pattern of allocyclicity. The role of foreland-basin flexure is to produce the relief (variation in accommodation space) which profoundly influences local stacking patterns and variations in local patterns of stratigraphic preservation.

ACKNOWLEDGMENTS

We wish to thank several people who have contributed to this project with new field data or discussions including Edwin Anderson, Robert Blair, Donald Crowley, Mark Domagalia, William Duke, Peter Goodwin, Stephen LoDuca, Lawrence Rickard and Dorothy Tepper. This paper has benefited significantly from reviews by Edward Cotter, John Dennison, and David Lehmann. Heidi Kimble typed several drafts of this manuscript. Our research has been supported by grants from the Petroleum Research Fund (American Chemical Society), New York State Museum, Pennsylvania State Geological Survey, Sigma Xi, and Envirogas Corporation.

REFERENCES

AIGNER, T., 1985, Storm depositional systems: Dynamic stratigraphy in modern and ancient shallow-marine sequences, in Lecture Notes in the Earth Sciences, v. 3: New York, Springer-Verlag, 174 p.

ANDERSON, E. J., AND GOODWIN, P. W., 1992, What distinguishes PACs from parasequences? (abs.): Northeastern Section Geological Society of America, Abstracts with Programs, v. 24, p. 3.

BATHURST, R. G. C., 1987, Diagenetically enhanced bedding in argillaceous platform limestones; stratified cementation and selective compaction: Sedimentology, v. 34, p. 749-778.

BEAUMONT, C., QUINLAN, B., AND HAMILTON, J., 1988, Orogeny and stratigraphy: numerical models of the Paleozoic in the eastern interior of North America: Tectonics, v. 7, p. 389-416.

BJORLYKKE, K., 1985, Glaciations, preservation of their sedimentary record and sea-level changes—a discussion based on the Late Precambrian and Lower Paleozoic sequence in Norway: Paleogeography, Paleoclimatology, and Paleoecology, v. 51, p. 197-207.

BLAIR, T. C., AND BILODEAU, W. L., 1988, Development of tectonic cyclothems in rift, pull-apart, and foreland basins: sedimentary response to episodic tectonism: Geology, v. 16, p. 517-520.

BOLTON, T. E., 1957, Silurian stratigraphy and paleontology of the Niagara Escarpment in Ontario: Ottawa, Geological Survey of Canada Memoir 289, 145 p.

BOUCOT, A. J., 1975, Evolution and Extinction Rate Controls: Amsterdam, Elsevier, 427 p.

BRETT, C. E., 1983, Sedimentology, facies relations, and depositional environments of the Silurian (Wenlockian) Rochester Shale: Journal of Sedimentary Petrology, v. 53, p. 947-972.

BRETT, C. E., in press, Wenlockian fossil communities in New York State and adjacent areas, in Boucot, A. J., and Lawson, J., eds., Final report of Project Ecostratigraphy: Cambridge University Press.

BRETT, C. E., AND BAIRD, G. C., 1986, Symmetrical and upward shallowing cycles in the Middle Devonian of New York State and their implications for the punctuated aggradational cycle hypothesis: Paleoceanography, v. 1, p. 431-445.

BRETT, C. E., GOODMAN, W. M., AND Lo DUCA, S. T., 1990a, Sequence stratigraphy of the type Niagaran Series (Silurian) of western New York and Ontario, in Lash, G., ed., New York State Geological Association 62nd Annual Meeting Field Trip Guidebook: Fredonia, New York State Geological Association, p. C1-C71.

BRETT, C. E., GOODMAN, W. M., AND Lo DUCA, S. T., 1990b, Sequences, cycles and basin dynamics in the Silurian of the Appalachian Foreland Basin: Sedimentary Geology, v. 69, p. 191-244.

BUSCH, R. G., AND ROLLINS, H. B., 1984, Correlation of Carboniferous strata using a hierarchy of transgressive-regressive units: Geology, v. 12, p. 471-474.

CATHLES, L. M., AND HALLAM, A., 1991, Stress-induced changes in plate density, Vail sequences, epeirogeny, and short-lived, global sea-level fluctuations: Tectonics, v. 10, p. 659-671.

CHANG, K. H., 1975, Unconformity-bounded stratigraphic units: Geological Society of America Bulletin, v. 86, p. 1544-1552.

COLLINGS, A. V. J., 1989, A major regressive event at the top of the Much Wenlock Limestone Formation in the west midlands (abs.): Keele, The Murchison Symposium Programme and Abstracts, University of Keele, p. 40.

COTTER, E., 1983, Silurian depositional history, in Nickelson, R. P., and Cotter, E., eds., Silurian Depositional History and Alleghanian Deformation in the Pennsylvania Valley and Ridge, Danville, Pennsylvania: Guidebook 48th Annual Field Conference of Pennsylvania Geologists, p. 3-27.

COTTER, E., 1988, Hierarchy of sea-level cycles in the medial-Silurian siliciclastic succession of Pennsylvania: Geology, v. 16, p. 242-245.

COTTER, E., 1990, Storm effects on siliciclastic and carbonate shelf sediments in the medial Silurian succession of Pennsylvania: Sedimentary Geology, v. 69, p. 245-258.

COTTER, E., 1992, Diagenetic alteration of chamositic clay minerals to ferric oxide in oolitic ironstone: Journal of Sedimentary Petrology, v. 62, p. 54-60.

DENNISON, J. M., AND HEAD, J. W., 1975, Sea-level variations interpreted from the Appalachian Basin Silurian and Devonian: American Journal of Science, v. 275, p. 1089-1120.

DUKE, W. L., 1991, The Lower Silurian Medina Group in New York State and Ontario, in Cheel, R. J., ed., Sedimentology and Depositional Environments of Silurian Strata of the Niagara Escarpment in Ontario and New York: Toronto, Geological Association of Canada, Mineralogical Society of Canada, Society of Economic Geologists, Joint Annual Meeting, Field Trip B4, p. 35-61.

ECKERT, J. D., 1984, Early Llandovery crinoids and stelleroids from the Cataract Group (Lower Silurian) in southern Ontario, Canada: Toronto, Royal Ontario Museum Life Sciences Contribution 127, 83 p.

ECKERT, B. Y., AND BRETT, C. E., 1989, Bathymetry and paleoecology of Silurian benthic assemblages, Late Llandoverian, New York State: Paleogeography, Paleoclimatology, Paleoecology, v. 74, p. 297-326.

EINSELE, G., AND BAYER, U., 1991, Asymmetry in transgressive-regressive cycles in shallow seas and passive continental margin settings, in Einsele, G., Ricken, W., and Seilacher, A., eds., Cycles and Events in Stratigraphy: Berlin, Springer Verlag, p. 660-681.

EINSELE, G., RICKEN, W., AND SEILACHER, A., 1991, Cycles and events in stratigraphy—basic concepts and terms, in Einsele, G., Ricken, W., and Seilacher, A., eds., Cycles and Events in Stratigraphy: Berlin, Springer-Verlag, p. 1-22.

EINSELE, G., AND SEILACHER, A., eds., 1982, Cyclic and Event Stratification: Berlin, Spring-Verlag, 536 p.

FISCHER, A. G., 1963, The Lofer cyclothems of the Alpine Triassic: Kansas Geological Survey Bulletin, v. 169, p. 107-149.

FISCHER, A. G., 1982, Long-term climatic oscillations recorded in stratigraphy, *in* Berger, W. H., and Crowell, J. C., eds., Climate in Earth History: Washington, D. C., National Academy Press, p. 97-104.

FISCHER, A. G., 1991, Orbital cyclicity in Mesozoic strata, *in* Einsele, G., Ricken, W., and Seilacher, A., eds., Cycles and Events in Stratigraphy: Berlin, Springer-Verlag, p. 48-62.

FRIEDMAN, G. M., 1987, Vertical movement of the crust: case histories from the northern Appalachian basin: Geology, v. 15, p. 1130-1133.

GALE, N. H., 1985, Numerical calibration of the Paleozoic time-scale: Ordovician, Silurian, and Devonian Periods, *in* Snelling, N. J., ed., The Chronology of the Geological Record: London, Geological Society Memoir 10, p. 81-88.

GILLETTE, T. W., 1947, The Clinton of western and central New York: Albany, New York State Museum Bulletin 341, 191 p.

GOODWIN, P. W., AND ANDERSON, E. J., 1985, Punctuated aggradational cycles: a general hypothesis of episodic stratigraphic accumulation: Journal of Geology, v. 93, p. 515-533.

GROTZINGER, J. P., 1986a, Cyclicity and paleoenvironmental dynamics, Rocknest Platform, northwest Canada: Geological Society of America Bulletin, v. 97, p. 1208-1231.

GROTZINGER, J. P., 1986b, Upward shallowing platform cycles—a response to 2.2 billion years of low-amplitude, high-frequency (Milankovitch band) sea-level oscillations: Paleoceanography, v. 1, p. 403-416.

HAMBREY, M. J., 1985, Late Ordovician-Early Silurian glacial period: Palaeogeography, Palaeoclimatology, Palaeoecology, v. 51, p. 273-289.

HAQ, B. U., HARDENBOL, J., AND VAIL, P. R., 1988, Mesozoic and Cenozoic chronostratigraphy and cycles of sea-level change, *in* Wilgus, C. K., and others, eds., Sea-level Changes: An Integrated Approach: Tulsa, Society of Economic Paleontologists and Mineralogists Special Publication 42, p. 71-108.

HECKEL, P. H., 1977, Origin of phosphatic black shale facies in Pennsylvanian cyclothems of Midcontinent North America: American Association of Petroleum Geologists Bulletin, v. 61, p. 1045-1068.

HECKEL, P. H., 1990, Evidence for global (glacial-eustatic) control over upper Carboniferous (Pennsylvanian) cyclothems in midcontinent North America, *in* Hardman, R. F. P., and Brooks, J., eds., Tectonic Events Responsible for Britain's Oil and Gas Reserves: London, Geological Society Special Publication 55, p. 35-47.

HECKEL, P. H., BRADY, L. L., EBANKS, W. J., JR., AND PABIAN, R. K., 1979, Field guide to Pennsylvanian cyclic deposits in Kansas and Nebraska: Kansas Geological Survey Guidebook Series 4, Ninth International Congress of Carboniferous Stratigraphy and Geology, 79 p.

HELLER, P. L., ANGEVINE, C. L., WINSLOW, N. S., AND PAOLA, C., 1988, Two phase stratigraphic model for foreland basin sequences: Geology, v. 16, p. 501-504.

HEWITT, M. C., AND CUFFEY, R. M., 1985, Lichenaliid fistuliporoid crust-mounds (Silurian, New York-Ontario): Typical early Paleozoic bryozoan reefs: Sydney, Proceedings, 5th International Coral Reef Symposium, p. 599-604.

HUNTER, R. E., 1970, Facies of iron sedimentation in the Clinton Group, *in* Fisher, G. W., ed., Studies of Appalachian Geology, Central and Southern: New York, Wiley Interscience, p. 101-124.

JOHNSON, M. E., 1987, Extent and bathymetry of North American platform seas in the Early Silurian: Paleoceanography, v. 2, p. 185-211.

JOHNSON, M. E., RONG, J. Y., AND YNAG, X. C., 1985, International correlation by sea-level events in the Early Silurian of North America and China (Yangtze Platform): Geological Society of America Bulletin, v. 96, p. 1384-1397.

JOHNSON, M. E., KALJO, D., AND RONG, J. Y., 1991, Silurian eustacy, *in* Bassett, M. E., ed., Murchison Symposium: London, Special Papers in Paleontology 44, p. 145-163.

KEMP, A. R. S., 1991, Silurian pelagic and hemipelagic sedimentation and paleogeography, *in* Bassett, M. E., ed., Murchison Symposium: London, Special Papers in Paleontology 44, p. 261-299.

KIDWELL, S. M., 1991, Condensed deposits in siliciclastic sequences: observed and expected features, *in* Einsele, G., Ricken, W., and Seilacher, A., eds., Cycles and Events in Stratigraphy: Berlin, Springer-Verlag, p. 682-695.

KLEIN, G. DEV., AND WILLARD, D. A., 1989, Origin of Pennsylvanian coal-bearing cyclothems of North America: Geology, v. 17, p. 152-155.

LIN, B. Y., AND BRETT, C. E., 1988, Stratigraphy and disconformable contacts of the Williamson-Willowvale interval: revised correlations of the Late Llandoverian (Silurian) in New York State: Northeastern Geology, v. 10, p. 241-253.

LO DUCA, S. T., AND BRETT, C. E., 1990, Stratigraphic relations of lower Clinton hematites (abs.): Geological Society of America, Abstracts with Programs, v. 22, p. 31.

LO DUCA, S. T., AND BRETT, C. E., 1991, Placement of the Wenlockian/Ludlovian boundary in New York State: Lethaia, v. 24, p. 255-264.

LO DUCA, S. T., AND BRETT, C. E., 1994, Revised stratigraphic and facies relationships of the lower part of the Clinton Group (Middle Llandoverian) of western New York State, *in* Landing, E., ed., Studies in Stratigraphy and Paleontology in Honor of Donald W. Fisher: Albany, New York State Museum Bulletin, p. 161-182.

LUKASIK, M., 1988, Lithostratigraphy of Silurian rocks in southern Ohio and adjacent Kentucky and West Virginia: Unpublished Ph.D. Dissertation, University of Cincinnati, Cincinnati, 313 p.

MCKERROW, W. S., 1979, Ordovician and Silurian changes in sea-level: Journal of the Geological Society of London, v. 136, p. 137-145.

MCKERROW, W. S., LAMBERT, R. ST. J., AND COCKS, L. R. M., 1985, The Ordovician, Silurian and Devonian Periods, *in* Snelling, N. J., ed., The Chronology of the Geological Record: London, The Geological Society Memoir 10, p. 73-80.

MIALL, A. D., 1991, Stratigraphic sequences and their chronostratigraphic correlation: Journal of Sedimentary Petrology, v. 61, p. 497-505.

MIDDLETON, G. V., RUTKA, M., AND SALAS, C. J., 1987, Depositional environments in the Whirlpool Sandstone Member of the Medina Formation, *in* Duke, W., ed., Sedimentology, Stratigraphy and Ichnology of the Lower Silurian Medina Formation: Niagara Falls, Guidebook for Society of Economic Paleontologists and Mineralogists, Eastern Section, p. 31-45.

MOORE, R. C., 1936, Stratigraphic classification of the Pennsylvanian rocks of Kansas: Kansas Geological Survey Bulletin, v. 22, p. 1-256.

PARKINSON, N., AND SUMMERHAYES, C., 1985, Synchronous global sequence boundaries: American Association of Petroleum Geologists Bulletin, v. 69, p. 685-687.

PITMAN, W. C. III, 1978, Relationship between eustacy and stratigraphic sequences of passive margins: Geological Society of America Bulletin, v. 89, p. 1389-1403.

PRAVE, A. R., ALCALA, M. L., AND EPSTEIN, J. B., 1989, Stratigraphy and sedimentology of Middle and Upper Silurian rocks and an enigmatic diamicitite, southwestern New York: Albany, 61st Annual Meeting New York State Geologic Association Field Trip Guidebook, p. 121-140.

QUINLAN, G. M., AND BEAUMONT, C., 1984, Appalachian thrusting, lithospheric flexure, and the Paleozoic stratigraphy of the eastern interior of North America: Canadian Journal of Earth Science, v. 21, p. 973- 996.

RAMSBOTTOM, W. H. C., 1979, Rates of transgression and regression in the Carboniferous of northwestern Europe: Journal of the Geological Society of London, v. 136, p. 147-153.

READ, J. F., GROTZINGER, J. P., BOVA, J. A., AND KOERSCHNER, W. F., 1986, Models for generation of carbonate cycles: Geology, v. 14, p. 107-110.

RICKARD, L. V., 1969, Stratigraphy of the Upper Silurian Salina Group, New York, Pennsylvania, Ohio, Ontario: Albany, New York State Museum and Science Services Map and Chart Series 12.

RICKARD, L. V., 1975, Correlation of the Silurian and Devonian rocks of New York State: Albany, New York State Museum and Science Service, Map and Chart Series 24.

ROSS, C. A., AND ROSS, J. R. P., 1988, Late Paleozoic transgressive-regressive deposition, in Wilgus, C. K., and others eds., Sea-level Changes: An Integrated Approach: Tulsa, Society of Economic Paleontologists and Mineralogists Special Publication 42, p. 227-247.

SANFORD, B. V., THOMPSON, F. J., AND MCFALL, G. H., 1985, Plate tectonics—a possible controlling mechanism in the development of hydrocarbon traps in southwestern Ontario: Bulletin of Canadian Petroleum Geology, v. 33, p. 52-71.

SARG, J. F., 1988, Carbonate sequence stratigraphy, in Wilgus, C. K., and others, eds., Sea-level Changes: An Integrated Approach: Tulsa, Society of Economic Paleontologists and Mineralogists Special Publication 42, p. 155-181.

SIVITER, D. J., OWENS, R. M., AND THOMAS, A. T., 1989, The northern Wenlock Edge Area: shelf muds and carbonates on the midland platform, in Bassett, M. G., ed., Silurian Field Excursions: A Geotraverse Across Wales and the Welsh Borderlands: Cardiff, National Museum of Wales Geological Series 10, 133 p.

SUMMERHAYES, C. P., 1986, Sea-level curves based upon seismic stratigraphy: Their chronostratigraphic significance: Palaeogeography, Palaeoclimatology, Palaeoecology, v. 57, p. 27-42.

SWARTZ, C. K., 1923, Stratigraphic and paleontologic relations of the Silurian strata of Maryland: Baltimore, Maryland Geological Survey, Silurian, p. 25-50.

SWARTZ, F. M., 1934, Silurian sections near Mt. Union, central Pennsylvania: Bulletin of the Geological Society of America, v. 45, p. 81-134.

TANKARD, A. J., 1986, On the depositional response to thrusting and lithospheric flexure: examples from the Appalachian and Rocky Mountain basins: Special Publications of the International Association of Sedimentologists 8, p. 369-392.

VAIL, P. R., MITCHUM, R. M., JR., AND THOMPSON, S., 1977, Seismic stratigraphy and global changes in sea-level, Parts 2-4, in Payton, C. E., ed., Seismic Stratigraphy—Applications to Hydrocarbon Exploration: Tulsa, American Association of Petroleum Geologists Memoir 26, p. 83-97.

VAIL, P. R., AUDEMARD, F., BOWMAN, S. A., EISNER, P. N., AND PEREZ-CRUZ, C., 1991, The stratigraphic signatures of tectonics, eustacy, and sedimentology—an overview, in Einsele, G., Ricken, W., and Seilacher, A., eds., Cycles and Events in Stratigraphy, Berlin, Springer-Verlag, p. 617-659.

VAN WAGONER, J. C., POSAMENTIER, H. W., MITCHUM, R. M., VAIL, P. R., SARG, J. F., LOUTIT, T. S., AND HARDENBOL, J., 1988, An overview of the fundamentals of sequence stratigraphy and key definitions, in Wilgus, C. K., and others, eds., Sea-level Changes: An Integrated Approach: Tulsa, Society of Economic Paleontologists and Mineralogists Special Publication 42, p. 39-46.

WATTS, A. B., 1982, Tectonic subsidence, flexure, and global changes in sea-level: Nature, v. 297, p. 469-474.

WIGLEY, T. M. L., AND RAPER, S. C. B., 1987, Thermal expansion of water associated with global warming: Nature, v. 330, p. 127-131.

WINDER, C. G., AND SANFORD, B. V., 1972, Stratigraphy and palaeontology of the Palaeozoic rocks of southern Ontario, in Glass, D. J., ed., Guidebook 24th International Geological Congress: Montreal, p. A.45-C.45.

ZENGER, D. W., 1965, Stratigraphy of the Lockport Formation (Middle Silurian) in New York State: Albany, New York State Museum Bulletin 404, 210 p.

ZENGER, D. W., 1971, Uppermost Clinton (Middle Silurian) stratigraphy and petrology, east-central New York: Albany, New York State Museum Bulletin 417, 58 p.

RECOGNITION OF REGIONAL (EUSTATIC?) AND LOCAL (TECTONIC?) RELATIVE SEA-LEVEL EVENTS IN OUTCROP AND GAMMA-RAY LOGS, ORDOVICIAN, WEST VIRGINIA

RICHARD J. DIECCHIO AND BRETT T. BRODERSEN[1]

Department of Geography and Earth Systems Science, George Mason University, Fairfax, Virginia 22030-4444

ABSTRACT: Plots of the cumulative aggradation of cyclically repetitive strata were generated by a method similar to that for Fischer plots. Plots were generated for three outcrop sections of the upper part of the Upper Ordovician Juniata Formation, eleven wells penetrating the Juniata Formation, and four wells penetrating the entire Ordovician System. The resultant plots are inferred to represent relative changes in sea level for each locality over the time interval represented.

Cycles of relative sea-level change with a periodicity of approximately 5 my are apparent throughout the Ordovician. These cycles are superimposed upon a larger scale relative sea-level curve that reflects the sea-level pattern associated with the Creek holostrome (Tippecanoe I sequence). Each of these sea-level cycles can be correlated from well to well and, because of their regional extent, are possibly the effect of an eustatic cause.

Other relative shallowing and deepening events are apparent in some of the wells, and at various stratigraphic positions. These local events are interpreted as tectonic uplifts and downwarps. Many of these localized sea-level events are more long-lived than the regional 5-my sea-level signal, but have a shorter duration than the Creek holostrome.

INTRODUCTION

The use of Fischer plots (Fischer, 1964) to generate relative sea-level curves for cyclical peritidal successions is now commonplace (Goldhammer and others, 1987, 1990; Read and Goldhammer, 1988; Koerschner and Read, 1989; Eriksson and Simpson, 1990). In the present paper, this method is applied first to cyclically-bedded outcrop sections of probable peritidal strata, then to cyclical subsurface successions of the same formation as discernible on gamma-ray logs. This method is extended further by applying it to Ordovician sandstone/shale or limestone/shale "cycles" that can be delineated on gamma-ray logs, but which cannot necessarily be documented as peritidal. The resultant plots appear to represent reasonable relative sea-level curves for the Ordovician, but their validity remains to be demonstrated.

The relative sea-level curves are presented with a vertical axis that represents time, thus preserving stratigraphic perspective. Because of this, and because the method is not applied strictly to peritidal strata, the resultant plots cannot be referred to as Fischer plots. Instead, they will be referred to as cumulative aggradation plots or CAPs.

Ordovician sea-level curves have been proposed by numerous authors (Vail and others, 1977; McKerrow, 1979; Fortey, 1984; Dennison, 1989). There is a general agreement that two worldwide lowstands occurred during the Ordovician; one during the late Early Ordovician or early Middle Ordovician, and the other near the end of the Ordovician. The former is associated with the Knox or Beekmantown unconformity in the Appalachians (Dennison, 1989), and the latter is associated with the Juniata redbeds (Dennison, 1976) and the Taconic hiatus, although tectonic uplift also contributed to this hiatus (Rodgers, 1971). A localized Trenton-Black River (sub-Bays) unconfor-

mity in parts of the Appalachians (Hergenroder, 1966; Haynes, 1992) is probably related to a Middle Ordovician eustatic sea-level drop.

Conversely, each of the hiatuses mentioned have been attributed by previous authors to tectonic uplift. Mussman and Read (1986) and Ettensohn (1991) have attributed the Knox unconformity to uplift associated with the end of the passive margin and the onset of convergence in the central Appalachians, possibly in conjunction with eustatic lowering. Ettensohn (1991) suggests that a pre-Rocklandian and Kirkfieldian (pre-Trentonian), post-Canadian and Black Riverian unconformity is associated with uplift accompanying thrust loading. The Taconic angular unconformity (Rodgers, 1971) certainly documents that deformation occurred close to the time of eustatic sea-level lowering during the later part of the Late Ordovician.

The purpose of this paper is to distinguish local (possibly tectonic) events from regional (possibly eustatic) events by detailed comparison of the relative sea-level history of several stratigraphic sections within one portion of the Appalachian basin (Fig. 1). This method is potentially applicable on a much larger scale, geographically and temporally.

METHODS

Figure 2 is a plot of cumulative aggradation versus time for a hypothetical stratigraphic sequence containing ten cycles, and which can be used to illustrate the analytical techniques used in this study. These cycles can be sandstone/shale or limestone/shale repetitions, as will be the case for the strata considered later in this report. The use of CAPs for modeling relative changes in sea level is based upon two necessary assumptions that controlled the thickness of the cycles: (1) that the cycles are the result of sea-level fluctuations that occurred with a constant period; and (2) that subsidence was occurring at a constant rate.

In Figure 2, the average thickness of the ten cycles is calculated and assumed to be the constant amount of subsidence

---

[1]Current address: Department of Geology, Portland State University, Portland, OR 97207

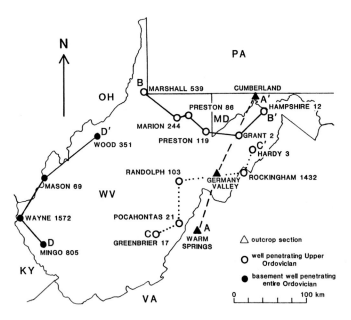

FIG. 1.—Map of study area showing location of outcrops and wells, and lines of cross-sections. Wells are designated by county and permit number.

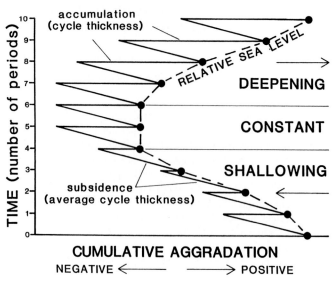

FIG. 2.—Hypothetical cumulative aggradation plot (CAP) (refer to text for explanation). Dots represent top of sediment or sediment-water interface.

during each cycle period. The amount of aggradation for each cycle is the thickness of that cycle (accumulation) minus the subsidence during that cycle period (average cycle thickness). For each cycle, aggradation is positive if the thickness of that cycle exceeds the average cycle thickness, and negative if average cycle thickness exceeds the thickness of the individual cycle. The base of the section is assigned an arbitrary value which represents either the sediment-water interface, or simply the top of the sediment. The amount of aggradation for the lowest cycle is added to (or subtracted from, depending on the sign) the base value to establish a new base value. Working up-section, a new base value is established successively for each cycle. Negative aggradation (deflection to the left) signifies that the top of the sediment has decreased in elevation, presumably because subsidence has exceeded accumulation. Positive aggradation (deflection to the right) indicates that the top of the sediment has built up to a higher elevation, presumably because accumulation has exceeded subsidence. The CAP therefore indicates the change in elevation of the top of the sediment. For peritidal strata the top of the sediment represents mean relative sea level. Therefore, the CAP for peritidal strata (Fischer plot) can be interpreted as a relative sea-level curve where positive aggradation is accommodated by a rise in sea level. Negative aggradation is accommodated by a sea-level drop.

In Figure 2, the lowest four cycles are thinner than the average cycle, the middle two cycles are as thick as the average cycle, and the upper four cycles are thicker than the average cycle. The lower four cycles accumulated under a condition where the top of the sediment was being lowered, indicated by subsidence exceeding accumulation, presumably due to an interval of shallowing (relative lowering of sea level), thus reduc-

ing the space available to accommodate sediment. Constant relative sea level allowed the middle two cycles to accumulate as much as subsidence would allow, keeping the top of the sediment at a constant elevation. The upper four cycles accumulated as the top of the sediment was building up as accumulation exceeded subsidence, presumably during an interval of deepening (relative rise of sea level), allowing the accommodation of more strata than would simply be accommodated by subsidence alone. The dashed relative sea-level curve of Figure 2 will be represented in all subsequent CAPs. CAPs are constructed with time as a vertical axis, calibrated in periods, to preserve the stratigraphic perspective of the diagram. This is different from a Fischer plot in which the time axis is horizontal.

The CAP is actually a sea-level plot only for peritidal strata. This is the significance of Fischer plots. It is important to realize that these sea-level plots are relative, and do not necessarily reflect eustasy alone. They may be effected by eustatic sea-level changes, variation in subsidence rate, variable cycle periods (Fischer, 1964), or other parameters that affect the amount of sedimentation during each cycle. Careful interpretation of CAPs may therefore allow the recognition of sea-level changes, subsidence rates, or other parameters. This kind of interpretation is the major emphasis of this paper.

FIG. 3.—Correlation of CAPs for three outcrop sections. Top of each section is the top of the Juniata Formation. Lines of correlation represent shallowing and deepening events. Line O-13 is the datum, and corresponds with a regional shallowing event shown in Figures 6 and 7. Ticks along the left margin of each curve indicate individual cycles. Correlation of outcrop cycles is shown diagrammatically in inset (after Diecchio, 1985), with a datum at the top of the Juniata (refer to Fig. 1 for locations).

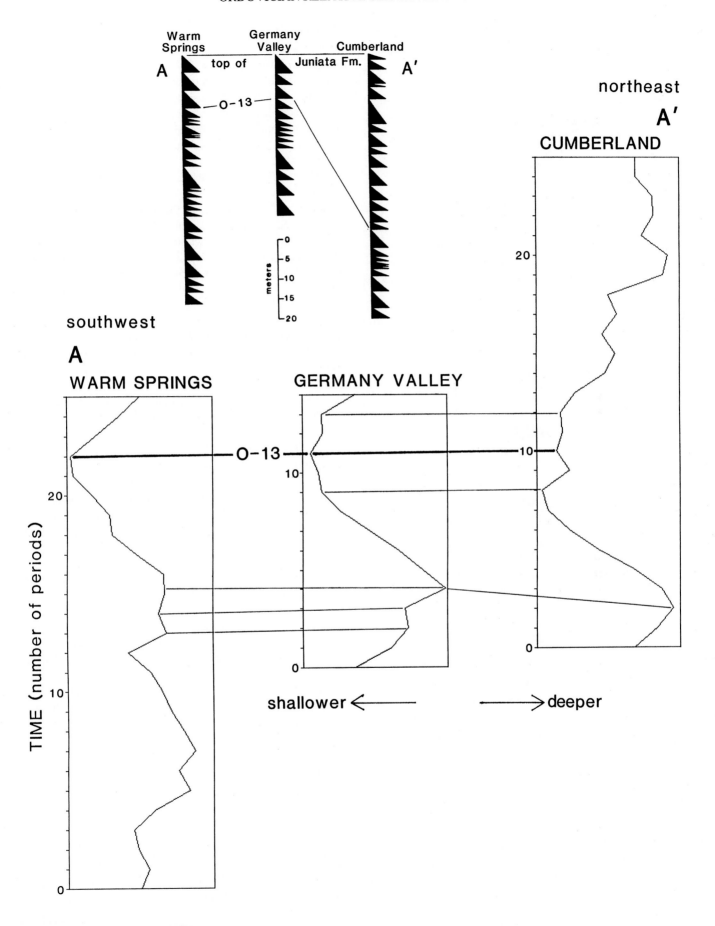

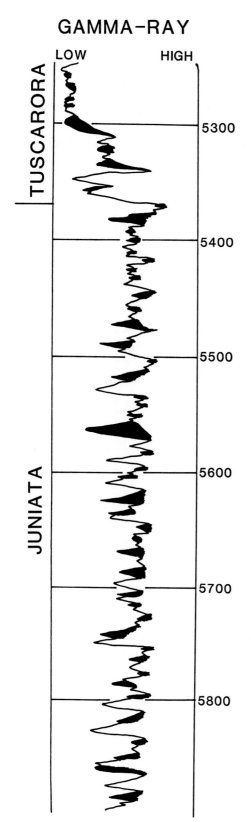

GAMMA-RAY

LOW   HIGH

TUSCARORA

5300

5400

5500

JUNIATA

5600

5700

5800

Fining-upward, siliciclastic, meter-scale, outcrop cycles of probable peritidal origin are described by Diecchio (1985) in the Upper Ordovician Juniata Formation of West Virginia and adjacent states. Three stratigraphic sections containing these cycles (indicated as triangles on Figure 1) provide the data used to generate relative sea-level curves in the form of CAPs (Fig. 3) for the upper part of the Juniata Formation. Because these strata are peritidal, the CAPs are interpreted as relative sea-level curves.

The fining-upward siliciclastic cycles within the Juniata Formation are also recognizable in gamma-ray well logs. Figure 4 illustrates the appearance of these cycles. Gamma-ray logs were chosen for this study simply because they are more available for Ordovician strata within the study area. Cycles were delineated on gamma-ray logs for each of eleven wells (open circles of Fig. 1) penetrating the Juniata Formation. The Oswego Formation, a localized facies of the lower part of the Juniata Formation (Diecchio, 1985) is included within the Juniata Formation in this report. Figure 5 is a log cross-section of these wells showing lithostratigraphic correlation of the tops of the Juniata and Reedsville Formations. The CAPs generated for these wells are shown and correlated in Figure 6. The CAPs are interpreted as relative sea-level curves because these cycles are the subsurface expression of cycles that are peritidal in outcrop.

The cyclical patterns that characterize the Juniata gamma-ray logs are not necessarily restricted to the Juniata Formation, but can be identified in the underlying formations. In fact, alternations of sandstone and shale, or of limestone and shale are apparent in gamma-ray logs through the entire Ordovician section. These do not all represent peritidal cycles, but they can be used to generate CAPs. This was done for each of four deep basement wells (closed circles of Fig. 1) that penetrate the entire Ordovician System. Figure 7 contains these CAPs. Table 1 contains the depth data for the formation tops in each well.

There is one error inherent in interpreting the CAPs of Figure 7 as relative sea-level curves. If the strata in question were peritidal (as is the case for Fischer plots), the repeated base value (reference datum representing the top of the sediment) is sea level, and cycle thickness is, therefore, controlled primarily by sea-level changes. In cases where the strata are not peritidal, the base value is some elevation above or below sea level. It is possible that within a formation containing a repeated set of facies, each cycle repeats similar paleobathymetric conditions. However, a formation does not necessarily repeat the same depth conditions as the overlying or underlying formations.

In an effort to minimize the complicating effect of facies changes on the CAPs generated for the deep wells (Fig. 6), portions of the well logs that have different gamma-ray signa-

FIG. 4.—Part of the gamma-ray log for the lower part of the Tuscarora Formation and the upper part of the Juniata Formation in the Hampshire 12 well. Gamma-ray intervals that were delineated and used to generate the CAP are indicated in alternating black and white.

FIG. 5.—Gamma-ray log cross-sections of wells penetrating the Juniata Formation. Dashed lines represent formation tops, modified slightly from Cardwell (1977) and Diecchio (1985). Datum is the top of the Juniata Formation. Vertical scale is depth in ft. Solid lines correspond to sea-level events delineated in Figure 6.

(Continued on p. 177)

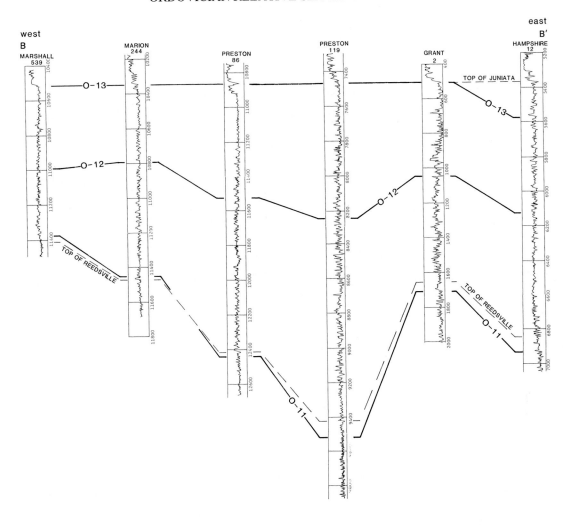

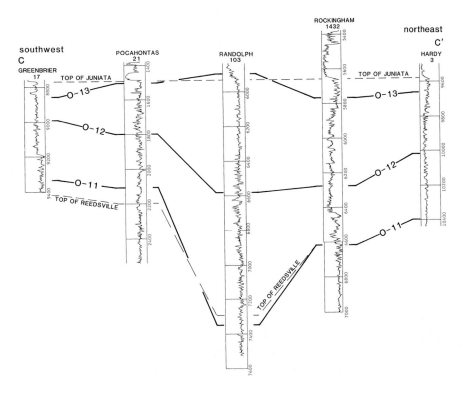

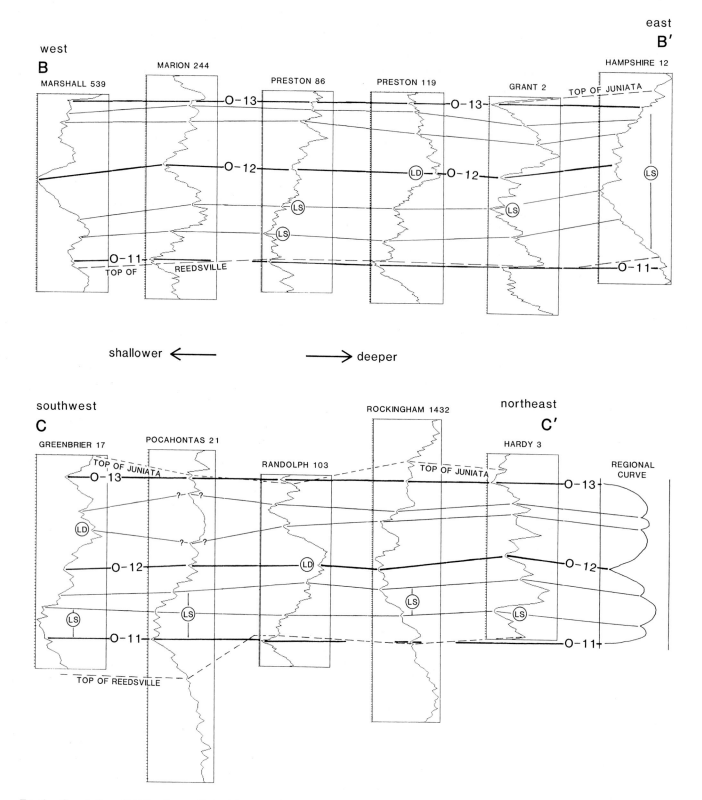

FIG. 6.—Correlation of CAPs of upper Ordovician strata in central and eastern part of study area based on gamma-ray logs. Interval includes all strata from the top of the Juniata Formation (upper dashed line) to the top of the Reedsville Formation (lower dashed line), as indicated in Figure 5. Well locations are shown in Figure 1. Lines of correlation (solid lines) represent shallowing events of regional extent. Heavy correlation lines correspond with lines of correlation of Figure 7. Regional curve on lower right is based on "average" trend determined from all plots. Circled symbols are as follows: LS— local shallowing event; LD— local deepening event. Vertical scales of CAPs have been adjusted so that time interval between O-11 and O-13 is constant, hence, vertical scale is time. O-11 and O-13 can each be considered a datum for this diagram.

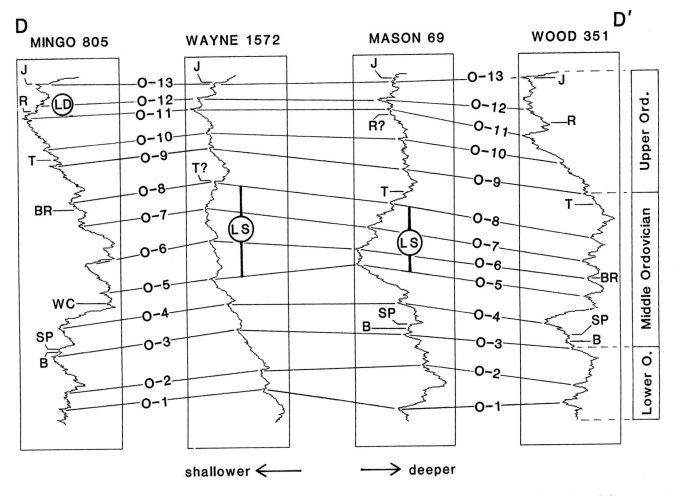

FIG. 7.—Correlation of CAPs of the entire Ordovician in western part of study area based on gamma-ray logs. Lines of correlation represent shallowing events of regional extent. Vertical scale is time, and represents the entire Ordovician Period for each well. Circled symbols are: LS— local shallowing event, LD— local deepening event. Formation tops are: J—Juniata, R—Reedsville, T— Trenton, BR— Black River, WC— Wells Creek, SP— St. Peter, and B— Beekmantown. Formation picks are after Cardwell (1977) for all wells except Wayne 1572. Datum is O-13.

tures were analyzed independently of each other. This was done by using a different subsidence rate (average cycle thickness) for each part of the log that had cycles with significantly different thicknesses or cycles with a significantly different gamma-ray intensity. For example, the Tuscarora interval (above 5312-ft depth) of Figure 4 was calculated with a greater subsidence rate (because the cycles have a greater average thickness) than the Juniata interval (below 5312 ft). Accordingly, for each of the wells that penetrate the entire Ordovician, three different intervals were defined: the predominantly siliciclastic (shale with some sandstone) interval between the top of the Juniata/Queenston and the top of the "Trenton" limestones; the carbonate section (limestone with minor shale) between the top of the "Trenton" and the top of the Wells Creek Limestone; and the carbonate section with a greater amount of shale between the top of the Wells Creek and the top of the Cambrian. A different average cycle thickness was calculated for each interval.

Data for the CAPs were manipulated using Lotus 1-2-3 software, which was also used to generate the plots. Data sets for

the gamma-ray logs were created by digitizing the actual well logs, creating X-Y data sets which were suitable for use in Lotus.

The CAPs of Figure 6 have been adjusted so that the vertical axis, assumed to be time, is the same for each CAP in the figure. In other words, the vertical axis of each CAP in Figure 6 has been adjusted so that the regional events assumed to be time lines (see section on Regional Sea-level Events) are as horizontal as possible, as is the case for stratigraphic chrono-correlation diagrams. Adjustment of the vertical CAP axes was not necessary for Figure 3, where a constant Y-axis scale results in horizontal event lines. For Figure 7, the vertical axis of each CAP is the duration of the Ordovician Period.

RELATIVE SEA-LEVEL CURVES

Cumulative aggradation plots for the outcrop sections (Fig. 3) reveal relative sea-level events that correlate among each of the sections. The top of each section is the top of the Juniata Formation. The plots suggest that the top of the Juniata Forma-

TABLE 1.—DEPTHS, IN FEET, OF FORMATION TOPS (MINGO, MASON AND WOOD, AFTER CARDWELL, 1977) AND SEA-LEVEL EVENTS INTERPRETED IN FIGURE 7 FOR EACH OF THE BASEMENT WELLS. (DETAILED WELL RECORDS ARE KEPT ON FILE AT THE WEST VIRGINIA GEOLOGICAL AND ECONOMIC SURVEY.)

	MINGO 805	WAYNE 1572	MASON 69	WOOD 351
Formation Tops				
Juniata	4986	4010	4501	7811
Reedsville	5385		4765	8421
Trenton	5963	5210?	5748	9528
Black River	6608			9899
Wells Creek	7760			
St. Peter	8494		6646	10675
Beekmantown	8522		6682	10708
Sea-level Events				
O-13	5012	4021	4511	7764
O-12	5281	4224	4699	8167
O-11	5490	4365	4850	8480
O-10	5872	4598	5041	8750
O-9	6075	4786	5232	9034
O-8	6523	5131	5633	9520
O-7	6820	5415	5876	9785
O-6	7334	5697	6077	9967
O-5	7675	6045	6197	10170
O-4	8211	6249	6468	10567
O-3	8600	6488	6720	10831
O-2	9052	6837	6954	11321
O-1	9285	7048	7365	11556

tion is diachronous, being significantly younger at Cumberland. Cumulative aggradation plots for the gamma-ray logs of wells penetrating the Juniata Formation (Fig. 6) show three major (O-11, O-12, O-13) and four minor shallowing events that correlate from well to well. The upper shallowing event (O-13) is also recognizable in the plots derived from the outcrop sections (Fig. 3). The diachronous nature of the top of the Juniata Formation, as well as the top of the Reedsville Formation, is also apparent in Figure 6.

The vertical axis of each CAP in Figure 6 had to be adjusted to attain chrono-correlation. This emphasizes the fact that inferred distinct time intervals contain different numbers of cycles in different wells. This observation suggests that not all cycles are preserved everywhere, and that wells in Figure 6 where fewer cycle periods are contained between sea-level events represent places where the stratigraphic section is less complete than elsewhere.

The CAPs generated for the basement wells are shown in Figure 7. Individual shallowing events are correlated and numbered.

Thirteen distinct Ordovician shallowing events are recognized in each of the basement wells (Fig. 7). The depths in each well corresponding to each event are listed in Table 1. These thirteen events, occurring over the duration of the Ordovician Period, about 65 my, delineate sea-level fluctuations with a period averaging about 5 my. These cycles have periods similar to Early and Middle Ordovician sea-level fluctuations determined using Fischer plots for meter-scale carbonate outcrop cycles in southwestern Virginia (Read and Goldhammer, 1988), suggesting that these outcrop cycles may represent the same sea-level events as those recognized in the deep wells. The widespread extent of these events, combined with their regular frequency of occurrence, are the criteria that we use to suggest these may be due to eustatic fluctuations rather than local subsidence or to tectonism. However, without the true test of eustasy, confirmation by comparison with other cratons, one cannot completely rule out the possibility of these events being caused by regional uplift or subsidence. These cycles correspond in period with third-order sea-level cycles (1-10 my) of Vail and others (1977).

The correlation of these thirteen events (Fig. 7) is subjective and, therefore, tentative in places. The three youngest of these events are recognizable throughout the study area (Fig. 6 and 7), except where they are masked by a local effect (such as O-12 in the Preston 119 well of Fig. 6). The youngest (O-13) is also recognized in outcrop (Fig. 3). If these are eustatic events, the event correlation lines of Figures 6 and 7 would be time lines. This is the basis for adjusting the vertical axes of the CAPs of Figure 6 to accomplish chrono-correlation.

Shorter period events are also recognizable. The thin correlation lines of Figure 6 indicate higher-frequency regional sea-level fluctuations. These are not necessarily correlatable throughout the study area.

Shallowing event O-11 is discernible in the Juniata or the Reedsville ("Martinsburg") Formations in Figures 6 and 7. This suggests that these two formations are facies of each other, at least in part. The Reedsville Formation is probably not peritidal. Recognition of the same event in two different facies may support the validity of using CAPs to determine relative sea-level changes for strata that are not peritidal.

Two large-scale sea-level events are also recognizable in each basement well (Fig. 7). A pronounced lowering of sea-level (O-3 to O-4) is recognizable near the top of the Knox or Beekmantown Group. Another large-scale shallowing event (O-12 to O-13) occurs near the Ordovician-Silurian boundary. The coincidence between these shallowing events and the lower and upper limits of the Creek holostrome (Wheeler, 1963) or Tippecanoe I sequence (Sloss, 1982) supports an eustatic cause.

The shallowing event marked O-8 in Figure 7 may correspond with the Trenton-Black River sea-level drop (Dennison, abstract, this volume). This suggests that this event had an eustatic influence.

LOCAL SEA-LEVEL EVENTS

The CAPs for the well sections (Figs. 6, 7) indicate relative sea-level events that occur only at some localities. These events are marked with either a circled LS for local shallowing or a circled LD for local deepening. On Figure 6, the regional curve in the lower right is intended to represent the "average" change in sea-level, estimated from all eleven curves. Significant departures from this regional curve are the basis for delineating the local sea-level changes on this diagram.

The sporadic occurrence of these local events rules out the possibility of regional or global eustatic events. These local events are interpreted to be the effects of local tectonism within the Appalachian basin. On Figures 6 and 7, the local sea-level events are generally of longer duration than the regional sea-level events.

These local sea-level events may be completely unrelated in time and space, and may simply represent localized uplifts or downwarps that occurred at different times. On the other hand, some of these events may be related to a single event that effected the basin differently in different places. For example, the localized effects recorded in wells penetrating the Upper Ordovician (Fig. 6) indicate that shallowing occurred in what is now the eastern panhandle of West Virginia at approximately the same time that deepening was occurring immediately to the west (Fig. 8). Based primarily on Upper Ordovician isopach data, Diecchio (1985) identified a positive axis located in the western part of the Valley and Ridge outcrop belt (coincident with the area of Upper Ordovician shallowing of Fig. 8), a western axis of subsidence in the central part of the Appalachian basin (coincident with the area of Upper Ordovician deepening of Fig. 8), and an eastern axis of subsidence in northern Virginia (east of the area of Upper Ordovician shallowing of Fig. 8). It is interesting that this geometry would be expected if the area east of the present study area (the true Martinsburg flysch basin) were a flexural moat subsiding under the load of Taconian thrust sheets, and the positive axis or area of shallowing of Figure 8 were part of a peripheral bulge. This model was suggested by Beaumont and others (1988), but their thrust-load flexural moat (Appalachian basin) and peripheral bulge (Cincinnati arch) were 100 km farther west.

The CAPs for the four basement wells indicate a local sea-level event during the Middle Ordovician. As indicated on Figure 7, a local, long-duration, shallowing event occurred in the area of the Wayne and Mason wells during that time. An alternate interpretation is that these two wells indicate the regional trend, and that the Mingo and Wood wells indicate a deepening event during the same interval. This localized shallowing (or deepening) is probably due to flexure (or differential subsidence) associated with Taconian tectonism, and may be localized by preexisting faults associated with the Rome trough. The timing of this local event may coincide with the pre-Rocklandian and Kirkfieldian flexural-induced unconformity of Ettensohn (1991). What is intriguing is that the correlation of the CAPs allows the distinction between the probable tectonic

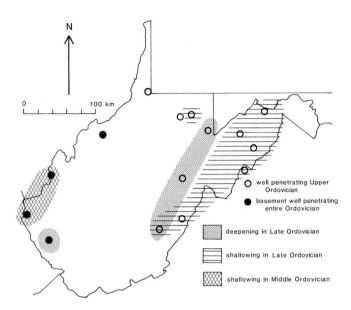

FIG. 8.—Geographic extent of local shallowing and deepening events, as depicted on Figures 6 and 7. Well locations are shown on Figure 1.

effects and the probable eustatic effects during Middle Ordovician time.

CONCLUSIONS

The CAPs presented in this report suggest that regionally correlative Ordovician bathymetric events can be delineated within the study area. These plots enable the refinement of stratigraphic correlations to 5-my intervals or less. Based on the comparison between many of these events with events and periodicities that others have attributed to sea-level fluctuations, we suggest that these plots reflect relative changes in sea level. At least, this application of CAPs warrants further investigation.

Using CAPs for outcrop and well data enables the recognition of regional and local "sea-level" events of various scales and durations, and thus helps differentiate between eustatic and tectonic controls. Generally speaking, regional (probably eustatic) sea-level changes seem to be shorter duration events that occur with a more regular frequency than local (probably tectonic) sea-level changes.

ACKNOWLEDGMENTS

The authors thank the donors of the Petroleum Research Fund of the American Chemical Society (grant number 19220-B2 to Diecchio) for supporting this research. This manuscript has benefited from comments by Jack Beuthin, John Dennison, and Lee Avary.

REFERENCES

BEAUMONT, C., QUINLAN, G., AND HAMILTON, J., 1988, Orogeny and stratigraphy: Numerical models of the Paleozoic in the Eastern Interior of North America: Tectonics, v. 7, p. 389-416.

CARDWELL, D. H., 1977, West Virginia gas developments in Tuscarora and deeper formations: Morgantown, West Virginia Geological and Economic Survey, Mineral Resources Series No. 8, 34 p.

DENNISON, J. M., 1976, Appalachian Queenston delta related to eustatic sea-level drop accompanying Late Ordovician glaciation centered in Africa, in Bassett, M. G., ed., The Ordovician System: Cardiff, Proceedings of a Palaeontological Association Symposium, Birmingham, England, 1974, University of Wales Press and National Museum of Wales, p. 107-120.

DENNISON, J. M., 1989, Paleozoic Sea-level Changes in the Appalachian basin: Washington, D.C., Field Trip Guidebook T354, 28th International Geological Congress, 56 p.

DIECCHIO, R. J., 1985, Post-Martinsburg Ordovician stratigraphy of Virginia and West Virginia: Charlottesville, Virginia Division of Mineral Resources Publication 57, 77 p.

ETTENSOHN, F. R., 1991, Flexural interpretation of relationships between Ordovician tectonism and stratigraphic sequences, central and southern Appalachians, U.S.A., in Barnes, C. R., and Williams, S. H., eds., Advances in Ordovician Geology: Ottawa, Geological Survey of Canada Paper 90-9, p. 213-224.

ERIKSSON, K. A., AND SIMPSON, E. L., 1990, Recognition of high-frequency sea-level fluctuations in Proterozoic siliciclastic tidal deposits, Mount Isa, Australia: Geology, v. 18, p. 474-477.

FISCHER, A. G., 1964, The Lofer cyclothems of the Alpine Triassic: Geological Survey of Kansas Bulletin, v. 169, p. 107-149.

FORTEY, R. A., 1984, Global earlier Ordovician transgressions and regressions and their biological implications, in Bruton, D. L., ed., Aspects of the Ordovician System: Oslo, Palaeontological Contributions, University of Oslo, No. 295, Universitetsforlaget, p. 37-50.

GOLDHAMMER, R. K., DUNN, P. A., AND HARDIE, L. A., 1987, High frequency glacio-eustatic sea level oscillations with Milankovitch characteristics recorded in Middle Triassic platform carbonates in northern Italy: American Journal of Science, v. 287, p. 853-892.

GOLDHAMMER, R. K., DUNN, P. A., AND HARDIE, L. A., 1990, Depositional cycles, composite sea-level changes, cycle stacking patterns, and the hierarchy of stratigraphic forcing: Examples from Alpine Triassic platform carbonates: Geological Society of America Bulletin, v. 102, p. 535-562.

HAYNES, J. T., 1992, Reinterpretation of Rocklandian (Upper Ordovician) K-bentonite stratigraphy in southwest Virginia, southeast West Virginia, and northeast Tennessee with a discussion of the conglomeratic sandstones in the Bays and Moccasin Formations: Charlottesville, Virginia Division of Mineral Resources Publication 126, 58 p.

HERGENRODER, J. D., 1966, The Bays Formation (Middle Ordovician) and related rocks of the southern Appalachians: Unpublished Ph.D. Dissertation, Virginia Polytechnic Institute and State University, Blacksburg, 323 p.

KOERSCHNER, W. F., III, AND READ, J. F., 1989, Field and modelling studies of Cambrian carbonate cycles, Virginia Appalachians: Journal of Sedimentary Petrology, v. 59, p. 654-687.

MCKERROW, W. S., 1979, Ordovician and Silurian changes in sea level: Journal of the Geological Society of London, v. 136, p. 137-145.

MUSSMAN, W. J., AND READ, J. F., 1986, Sedimentology and development of a passive- to convergent-margin unconformity: Middle Ordovician Knox unconformity, Virginia Appalachians: Geological Society of America Bulletin, v. 97, p. 282-295.

READ, J. F., AND GOLDHAMMER, R. K., 1988, use of Fischer plots to define third-order sea-level curves in Ordovician peritidal carbonates, Appalachians: Geology, v. 16, p. 895-899.

RODGERS, J., 1971, The Taconic orogeny: Geological Society of America Bulletin, v. 82, p. 1141-1178.

SLOSS, L. L., 1982, The Midcontinent Province: United States, in Palmer, A. R., ed., Perspectives in Regional Geological Synthesis: Boulder, Geological Society of America Decade of North American Geology Special Publication 1, p. 27-39.

VAIL, P. R., MITCHUM, R. M., AND THOMPSON, S., III, 1977, Seismic stratigraphy and global changes of sea-level, part 4: Global cycles of relative changes of sea-level, in Payton, C. E., ed., Seismic Stratigraphy; Applications to Hydrocarbon Exploration: Tulsa, American Association of Petroleum Geologists Memoir 36, p. 83-97.

WHEELER, H. E., 1963, Post-Sauk and Pre-Absaroka Paleozoic stratigraphic patterns in North America: American Association of Petroleum Geologists Bulletin, v. 47, p. 1497-1526.

TECTONIC AND EUSTATIC INFLUENCES UPON THE SEDIMENTARY ENVIRONMENTS OF THE UPPER ORDOVICIAN STRATA OF NEW YORK AND ONTARIO

DAVID LEHMANN[1], CARLTON E. BRETT, AND RONALD COLE

Department of Earth and Environmental Sciences, University of Rochester, Rochester, New York 14627

ABSTRACT: The Upper Ordovician stratigraphic succession in New York and Ontario is superficially similar to a very large eustatically-controlled sequence. These strata are bounded by unconformities, and analogs of systems tracts are present. Submarine fan siltstones, sandstones, and shales comprise the lowstand systems tract analog; the transgressive systems tract analog is represented by a stratigraphically condensed section, and the highstand systems tract analog is characterized by an upward-coarsening, and generally upward-shallowing, succession of strata. The diachroneity of the systems tract analogs and of sequence-bounding unconformities suggest, however, that this stratigraphic succession was most strongly influenced by tectonic forces associated with the Taconic Orogeny.

In the Taconic foreland basin of New York and Ontario, thrust- and sediment-loading drove subsidence so that the Middle to Late Ordovician Trenton carbonate ramp progressively oversteepened and collapsed. In New York, the oversteepening is represented by a stratigraphic succession in which carbonate-dominated deposits are disconformably overlain by flysch. In the Toronto, Ontario region, which was more distal to the thrust sheets and, presumably, more proximal to the Late Ordovician tectonic hinge, the disconformable relationship between the underlying carbonates and the overlying siliciclastics grades to conformity. In New York and Ontario, progressive southeast to northwest oversteepening of the carbonate ramp resulted in a geographically diachronous shift from carbonate-dominated deposition to organic-rich mud deposition; a sediment-starvation surface (a condensed section) is often associated with this shift. The siliciclastic strata that overlie the condensed section record the prograding Queenston clastic wedge. Paleocurrent and stratigraphic data suggest that movement along Taconic, or older rejuvenated, basement normal faults was an important control on basin subsidence and filling.

Smaller scale isochronous stratigraphic changes that cross-cut facies patterns *may* record eustatic events. These possible eustatic events include: (1) an Early Maysvillian deepening event, recorded by the Collingwood Formation in Ontario; (2) a basal-Pulaski progradational event; and (3) a mid-Queenston transgression.

INTRODUCTION

Over the last three decades, two broad-based models, sequence stratigraphy and plate tectonics, have revolutionized the way geologists observe, analyze, and interpret stratigraphic data. Unquestionably, both eustatic and tectonic forces, by affecting relative sea level and sedimentation rate, can strongly influence stratigraphic successions.

In this study, we examine the Upper Ordovician stratigraphic succession of north-central New York and of central and western Ontario (Figs. 1, 2) and attempt to differentiate between eustatic and tectonic controls upon deposition of the strata. The Late Ordovician was a time of active tectonism in eastern North America (the Taconic Orogeny) during which a peripheral foreland basin was developing along the eastern continental margin. Whereas tectonic models are often invoked to explain stratigraphic successions of active foreland basins (Allen and Homewood, 1986; Allen and Allen, 1990), eustatic models may best explain stratigraphic successions of passive continental margins (see Kendall and Lerche, 1988, for historical review). Most of the strata which we will discuss occur distally from (west of) the most actively subsiding portion of the Taconic foreland basin. (We estimated subsidence using stratigraphic thickness and biostratigraphic data.) Therefore, the study area occupies a critical transition interval between the active foreland

basin (in which tectonic forces are believed to dominate) and the platform (on which eustatic events may exert greater control of the stratigraphic succession).

To recognize and differentiate eustatic and tectonic events, we employed sequence stratigraphy, in addition to biostratigraphy and event stratigraphy. Although unconformity-bounded stratigraphic sequences are typically reported to be the result of fluctuations in eustatic sea level (Jervey, 1988), tectonic forces can develop stratigraphic successions which mimic eustatically-driven sequences (Galloway, 1989; Miall, 1991), as outlined in Table 1. Upper Ordovician strata of New York and Ontario comprise a "pseudo-sequence" primarily attributed to tectonic influences as opposed to eustatic control. This paper will characterize a tectonic "pseudo-sequence" within a foreland basin and define tectonic analogs for eustatic depositional systems tracts.

THE SEQUENCE STRATIGRAPHIC MODEL

A stratigraphic sequence has lower- and upper-bounding unconformities and their correlative conformities (Vail and others, 1977; Van Wagoner and others, 1988). Sequence-bounding unconformities can be the result of eustatic sea-level fall (Vail and others, 1977) or tectonic uplift (Cloetingh, 1988; Bond and Kominz, 1991; Miall, 1991; Sloss, 1991). Between these unconformities, there are a number of systems tracts, each corresponding to specific eustatic conditions (Posamentier and Vail, 1988).

---

[1] Present address: Huntingdon Engineering and Environmental, 535 Summit Point Drive, Henrietta, New York 14467

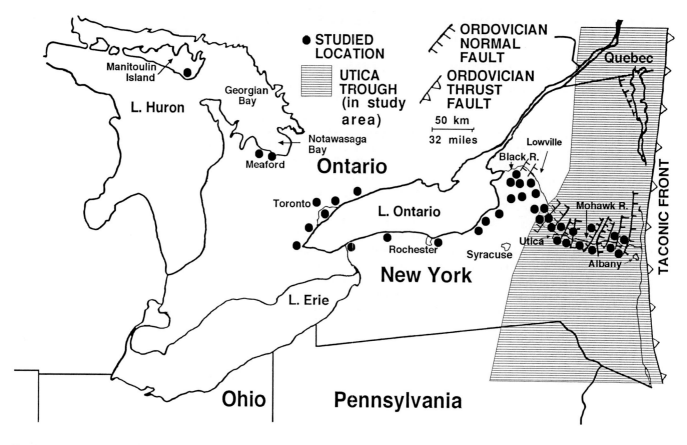

FIG. 1.—Locality map showing some of the outcrops studied. The approximate position of the Utica trough parallels the strike of normal faults which were active during the Ordovician (Cisne and others, 1982; Mehrtens, 1988; Bradley and Kidd, 1991). The present western margin of the Taconic Front (allochthon) is probably east of the Late Ordovician Taconic margin due to later erosion.

In many cases (Type 1 sequences of Van Wagoner and others, 1988), the lower-bounding unconformity (or correlative conformity) is overlain by the Lowstand Systems Tract (LST). The LST may include a lowstand submarine fan of sediment which is best developed toward the basin center (Posamentier and others, 1988). Lowstand fans can be the result of erosion along the shelf edge and from incised river valleys during relative lowering of sea level (Vail and others, 1977; Mutti, 1985; Kolla and Macurda, 1988). Alternately, similar submarine fans may develop, independent of eustasy, following shelf edge or platform collapse or following tectonic uplift in the hinterland of an active foreland basin (Mutti and Ricci Lucchi, 1978).

When present, the LST is overlain by the Transgressive Systems Tract (TST) which comprises deposits marking the landward migration of the strandline (Vail and others, 1977; Baum and Vail, 1988). In many shallow-shelf deposits, lowstand deposits are absent, and the transgressive surface (the base of transgressive deposits) and basal sequence boundary are merged into a single disconformity. At its top, the TST also includes a condensed section, which is best developed away from the sediment source (Nummedal and Swift, 1987; Posamentier and others, 1988). The condensed section contains

abundant authigenic mineral grains and reflects relative sediment starvation (Loutit and others, 1988). Because nearshore transgressive deposits may be removed by erosion associated with the upper-bounding sequence unconformity, the condensed section has proven to be one of the most recognizable and traceable elements of stratigraphic sequences. Condensed sections can reflect sediment starvation of the basin center as a result of estuarine deposition during a rapid eustatic sea-level

FIG. 2.—(A) Lithostratigraphy of the study area. Argillaceous carbonates underlying the siliciclastic strata were erosionally removed during the Late Ordovician in, and to the west of, the Utica trough. To the west, this disconformity is replaced by a condensed interval (CI). Siltstones capping the Utica trough (Hasenclever Member of the Frankfort Formation) represent a short-lived supply of coarser siliciclastic sediment. Note that the Silurian cut-out surface, marking the Cherokee Unconformity, roughly parallels the Utica trough, suggesting that motion along faults which controlled subsidence during the Ordovician reversed during the latest Ordovician to the Early Silurian. Lithostratigraphic relations across Lake Ontario are somewhat conjectural. (B) Time-stratigraphic relations of strata which are discussed in the text. Paleocurrent arrowheads point downcurrent. Strata were correlated using biostratigraphy, event beds, and facies relations. Abbreviations include Rock.: Rocklandian; Kirk.: Kirkfieldian.

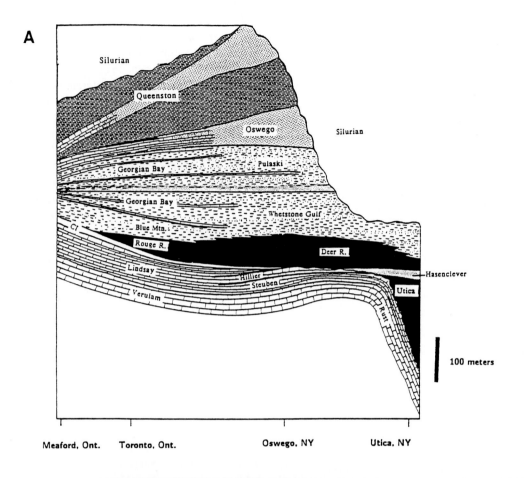

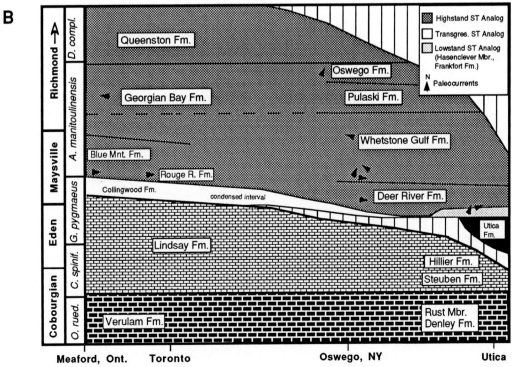

TABLE 1.— COMPARISONS OF SYSTEMS TRACTS AND SYSTEMS TRACT ANALOGS.

	Eustatic Systems Tract (Posamentier and Vail, 1988)	Tectonic Systems Tract Analog (this study)
Lowstand		
Deposits	Submarine fans in basin center, submarine canyon fill in incised portions of old shelf, and progradational delta-front turbidites.	Axial submarine fan systems in elongate foreland basins.
Position in sequence	Sharp to disconformable contact with underlying Highstand Systems Tract; overlain by Transgressive Systems Tract.	Random; may overlie condensed section of Transgressive Systems Tract Analog.
Causes	Rate of eustatic fall is greater than rate of subsidence at or near shoreline.	Increased sediment supply due to tectonic uplift and/or new accommodation space for siliciclastics due to shelf-edge collapse.
Transgressive		
Deposits	Retrogradational transgressive lag deposits in near-shore settings; stratigraphically-condensed section in basinal settings.	Stratigraphically-condensed section in basinal settings.
Position in sequence	Separated from overlying Highstand deposits by a surface of maximum sediment starvation.	Overlies carbonate-dominated strata; separated from overlying Highstand deposits by disconformable to conformable surface.
Causes	Eustatic rise leads to flooding of the shelf; siliciclastics trapped in estuaries (old incised stream channels).	Carbonate ramp/platform subsides; oversteepening results in little accumulation of siliciclastic sediment; carbonates dissolved by bottom waters; sediment starvation.
Highstand		
Deposits	Aggradational to progradational strata.	Aggradational to progradational strata.
Position in sequence	Disconformable surface at top.	Disconformable surface at top.
Causes	Sediment supply is greater than accommodation space as eustatic rise ceases and during initial eustatic fall.	Clastic wedge progradation.

rise (Jervey, 1988). Alternately, condensed sections may be the result of collapse or subsidence of a carbonate platform (Bally and Snelson, 1980). If a carbonate platform subsides and siliciclastic sediment does not reach the subsided platform, sediment starvation will occur.

Interestingly, the relationship between a LST and a TST can be an important criterion for differentiating a eustatic sea-level rise from tectonic basin subsidence. In a passive margin, eustatically-controlled, Type I sequence, the TST will overlie the LST, whereas in a tectonically-controlled foreland basin succession, a submarine fan may occur at various positions within the pseudo-sequence, often overlying a condensed section reflecting subsidence followed by sedimentation.

The TST, in turn, is overlain by the Highstand Systems Tract (HST), an aggradational to progradational succession of strata. The HST records times of gentle eustatic sea-level rise and initial fall during which sedimentation keeps pace with, and may even outpace, basin subsidence. Aggradational to progradational clastic wedges (molasse) are also characteristic of foreland-basin fill and late orogenic platform sedimentation (Miall, 1991) and, similarly, reflect sedimentation outpacing basin subsidence.

GEOLOGIC AND STRATIGRAPHIC SETTING

The Taconic Orogeny transformed eastern North America from a predominantly shallow-marine, carbonate platform into a siliciclastic-dominated peripheral foreland basin. This orogeny probably resulted from the collision of eastern North America and a series of island arcs (Rowley and Kidd, 1981; Cisne and others, 1982; Hay and Cisne, 1988), although recent work indicates microcontinents may also have been involved in this collisional orogeny (Delano and others, 1990). During Middle Ordovician time, slices of an accretionary prism (the Taconic allochthon) were thrust westward onto the North American platform as a result of the subduction of the eastern edge of the North American plate beneath the Iapetus plate. This thrusting culminated with uplifts of Precambrian North American basement rocks (the Green Mountain and Berkshire Massifs) in the hinterland of the orogen (Stanley and Ratcliffe, 1985).

Lithospheric flexural subsidence during and following the thrust-load emplacement created the Taconic peripheral foreland basin (Rowley and Kidd, 1981; Quinlan and Beaumont, 1984; Tankard, 1986; Lash, 1988; Bradley and Kidd, 1991). Cisne and others (1982) and Bradley and Kidd (1991) have shown that the development of the eastern portion of the foreland basin in New York, the Utica trough, was at least in part the result of movement along high-angle faults forming grabens and half-grabens which dropped to the east. These faults generally trend NNE-SSW and are collateral with the strike of the foreland basin (Fig. 1). Progressive (east to west) movements along these faults were coincident with the westward progressive change from carbonate-dominated to siliciclastic sedimentation within the foreland basin (Bradley and Kidd, 1991).

The shift from a carbonate-platform to a foreland-basin setting resulted in drastic changes in sedimentary deposits (Fig.

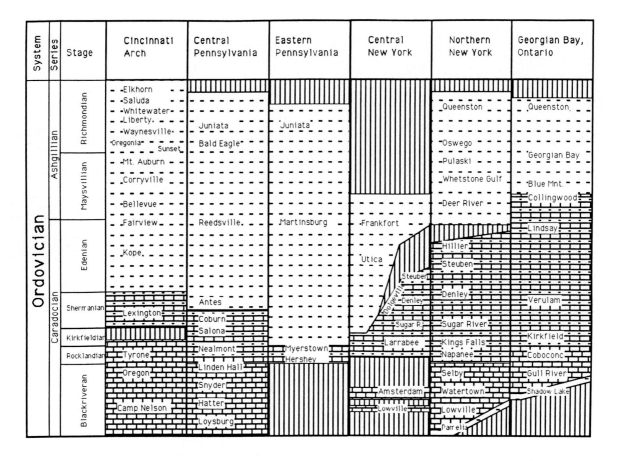

FIG. 3.— Caradocian and Ashgillian strata of northeastern and northcentral North America. Only formational names are included on this diagram. Note the three very generalized packages of strata: the siliciclastic-free carbonates (shown in a brick pattern), the argillaceous carbonates (shown as intermittent bricks and dashed lines), and the clastic wedge (shown as dashed lines). Hiatuses are shown as vertical lines. The Blackriveran clastics in New York and Ontario are basal arkoses and related siliciclastics which are present between underlying Precambrian basement strata and overlying carbonates of the Black River Group. Includes data from MacLachlan, 1967; Liberty, 1978; Sweet and Bergstrom, 1976; Fisher, 1979; Bergstrom and Mitchell, 1986; Keith, 1988; Holland, 1989; Lehmann and Brett, 1991.

3). In general, the Middle and Upper Ordovician rocks that reflect this shift belong to the Creek Holostrome (Dennison and Head, 1975) of northeastern North America and can be described as comprising three informal packages: (1) siliciclastic-free carbonates, (2) argillaceous carbonates, and (3) siliciclastics.

Siliciclastic-free carbonates include micrites and biomicrites belonging to the Black River Group in New York and Ontario. This package comprises an upward-deepening progression from desiccation-cracked, supratidal micrites to shallow shelf biomicrites (Anderson and others, 1978). Although lack of siliciclastic detritus in these rocks seems to indicate tectonic quiescence, metabentonites are present (Kay, 1935; Kunk and Sutter, 1984), foreshadowing the effects of tectonism evident in overlying strata.

Argillaceous carbonates include marine limestones and argillaceous limestones which are interbedded with shales (Trenton and Cobourg Groups of New York and Ontario). The rocks in this package seem to represent relative basin deepening, followed by an interval of shallowing, and culminating in an

abrupt relative deepening event (Titus, 1986, 1988). The upper boundary of this package is diachronous; from east to west, the argillaceous carbonates thicken and represent progressively more time (Lehmann and Brett, 1991). An inverse relationship exists between the thickness of argillaceous carbonates and the thickness of overlying siliciclastics (Fisher, 1977; Liberty, 1969).

A sediment-starved transition interval typically occurs between the underlying argillaceous carbonates and the overlying siliciclastics. This sediment-starved transition interval is characterized by early diagenetic cementation of limestones, abundant organic material (Russel and Telford, 1983), oil shales, and phosphatic pebbles and molds of fossils. The transition interval, represented by the Collingwood Formation in Ontario, is approximately 10 m thick along the Georgian Bay in Ontario and thins progressively towards the southeast (Fig. 2). Carbonates below the transition interval are typically part of a "deepening-upwards" package of strata. The deepening upwards succession occurs abruptly in the Utica trough but occurs gradually west of the trough. In the Utica trough, a diachronous, discontinuous,

FIG. 4.— (A) Contact between the Steuben Limestone and turbidites of the Hasenclever Member of the Frankfort Formation (Wells Creek, Frenchville, NY). The daypack is resting on the corroded upper surface of the Steuben Limestone. (B) Turbidites showing Bouma T_b through T_e with well developed climbing ripple cross lamination.

transition interval rests unconformably on Trenton carbonates.

Siliciclastics include rocks representing a wide range of depositional environments, from anoxic marine deposits to nonmarine delta-platform deposits. In general, these rocks represent upward-shallowing conditions, culminating in non-marine Queenston/Juniata red beds. In New York, the oldest autochthonous siliciclastic rocks (Snake Hill and Martinsburg Formations) underlie the Taconic allochthon (Fisher, 1977). The base of the siliciclastic package becomes progressively younger to the west. As will be discussed, this trend of diachronous westward onlap is the result of basin thalweg (which we define as the deepest part of the basin) migration, which in turn, was presumably driven by tectonic subsidence and delta progradation.

This study focuses on rocks belonging to the siliciclastic package in the westernmost portion of the Utica trough and strata northwest of the trough. The Upper Ordovician siliciclastic strata in the study area are bounded by unconformities and their correlative conformities (Fig. 2). The upper unconformity, above the Queenston Formation and below Lower Silurian strata is unquestionably subaerial and is partly the result of a Late Ordovician glacio-eustatic sea-level fall. However, this contact between the Queenston Formation and overlying Silurian strata is a regionally angular unconformity indicating that the erosional hiatus may have been enhanced by tectonic or isostatic uplift. The lower unconformity separates Late Ordovician siliciclastics from underlying argillaceous carbonates of the Trenton and Cobourg Groups in the Utica trough. The origin of this lower unconformity is more problematic, although Titus (1986) has suggested that, at least in part, this unconformity may be the result of local uplift and subaerial exposure. Between these unconformities, there are three basic stratigraphic packages, analogous to systems tracts, which are discussed below.

SYSTEMS TRACT ANALOGS

Three "systems-tract analogs" are present in the strata which we studied. These systems-tract analogs differ from true systems tracts in that the analogs are as much (or more) a result of tectonic forces as they are a result of eustasy (Duke, 1991). However, as Figure 2 depicts, the lateral and vertical arrangement of the systems-tract analogs closely mimic the arrangements of true systems tracts within eustatically-controlled sequences occurring in passive margin settings.

Lowstand Systems Tract Analog (LSTA)

Turbiditic siltstones, sandstones, and interbedded shales of the Hasenclever Member, Frankfort Formation overlie black shales of the Utica Formation in the most rapidly subsided portion of the foreland basin, the Utica trough (Fig. 2). To the northwest of the Utica trough, in north-central New York, these turbidites overlie argillaceous carbonates of the Trenton Group (Fig. 4A). These turbidites are part of a distal submarine fan and form the LSTA.

Strata representing the LSTA are approximately 20-m thick in the central portion of the Utica trough and thin to approximately 15 m along the western margin of the trough. To the northwest of the trough, in north-central New York, the turbiditic siltstones and sandstones laterally grade into black shales.

The interbedded siltstones, sandstones, and shales of the LSTA are arranged as classic Bouma sequence turbidites (after Bouma, 1962), typically with units T_b through T_e or T_c through T_e present (Fig. 4B). Individual Bouma sequences have relatively uniform thickness and can be traced over hundreds of meters. No channels have been recognized in this lithofacies.

The bases of individual turbidites are erosional. Uppermost mudstones underlying bases of each successive Bouma se-

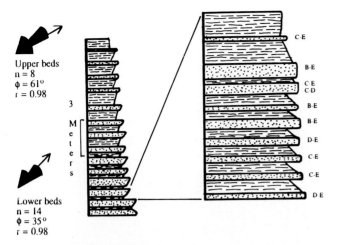

Upper beds
n = 8
φ = 61°
r = 0.98

3
M
e
t
e
r
s

Lower beds
n = 14
φ = 35°
r = 0.98

FIG. 5.—Composite section of LSTA from Frenchville and Northwestern, NY with enlargement of a upward-coarsening succession of turbidites. Turbiditic siltstones increase in thickness upwards in each progradational succession. Letters next to siltstones refer to Bouma divisions present. Where two sets of letters appear next to a siltstone, turbidites are amalgamated. Paleocurrent rose diagrams for the lower and upper strata are shown to the left of the composite section.

quence are capped by laminae of graptolite (*Geniculograptus pygmaeus*) and trilobite *(Triarthrus)* debris, suggesting that turbidity currents initially scoured the muddy sea floor, leaving very thin winnowed lag deposits. Siltstones at the bases of Bouma sequences rarely show flute casts or parting lineations. T_b is typically very thin (less than 1 cm) or absent, whereas T_c is typically well developed and can contain climbing-ripple cross-laminations with depositional stoss and lee sides preserved (Fig. 4B). Sectioning of cross-laminated siltstone slabs suggests that the generating ripple forms were 3-dimensional and linguloid in shape. Ripple forms are generally between 10 and 20 cm in length and have up to 3 cm of relief. Because of the relief of the ripple forms, T_d is often draped over T_c, giving the more resistant part of the Bouma sequence a hummocky appearance.

Bouma sequences appear to be arranged in approximately 1-m thick upward-coarsening packages in which each progressive sequence becomes more silt-dominated (Fig. 5). Lower Bouma sequences in these upward-coarsening packages typically do not contain T_b, and cross-lamination in T_c generally lacks depositional-stoss ripples. These lower Bouma sequences are characteristically capped by black shales. The lower shale-rich Bouma sequences grade upward into turbidites with more complete Bouma sequences containing depositional-stoss cross-lamination. These upper turbidites containing nearly complete Bouma sequences include thicker silty deposits and are capped by a smaller thickness of shales, and the shales are gray—not black. The silty units of Bouma sequences (T_c and T_d) are occasionally amalgamated in the uppermost portions of the 1 m packages, suggesting that greater scouring occurred.

Although the LSTA contains discrete meter-scale coarsening-upward packages, this lithofacies assemblage, as a whole, fines upward (Fig. 5). Each successive package of Bouma

sequences is slightly more clay-rich and silt-poor than the preceding package. The uppermost part of the LSTA contains only widely-spaced silt laminae interbedded with gray to black shales. These beds, in turn, grade upward into black shales. The shift from gray shales with turbiditic siltstones to black shales is, therefore, gradational.

Dip directions of ripple foresets in T_c show strongly preferred orientations which progressively shift upsection. In the lower, coarsest interval of this lithofacies assemblage, ripple foresets dip to the NNE (Fig. 5), but in the upper interval of this lithofacies assemblage, ripple foresets dip to the ENE (Fig. 5). (All rose diagrams included in this paper were constructed using methods advocated by Andreassen, 1990.) This change in dip directions is gradational; within relatively thin stratigraphic intervals (less than 3 m), azimuths of paleocurrent indicators are extremely uniform with high r-values.

The thin nature of T_b, depositional stoss ripple forms, and draping of lower-flow-regime plane beds (T_d) suggest that these Bouma sequences represent distal turbidity flows (Mutti and Ricci Lucchi, 1978; Pickering and others, 1989). These turbidity flows deposited sheet-like blankets of sediments on outer fan lobes or on the lower fan. Sheet-like, moderate concentration turbidites are characteristic of fan lobe-fringe sediments deposited during relative sea-level lowstands (Type 1 of Mutti, 1985).

The paleocurrent indicators suggest that these strata are a result of turbidity currents flowing parallel to, but west of, the basin thalweg. The principal sediment source was probably to the southeast (Zerrahn, 1978), but dip directions of ripple foresets within Bouma sequences suggest that fan sediments were deposited by turbidity currents directed to the NNE (Fig. 5). This direction parallels the strike of normal faults present in the Utica trough. As turbiditic siltstones thin upsection within this lithofacies, paleocurrent indicators become increasingly deflected toward the ancient basin thalweg, to the east (Fig. 5).

Transgressive Systems Tract Analog (TSTA)

Along the western margin of the Utica trough, turbidites of the LSTA are overlain by the black shales of the Deer River Formation (of Ruedemann, 1925). To the northwest of the trough, in north-central New York, Deer River black shales are separated from uppermost Trenton carbonates by phosphatic shales containing concretionary limestones. The phosphatic interval is the condensed section which comprises the TSTA (Fig. 2). These phosphatic strata form a diachronous TSTA which grades westward into the stratigraphically higher, interbedded carbonates and oil shales of the Collingwood Formation in central and western Ontario. No such condensed interval separates the Hasenclever LSTA from overlying Deer River shale to the southeast.

In north-central New York, the condensed section is generally poorly exposed. Along Gulf Stream, near Rodman, however, the condensed section crops out and can be traced for a few hundred meters (Fig. 6A). At this locality, the condensed section disconformably overlies nodular carbonates of the Hillier For-

FIG. 6.— (A) Condensed section at Gulf Stream, Rodman, NY. The condensed section, containing concretionary limestone ellipsoids and phosphatic shale, occurs between underlying argillaceous carbonates of the Hillier Formation and overlying black shales of the Deer River Formation. (B) Condensed section (Collingwood Formation) at Craigleith, Ont., containing cyclically bedded phosphatic limestones and oil shales.

mation and underlies black shales of the Deer River Formation. The condensed section pinches and swells, and ranges in thickness from a few centimeters up to about 0.5 meters. Unlike the irregularly-shaped limestone nodules of the Hillier Formation, carbonate concretions within the condensed interval are smooth and ellipsoidal. These concretions contain well preserved, uncompressed fossils, indicating early diagenesis. The concretions and the phosphatic shale contain similar faunas including: inarticulate brachiopods, trilobites (calymenids, ceraurids, and asaphids), and conulariids. Although the exact biostratigraphic position of the condensed section is uncertain, it is directly overlain by black shales which contain the graptolite *Geniculograptus pygmaeus*. Furthermore, the condensed section is approximately 65 m below the first occurrence of *Geniculograptus pygmaeus magnificus*, a key index fossil of the uppermost *G. pygmaeus* zone (Riva, 1969). In north-central New York, therefore, the condensed section probably belongs to the middle *G. pygmaeus* zone.

In central and western Ontario, a condensed section up to 10 m in thickness (the Collingwood Formation) occupies a similar lithostratigraphic position. The Collingwood Formation (Fig. 2) overlies uppermost Cobourg argillaceous carbonates and underlies organic-rich, black to dark gray shales of the Rouge River and Blue Mountain Formations. The Collingwood Formation comprises alternating beds of oil shale and phosphatic biomicrite to micrite. The shale and carbonate beds are arranged in 0.5- to 1.5-m thick symmetrical cycles with thin marly intervals occupying intermediate positions between the shales and the

limestones (Fig. 6B). As noted above, the Collingwood Formation is younger than the condensed section in New York. In western Ontario, Collingwood shales contain the graptolite *Geniculograptus pygmaeus magnificus,* and overlying shales of the Blue Mountain Formation contain the graptolite *Amplexograptus manitoulinensis*, indicating that the Collingwood Formation in western Ontario belongs to the uppermost part of the *G. pygmaeus* zone (Senior, 1991).

The condensed section in north-central New York overlies a corroded disconformable surface, but the Collingwood Formation seems to represent a conformable transition between carbonate and siliciclastic deposition. Russel and Telford (1983), using geochemical evidence, argued that there may be a slight disconformity at the top of the Collingwood Formation. This could be interpreted as a surface of maximum flooding and sediment starvation. Examination of their data in conjunction with examination of core samples and outcrop exposures suggests that the disconformity is patchy and does not form a continuous surface. Where the disconformity occurs, the uppermost shales of the Collingwood Formation are slightly pyritic and the lowermost shales of the overlying Blue Mountain Formation contain shale rip-up clasts. This discontinuous disconformable surface may represent the surface of maximum sediment starvation, which is the upper boundary of the TSTA.

The condensed section does not occur above the LSTA (as would be expected if this "sequence" were purely eustatically controlled), but a disconformity occurs below the LSTA. The hiatus represented by this surface is probably in part time-correlative with the disconformity below the condensed interval present in north-central New York. Where silty, distal turbidites unconformably rest on carbonates of the upper Trenton Group, the upper surface of the carbonates is pitted and shows evidence of solution.

Along the western margin of the Utica trough, similar solution features are present on the uppermost surface of Trenton carbonates and below the Utica Formation (lowermost *G. pygmaeus* zone). At some localities, this surface is covered by a thin pyritic crust and a lag of phosphatic pebbles. Further east still, in the Utica trough, a thin phosphatic breccia bed (TSTA) occurs between middle Trenton strata (*Orthograptus ruedemanni* zone) and lower Utica strata (*Climacograptus spiniferus* zone). In summary, the condensed section (representing the TSTA) and the carbonate solution surface that correlates with its base span at least a complete graptolite zone; the solution surface is oldest in the Utica trough and is progressively younger to the northwest.

Although the condensed section represents a diachronous interval, it occurs consistently above the carbonate and below the siliciclastic packages. On a seismic profile, this would show as an essentially continuous reflective surface. The diachroneity of this interval indicates that this condensed section is not simply the result of an abrupt eustatic sea-level rise. The condensed section represents conditions in which both carbonate accumulation and siliciclastic sedimentation were low. We suspect that these conditions arose as the Trenton/Cobourg carbonate ramp was progressively oversteepened. This oversteepened ramp became a region of sediment bypass and pycnoclinal erosion. Much of the carbonate sediment deposited towards the base of the ramp was dissolved under conditions of carbonate undersaturation or low pH in bottom waters, as outlined by Baird and Brett (1991). In this respect, phosphatic and pyritic intervals within the condensed section represent geochemical lag deposits and post-solution authigenic mineralization. Limestones within the condensed section record bottom water conditions of carbonate saturation and/ or slightly elevated pH.

Highstand Systems Tract Analog (HSTA): Early Phase

Aggradational to progradational siliciclastics of the HSTA range in thickness from 650 m (west of Oswego, New York) to approximately 300 m (along the Georgian Bay in western Ontario) (Fig. 2). This systems-tract analog comprises the Deer River, Whetstone Gulf, Pulaski, Oswego, and Queenston Formations of New York and the Rouge River, Blue Mountain, Georgian Bay, and Queenston Formations of Ontario.

The Deer River Formation (*Geniculograptus pygmaeus* zone) in north-central New York and the Rouge River and lower Blue Mountain Formations (lower *Amplexograptus manitoulinensis* zone) in Ontario contain black to dark gray shales. Black shales occur above the condensed section of the TSTA in the Black River Valley of New York and in much of the Upper Ordovician outcrop belt in central and western Ontario. In the upper Mohawk River Valley, black shales of the Deer River Formation

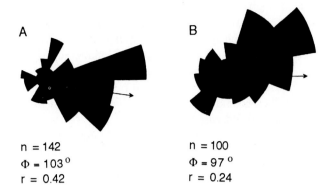

A   B

n = 142   n = 100
Φ = 103°   Φ = 97°
r = 0.42   r = 0.24

FIG. 7.— (A) Rose diagram of straight-shelled nautiloid orientation in the Deer River Formation (Constableville, NY). Note that orthoconic shells point into currents; rose diagrams show adapical orientation. The dominant paleocurrent direction during deposition of muds which formed the Deer River Formation was WNW to ESE. (B) Rose diagram of orthoconic nautiloids in the Blue Mountain Formation (Camperdown, Ont.). During deposition of muds which formed the Blue Mountain Formation, dominant paleocurrent direction was from west to east.

also occur above rocks of the LSTA, represented by the Hasenclever Member of the Frankfort Formation. The black shales are evenly laminated and organic-rich, with subordinate quartz silt. The black shales grade stratigraphically upsection into gray shales which contain subordinate thin (2 to 7 cm) siltstones (the Whetstone Gulf and lower Georgian Bay Formations). The interbedded siltstones and shales are similar to, but more clay-rich than, deposits of the LSTA and, in part, may represent renewed progradation of a submarine fan system.

The black shales contain a low to moderate diversity fossil fauna which is dominated by graptolites, trilobites, and nautiloid cephalopods (Lehmann and Brett, 1990; Lehmann and others, 1990). Although fossils occur throughout the black shales, most fossils are concentrated on bedding planes. These fossil-rich beds are separated by rather barren shale intervals. There are no indications that any of the fossil organisms were transported a great distance after death. In fact, the occurrence of moults and partly disarticulated complete trilobites, nearly equal proportions of convex-up and convex-down trilobite cranidia, clams in butterflied positions, and fossils of delicate nonmineralized organisms (such as the hydroid, *Mastigograptus*) all argue for relatively quiet background conditions with little postmortem transport. Generally, the shales are not bioturbated, although small horizontal burrows occur at some horizons. The occurrence of a moderately diverse (12 genera), but high-dominance, benthic fauna and the relative scarcity of bioturbation indicates that the black shales represent dysoxic environments.

Unraveling the sedimentary history of black shales can be frustrating; the absence of sedimentary structures hampers the interpretation of depositional setting. Fossils (as sedimentary particles) can give important clues to depositional settings in such strata. For example, orthoconic cephalopod shell orientation may record paleocurrent directions; in a current, orthoconic cephalopod shells pivot on the shell apex and point upcurrent (Reyment, 1971). The black shales of the HSTA contain abundant fossil cephalopods, many of which are concentrated on discrete bedding planes. Cephalopod orientation in the Deer River Formation indicates paleocurrents directed towards the ESE (Fig. 7A). Likewise, cephalopod orientation in the slightly younger Blue Mountain Formation also indicates paleocurrents directed towards the east (Fig. 7B).

We interpret these eastward-directed paleocurrents as being indicative of either down-slope directed gradient currents or down-slope deflected axial turbidity currents. The organic-rich muds may represent both pelitic tails of turbidites and hemipelagic deposits.

The presumably down-slope, eastward-directed currents indicate that the organic-rich muds were not deposited along a delta slope prograding from the east or axially along the basin thalweg. Paleocurrent indicators in shales of the Deer River and Blue Mountain Formations suggest that the organic-rich muds were typically deposited on an eastward dipping, subsided carbonate ramp.

In the Utica trough of New York, paleocurrent indicators suggest that muds which formed the shales of the Utica Formation were deposited on fault-bounded blocks of carbonate platform which were typically dropped down to the east (Cisne and others, 1982; Fisher, 1979; Hay and Cisne, 1988; Mehrtens, 1988). Small normal faults west of the Utica trough may have similarly influenced the depositional setting of the Deer River shales (discussed under "Tectonic Influences," below).

Within the HSTA, organic-rich shales grade upward into predominantly gray shales which are interbedded with siltstones and fine-grained sandstones (Fig. 2). These shales and siltstones comprise the Whetstone Gulf Formation (upper *G. pygmaeus* to lower *A. manitoulinensis* zone) in New York and the upper Blue Mountain and lower Georgian Bay Formations (*A. manitoulinensis* zone) in Ontario.

The lowermost 30 m of the Whetstone Gulf Formation contain 10- to 15-percent siltstones and 85- to 90-percent gray shales. Most siltstone beds are relatively thin (less than 4 cm), evenly-laminated, and laterally continuous over 100's of meters. Sparse thin Bouma sequences (typically less than 7 cm in thickness) are also present in these strata. Overturned ripple forms, flute casts, and climbing-ripple cross-laminations with depositional stoss and lee sides preserved indicate that turbidity currents were responsible for depositing some of the muds composing these rocks. The shales in this interval are typically unfossiliferous, although fossil "hash" beds containing molds of broken pieces of small orthid brachiopods, disarticulated trilobites (*Triarthrus*), graptolites, and inarticulate brachiopods occur in some dark gray shales immediately below siltstone beds. Small, pyritized *Chondrites* burrows also occur in some shales.

The lower Whetstone Gulf Formation seems to record the westward migration of the basin thalweg. Although unidirectional paleocurrent indicators are sparse in this interval, the available data (orientations of cross-laminations, flute casts, and straight-shelled nautiloids) indicate that sediments were deposited by first eastward, then northward, and finally, westward-directed currents (Fig. 2). These strata are transitional between organic-rich mud deposition occurring on an eastward-dipping slope (the underlying black shales) and silty mud deposition on a westward dipping slope (the overlying siltstones and shales). We have not been able to collect paleocurrent data from a correlative transitional interval of the Blue Mountain Formation in Ontario because of poor exposure.

The upper Whetstone Gulf Formation of New York and the upper Blue Mountain and lower Georgian Bay Formations of Ontario (Fig. 2) contain 10- to 30-percent siltstones and fine-grained sandstones and 70- to 90-percent gray to blue gray shales. Subordinate thin biosparites and biomicrosparites also occur in the lower Georgian Bay Formation. Some of the siltstone beds in the upper Whetstone Gulf Formation are continuous for hundreds of meters, but this interval is generally characterized by thin, relatively discontinuous siltstone beds. Some siltstone beds are amalgamated, and scour surfaces occur between amalgamated beds. Sedimentary structures within this interval include low-angle cross stratified siltstones, gutter casts, and hummocky cross stratification (HCS). Orientations of straight shelled nautiloids indicate paleocurrents were directed towards the WNW. Sedimentary structures and paleocurrent indicators, therefore, suggest that sediment forming these strata was deposited by gradient currents coming down a WNW sloping clastic wedge. Furthermore, the occurrence of HCS indicates that combined flow (hence, storm waves) influenced the seafloor.

These siltstones and shales show a moderate degree of bioturbation with *Teichichnus*, *Chondrites*, and vertical escape burrows being the most abundant trace fossils. A moderate diversity epifauna also occurs in this interval and includes inarticulate brachiopods, orthid and sowerbyellid brachiopods, ramose bryozoans, bellerophontids, small pelecypods, trilobites, and crinoids. Fossils occur as winnowed storm lag deposits at the bases of siltstones, within the shales, and at the bases of gutter casts. Thin biosparites (less than 4 cm) within the lower Georgian Bay Formation probably also represent storm lag deposits. The lag deposits typically contain only broken and disarticulated shell material, but well-preserved, articulated brachiopods, trilobites, and crinoids occur in some shales. The well-preserved fossils are not associated with shell lags, and these death assemblages may be the result of rapid burial by muddy tempestites and/or turbidites.

Highstand Systems Tract Analogy (HSTA): Progradational Phase

In New York, upper Whetstone Gulf strata are abruptly overlain by abundantly fossiliferous, coarser-grained deposits of the Pulaski Formation (Fig. 2, 8A). The Pulaski Formation contains approximately equal amounts of calcareous sandstone and of silty shale. Most sandstone beds are less than 25 cm in thickness, although sandstone beds up to 80 cm thick occur. Abundant gutter casts, symmetrical straight-crested ripple marks,

FIG. 8.— (A) Abrupt coarsening of siliciclastic rocks at the base of the Pulaski Formation at Barnes Corners, NY (marked by dashed line). (B) Abrupt shift from predominantly shales below to a thickly-bedded sandstone interval above in the lower Georgian Bay Formation at the cliffs between Camperdown and Meaford, Ontario (marked by dashed line).

interference ripple marks and HCS indicate that these sediments were deposited between normal wave base and storm wave base. Sandstone beds pinch and swell; most beds can be traced only tens of meters. Many of the sandstone beds represent single event deposits (or, more likely, single event reorganization of sedimentary grains) with coarse sandy shelly lags overlain by hummocky cross-stratified fine sands. Amalgamation of sandstone beds is common, and interbedded shales are erosionally scoured.

The shift from shale-dominated deposits in the Whetstone Gulf Formation to the more sandstone-rich deposits in the Pulaski Formation may represent rapid shoreline progradation. Whereas the Whetstone Gulf Formation includes offshore, muddy shelf deposits, the Pulaski Formation represents storm- dominated, shallow shelf deposits.

Evidence of a similar abrupt shallowing event occurs to the northwest in the approximately temporally correlative lower Georgian Bay Formation of western Ontario (Fig. 2). At and near Meaford, Ontario, strata abruptly coarsen at approximately twenty-six meters above the base of the Georgian Bay Formation (Fig. 8B). Three meters of predominantly thickly bedded (10 to 40 cm) siltstones and fine-grained, calcite-cemented sandstones overlie the predominantly shaley lower Georgian Bay Formation. This relatively coarse-grained interval, which forms resistant ledges and caps low cliffs, can be traced throughout the Notawasaga Bay region. These resistant beds typically have flat bases, contain low-angle (10 degrees) cross-laminations, and have wavy to undulatory upper surfaces. Some

siltstone and sandstone beds are amalgamated. Trace fossils and body fossils are rare within the siltstones and sandstones, but some body fossils (crinoid plates, pelecypods, and articulate brachiopods) occur in interbedded shales. This coarse interval is overlain by shale-rich strata which are similar to the lower Georgian Bay strata. The shale-rich interval is 18-m thick and contains increasingly higher proportions of limestone (biomicrite and biomicrosparite) towards its top. The strata above this shaley interval, in turn, generally coarsen upward to the top of the Georgian Bay Formation. Interestingly, Kerr and Eyles (1991) recognized an almost identical stratigraphic succession in a drill core from Wiarton, Ontario approximately 50 km to the northwest of Meaford. In the Wiarton core, the thickly bedded siltstone and sandstone interval occurs between 10- and 13-m above the base of the Georgian Bay Formation.

Kerr and Eyles (1991) also showed a similar, and possibly correlative, abrupt shift from shale-rich strata to predominantly sandstones occurring in the middle part of the Georgian Bay Formation (127 m above the base of the formation) near Ottawa. As in the Pulaski Formation of New York, the Georgian Bay Formation near Ottawa remains relatively coarse-grained following this progradational "kick." In addition, the Georgian Bay core contains a particularly shaley strata approximately 30 m above the base of the resistant beds. These shales may be correlative with the shaley interval which crops out in the Notawasaga Bay region.

In summary, evidence of an episode of rapid shallowing occurs at the base of the Pulaski Formation in New York, in the

FIG. 9.—(A) Flow-roll filled channel in storm-dominated shelfal deposits of the upper Pulaski Formation at Bennets Bridge, NY (base of the channel marked by white arrows). The channel cuts into HCS sandstone beds. (B) Channel sandstones in the uppermost Pulaski Formation at Ninemile Point, NY. Note the large trough cross-bedding in the upper left. This cross-bedding represents a migrating bar form. ((C) Epsilon cross beds of sandstone sandwiching interdistributary muds (near Salmon River Falls, NY). The upper Pulaski Formation displays an upward facies gradation from storm-dominated shallow shelf to distributary channel and bay complex.

middle Georgian Bay Formation in the Ottawa, Ontario region, and in the lower Georgian Bay Formation in western Ontario.

In New York, fossiliferous Pulaski strata grade upward into poorly fossiliferous, well-sorted, sand-rich tempestites and fluvial-deltaic deposits, tentatively assigned to the upper portion of the Pulaski Formation. HCS, epsilon cross-stratification, and soft sediment deformation (flow rolls and slump folds) are characteristic of this facies (Fig. 9). Dark-gray to black shales are interbedded with the sandstones and also are present as the lowest fill in large (tens to hundreds of meters wide) channel forms. These strata represent distributary channel and bay deposition.

The upper Pulaski strata are overlain by the massively bedded sandstones of the Oswego Formation (Fig. 2). Herringbone cross-stratification, abundant interference ripple marks, raindrop impact impressions (Patchen, 1978), desiccation cracks, and channel-shaped sandstone bodies indicate that these sediments were deposited in peritidal settings. Large foresets of trough cross-stratified sandstones in presumed tidal channels indicate a paleocurrent direction towards the NNE.

The red shales and sandstones of the Queenston Formation overlie the Oswego and Georgian Bay Formations (Fig. 2). In New York, much of the Queenston Formation is unfossiliferous and represents sandy braided stream (in north-central New York) to peritidal (in western New York) environments (Lehmann, in preparation). In the Toronto region, lowermost strata of the Queenston Formation contain a normal marine fauna, and bryozoan biostromes are present. Near Notawasaga Bay (western Ontario), strata low in the Queenston Formation contain leperditid ostracods, pelecypods, and rare strophomenid brachiopods, indicating at least marginal-marine conditions.

The Queenston Formation of western New York and Ontario contains a transgressive-regressive cycle. In the Notawasaga Bay region, a green calcareous tongue of the Queenston is sandwiched between gypsum-rich, red, desiccation-cracked shales. These green strata include biomicrites and biosparites containing a normal marine fauna. Using core samples, we have correlated this interval with an interval of green and red bioturbated sandstones in the Toronto, Ontario area. Red, bioturbated sandstones, possibly correlative with the green strata of Ontario, are present in the uppermost Queenston Formation near Rochester, New York.

PALEOENVIRONMENTAL RECONSTRUCTIONS

The persistent pattern of change from primarily northward directed paleocurrents represented in the silty LSTA turbidites (Hasenclever Member, Frankfort Formation) overlying the upper Trenton strata in the Mohawk River Valley shifting to primarily eastward directed currents in the laterally correlative and overlying Deer River Formation (of the HSTA) suggests the paleoenvironmental reconstruction for late Edenian time shown in Fig. 10A. The clastic wedge was prograding from the east. Axial drainage existed, and the basin thalweg sloped northward. During times of relative lowstands of sea level or uplift in the sediment source area, turbidity currents deposited silts and sands over great expanses of the basin to form large sheet systems. During times of relative highstands of sea level, silts and sands

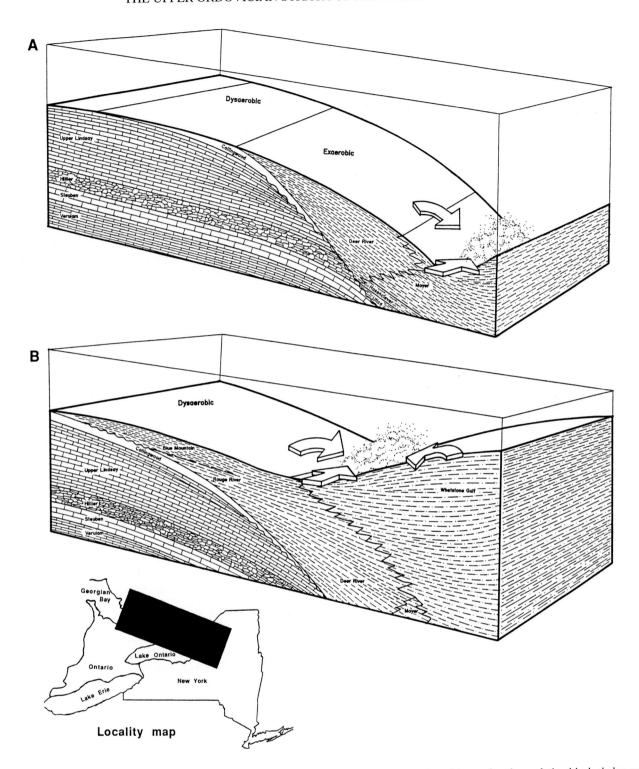

Fig. 10.— Paleoenvironmental reconstructions. (A) Late Edenian. Sediments forming a condensed interval and correlative black shales were deposited on a collapsed carbonate ramp. In the Utica trough, a distal submarine fan still occupied the basin axis. (B) Middle Maysvillian. General paleocurrent patterns persisted although the basin thalweg had migrated.

were not as laterally persistent. This basin geometry and drainage persisted even as the basin thalweg migrated to the west. Similar paleocurrent patterns occur in the slightly younger Blue Mountain and lower Whetstone Gulf Formations (Fig. 10B).

Sediments from the eastern clastic wedge probably supplied the basin thalweg. If so, westward-directed currents coming down the eastern slope of the basin were deflected to the north by persistent axial currents following the slope of the basin thalweg. A southern sediment source may have existed. The basin thalweg contained a distal submarine fan, yet no temporally correlative proximal fan deposits have been recognized in north-central New York or Ontario. Richmondian gravels deposited in central Pennsylvania (which formed the Bald Eagle Formation) grade into sands in central New York (the Oswego Formation) suggesting a southern sediment source for that time (Yeakle, 1962; Zerrahn, 1978), and paleocurrent data from the Oswego Formation indicate north-northeast-directed flow (Fig. 2). Paleocurrents documented by Cisne and others (1982) in the lower to middle Edenian rocks from the Utica trough of New York indicate drainage directions similar to those seen in the upper Edenian to lower Maysvillian rocks of New York and Ontario. Throughout the Edenian and Maysvillian Stages, the basin thalweg roughly paralleled normal faults present in the Utica trough—even when the thalweg was far to the west of the original Utica trough area. The axial drainage pattern, therefore, seems to represent a long-lasting, tectonic feature of the foreland basin.

The organic-rich muds forming the black-shale lithosome (Deer River, Rouge River, and lower Blue Mountain Formations) were deposited in dysoxic settings below storm-wave base. The black shales indicate typically low-energy conditions, relatively high-organic productivity, low carbonate productivity, and absence of a coarser siliciclastic supply (Hallam, 1967). The black-shale *magnafacies* does not always represent the deepest part of the basin, however. The presence of cyclothemic alternations between platform carbonates and oil shales in the Collingwood Formation of Ontario (Fig. 6B) (Liberty, 1969; Russel and Telford, 1983; Lehmann and Brett, 1990) suggests that organic-rich shales can form at moderate depths and that factors such as basin circulation (Fischer and others, 1985), bathymetry, restriction (Witzke, 1987), and depth of the oxygen-minimum zone (Fischer and Arthur, 1977) may be of primary importance in controlling the occurrence of organic-rich muds.

The organic-rich muds forming the black shale may have initially been deposited as the pelitic tails of axial fan turbidites or as suspended mud moving up onto the shelf. This organic-rich mud was deposited on a drowned carbonate ramp forming the western slope of the Taconic foreland basin—not on a delta slope or in the basin thalweg. Black shales in New York grade upslope into temporally equivalent carbonates in Ontario, so there is no upslope source of siliciclastic material. The eastward-directed currents were either down-slope deflected, axial turbidity currents or dilute gradient currents that merely reworked muds locally.

The black shales of the Deer River and Blue Mountain Formations grade stratigraphically upsection into gray shales which contain subordinate thin (2 to 7 cm) siltstones (the Whetstone Gulf and lower Georgian Bay Formations, respectively). These interbedded siltstones and shales represent renewed progradation of the submarine fan system and deep shelf deposits. The siltstones, representing distal low-concentration turbidites and/or distal tempestites, are typically either evenly laminated or cross-laminated, and some can be traced over entire field exposures (in some cases, over 100 m).

Orientations of ripple foresets, overturned soft-sediment folds, and orthoconic cephalopods in this gray shale and siltstone interval indicate shifts in paleocurrents associated with the prograding clastic wedge. Working stratigraphically upsection from the black shales into the interbedded shales and siltstones, paleocurrents shift from primarily eastward-directed to northward-directed and then to westward-directed (Fig. 2). The northward-directed paleocurrents are the result of axial drainage, but the westward-directed paleocurrents (also reported in the Whetstone Gulf Formation by Zerrahn, 1978) represent turbidity and gradient currents derived from the clastic wedge, which was prograding from east to west.

Siltstones and sandstones in the Pulaski Formation and the upper Georgian Bay Formation represent a storm-dominated, marine shelf. Ripple crest, gutter cast, and fossil orientation measurements from these deposits typically do not show preferred orientations, suggesting multidirectional paleocurrent patterns.

TECTONIC AND EUSTATIC INFLUENCES

Tectonic and eustatic effects can both influence the deposition and preservation of organic-rich muds and of prograding clastic wedges. Eustatic sea-level rises can drown carbonate platforms and in this respect change previously carbonate-dominated depositional regimes into siliciclastic depositional regimes (Sarg, 1988). Fine-grained siliciclastics and authigenic minerals (pyrite and phosphate) are typical deposits of eustatic sea-level rises because most coarser grained siliciclastic sedimentation occurs in newly developed estuaries occupying drowned river valleys (Loutit and others, 1988; Van Wagoner and others, 1988). Furthermore, oxygen-minimum zones may expand during sea-level rises and create low-oxygen environments—the sort of environments in which organic material does not quickly break down (see Allison and Briggs, 1991, for review).

Likewise, tectonic deepening can drown carbonate platforms and lead to low-oxygen environments. Bally and Snelson (1980) noted that tectonic subsidence usually leads to a three-part stratigraphic succession in foreland basins (from bottom to top, with corresponding Middle and Upper Ordovician stratigraphic units of this study): (1) "platform" sediments (Black River and Trenton Groups), (2) unconformity (pre-LSTA unconformity and the lower Deer River and Collingwood condensed sections), and (3) clastic wedge, typically beginning with fine-grained siliciclastics (Deer River and Blue Mountain through Queenston Formations). Our data indicate that for the Upper Ordovician clastic wedge of New York, initial muddy clastic wedge sediments grade into silty and sandy deposits approaching the orogen (LSTA). The tripartite tectonic stratigraphic succession typically onlaps cratonward (Bally and Snelson, 1980), as does the Middle to Upper Ordovician succession.

Orogen development influences circulation in epeiric seas and, thus, may cause low-oxygen conditions. The development of hinterland mountain barriers can sill basins, and delta development may also influence the sedimentary and geochemical environment of large portions of basins (Ettensohn, 1985). Freshwater deltaic runoff, being less dense than seawater, can cause stratification of the water column (Witzke, 1987). Because of water-column stratification, the lower water mass is oxygen deficient. Intuitively, organic-rich mud deposition should be most widespread when tectonic and eustatic influences (such as subsidence and maximum eustatic rise) are superimposed.

Tectonic Influences

Three lines of evidence suggest that the Late Ordovician siliciclastic succession was strongly influenced by tectonic events. First, the basin geometry and local bathymetry changed progressively from Rocklandian (*Climacograptus americanus* zone) to middle Maysvillian (upper *G. pygmaeus* zone) times. Second, normal faults on the "platform" were active during the Late Ordovician, and movements along these faults seem to be related to the transition from carbonate platform deposition to organic-rich mud deposition. Third, the large scale Late Ordovician pseudo-sequence, and systems tract analogs within the pseudo-sequence are diachronous, becoming progressively younger to the northwest.

The facies and systems tract analog progressions in north-central New York and Ontario extend into the Utica trough (Figs. 2, 11). Localized paleobathymetric changes of the Trenton ramp in the region of the developing Utica trough argue for tectonically influenced subsidence and uplift in that region (Fig. 11). In the Black River Valley and much of the Mohawk River Valley, oldest Trenton strata comprise shelfal micrites, biomicrites, and interbedded shales. These shelfal deposits are overlain by shoal biosparites (Kings Falls Formation). The contact between the Kings Falls and the underlying strata is sharp, and in the middle Mohawk River Valley region, conglomerates containing clasts of Precambrian metamorphic rocks and Cambrian dolostone locally mark the contact. Titus and Cameron (1976) suggested that the change in carbonate lithology at the base of the Kings Falls reflects shoal migration. However, the basal Kings Falls conglomerate suggests tectonic activity. Bradley and Kidd (1991) argued that clasts in this conglomerate represented local uplift and scarp erosion along normal faults in the middle Mohawk River Valley.

Through much of north-central New York, the middle Shermanian (basal Denmarkian) strata records a dramatic deepening event; tempestitic packstones and grainstones are overlain by deeper shelf micrites, biomicrites, and dark calcareous shales (Fig. 11). This lithologic contact can be traced from Lowville, New York (middle Black River Valley) to Little Falls, New York (middle Mohawk River Valley), a distance of approximately 160 kilometers. Northwest of the Lowville area, however, the Shermanian micrites laterally grade into bioclastic limestones (Chenowith, 1952). The abrupt vertical transition

from storm-dominated shelf deposits to deeper shelf deposits may record an episode of tectonically-induced subsidence which was accommodated by movement along faults in the middle Black River Valley (Fig. 11). Interestingly, lower Shermanian strata are slump folded in the middle Black River Valley suggesting a localized region of ramp oversteepening or subsidence.

Another episode of ramp oversteepening along the western margin of the Utica trough occurred during Cobourgian time. Slump folded strata representing this oversteepening event can be traced throughout the Black River and Mohawk River valleys (Fig. 11). Cobourgian strata in the Utica trough may show an extremely steep facies gradient; physical stratigraphy and K-bentonite correlation suggest that shallow shelf biosparites in the upper Mohawk River Valley are correlative with interbedded black shales and thin carbonate turbidites (Dolgeville Formation) present 10 km to the southeast (Baird and Brett, 1994; Mitchell and others, 1994). Thin, phosphatic, stratigraphically-condensed intervals define the bases of the overlying lower and upper Utica strata. These condensed intervals also represent times of sediment starvation following subsidence of portions of the Trenton ramp.

Following deposition of carbonate sediments which formed the biosparites of the Steuben Formation; in north central New York, relative sea-level rose and, by the late Edenian, drowned the carbonate platform. The middle Edenian through the early Maysvillian time may also represent a time of rising relative sea level in central and western Ontario, when argillaceous carbonate sediments of the Lindsay Formation were deposited. However, in central and western Ontario, shallow-marine conditions did reoccur as is evidenced by grainstone shoal deposits in the middle Lindsay Formation. In central and western Ontario, maximum sea-level rise (represented by TSTA deposits of the Collingwood Formation) occurred during early to middle Maysvillian time. This pattern of a northwestward migration of maximum relative sea-level rise indicates that tectonic subsidence was the primary control on the stratigraphic succession of New York and Ontario. Maxium eustatic sea-level rises are nearly isochronous (Loutit and others, 1988) and, in this respect, differ from the pattern of relative sea-level rise recorded in our study area.

Tectonically induced subsidence may have greatly influenced the bathymetry and sedimentary environments of the eastern and central North American platform (Lash, 1988). The Michigan basin became an active depocenter during the Chazyan Epoch, possibly as a result of early Taconic plate collision and stresses (Howell and van der Pluijm, 1990). Furthermore, during Shermanian (middle to late Caradocian) time, there were widespread episodes of subsidence and flexures of the carbonate platform in the midcontinent (Cressman, 1973; Witzke, 1980; Bergstrom and others, 1990; Wickstrom, 1990). In New York, Cobourgian ramp oversteepening is manifested by the slump-folded carbonate turbidites of the Dolgeville Formation (Fisher, 1979).

The relationship between basin-thalweg alignment in the

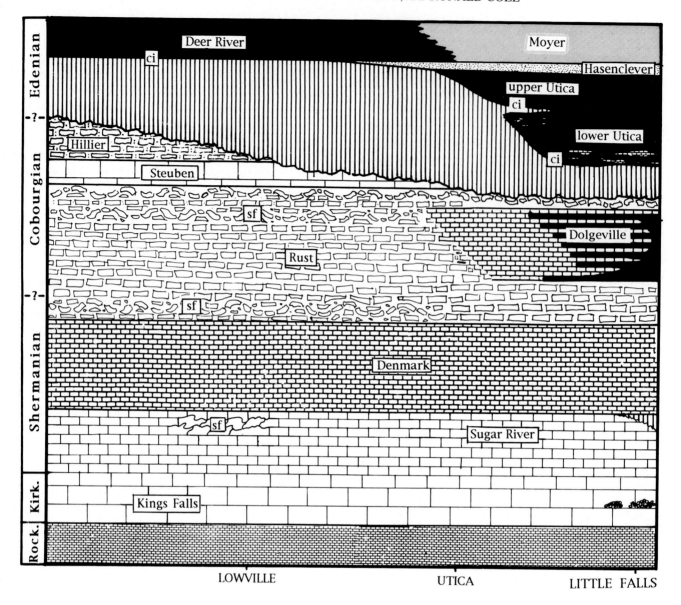

FIG. 11.—Stratigraphy of the Trenton Group and correlative black shales in and to the northwest of the Utica trough (based partly on Baird and others, 1992; C. Mitchell, personal communication, 1992). Note that progressive northwestward collapse of the carbonate ramp to the northwest of the Utica trough (see Fig. 2) is merely a continuation of earlier, yet similar, progressive ramp collapse in the developing trough. Stratigraphically-condensed intervals (CI) occur just above the bases of the lower and upper Utica strata. Slump folded strata (sf) correlates to times of ramp-oversteepening events. The age of the Rust through Hillier strata is somewhat conjectural; biostratigraphic control for this interval is poor.

field area and normal fault orientation in the Utica trough suggests a relationship between tectonically-induced subsidence and basin geometry. The north-northeast trend of the Taconic foreland axis in the study area was probably influenced by a combination of variables including the distribution of the structural and sedimentary loads imposed on the continental margin during the Taconic orogeny (Beaumont, 1981; Rowley and Kidd, 1981; Stanley and Ratcliffe, 1985; Tankard, 1986; Bradley and Kidd, 1991), the alignment of the Taconic subduction zone, and the locations and alignments of promontories along the continental margin (Thomas, 1977). The orientation

of the basin, generally parallel to the orogenic belt, along with the large-scale basin shape, an asymmetric trough steepest adjacent to the orogen, are typical of modern and ancient peripheral foreland basins (Quinlan and Beaumont, 1984; Covey, 1986; Houseknecht, 1986; Tankard, 1986). This overall geometry provided a primary control on the distribution of sediment which forms the black-shale *magnafacies* and related deposits. According to Karner (1986), Allen and Allen (1990, p. 161), and Bond and Kominz (1991), in-plane stresses imposed at plate margins amplify foreland-basin subsidence and can be transmitted far into plate interiors; correlations between timing of sea-level changes along active continental margins and continental interiors may thus be explained without resorting to global eustatic controls (Sloss, 1991).

The foreland-basin shape and subsidence history may reflect preexisting heterogeneities in the continental crust (Allen and Allen, 1990). As alluded to earlier in this paper and as described in Cisne and others (1982), smaller-scale structural elements were superimposed on the large-scale Taconic foreland geometry. These secondary features are north-northeast trending normal faults, possibly formed during Precambrian rifting and then reactivated during the Taconic orogeny.

The influence of reactivated basement structures on sedimentation has been documented within other portions of the Appalachian basin (Wagner, 1976; Root, 1980; Sanford and others, 1985; Hiscott and others, 1986) and within other peripheral foreland basins (Ricci Lucchi, 1975, 1984; Houseknecht, 1986). In each of these examples, reactivation of preexisting basement faults during an orogeny presumably resulted in internal partitioning of a foreland basin and subsequent control on sediment dispersal patterns.

Movement of normal fault-bounded blocks in the Utica trough and possibly west of the trough is indicated by extensive slump folds in ribbon carbonates of the uppermost Dolgeville Formation (Fisher, 1979) and possibly by the persistent east-southeast-directed paleocurrent trends in rocks of the black shale *magnafacies* documented in this paper and by Lehmann and others (1990). Overturned ripples in the Frankfort and Whetstone Gulf Formations, as well as slump folds in ribbon carbonates of the uppermost Dolgeville Formation, suggest that seismic activity may have been related to movement along the north-northeast trending faults. The surface expression of these faults may have been subtle; steep Ordovician fault scarps have not been reported west of the Utica trough. Proprietary seismic lines run by Geodata, Inc., however, show small normal faults, with maximum displacements of only 20 meters, in the subsurface of central New York; these faults are present well west of the Utica trough. The faults displace Middle to Upper Ordovician Trenton rocks, but uppermost Ordovician Queenston strata are not disturbed over these faults. Rather abrupt transitions from carbonate-platform sedimentation (Trenton) to deeper water, organic-rich mud deposition probably reflect movements along these faults.

Local tectonic influences should affect only basins developed within or adjacent to active orogenic belts. Evidence of eustatic sea-level rise should be present interbasinally and worldwide. Previous attempts to match Late Ordovician cyclicity in North America with cyclicity occurring in other continents (Johnson and others, 1989) have not been entirely successful. This poor correlation may reflect a lack of precise stratigraphic data from outside of North America or, alternatively, may indicate that Late Ordovician depositional settings in North America were influenced more by tectonic activity than by eustasy. In much of the world (excluding North America), relative sea-level changes are correlative and appear to reflect

eustasy (Barnes and others, 1981; Dean, 1980; Fortey, 1984; Hamman and others, 1982; Shen-Fu, 1980; Webby and others, 1981; Williams and others, 1973). Ordovician seas were most widespread globally during the early to middle Caradoc (Kirkfield through late Eden stages of North America) and regressed following the Caradocian high stand. Renewed transgression occurred during the early Ashgill (Maysville Stage of North America) but the Ashgillian transgression was not nearly as widespread as the Caradocian transgression.

Correlating presumed eustatically-controlled Upper Ordovician sedimentary sequences has proven problematical in North America. In the Cincinnati, Ohio region (the Upper Ordovician standard for North America), water depth apparently reached its Ordovician maximum during the early Edenian Stage (Late Caradocian, represented by the Kope Formation) (Hay, 1981; Holland, 1989). Farther west, in Iowa, the late Maysvillian (Ashgillian) Maquoketa Formation is believed to represent maximum water depth for the Ordovician Period (Witzke and Glenister, 1987; Johnson and Jia-Yu, 1989; Johnson and others, 1989). Witzke (pers. commun.) believes that the phosphatic, organic-rich, lower Maquoketa shales represent the Ashgillian transgression recognized in other parts of the world. The Ashgillian transgression should follow a late Caradocian regression, however, and evidence of this regression is lacking in the western midcontinent of North America. The late Caradocian eustatic sea-level fall recognized by Fortey (1984) is approximately coincident with the time of widespread carbonate platform collapse in eastern and central North America; tectonic subsidence may have largely overprinted this sea-level fall.

Eustatic Influences

Tectonic and orogenic effects had strong influences upon the orientation and paleobathymetry of the Middle and Late Ordovician foreland basin, but evidence of possible Late Ordovician eustatic events also is present in the field area. The condensed section associated with the Collingwood Formation of Ontario, partly a response to tectonic subsidence and clastic wedge progradataion, may also represent a eustatic sea-level rise. Deepening-upward carbonate platform successions culminating in organic-rich mud deposition occur in temporally correlative strata (late *G. pygmaeus* zone) in the Hudson Bay Lowlands, Northwest Territories (Dewing and Copper, 1991) and in the upper Mississippi Valley of the midcontinent (Witzke and Glenister, 1987; Johnson and Jia-Yu, 1989; Johnson and others, 1989; Bergstrom and Mitchell, 1992).

Times of extremely rapid progradation may reflect eustatic sea-level drops, although rejuvenation of the sediment source may also have led to these rapid progradations. The abrupt occurrence of thickly bedded sandstones and siltstones in the lower Georgian Bay Formation of Ontario is approximately correlative with a similarly abrupt coarsening event in New York (Figs. 2, 8). As noted above, in New York, this change from predominantly shales to predominantly siltstones and sandstones marks the contact between the Whetstone Gulf and

Pulaski Formations. Similarly, the transgressive-regressive cycle in the Queenston Formation is traceable across a range of facies and may reflect eustatic fluctuations.

DISCUSSION

The Taconic Orogeny is largely viewed as a compressional event, but extension occurred well into the platform during the orogeny, possibly in response to lithospheric flexure, as outlined by Bradley and Kidd (1991). Regardless of the causes of extension, the largest scale sequence stratigraphic pattern in New York and Ontario is the result of Middle to Late Ordovician orogenesis and associated basin subsidence. Tectonically-induced subsidence, via normal faulting and possibly lithospheric flexure, drowned portions of the carbonate ramp and created a sediment-starved condition. With uplift and erosion in the hinterland, sedimentation began to outpace subsidence.

Upper Ordovician strata of New York and Ontario, in addition to being strongly influenced by tectonism may also contain evidence of eustatic events. These include the "Collingwood" condensed section, the abrupt Pulaski-Georgian Bay progradation, and the transgressive-regressive cycle in the Queenston Formation. We stress that the sequence stratigraphic model has developed primarily through observations in passive margin settings. Eustatic fluctuations may be manifested differently in foreland-basin settings for a number of reasons. In foreland-basin settings, the tectonic hinge (as defined by Posamentier and others, 1988) and the primary siliciclastic sediment source will be on opposite sides of the basin; but in passive-margin settings, the tectonic hinge and the primary sediment source will be located on the same side of the basin. Furthermore, subsidence histories within foreland basins may be more complex and less predictable than in passive-margin settings. In orogenically active regions, rate of uplift and time-dependent lithospheric responses to loading may exert strong influences on the amount of sediment supplied to a basin (Miall, 1991).

Both tectonic and eustatic events can produce strata resembling stratigraphic sequences. By combining sequence stratigraphy with event stratigraphy, biostratigraphy, and chronostratigraphy; eustatically- and tectonically-driven sequences can be differentiated. Eustatically-driven sequences are largely isochronous. The evidence of eustatic events in tectonically active areas may be cryptic. However, eustatic events cause changes which crosscut lithofacies. Alternatively, tectonically-driven pseudo-sequences are diachronous; these become progressively younger away from orogenic fronts.

ACKNOWLEDGMENTS

Financial support for this research has been provided by grants to David Lehmann from the Geological Society of America, Sigma Xi, Northeastern Science Foundation, New York State Museum, and the Paleontological Society and by a Petroleum Research Fund (American Chemical Society) to Carlton Brett. Sanford Britt, Wendy Taylor, and Pat Voglesang of the University of Rochester and Steve Mayer aided in field work. Talia Sher, Laura Abbot and Sharon Ingram (University of Rochester) helped draft diagrams. Gordon Baird (SUNY at Fredonia) aided in fieldwork and posed many thought-provoking questions to us. Charles Mitchell and Dan Goldman greatly assisted the senior author in identifying key graptolites. We thank all of these people and organizations for their support.

REFERENCES

ALLEN, P. A., AND ALLEN, J. R., 1990, Basin Analysis: Principles and Applications: Oxford, Blackwell Scientific Publications, 451 p.

ALLEN, P. A., AND HOMEWOOD, P., 1986, Foreland Basins: Oxford, International Association of Sedimentologists Special Publication 8, Blackwell Scientific, 453 p.

ALLISON, P. A., AND BRIGGS, D., 1991, The taphonomy of soft-bodied animals, in Donovan, S. K., ed., The Process of Fossilization: London, Belhaven Press, p. 120-140.

ANDERSON, E. J., GOODWIN, P. W., AND CAMERON, B., 1978, Punctuated aggradational cycles (PACS) in Middle Ordovician and Lower Devonian sequences, in Merriam, D. F., ed., Field Trip Guidebook: Syracuse, New York State Geological Association, 50th Annual Meeting, p. 204-224.

ANDREASSEN, C., 1990, A suggested standard procedure for the construction of unimodal current rose-diagrams: Journal of Sedimentary Petrography, v. 60, p. 628-629.

BAIRD, G. C., AND BRETT, C., 1991, Submarine erosion on the anoxic sea floor: stratinomic, paleoenvironmental, and temporal significance of reworked pyrite-bone deposits, in Tyson, R. V., and Pearson, T. H., eds., Modern and Ancient Continental Shelf Anoxia: London, Geological Society Special Publication 58, p. 233-257.

BAIRD, G. C., AND BRETT, C. E., 1994, Revised correlation for Late Ordovician shelf-slope deposits, Mohawk Valley, N.Y.: Implications for depositional dynamics in a foreland basin (abs.): Geological Society of America, Abstracts with Programs, v. 27, p. 4.

BAIRD, G. C., BRETT, C. E., AND LEHMANN, D., 1992, The Trenton- Utica problem revisited: new observation and ideas regarding Middle-Late Ordovician stratigraphy and depositional environments in central New York, in Pinet, P., ed., Field Trip Guidebook: Hamilton, New York State Geological Association, 64th Annual Meeting, p. 1-40.

BALLY, A. W., AND SNELSON, S., 1980, Realms of subsidence, in Miall, A. D., ed, Facts and Principles of World Petroleum Occurrence: Calgary, Canadian Society of Petroleum Geologists Memoir 6, p. 9-94.

BARNES, C. R., NORFORD, B. S., AND SKEVINGTON, D., 1981, The Ordovician System in Canada: Paris, International Union of Geological Sciences Publication 8, 27 p.

BAUM, G. R., AND VAIL, P. R., 1988, Sequence stratigraphic concepts applied to Paleogene outcrops, Gulf and Atlantic Basins, in Wilgus, C. K., and others, eds., Sea-level Changes: An Integrated Approach: Tulsa, Society of Economic Paleontologists and Mineralogists Special Publication 42, p. 309-328.

BEAUMONT, E. A., 1981, Foreland basins: Geophysical Journal of the Royal Astronomical Society, v. 65, p. 291-329.

BERGSTROM, S., AND MITCHELL, C., 1986, The graptolite correlation of the North American Upper Ordovician standard: Lethia, v. 19, p. 247-266.

BERGSTROM, S., AND MITCHELL, C., 1992, The Ordovician Utica Shale in the eastern midcontinent region: Age, lithofacies, and regional relationships, *in* Chaplin, J. R., and Barrick, J. E., eds., Special Papers in Paleontology and Stratigraphy: A Commemorative Volume for Thomas W. Amsden: Tulsa, Oklahoma Geological Survey Bulletin, v. 145, p. 67-89.

BERGSTROM, S., MITCHELL, C., SCHUMACHER, G., AND SWINFORD, E. M., 1990, The Sebree Project II: Upper Middle and lower Upper Ordovician biostratigraphy in the Oxford, Ohio core and coeval successions in southwestern Ohio and easternmost Indiana cores (abs.): Geological Society of America, Abstracts with Programs, v. 22, p. 3.

BOND, G. C., AND KOMINZ, M. A., 1991, Disentangling Middle Paleozoic sea level and tectonic events in cratonic margins and cratonic basins of North America: Journal of Geophysical Research, v. 96, p. 6619-6639.

BOUMA, A. H., 1962, Sedimentology of Some Flysch Deposits: A Graphic Approach to Facies Interpretation: Amsterdam, Elsevier, 168 p.

BRADLEY, D. C., AND KIDD, W. S., 1991, Flexural extension of the upper continental crust in collisional foredeeps: Geological Society of America Bulletin, v. 103, p. 1416-1438.

CHENOWITH, P. A., 1952, Statistical methods applied to Trenton stratigraphy in northwestern New York: Geological Society of America Bulletin, v. 63, p. 521-560.

CISNE, J. L., KARIG, D. E., RABE, B. D., AND HAY, B. J., 1982, Topography and tectonics of the Taconic outer trench slope as revealed through gradient analysis of fossil assemblages: Lethia, v. 15, p. 229-246.

CLOETINGH, S., 1988, Intraplate stresses: a tectonic cause for third-order cycles in apparent sea-level, *in* Wilgus, C. K., and others, eds., Sea-level Changes: An Integrated Approach: Tulsa, Society of Economic Paleontologists and Mineralogists Special Publication 42, p. 19-30.

COVEY, M., 1986, The evolution of foreland basins to steady state: evidence from the western Taiwan foreland basin, *in* Allen, P. A. and Homewood, P., eds., Foreland Basins: Oxford, International Association of Sedimentologists Special Publication 8, Blackwell Scientific, p. 77-90.

CRESSMAN, E. R., 1973, Lithostratigraplhy and depositional environments of the Lexington Limestone (Ordovician) of central Kentucky: Washington D. C., Unites States Geological Survey, Professional Paper 768, 61 p.

DEAN, W. T., 1980, The Ordovician System in the Near and Middle East: Ottawa, International Union of Geological Sciences Publication 2, 22 p.

DELANO, J. W., SHIRNICK, C., BOCK, B., KIDD, S. F., HEIZLER, T., PUTNAM, G. W., DELONG, S. E., AND OHR, M., 1990, Petrology and geochemistry of Ordovician K-bentonites in New York State: Constraints on the nature of a volcanic arc: Journal of Geology, v. 98, p. 157-170.

DENNISON, J. M., AND HEAD, J. W., 1975, Sea level variations interpreted from the Appalachian Basin, Silurian and Devonian: American Journal of Science, v. 275, p. 1089-1120.

DEWING, K., AND COPPER, P., 1991, Upper Ordovician stratigraphy of Southampton Island, Northwest Territories: Canadian Journal of Earth Sciences, v. 28, p. 283-291.

DUKE, W. L., 1991, The Lower Silurian Medina Group in New York and Ontario, *in* Cheel, R. J., ed., Sedimentology and Depositional Environments of Silurian Strata of the Niagara Escarpment, Ontario and New York: Toronto, Geological Association of Canada, p. 35-61.

ETTENSOHN, F., 1985, Controls on development of Catskill Delta complex basin-facies, *in* Woodrow, D., and Sevon, W., eds., The Catskill Delta: Boulder, Geological Society of America Special Paper 201,

p. 51-63.

FISCHER, A. G., AND ARTHUR, M. A., 1977, Secular variations in the pelagic realm, *in* Cook, H. E., and Enos, P., eds., Deep-water Carbonate Environments: Tulsa, Society of Economic Paleontologists and Mineralogists Special Publication 25, p. 19-50.

FISCHER, A. G., HERBERT, T., AND PREMOLI SILVA, I., 1985, Carbonate bedding cycles in Cretaceous pelagic and hemipelagic sequences, *in* Pratt, L. M., and others, eds., Fine Grained Deposits and Biofacies of the Western Interior Sea Way: Evidence for Cyclic Sedimentary Processes: Tulsa, Society of Economic Paleontologists and Mineralogists, p. 1-10.

FISHER, D. W., 1977, Correlation of the Hadrynian, Cambrian and Ordovician rocks in New York State: New York State Museum Map and Chart Series 25, 75 p.

FISHER, D. W., 1979, Folding in the foreland, Middle Ordovician Dolgeville facies, Mohawk Valley, New York: Geology, v. 7, p. 455-459.

FORTEY, R. A., 1984, Global earlier Ordovician transgressions and regressions and their biological implications, *in* Bruton, D. L., ed., Aspects of the Ordovician System: Olso, Univeritetsforlaget, p. 37-50.

GALLOWAY, W., 1989, Genetic stratigraphic sequences in basin analysis I: architecture and genesis of flooding-surface bounded depositional units: American Association of Petroleum Geologists Bulletin, v. 73, p. 125-154.

HALLAM, A., 1967, Depth significance of shale with bituminous laminae: Marine Geology, v. 5, p. 481-493.

HAMMAN, W., ROBARDET, M., AND ROMANO, M., 1982, The Ordovician System in Southwestern Europe: Paris, International Union of Geological Sciences Publication 11, 47 p.

HAY, B. J., AND CISNE, J. L., 1988, Deposition in the Oxygen-deficient Taconic Foreland Basin, Late Ordovician, *in* Keith, B. D., ed., The Trenton Group (Upper Ordovician Series) of Eastern North America: Tulsa, American Association of Petroleum Geologists Studies in Geology 29, p. 113-134.

HAY, H. B., 1981, Lithofacies and formations of the Cincinnatian Series (Upper Ordovician), southeastern Indiana and southwestern Ohio: Unpublished Ph.D. Dissertation, Miami University, Oxford, 234 p.

HISCOTT, R. N., PICKERING, K. T., AND BEEDEN, D. R., 1986, Progressive filling of a confined Middle Ordovician foreland basin associated with the Taconic Orogeny, Quebec, Canada, *in* Allen, P. A., and Homewood, P., eds., Foreland Basins: Oxford, International Association of Sedimentologists Special Publication 8, Blackwell Scientific, p. 309-326.

HOLLAND, S., 1989, Sequence-based correlation: Sorting through the complicated stratigraphy of the eastern United States (abs.): Geological Association of America, Abstracts with Programs, v. 21, p. 80.

HOUSEKNECHT, D. W., 1986, Evolution from passive margin to foreland basin: the Atoka Formation of the Arkoma Basin, south-central U.S.A., *in* Allen, P. A., and Homewood, P., eds., Foreland Basins: Oxford, International Association of Sedimentologists Special Publication 8, Blackwell Scientific, p. 327-345.

HOWELL, P. D., AND VAN DER PLUIJM, B. A., 1990, Early history of the Michigan basin: Subsidence and Appalachian tectonics: Geology, v. 18, p. 1195-1198.

JERVEY, M. T., 1988, Quantitative geological modeling of siliciclastic rock sequences and their seismic expression, *in* Wilgus, C. K., and others, eds., Sea-level Changes: An Integrated Approach: Tulsa, Society of Economic Paleontologists and Mineralogists Special

Publication 42, p. 47-70.

JOHNSON, M. E., AND JIA-YU, R., 1989, Middle to Late Ordovician rocky shores from the Manitoulin Island area, Ontario: Canadian Journal of Earth Science, v. 26, p. 642-653.

JOHNSON, M. E., JIA-YU, R., AND FOX, W. T., 1989, Comparison of Late Ordovician epicontinental seas and their relative bathymetry in North America and China: PALAIOS, v. 4, p. 43-50.

KARNER, G. D., 1986, Effects of lithospheric in-plane stress on sedimentary basin stratigraphy: Tectonics, v. 5, p. 573-588.

KAY, M., 1935, Distribution of Ordovician altered volcanic materials and related clays: Geological Society of America Bulletin, v. 46, p. 225-244.

KAY, M., 1953, Geology of the Utica Quadrangle, New York: New York State Museum Bulletin, v. 347, 126 p.

KEITH, B. D., 1988, Regional facies of the Upper Ordovician Series of eastern North America, in Keith, B. D., ed., The Trenton Group (Upper Ordovician Series) of Eastern North America: Tulsa, American Association of Petroleum Geologists Studies in Geology 29, p. 1-16.

KENDALL, C. G. ST. C, AND LERCHE, I., 1988, The rise and fall of eustasy, in Wilgus, C. K., and others, eds., Sea-level Changes: An Integrated Approach: Tulsa, Society of Economic Paleontologists and Mineralogists Special Publication 42, p. 3- 16.

KERR, M., AND EYLES, N., 1991, Storm-deposited sandstones (tempestites) and related ichnofossils of the Late Ordovician Georgian Bay Formation, southern Ontario, Canada: Canadian Journal of Earth Sciences, v. 28, p. 266-282.

KOLLA, V., AND MACURDA, D. B., 1988, Sea-level changes and timing of turbidity-current events in deep-sea fan systems, in Wilgus, C. K., and others, eds., Sea-level Changes: An Integrated Approach: Tulsa, Society of Economic Paleontologists and Mineralogists Special Publication 42, p. 381-392.

KUNK, M. J., AND SUTTER, J. F., 1984, $^{40}Ar/^{39}Ar$ spectrum dating of biotite from Middle Ordovician bentonites, eastern North America, in Bassett, M. G., ed., The Ordovician System: Candiff, University of Wales Press, p. 121-152.

LASH, G. G., 1988, Middle and Late Ordovician shelf activation and foredeep evolution, central Appalachian orogen, in Keith, B. D., ed., The Trenton Group (Upper Ordovician Series) of Eastern North America: Tulsa, American Association of Petroleum Geologists Studies in Geology 29, p. 37- 54.

LEHMANN, D., AND BRETT, C. E., 1990, Paleoecology of the Late Ordovician foreland basin in New York and Ontario: Faunal assemblages controlled by sedimentation rate (abs.): Geological Society of America, Abstracts with Programs, v. 23, p. 58.

LEHMANN, D., AND BRETT, C. E., 1991, Tectonic and eustatic influences upon the sedimentary environments of the Upper Ordovician strata of New York and Ontario (abs.): Geological Society of America, Abstracts with Programs, v. 24, p. 29.

LEHMANN, D., BRETT, C. E., BRITT, S., PARSONS, M., AND RYAN, D., 1990, Fossil concentrations in black shale environments: the role of gradient currents (abs.): Geological Society of America, Abstracts with Programs, v. 23, p. 227.

LIBERTY, B. A., 1969, Paleozoic Geology of the Lake Simcoe Area, Ontario: Ottawa, Geological Survey of Canada Memoir 355, 201 p.

LIBERTY, B. A., 1978, Ordovician nomenclature of Manitoulin Island, in Sanford, J. T., and others, eds., Geology of the Manitoulin Area Including the Road Log to the Michigan Basin Geological Society Field Trip: Kalamazoo, Michigan Basin Geological Society Special Paper 3, p. 43-45.

LOUTIT, T. S., HARDENBOL, J., VAIL, P. R, AND BAUM, G. R., 1988, Condensed sections: the key to age dating and correlation of continental margin sequences, in Wilgus, C. K., and others, eds., Sea-level Changes: An Integrated Approach: Tulsa, Society of Economic Paleontologists and Mineralogists Special Publication 42, p. 183-215.

MACLACHLAN, D. B., 1967, Structure and stratigraphy of the limestones and dolomites of Dauphin Counlty, Pennsylvania: Pennsylvania Topographic and Geologic Survey, General Geology Report G44, 168 p.

MEHRTENS, C. J., 1988, Bioclastic turbidites in the Trenton Limestone: significance and criteria for recognition, in Keith, B. D., ed., The Trenton Group (Upper Ordovician Series) of Eastern North America: Tulsa, American Association of Petroleum Geologists Studies in Geology 29, p. 87-112.

MIALL, A., 1991, Stratigraphic sequences and their chronostratigraphic correlation: Journal of Sedimentary Petrology, v. 61, p. 497-505.

MITCHELL, C., GOLDMAN, D., AND DELANO, J., 1994, Correlation of the Denley Limestone and Dolgeville Formation, New York State, based on K-bentonite and graptolite chronostratigraphy (abs.): Geological Society of America, Abstracts with Programs, v. 27, p. 63.

MUTTI, E., 1985, Turbidite systems and their relations to depositional sequences, in Zuffa, G., ed., Provenance of Arenites: Hingham, Reidel Publishing, p. 65-93.

MUTTI, E., AND RICCHI LUCCHI, F., 1978, Turbidites of the northern Appennines: Introduction to facies analysis: International Geological Review, v. 20, p. 125-166.

NUMMEDAL, D., AND SWIFT, D. J. P., 1987, Transgressive stratigraphy at sequence-bounding unconformities: Some principles derived from Holocene and Cretaceous examples, in Nummedal, D., and others, eds., Sea Level Fluctuation and Coastal Evolution: Tulsa, Society of Economic Paleontologists and Mineralogists Special Publication 41, p. 241-260.

PATCHEN, D. G., 1978, Depositional environment of the Oswego Sandstone, in Merriam, D. F., ed., Field Trip Guidebook: Syracuse, New York State Geological Association, 50th Annual Meeting, p. 368-385.

PICKERING, K. T., HISCOTT, R. N., AND HEIN, F. J., 1989, Deep Marine Environments: Clastic Sedimentation and Tectonics: Boston, Unwin Hyman, 416 p.

POSAMENTIER, H. W., JERVEY, M. T., AND VAIL, P. R., 1988, Eustatic controls on clastic deposition- conceptual framework, in Wilgus, C. K., and others, eds., Sea-level Changes: An Integrated Approach: Tulsa, Society of Economic Paleontologists and Mineralogists Special Publication 42, p. 109-124.

POSAMENTIER, H. W., AND VAIL, P. R., 1988, Eustatic controls on clastic deposition-sequence and systems tract models, in Wilgus, C. K., and others, eds., Sea-level Changes: An Integrated Approach: Tulsa, Society of Economic Paleontologists and Mineralogists Special Publication 42, p. 125-154.

QUINLAN, G. M., AND BEAUMONT, C., 1984, Appalachian thrusting, lithospheric flexure, and the Paleozoic stratigraphy of the eastern interior of North America: Canadian Journal of Earth Sciences, v. 21, p. 973-996.

REYMENT, R., 1971, Introduction to Quantitative Paleoecology: Amsterdam, Elsevier, 227 p.

RICCI LUCCHI, F., 1975, Miocene palaeogeography and basin analysis in the Periadriatic Appennines, in Squyres, C., ed., Geology of Italy, Pt. II: Tripoli, Prentice Hall, p. 129-236.

RICCI LUCCHI, F., 1984, Deep-sea fan deposits on the Miocene Marnosoarenacea Formation, northern Apennines: Geo-Marine Letters, v. 3, p. 203- 210.

RIVA, J., 1969, Middle and Upper Ordovician graptolite faunas of the St. Lawrence Lowlands of Quebec, and of Anacosti Island, *in* Kay, M., ed., North Atlantic Geology and Continental Drift, A Symposium: Tulsa, American Association of Petroleum Geologist Memoir 12, p. 513-556.

ROOT, S. I., 1980, Possible recurrent basement faulting, Pennsylvania: Part I, geologic framework: Harrisburg, Pennsylvania Geological Survey, Open-file Report, 23 p.

ROWLEY, D. B., AND KIDD, W. S. F., 1981, Stratigraphic relationships and detrital composition of the Medial Ordovician flysch of western New England: Implications for the tectonic evolution of the Taconic Orogeny: Journal of Geology, v. 89, p. 199-218.

RUEDEMANN, R., 1925, The Utica and Lorraine Formations of New York, Part I: Stratigraphy: New York State Museum Bulletin 258, 176 p.

RUSSEL, D., AND TELFORD, P., 1983, Revisions to the stratigraphy of the Upper Ordovician Collingwood Beds of Ontario—a potential oil shale: Canadian Journal of Earth Sciences, v. 20, p. 1780-1790.

SARG, J. F., 1988, Carbonate sequence stratigraphy, *in* Wilgus, C. K., and others, eds., Sea-level Changes: An Integrated Approach: Tulsa, Society of Economic Paleontologists and Mineralogists Special Publication 42, p. 155- 181.

SANFORD, B. V., THOMPSON, F. J., AND MCFALL, G. H., 1985, Plate tectonics—a possible controlling mechanism in the development of hydrocarbon traps in southwestern Ontario: Bulletin of Canadian Petroleum Geology, v. 33, p. 52-71.

SENIOR, S., 1991, A new species of graptoloid, *Dicellograptus uncatus* n. sp., from the Blue Mountain Formation of Southern Ontario, Canada: Canadian Journal of Earth Sciences, v. 28, p. 822-826.

SHEN-FU, S. 1980, The Ordovician System in China: Ottawa, International Union of Geological Sciences Publication 1, 7 p.

SLOSS, L. L., 1991, The tectonic factor in sea level change: a countervailing view: Journal of Geophysical Research, v. 96, p. 6609- 6617.

STANLEY, R. S., AND RATCLIFFE, N. M., 1985, Tectonic synthesis of the Taconic orogeny in western New England: Geological Society of America Bulletin, v. 96, p. 1227-1250.

SWEET, W. C., AND BERGSTROM, S. M., 1976, Conodont biostratigraphy of the Middle and Upper Ordovician of the United States Midcontinent, *in* Bassett, M. G., ed., The Ordovician System: Candiff, University of Wales Press, p. 121-152.

TANKARD, A. J., 1986, On the depositional response to thrusting and lithospheric flexure: examples from the Appalachian and Rocky Mountain basins, *in* Allen, P. A., and Homewood, P., eds., Foreland Basins: Oxford, International Association of Sedimentologists Special Publication 8, Blackwell Scientific, p. 369-392.

THOMAS, W. A., 1977, Evolution of the Appalachian-Ouachita salients and recesses from reentrants and promentories in the continental margin: American Journal of Science, v. 277, p. 1233-1278.

TITUS, R., 1986, Fossil communities of the upper Trenton Group (Ordovician) of central and northwestern New York State: Journal of Paleontology, v. 50, p. 1209-1225.

TITUS, R., 1988, Facies of the Trenton Group of New York, *in* Keith, B. D., ed., The Trenton Group (Upper Ordovician Series) of Eastern North America: Tulsa, American Association of Petroleum Geologists Studies in Geology 29, p. 77-86.

TITUS, R., AND CAMERON, B., 1976, Fossil communities of the lower Trenton Group (Middle Ordovician) of central and northwestern New York State: Journal of Paleontology, v. 60, p. 805-814.

VAIL, P. R., MITCHUM, R. M., AND THOMPSON, S., 1977, Global cycles of relative changes of sea level, *in* Payton, C. E., ed., Seismic Stratigraphy— Applications to Hydrocarbon Exploration: Tulsa, American Association of Petroleum Geologists Memoir 26, p. 83-97.

VAN WAGONER, J. C., POSAMENTIER, H. W., MITCHUM, R. M., VAIL, P. R., SARG, J. F., LOUTIT, T. S., AND HARBENDOL, J., 1988, An overview of the fundamentals of sequence stratigraphy and key definitions, *in* Wilgus, C. K., and others, eds., Sea-level Changes: An Integrated Approach: Tulsa, Society of Economic Paleontologists and Mineralogists Special Publication 42, p. 39-46.

WAGNER, W. R., 1976, Growth faults in Cambrian and Lower Ordovician rocks of western Pennsylvania: American Association of Petroleum Geologists Bulletin, v. 60, p. 414-427.

WEBBY, B. D., 1981, The Ordovician System in Australia, New Zealand and Antartica: Ottawa, International Union of Geological Sciences Publication 6, 64 p.

WICKSTROM, L., 1990, A new look at Trenton structure in northwestern Ohio: Northeastern Geology, v. 12, p. 103-113

WILLIAMS, A., STRACHAN, I., BASSETT, D. A., DEAN, W. T., INGHAM, J. K., WRIGHT, A. D., AND WHITTINGTON, H. B., 1973, A correlation of Ordovician rocks in the British Isles: Geological Society Special Report 3, 74 p.

WITZKE, B. J., 1980, Middle and Upper Ordovician paleogeography of the region bordering the Transcontinental Arch, *in* Fouch, T. D., and Magathan, E. R., eds., Paleozoic Paleogeography of West-Central United States: Tulsa, Society of Economic Paleontologists and Mineralogists, p. 1-15.

WITZKE, B. J., 1987, Models for circulation patterns in epicontinental seas applied to Paleozoic facies of North America craton: Paleoceanography, v. 2, p. 229-248.

WITZKE, B. J., AND GLENISTER, B. F., 1987, Upper Ordovician Maquoketa Formation in the Graf area, eastern Iowa, *in* Biggs, D. L., ed., North-Central Section of the Geological Society of America, Centenial Field Guide: Boulder, Geological Society of America, p. 103-107.

YEAKLE, L. S., 1962, Tuscarora, Juniata, and Bald Eagle paleocurrents and paleogeography in the Central Appalachians: Geological Society of America Bulletin, v. 73, p. 1515-1540.

ZERRAHN, G. J., 1978, Ordovician (Trenton to Richmondian) depositional patterns of New York State and their relationships to the Taconic orogeny: Geological Society of America Bulletin, v. 89, p. 1751-1760.

FOUNDERING OF THE CAMBRO-ORDOVICIAN SHELF MARGIN: ONSET OF TACONIAN OROGENESIS OR EUSTATIC DROWNING

PAUL A. WASHINGTON[1] AND STEVEN A. CHISICK[2]

[1]1385 South West Broad Street, Southern Pines, North Carolina 28387

[2]Sevenson Environmental Services, 9245 Calumet Avenue, Munster, Indiana 46321

ABSTRACT: Onset of the Taconic Orogeny is generally dated by the age of the last shelf strata beneath Ordovician pelitic sequences. Recent stratigraphic analysis of the shelf sequence in the Champlain Valley, however, suggests that timing of the end of shelf sedimentation in a given area was controlled by eustasy as well as tectonics. Following major eustatic lowstands in Cambrian and Early-Middle Ordovician time, the shelf margin abruptly jumped cratonward and large areas of outer shelf foundered. Only by analyzing subsidence (stratigraphic accumulation) rates can the timing of the orogenic event be determined. Analysis of subsidence rates indicate a simultaneous start of the initial phase of orogenic subsidence in late Early Ordovician time for the entire shelf from Newfoundland to Pennsylvania. Arrival of the orogenic wedge varied from Middle Ordovician (Newfoundland) to early Late Ordovician (Champlain Valley) time.

INTRODUCTION

Timing of the beginning of the Taconic orogeny has long been a matter of controversy because the primary evidence for the onset of orogenesis is stratigraphic markers that may be unrelated to orogenesis. Based on the local timing of the changeover from carbonate to pelitic sedimentation on the outer portions of the Cambro-Ordovician shelf (Rowley and Kidd, 1981; González-Bonorino, 1990), most geologists currently consider the onset of the Taconic orogeny to have been diachronous, varying with position along the margin of proto-North America (Hall and Roberts, 1988; Drake and others, 1989; Bradley, 1989). This carbonate-to-pelite transition is generally considered to mark the tectonic foundering of the shelf (Fig. 1), but we believe that in most cases the transition was caused by prolonged eustatic events during which sea level was significantly lowered (Chisick and Washington, 1989; Washington, 1992a).

In a sense, nomenclature has anticipated this conclusion. The timing of this stratigraphic event (the changeover) varies by at least 30 million years (earliest Early Ordovician to Late Ordovician) along the length of the Appalachians and is often temporally distinct from the primary deformational event. Alternate names (e.g. Penobscottian and Blountian) have been applied to various temporal groupings of changeovers in the belief that they represent distinct tectonic events (see Rast and others, 1988; Hall and Roberts, 1988; Drake and others, 1989). Most recent authors, however, consider the various named events to constitute episodes within a larger tectonic event. Currently, the popular explanations for the varied timing of these episodes are tectonic migration along strike (Rodgers, 1971; Bradley, 1989) and prolonged tectonic convergence (Rowley and Kidd, 1981; Stanley and Ratcliffe, 1985).

Over the last several years, we have been studying the structure and stratigraphy of the Champlain Valley of Vermont and New York, the last portion of the Taconic outer shelf system to experience the carbonate-to-pelite transition (Rodgers, 1971; Bradley, 1989). Our analysis (Chisick and Washington, 1989; Washington, 1992a) suggests that this transition is not a valid marker for the onset of orogenic effects, and that the Champlain Valley experienced the initial effects of Taconian tectonism at the same time as the rest of the orogen. A key piece of evidence leading to this conclusion is our discovery of several distinct carbonate-pelite changeover events capping large portions of Cambrian and Ordovician shelf strata. Each of these events resulted from an abrupt landward retreat of the shelf margin. The drowned shelf-segments are directly overlain by deeper water shales or pelitic carbonates, which continued to Late Ordovician time. These changeover events are distributed throughout Cambrian and Early and Middle Ordovician time.

Abrupt step-back events along the shelf margin correlate with episodes of subaerial exposure and accompanying hiatuses in the otherwise continuous shelf sequence. In the Champlain Valley, subaerial exposure events within the carbonate-shelf sequence are primarily expressed as major erosional events marked by estuarine valley fill at the mouths of craton-draining river systems (Washington, 1992a). We attribute most of these step-backs and correlative exposure events to global eustatic events (Barnes, 1984; Fortey, 1984; Erdtmann, 1986; Chen, 1988) apparently unrelated to regional tectonics (Washington and Chisick, 1989; Washington, 1992a), although the effects of later events may have been aggravated by the onset of orogenesis. The last transition immediately precedes the deformational phase of the Taconic orogeny, and so is probably directly related to the tectonism.

The repeated eustatic events resulted in permanent dissection of the outer shelf by lowstand river valleys, which segmented the shelf into distinct sedimentologic blocks separated by submarine canyons (Chisick and Washington, 1989; Washington, 1992a). Each of these blocks developed a unique lithostratigraphy which is not easily correlated with adjacent blocks. However, the major stratigraphic boundaries within the blocks are coeval; we believe this is because the boundaries were caused by major eustatic events. This segmentation of the shelf stratigraphy has plagued Appalachian geologists for more than a century.

This paper will investigate the evidence for eustatic and

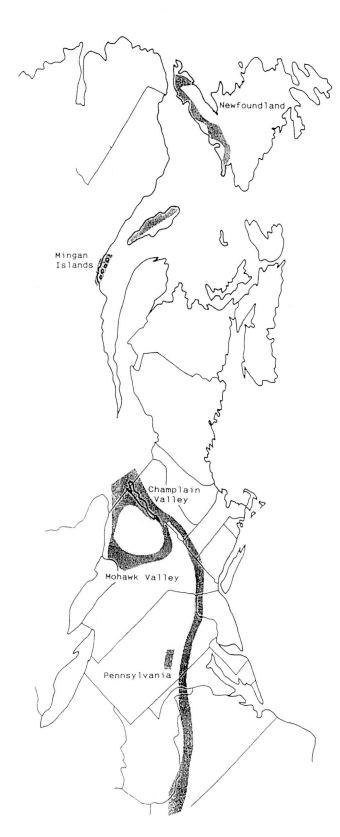

FIG. 1—Areas of outcrop of Cambrian through lower Upper Ordovician shelf rocks in the northern and central Appalachians.

tectonic controls on the carbonate-to-pelite transitions at the top of the Cambro-Ordovician shelf sequence and reevaluate the timing of the onset of Taconian orogenesis. Toward this end, we will compare the processes and results of shelf foundering and carbonate-to-pelite transitions caused by tectonic and eustatic events, review evidence from the Champlain Valley, and review and reevaluate tectonic and nontectonic stratigraphic indicators in the northern and central Appalachians.

EFFECTS OF EUSTASY AND TECTONISM

Passive margins slowly subside over time in response to mantle cooling and sediment accumulation. Mantle cooling results in a subsidence of more than 3 km over the first 100 my following a curve approximated by $k_1 + k_2 t^{0.5}$, where k_1 and k_2 may vary along stratigraphic strike depending on the local amount of crustal extension, mantle heating, and sedimentation (Sleep, 1971; Bott, 1992). Isostatic response to sediment deposition depends on the rate of accumulation, the relation of the top of sediment to normal sea level, the density of the sediments, and, for passive-margin systems, the relative position and nature of the continental-oceanic crustal boundary. Fortunately, shelf carbonates have a relatively constant density and carbonate deposition keeps pace with subsidence; thus, the resulting stratigraphic accumulation curve follows the mantle-cooling subsidence curve (except that k_2 is increased by a constant multiplier) as long as the shelf margin stays relatively stationary. One of the problems encountered when analyzing the accumulation of the Cambro-Ordovician shelf sequence in the Appalachians is that the shelf margin did not remain stationary.

Eustatic Events

During major eustatic lowstands, shelf surfaces generally lie above sea level, interrupting shelf sedimentation and exposing the shelf to erosion. Interruption of the sedimentation reduces the subsidence rate for that period because only cooling subsidence is active; at the same time, erosion may further reduce the subsidence. Nevertheless, when sea level rebounds, the shelf surface will have subsided from where it was when the eustatic event began, so sedimentation attempts to catch up, increasing the subsidence rate to greater than normal until total isostatic adjustment catches up with the steady-state curve (Fig. 2). Thus, total sediment accumulation should be the same with or without intermittent eustatic interruptions.

This relation breaks down, however, if the interruptions are long enough that the post-eustatic depth of the shelf becomes so great that subsequent sedimentation can not restore the previous shelf morphology (i.e., some of the shelf area is lost). Carbonate sedimentation is quite sensitive to water depth, so long eustatic lowstands should result in the outer edges of the shelf (which subside fastest) becoming too deep for rapid accumulation. Where the depth is too great, sediment accumulation rates are slower than subsidence rates, and that portion of the shelf is permanently drowned (Fig. 3). A new shelf margin is ultimately

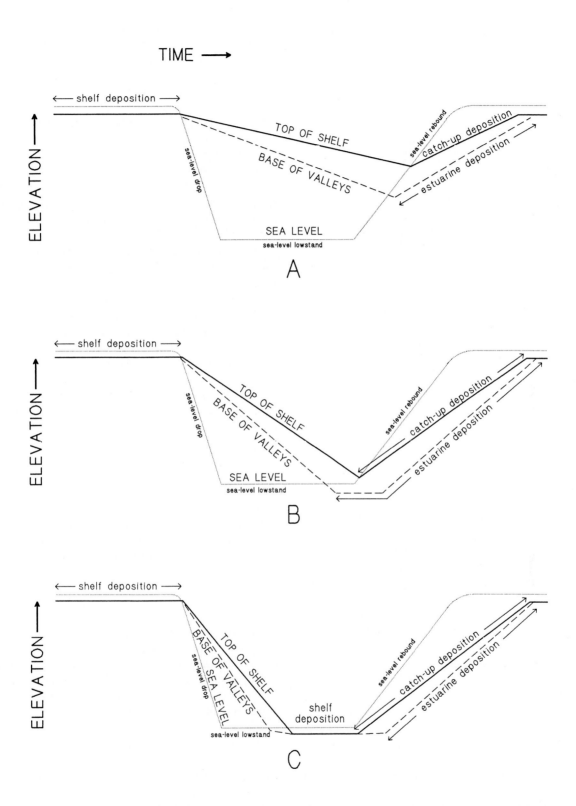

FIG. 2—Sedimentation/Erosion vs. Time curves for shelves affected by major eustatic events. The vertical position of the various curves within each diagram represents relative changes in elevation (normalized to highstand sea level) of critical portions of the shelf through time; the different patterns can be produced by varying either event duration or the amount of sea-level drop. (A) event during which shelf and river valleys continue to erode until reflooded during sea-level rebound. (B) event during which shelf subsidence allows deposition in the outer ends of erosional valleys, but not on the shelf surface, prior to sea-level rebound. (C) event during which shelf subsidence allows deposition to resume within the river valleys and on the shelf surface prior to sea-level rebound.

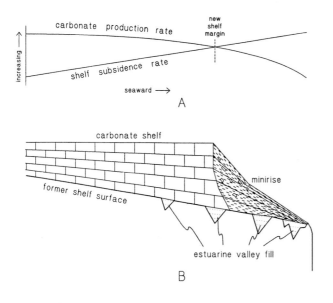

Fig. 3—Following sea-level rebound at the end of an eustatic event, the outer edges of the shelf are much deeper than the inner portions of the shelf, so carbonate production will be significantly slower on the edges of the shelf. Because the outer edge of the shelf is also the area of the shelf that is generally subsiding most rapidly, it is common that subsidence will outstrip carbonate deposition at least locally. A new shelf edge will be established where carbonate production rate during that initial period equals subsidence rate.

established landward of this foundered portion of shelf, and the foundered portion either forms a submarine plateau (such as the Blake plateau) or is buried beneath subsequent slope-rise sequence.

The sensitivity of carbonate deposition to water depth should cause carbonate-shelf loss in this manner whenever there is a major eustatic lowstand. Furthermore, any stream valley of significant depth developed on a carbonate shelf during a eustatic lowstand will be preserved. On the Cambro-Ordovician shelf of the Champlain Valley, clastic carbonate washed into such valleys from the adjacent shelf was insufficient to fill the valleys and restore shelf continuity. Only where significant fluvio-clastic sediment was delivered to the shelf from the continental interior was shelf continuity restored.

Upon sea-level rebound, fluvio-clastic sediment is trapped in drowned stream valleys (estuaries) where it may provide enough additional sediment to fill the valleys. The stratigraphic architecture of the estuarine fill depends on the relative volumes and input patterns of the fluvial siliciclastics and marine clastic carbonates (Dalrymple and others, 1992). Estuarine deposits in the Cambro-Ordovician shelf strata of the Champlain Valley are generally dominated by siliciclastic sands. Within the estuaries, the change from siliciclastic-dominated estuarine sediments to carbonate-dominated shelf sediments occurred when the estuarine system was nearly filled and the fluvio-clastic sediments were diverted onto the adjacent shelf, where they were trapped along the coast (Fig. 2).

Because estuarine environments trap fluvial sediments, estuarine deposition can easily keep up with thermal subsidence. Therefore, thermal subsidence during eustatic lowstands causes siliciclastic sediments to build up at the ends of deep river valleys.

When shelf subsidence during an extended eustatic event is sufficient to reflood the outer edge of the shelf before sea-level rebound occurs (Fig. 2C), carbonates are briefly reestablished. Following sea-level rebound, clastic outwash from estuaries that is usually carried to deep water through the drowned valley ends is diverted to the outer shelf.

During prolonged eustatic lowstands, river systems may erode deep valleys into the shelf without carrying significant sediment loads from the continental interior. Where the fluvial sediment input is insufficient, deep river valleys are preserved as embayments (such as Tongue of the Ocean in the Bahamas) or submarine canyons. Shelf margins develop atop the valley walls, and subsequent accumulation on the shelf preserves and accentuates embayments and canyons.

So, how long must a eustatic lowstand last for these effects to develop? Whether these effects occur depends on thermal subsidence rate, magnitude of the sea-level drop, and duration of the eustatic event (Fig. 2). Total subsidence during an event (subsidence rate ξ duration) is the critical factor controlling the amount of carbonate shelf loss, with more loss accompanying rapid thermal-subsidence phases and long-duration eustatic events. Development of erosional features, including river valleys incised into the shelf, is controlled by magnitude of sea-level drop and duration of the event. Erosion rates increase as base level is lowered, and total erosion is a function of rate and duration.

Flux of meteoric water during subaerial exposure of the shelf generally results in karst and carbonate diagenesis. These effects are most pronounced near the shelf edge where higher relief results in higher groundwater flux, but prolonged exposure will result in noticeable effects even in the interior of the shelf.

Tectonic Events

At first glance, tectonic events resemble eustatic events in many ways. As tectonism proceeds from out-board to in-board, it is usually assumed that the shelf edge will retreat ahead of the advancing orogenic wedge (see Rowley and Kidd, 1981; Jacobi, 1981). Because carbonate-production rates are potentially very high, it is unlikely that tectonic subsidence, whether regional or local (fault blocks), will cause foundering of a carbonate shelf, despite such claims for many recent shelves (see Mullins and others 1991). Rather, retreat will occur fairly rapidly as the advancing tectonic wedge approaches the edge of the shelf and mud eroded from the orogenic wedge begins to interfere with carbonate production. This initial retreat may mimic the retreat accompanying a eustatic event if it is fast enough to appear virtually instantaneous. An advancing tectonic wedge, however, should force the shelf edge to retreat in a nearly continuous manner, whereas a eustatic event results in relatively abrupt retreats interspersed with static or slowly prograding shelf edges. Where both tectonic and eustatic events operate simultaneously, the abrupt retreats will be much greater than would otherwise be expected because subsidence rates on the outer shelf are greater; whether there is retreat between eustatic events will depend on the rate of tectonic advance and the temporal spacing of the eustatic events.

In addition, there is often some subaerial exposure on the foreland bulge preceding the orogenic wedge (see Bradley and Kidd, 1991). As the bulge migrates across the shelf, it raises portions of the shelf above sea level, producing a migrating zone that experiences karstic and diagenetic events. Often the shelf edge locates itself near the outboard side of the foreland bulge, so the location of these events mimics that produced by eustasy. The timing of exposure, however, is time transgressive when caused by a tectonic event but is temporally uniform when caused by a eustatic event. There are also no post-exposure channel-fill units associated with tectonic exposure.

Carbonate-clastic transitions produced by tectonic and eustatic events look very similar. Both mechanisms result in shelf carbonates being directly overlain by pelites or deep-water carbonates. The difference is that eustatic retreat produces a draping unit (we call this a minirise) of long duration where sedimentation rate decreases upward, whereas tectonic retreat produces a short-duration drape that is quickly overrun by the foreland fluvio-clastics of the advancing wedge so that sedimentation rate increases upward. From a simple physical perspective, post-eustatic drapes should have relatively consistent grain-sizes, except during and immediately after subsequent eustatic events, but tectonic drapes should have upward coarsening grain-sizes. Unfortunately, the nature of the exposed off-shelf clastics in most of the Taconian orogen do not readily lend themselves to analysis of this sort.

Dating Tectonic Onset

Probably the most diagnostic feature to look for when trying to determine onset of tectonism is the rate of shelf-edge retreat. Eustatic lowstands occur fairly frequently, and the amount of retreat in each eustatic event will generally decrease as the thermal subsidence rate decreases. Except for isolated long-duration events, there should be little or no shelf loss accompanying eustatic lowstands affecting mature, slowly subsiding shelves. The onset of tectonism, however, is usually accompanied by a significant increase in subsidence rate, so shelf loss accompanying eustatic lowstands increases dramatically. Even when there are no eustatic events, the approach of a tectonic wedge will result in significant shelf loss. Thus, the beginning of a pattern of regional shelf edge retreat marks the onset of tectonic subsidence, whether or not a tectonic wedge has arrived. If frequent eustatic events occur during the approach of tectonism, only those areas with significant fluvio-clastic input to the shelf will tend to survive the initial eustatic effects to be drowned tectonically.

EVIDENCE FROM THE CHAMPLAIN VALLEY

Outer portions of the Cambro-Ordovician continental shelf are exposed in the structurally complex Taconic thrust belt of western Vermont, eastern New York, and southern Quebec (Fig. 4). Although this area has been studied *ad nauseam* for over 170 years (beginning with James, 1820), the complexity of the structural framework has obscured certain key stratigraphic relations from prior workers. Our recent detailed structural and stratigraphic reanalysis of this belt (Washington, 1982, 1985, 1987a, 1987b, 1992a, 1992b; Washington and Chisick, 1986; Bosworth and Chisick, 1987; Chisick and others, 1987) has unravelled many of the pre-orogenic relations among exposed portions of the shelf sequence. As a result, it is now possible to reconstruct the gross morphology of the pre-orogenic shelf margin and begin integrating sedimentologic data from the exposed stratigraphic section into a unified evolutionary model for the shelf. Throughout the following discussion, we will follow the terminology for Ordovician stages commonly used in the Champlain Valley (Table 1) because of their greater resolution than the equivalent European stages (Harland and others, 1990).

Shelf Stratigraphy

As in much of the Appalachians (Schwab and others, 1988; Rankin and others, 1989), the shelf strata of the Champlain Valley (Fig. 5) are broadly partitioned into three groups based on areal, temporal, and tectonic relations: (1) an Eocambrian clastic sequence with interbedded volcanic material at the base and shelf carbonates near the top; (2) a Lower to Middle(?) Cambrian clastic and carbonate (dominantly dolostone) sequence; and (3) an Upper Cambrian to Upper Ordovician carbonate and clastic sequence. The Eocambrian clastic sequence has generally been considered to represent synrifting sediments associated with the initial opening of a proto-Atlantic ocean (Cady, 1968; Rodgers, 1970), although the upper parts have recently been shown to represent early shelf deposits (Doolan, 1988); it is confined to more internal portions of the orogen and will be ignored in this paper. The Lower-Middle Cambrian sequence consists of the Cheshire and Monkton quartzites and accompanying Dunham and Winooski dolostones; these strata are limited to the structurally highest thrust sheets in the Champlain Valley. Excellent exposures of the shelf margin for the later portions of this sequence are found just north of Burlington, Vermont, where they have been studied intensively by Mehrtens (1985, 1989). The Upper Cambrian to Middle Ordovician sequence includes the Potsdam, Beekmantown, and Chazy-Trenton sections. This sequence is primarily found in the structurally lower thrust sheets and in the foreland to the west (Washington and Chisick, 1986; Chisick and others, 1987; Washington, 1992a). The shelf was smothered during Late Ordovician time by synorogenic shale associated with the Taconic orogeny (Rowley and Kidd, 1981; Stanley and Ratcliffe, 1985).

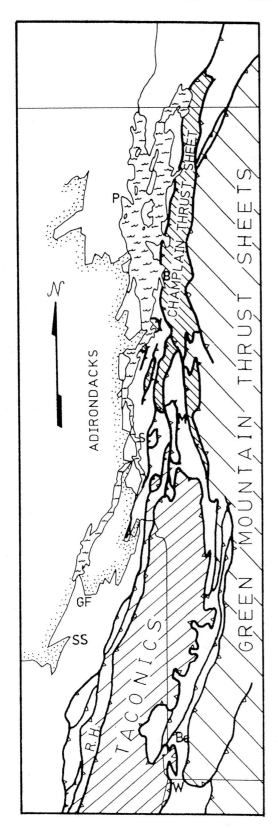

FIG. 4—Structure map of the Taconic thrust belt in and around the Champlain Valley. B - Burlington, VT; Be - Bennington, VT; GF - Glens Falls, NY; P -Plattsburg, NY; S - Shoreham, VT; SS -Saratoga Springs, NY; W - Williamstown, MA.

The Lower-Middle Cambrian and Upper Cambrian-Upper Ordovician sequences can be subdivided into stratigraphic packages which progress from an initial pulse of siliciclastic sediment to a more stable carbonate shelf phase. Detailed biostratigraphic work (Chisick and others, 1987; Washington, 1992a) has shown that the sand pulses always occur immediately after major breaks in stratigraphic continuity. These breaks are associated with erosional, diagenetic, and karstic events and, where high-quality data is available, by paleontologic discontinuities indicative of major depositional hiatuses. Seven such events are identified in the stratigraphic section (Fig. 6). For four of the events (Early Cambrian, Middle Cambrian, early Demingian, Cassinian), thickness of estuarine deposits and/or depth of diagenesis provides clear evidence of sea-level stands more than 30 m below normal. Similarities with the other events lead us to infer that the other hiatuses were also accompanied by long periods of significantly lowered sea level.

Following each of the eustatic events, there was a period of siliciclastic sedimentation, primarily in estuarine ends of lowstand river valleys, until carbonate shelf sedimentation was reestablished. During each subsequent stable shelf phase there was little clastic deposition on the shelf. In two cases (during Early and Middle Cambrian time), some shelf carbonate separates the depositional hiatus from the main body of the clastic pulse outside of the filled river valleys.

Shelf Margin

The shelf margin in the Champlain Valley region receded cratonward throughout Cambrian and Ordovician time. Migration of the margin was episodic, with the shelf edge remaining fixed for long periods of time and stepping back abruptly along regionally correlatable stratigraphic boundaries. The step-back boundaries coincide with the major stratigraphic hiatuses recognized in the shelf strata. Following each regression of the margin, the drowned former outer edge of the shelf caught sediment that normally would have been destined for the rise and abyssal plain. Once the sediments built up to a critical slope, subsequent sediments were directed to the deep ocean. These perched deep-water sediment wedges (we call them "minirises") are temporally correlative with the post-break shelf strata.

TABLE 1—STAGE EQUIVALENTS

CHAMPLAIN VALLEY	EUROPEAN
Trenton	Caradoc
Black River	
Chazy	Llandeilo
Whiterock	Llanvirn
Cassinian	Arenig
Jeffersonian	
Deminigian	Tremadoc
Gasconadian	

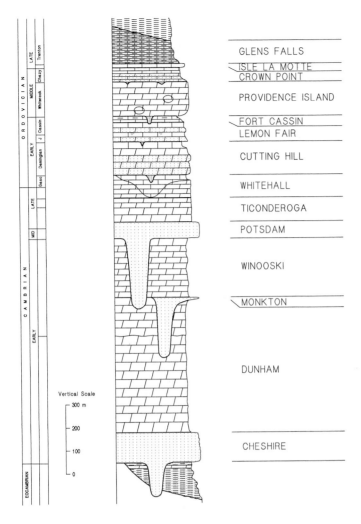

FIG. 5—Composite stratigraphic column of the Cambro-Ordovician shelf sequence in the Champlain Valley. Downward extension of sandstone bodies indicates depth of erosional valleys filled by siliciclastic sediments.

Three minirises are readily distinguished in the Champlain Valley. The mid-Early Cambrian eustatic event was followed by the deposition of the Georgia shales (Shaw, 1958; Mehrtens, 1985) and the New Haven mélange (Washington, 1992a) atop Dunham dolostones. The middle Jeffersonian event was followed by deposition of the Taconic mélange atop lower Lemon Fair shelf carbonates. The Whiterock-Chazy event was followed by the deposition of the Middlebury Limestone (deeper water) atop shelf carbonates of the Providence Island Formation. In shale minirises, only the presence of carbonate breccias and olistoliths from the adjacent shelf confirm the proximity of each of these sequences to the shelf margin. The Middlebury Limestone carries more subtle echoes of the adjacent shelf stratigraphy (Washington, 1987a). We believe that the main Taconic mass is the more distal portion of the late Early Cambrian minirise system overlain by the excess material shed off later minirise systems.

Most minirise systems (the Middlebury Limestone being a

notable exception) are now separated from both underlying and coeval shelf strata by thrust faults, making correlation difficult. This is especially true for the so-called Taconic mélange (Zen, 1968; Vollmer and Bosworth, 1984; Bosworth and Kidd, 1985) bounding the west edge of the Taconic allochthons. Only recent detailed analysis of strings of large blocks of shelf strata imbedded in the shale sequence (Fisher, 1985; Chisick and others, 1987) has enabled the identification of an internal stratigraphy (proving it is not a tectonic mélange) and the correlation with the coeval shelf strata.

Foundered areas of the shelf can be quite large or relatively small. The areas lost during the Early and Middle Cambrian eustatic events were large. There is a near total lack of post-break shelf strata on major thrust sheets (including the Green Mountain and Champlain thrust sheets) dominated by pre-break shelf strata. The Taconic allochthons are mostly made up of material that appears to be a late Early Cambrian minirise sequence (Baldwin, 1983) stripped off of foundered portions of early Early Cambrian shelf. The Georgia shales north of Burlington (Mehrtens, 1985, 1989) and the mélange in eastern New Haven, Vermont, (Washington, 1987b, 1992a) appear to be parts of this same minirise that have remained attached to the adjacent shelf margin.

The shelf area lost during the first two Early Ordovician breaks was apparently small; in fact, it is not clear that any shelf at all was lost in the first of these. Occurrences of pre-break strata without post-break strata are rare.

Large areas of shelf foundered during the middle Jeffersonian break, including most of the Vermont Valley and the north edge of the Champlain Valley shelf sequence. In both areas the youngest shelf strata are apparently lower Lemon Fair strata. The top of the Vermont Valley sequence may have served as the original location of the Taconic mélange, which appears to be a minirise system of late Jeffersonian through Chazy age accreted to the front of the advancing Taconic thrust sheets (Chisick and others, 1987; Washington, 1992a, 1992b). The absence of post Jeffersonian strata in the northern Champlain Valley is well illustrated in the incomplete Beekmantown Group section at Beekmantown, New York, (the name locality; the reference section is at Shoreham, Vermont) where the carbonate-pelite transition defining the top of the section corresponds to the middle Jeffersonian hiatus found in the shelf sequence farther south.

Apparently little shelf was lost during the late Cassinian hiatus, although critical sections are hidden beneath the Taconics. On the other hand, the structural importance of this boundary within the shelf strata should not be ignored — it forms the most consistent detachment level in the central and southern Champlain Valley. The minimal thickness of the basal sand unit (Weybridge member) suggests that this was a relatively short-lived event.

The Whiterock-Chazy event caused another major section of shelf to be lost. The subsequent shelf section represents the last stable shelf area in the Champlain Valley, and the last stable shelf area near the cratonic margin for the entire orogen. The Middlebury Limestone is the off-shelf pelitic carbonate se-

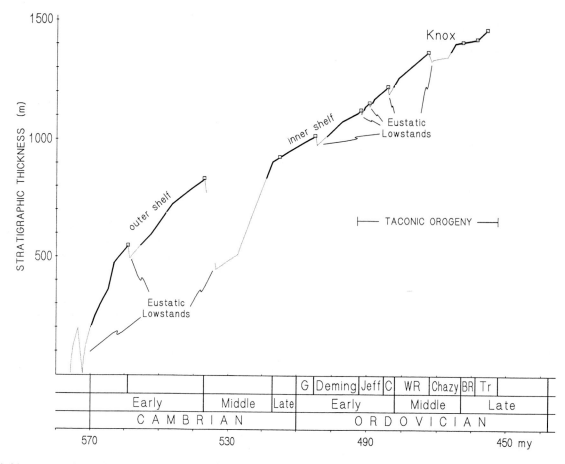

FIG. 6—Subsidence curve for the Cambro-Ordovician shelf sequence in the Champlain Valley showing the timing and approximate sea-level drops of the major eustatic events. Time values are based on Harland and others (1990). Knox denotes the lowstand that created the Knox unconformity.

quence formed atop the foundered shelf; the post-break shelf margin lies only a few kilometers west of Middlebury.

Anomalous Zones

Zones of missing or anomalous strata cut across the shelf system. Two of the zones in the Champlain Valley region are seminal to this discussion. All of the shelf sands are extremely thick in the central Champlain Valley (Shoreham-Middlebury area). Each overthickened section sits within a local depression in the top of the underlying carbonate unit. All of the sands correlate with the return of sedimentation to the shelf surface following hiatuses. We interpret these thick sands to be estuarine deposits filling the mouths of river valleys incised into the shelf during sea-level lowstands. Similar, though thinner, coeval sand units are found in the Great Meadows block south of Whitehall, New York (Flower, 1964; Fisher and Mazzullo, 1976; Fisher, 1985).

The other zone is the Kayaderosseras stratigraphic break between the Glens Falls area (Champlain Valley sequence) and the Saratoga Springs area (Mohawk Valley sequence) in eastern New York. This break consists of a 12-km wide zone (South Glens Falls to Wilton) with thick Upper Ordovician shales but no

exposures of shelf strata (Fisher, 1985). Within this zone, Upper Ordovician shales directly overlie Cambrian Potsdam sands and Precambrian basement gneisses. The lack of lithostratigraphic continuity across this break has long confounded geologists in the region.

A similar stratigraphic discontinuity is encountered in Pownal, Vermont, between the Champlain Valley/Vermont Valley sequence at Bennington, Vermont, and the Stockbridge limestone/ marble sequence (= Mohawk Valley sequence) at Williamstown, Massachusetts. Higher metamorphic grades and structural complexity have obscured this discontinuity and discouraged the detailed stratigraphic analysis that has been done in the Saratoga Springs-Glens Falls area. The discontinuity around Pownal does not appear to cut across the Cheshire Quartzite (basal Cambrian), but no overlying formations continue across the break. The stratigraphic sequences are notably different after Middle Cambrian time (possibly middle Early Cambrian), and the zone is filled with thick Upper Ordovician shales.

We correlate the Kayaderosseras and Pownal breaks and interpret them as marking a major submarine canyon that dissected the shelf. We hypothesize that this canyon developed during an Early or Middle Cambrian eustatic event and that its presence allowed the Champlain Valley/Vermont Valley and

Mohawk Valley/Stockbridge stratigraphic sequences to evolve into lithologically and faunally-distinct sequences. This canyon was apparently filled by syntectonic flysch during the Taconic orogeny.

Local lithologic- and stratigraphic-thickness variations cause many structural complications that have long befuddled geologists working in the Champlain Valley. Besides the shelf margins and anomalous zones discussed above, other erosional and estuarine features cut across the shelf margin in western Vermont at Dorset (Thompson, 1967), Sudbury, and Milton (Mehrtens, 1989). Other major stratigraphic boundaries interrupt the Cambro-Ordovician shelf sequence at the Quebec reentrant and across an enigmatic zone south of Albany, New York, and Stockbridge, Massachusetts.

Subsidence Record

Subsidence of the Cambro-Ordovician shelf in the Champlain Valley followed a standard passive margin subsidence pattern (Fig. 6) throughout Cambrian time and into Early Ordovician time. Deviations from the expected curve can be attributed to the cratonward migration of the shelf margin and the use of a composite column that included shelf strata from more distal (Early and Middle Cambrian) and more proximal (after Middle Cambrian) portions of the shelf system. Beginning just before the early Jeffersonian hiatus, however, subsidence rates increased markedly. This rapid subsidence continued until the Whiterock hiatus (the "real" Knox unconformity). Subsidence throughout the Chazy was very slow, probably less than expected for thermal subsidence of a mature shelf. At the beginning of the Black River, subsidence increased abruptly with rapid subsidence continuing until the final foundering of the shelf. Localized deposits of chaotic olistoliths (Washington, 1992b) in the Trenton shales overlying the carbonates indicate that these shales were deposited as the shelf was overridden by the tectonic wedge.

The amount of shelf loss during each eustatic lowstand generally reflects the shelf subsidence rate, except that the late Cassinian and Whiterock-Chazy events reflect their duration (the former must have been quite short, whereas the latter was quite long). The amount of sea-level drop affected the amount of erosion, thereby controlling the thickness of the estuarine deposits. Diagenetic effects were mostly limited to the sides of erosional valleys, whereas limited karst effects are found mainly near the outer shelf margin.

CORRELATIVE EVIDENCE
Newfoundland

High-quality work in Newfoundland over the last 150 years, from the early work of the Canadian Geological Survey to recent work by Noel James and contemporaries, has discovered similar features to those found in the Champlain Valley. Shelf marginal elements are well exposed in a number of thrust sheets.

Beginning in earliest Early Ordovician time, a series of thick flysch units were deposited on shelf strata. Each successive flysch depocenter lies west of the former depocenter. This westward progression has been cited (Rodgers, 1971; Bradley, 1989; González-Bonorino, 1990) as evidence for the arrival of the Taconian orogenic wedge, with the flysch considered to be foreland-basinal deposits. We believe that this progression could just as easily be attributed to westward migration of the shelf margin in response to the eustatic events recognized in western Vermont, most of which have already been documented in the western Newfoundland shelf strata (Barnes, 1984; James and others, 1989; Knight and others, 1991).

As in the Champlain Valley, foundering of large portions of the outer carbonate platform occurred during Early and Middle Cambrian time, but not during Late Cambrian or early Early Ordovician time. Large-scale shelf loss resumed in late Early Ordovician time. The carbonate-to-pelite transitions appear to have occurred at the same time over large areas; shelf loss and carbonate-to-pelite transitions correlate with recognized eustatic events. Rapid subsidence in excess of normal passive margin subsidence begins in late Early Ordovician time (Jeffersonian-Cassinian) and continued until the shelf foundered. Final foundering occurred in late Whiterock time (early Middle Ordovician), and definitive synorogenic flysch deposition began shortly thereafter during Chazy time.

Quebec

Except for the continuation of Champlain Valley strata into southwestern Quebec, the only significant exposures of the carbonate bank in Quebec are found on Anticosti Island and along nearby coastlines of the mainland; Upper Cambrian sandstones are found several places along the Laurentian front. Anticosti Island contains Mid-Upper Ordovician strata that offer little for the present analysis. However, Lower Ordovician carbonates along the north shore of the St. Lawrence northeast of Mingan contain clear evidence of erosional and karstic features (Desrochers and James, 1988). These features apparently developed during subaerial exposure in Whiterock time (early Middle Ordovician), which correlates with a major eustatic event recognized in the Champlain Valley (Chisick and others, 1987), Newfoundland (Knight and others, 1991), and throughout the entire Appalachians. This event is generally considered to be the real "Knox Unconformity" (Mussman and others, 1988).

Mohawk Valley

The Mohawk Valley of New York was the first area where Cambro-Ordovician strata were studied in any detail (Eaton, 1824) and has subsequently been studied by many of the most eminent geologists. Nevertheless, significant questions remain unresolved.

Continuing attempts to link the Mohawk Valley strata to the Champlain Valley strata has hampered stratigraphic investigations and caused confusion. The two sequences are lithologically distinct, and the Mohawk Valley sequence (*sensu stricto*) spans a much shorter time period. Recent detailed paleontologic work (Mazzullo and others, 1978) across the boundary between the two sequences (the Kayaderosseras break) has shown that the stratigraphic boundaries in the two sequences are coeval. After the Middle-Upper Cambrian Potsdam sandstone, however, there is no lithofacies continuity across the break despite virtually complete internal continuity within the sequences on either side.

The Champlain Valley and Mohawk Valley sequences differ greatly in the amount of sand, with Mohawk Valley strata containing very little sand as compared with Champlain Valley strata. The timing of diagenetic, karstic, and erosional events, however, is the same for both sequences.

The foundering of the Mohawk Valley portion of the carbonate shelf occurred quite abruptly in late Early Ordovician time (mid-Jeffersonian) (Fisher, 1966; Mazzullo, and others, 1978). Except in western end of the valley where carbonate deposition continued into Late Ordovician time, shales directly overlie the Lower Ordovician carbonates; Berry (1988) has interpreted these shales to be slope-rise equivalents rather than synorogenic flysch.

Along the north side of valley, there are several places where significant sections of carbonate are absent and shales directly overlie older strata or even Precambrian basement. These relations are often revealed adjacent to faults associated with the current uplift of the Adirondacks. This correlation led prior workers (Fisher, 1979; Jacobi, 1981; Rowley and Kidd, 1981; Bradley and Kidd, 1991) to interpret these as sites of localized erosion around normal faults (reactivated by Adirondack doming) developed on the foreland bulge of the Taconic orogen. We believe, however, that these mostly represent local indentations in the northern margin of the Mohawk Valley shelf segment along the edge of the Kayaderosseras paleocanyon, with only some minor syntectonic (Late Ordovician) normal faulting (Mehrtens, 1988). Corroborative evidence for the proximity of the canyon is provided by the unique carbonate sequence at Wells in the southern Adirondacks, which appears to represent carbonates deposited within the canyon during successive eustatic lowstands.

Central Appalachians

The Cambro-Ordovician shelf sequence of the central Appalachians in Pennsylvania has been mapped and studied by some of the best geologists of the last two centuries. Most of the outcrop is found in the Great Valley, but Cambro-Ordovician strata also emerge within the Juniata culmination. The carbonate strata within the Juniata culmination extend into Late Ordovician time, but the carbonate-pelite transition occurred much earlier elsewhere (Berg and others, 1986). Areal and temporal distribution of the transition suggests that the central Appalachian shelf also experienced abrupt shelf-edge regressions. In particular, the differences in timing of the transition between eastern (earlier) and central (later) Pennsylvania has caused

Rodgers (1971) and others to consider the Taconian orogenic effects diachronous, successively younger to the west. There are also indications of two paleocanyons and concomitant stratigraphic discontinuities.

The carbonate-pelite transitions in the Great Valley all occur at or prior to the Knox unconformity (Middle Ordovician). Northeast of the Lehigh River, the transition occurred during late Early Ordovician time, possibly in the Jeffersonian (Hobson, 1963). The thick section of Martinsburg in the vicinity of the Lehigh River correlates with a break in the stratigraphic sequence and an abrupt change in structural style; we interpret this to mark a paleocanyon.

Between the Lehigh River and Hershey, the carbonate transition is at or just below the Knox unconformity (Hobson, 1957, 1963). Neither this area nor area east of the Lehigh River have significant thicknesses of sand; the only sands are calcarenites and thin quartz sands of limited areal extent. The western edge of this segment correlates with abrupt stratigraphic and structural discontinuities across a zone running from Hershey to Bloomsburg. Local structural complications obscure whether the area contains a thicker section Martinsburg, but all other indicators suggest the presence of a paleocanyon.

The Juniata culmination has at least two transition levels. Carbonate deposition in the center of the culmination did give way to shales until early Trenton time (Berg and others, 1986). As in the central Champlain Valley, this sequence contains several thick sands. In strata along the Susquehanna river, however, the transition is no later than the Knox unconformity. Again, the structure mimics the stratigraphic change. The late lateral shelf margin is marked by a thick section of Martinsburg shale (a.k.a. Reedsville) involved in chevron folds at the transition. There is similar evidence of a late shelf margin along the southwest side of the culmination. Continuing southwest, the transition occurs no later than the Knox unconformity (Read, 1989).

Farther to the southwest, the transition becomes even older (Mussman and others, 1988; Hardie, 1989); whether this change represents another shelf margin or another paleocanyon has not yet been determined. Differences in the timing of the carbonate-pelite transition are to be found throughout the central and southern Appalachians, and there are several additional features which we would suggest mark paleocanyons. In eastern Tennessee, where the shelf and subsequent pelitic rocks have been studied extensively, the shelf foundered in late Early Ordovician time. Paleobathymetry of the pelitic sequence indicate rapid subsidence during the early Whiterock, slow subsidence during late Whiterock and Chazy, and the arrival of east-derived (orogenic) sediments in the late Chazy to Trenton (Shanmugam and Walker, 1978, 1980; Benedict and Walker, 1978).

DISCUSSION

Cambro-Ordovician shelf strata in the northern and central Appalachians exhibit many of features which would be expected on a carbonate shelf that had experienced intermittent eustatic

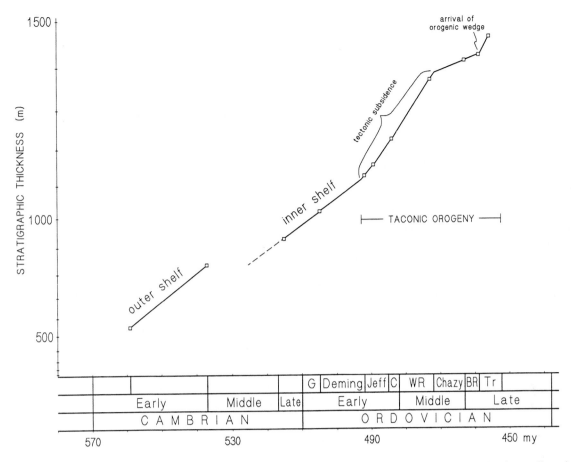

FIG. 7—Subsidence vs. time for the Cambro-Ordovician shelf sequence in the Champlain Valley where thickness has been adjusted for normal subsidence of a passive margin. Note that this representation clearly shows the beginning of tectonic subsidence. Squares represent stratigraphic horizons (generally formational boundaries) that can be accurately dated.

lowstands throughout its evolution. Among these features are distinct lithostratigraphic sequences developed over large areas with sharp, consistent boundaries between adjoining sequences, shelf edge step-backs occurring early and late in the history of the shelf, and regionally correlatable erosional, karstic, and diagenetic events. The problem, therefore, becomes distinguishing the effects produced by the onset of tectonism from the effects produced by the intermittent eustasy.

The stratigraphic record indicates at least five distinct major eustatic events during Early and Middle Ordovician time. Nowhere in the northern and central Appalachians was there significant shelf loss during the first of these events (early Demingian). Minor shelf loss accompanied the second event (earliest Jeffersonian) at several places along the shelf. Both of these events are marked by effects indicative of a significant eustatic event. The first was accompanied by a major diagenetic event (the most important diagenetic event to affect the Cambro-Ordovician section) in the Champlain and Mohawk Valleys, and there are signs of major erosion and solution. Based on subsequent sand thicknesses, some of the erosional channels were tens of meters deep. The second event was accompanied by only minor, localized diagenesis, but indications of erosion and small-scale karst are common; thicknesses of subsequent sands

indicate that erosional channels were not very deep. It is our interpretation that the first event probably lasted much longer than the second event, so it is interesting that there was virtually no shelf loss during the first event whereas there was noticeable, though minor, shelf loss during the second event.

It is the third event (mid Jeffersonian) that is accompanied by major shelf loss. In fact, most of the shelf loss during Early and Middle Ordovician time happened during this event. Only in the clastic depocenters of the central Champlain Valley and the Juniata culmination (central Pennsylvania) was shelf loss minimal. So, does this indicate the onset of Taconian orogenesis? We believe that this event does indeed signal that orogenic subsidence had begun, but let us examine this question at some length.

The subsidence rate, recorded as stratigraphic accumulation rate, is the key factor to be considered. As can be seen in Fig. 7, there is an abrupt increase in the rate of carbonate accumulation in the Champlain Valley at about this time, suggesting that indeed this is the time of orogenic onset in those areas. Newfoundland and central Pennsylvania experienced coeval increases in subsidence rate, and eastern Tennessee experienced a similar increase in subsidence rate only a few million years later during the Cassinian. In addition, a valley cut into shelf strata in near Dorset, Vermont, (Vermont Valley portion of the Champlain Valley block) is filled with volcanic ash (Thompson, 1967); this ash deposit sits directly atop shelf carbonates (of probable early Jeffersonian age), so it must have been deposited before the shelf was reflooded and subsequent sediments (in this case, slope-rise shales) could blanket the shelf strata. Finally, the erosion, solution, and diagenesis accompanying this event was relatively minor, indicating that this was a relatively short-lived event. Nevertheless, we believe that it lasted longer than the second event based on the somewhat greater thicknesses of subsequent estuarine sands in the Champlain Valley.

The fourth event involved significant erosion (sand thickness exceeds 30 m in the Champlain Valley) and the development of a major paleosol with accompanying minor karst. Although we consider this event to have lasted longer than the third event, there is only minor shelf loss. We attribute the lack of shelf loss to the prior loss of the most rapidly subsiding portions of the margin and a stationary hingeline for the subsidence.

The fifth event was accompanied by loss of much of the remaining shelf. The lowstand has been postulated as lasting 10 my (Knight and others, 1991), a value we would not strongly dispute. During this event, major karst developed on the shelf (Mussman and others, 1988; Desrochers and James, 1988) and all remaining shelf outside of the high-volume clastic depocenters was lost. This event has been called the "Knox Unconformity" because it occurs at the end of Beekmantown (a.k.a. Knox) deposition; the hiatus coincides with a fundamental change in sedimentary character (shelf carbonates to deep-marine pelites) in most areas of the Appalachians, and the stratigraphic level of the underlying sequence varies from place to place (Mussman and others, 1988). As has been stated herein, we interpret the temporal variation of the carbonate-pelite transition to be the result of shelf-edge regression events caused by eustasy rather than the result of an erosional surface as generally claimed by others.

Direct evidence of an overriding orogenic wedge does not appear until Middle Ordovician time (Whiterock) or later. In the central Champlain Valley the arrival of the orogenic wedge is marked by the appearance of small blocks of older strata in the shales atop the Trenton carbonates (Washington, 1992b) and a subsequent coarsening upward succession. In Newfoundland, the arrival is marked by the development of wildflysch and coarse clastics containing evidence of an allochthonous source (James and others, 1989; González-Bonorino, 1990). In the Mohawk Valley and southward, the arrival is marked by the growth of a molasse wedge (Zerrahn, 1978; Keith, 1988) atop the inner-shelf carbonate platform, not the erosional effects invoked by Vollmer and Bosworth (1984), Bradley and Kidd (1991), and many others.

CONCLUSIONS

Stratigraphic accumulation rate and amount of shelf loss provide the best evidence for the timing of orogenic onset in the Cambro-Ordovician shelf of the northern and central Appalachians. Variations in the effects of intermittent eustatic lowstands have caused confusion about the timing of orogenic onset. Foundering of large areas of outer shelf following major eustatic lowstands has been mistaken for the formation of a foreland basin in advance of an orogenic wedge. Interaction between sedimentation, thermal or tectonic subsidence, and eustasy easily explains the loss of shelf.

The Taconic orogeny, generally considered to be diachronous, is first expressed in middle Early Ordovician time (earliest Jeffersonian) by a simultaneous increase in subsidence rates (i.e. sediment accumulation rates) along the entire Cambro-Ordovician shelf from Newfoundland to Pennsylvania; the change in subsidence rate may have occurred slightly later in the southern Appalachians, though no later than latest Early Ordovician (late Cassinian). However, timing of the actual arrival of the orogenic wedge does vary along strike. Arrival of the tectonic wedge occurred first (Whiterock) in Newfoundland, where subsidence associated with wedge arrival cannot be distinguished from the initial tectonic subsidence. In the Champlain Valley, central Pennsylvania, and eastern Tennessee, there was a period with little subsidence between the initial subsidence beginning in middle Early Ordovician time and the arrival of the orogenic wedge in early Late Ordovician time. The arrival of the wedge in central Pennsylvania (and eastern Tennessee) occurred in earliest Late Ordovician time (Black River), whereas the wedge arrived slightly later (Trenton) in the Champlain Valley.

REFERENCES

BALDWIN, B., 1983, Sedimentation rates of the Taconic sequence and the Martinsburg formation: American Journal of Science, v. 283, p. 178-191.

BARNES, C. R., 1984, Early Ordovician eustatic events in Canada, *in* Bruton, D. L., ed., Aspects of the Ordovician System: Oslo, University of Oslo Palaeontologic Contributions, no. 295, p. 51-63.

BENEDICT, G. L. III, AND WALKER, K. R., 1978, Paleobathymetric analysis in Paleozoic sequences and its geodynamic significance: American Journal of Science, v. 278, p. 579-607.

BERG, T. M., McINERNEY, M. K., WAY, J. H., AND MacLACHLAN, D. B., 1986, Stratigraphic correlation chart of Pennsylvania (2nd ed.): Pennsylvania Geological Survey, General Geology Report 75.

BERRY, W. B. N., 1988, Deepkill and Schaghticoke shales: shelf margin remnants?: Socorro, New Mexico Bureau of Mines and Mineral Research Memoir 44, p. 81-85.

BOSWORTH, W., AND CHISICK, S., 1987, The Taconic allochthon frontal thrust: Bald Mountain and Schaghticoke gorge, eastern New York: Boulder, Geological Society of America, Centennial Field Guide-Northeastern Section, p. 141-146.

BOSWORTH, W., AND KIDD, W. S. F., 1985, Thrusts, melanges, folded thrusts and duplexes in the Taconic foreland: Saratoga Springs, New York State Geological Association, 1985 Guidebook, p. 117-147.

BOTT, M. H. P., 1992; Passive margins and their subsidence: Journal of the Geological Society of London, v. 149, p. 805-812.

BRADLEY, D. C., 1989, Taconic plate tectonic kinematics as revealed by fore-deep stratigraphy, Appalachian orogen: Tectonics, v. 8, p. 1037-1049.

BRADLEY, D. C., AND KIDD, W. S. F., 1991, Flexural extension of the upper continental crust in collisional foredeeps: Geological Society of America Bulletin, v. 103, p. 1416-1438.

CADY, W. M., 1968, The lateral transition from the miogeosynclinal to the eugeosynclinal zone in northwestern New England and adjacent Quebec, in Zen, E., and others, eds., Studies of Appalachian Geology: Northern and Maritime: New York, Wiley Interscience Publishers, p. 23-34.

CHEN, J.-Y., 1988, Ordovician changes of sea-level: Socorro, New Mexico Bureau of Mines and Mineral Resources Memoir 44, p. 387-404.

CHISICK, S. A., AND WASHINGTON, P. A., 1989, Evolution of the Cambro-Ordovician shelf in the Champlain valley: Interaction of an evolving carbonate bank with tectonic subsidence and sea-level oscillations (abs.), in Colpron, M., and Doolan, B., eds., Proceedings of the Quebec-Vermont Appalachian Workshop: Burlington, University of Vermont, p. 11.

CHISICK, S. A., WASHINGTON, P. A., AND FRIEDMAN, G. M., 1987, Carbonate-clastic synthems of the middle Beekmantown group in the central and southern Champlain Valley: **Northfield**, New England Intercollegiate Geological Conference, 1987 Guidebook, p. 406-442.

DALRYMPLE, R. W., ZAITLIN, B. A., AND BOYD, R., 1992, Estuarine facies models: conceptual basis and stratigraphic implications: Journal of Sedimentary Petrology, v. 62, p. 1130-1146.

DESROCHERS, A., AND JAMES, N. P., 1988, Early Paleozoic surface and subsurface paleokarst: Middle Ordovician carbonates, Mingan Islands, Quebec, in James, N. P., and Choquette, P. W., eds., Paleokarst: New York, Springer-Verlag Publications, p. 183-210.

DOOLAN, B., 1988, Stratigraphy and structure of the Camels Hump group along the Lamoille River transect, northern Vermont: Vermont Geology, v. 5, p. C1-C33.

DRAKE, A. A., JR., SINHA, A. K., LAIRD, J., AND GUY, R. E., 1989, The Taconic orogen, in Hatcher, R. D., and others, The Appalachian-Ouachita Orogen in the United States: Boulder, Geological Society of America, DNAG, The Geology of North America, v. F-2, p. 101-177.

EATON, A., 1824, A Geological and Agricultural Survey of the District adjoining the Erie Canal, in the State of New York: Albany, Packard & Van Benthuysen, 163 p.

ERDTMANN, B-D., 1986, Early Ordovician eustatic cycles and their bearing on punctuations in early nematophorid (planktic) graptolite evolution, in Walliser, O., ed., Global Bio-Events: Berlin, Springer-Verlag, Lecture Notes in Earth Sciences, v. 8, p. 139-152.

FISHER, D. M., 1966, Mohawk Valley strata and structure, Saratoga to Canajoharie: Albany, New York State Museum and Science Service, Educational Leaflet 18, p. 1-58.

FISHER, D. M., 1979, Folding in the foreland, Middle Ordovician Dolgeville facies, Mohawk Valley, New York: Geology, v. 7, p. 455-459.

FISHER, D. M., 1985, Bedrock geology of the Glens Falls-Whitehall region, New York: Albany, New York State Museum and Science Service, Map and Chart Series 35.

FISHER, D. M., AND MAZZULLO, S. J., 1976, Lower Ordovician (Gasconadian) Great Meadows formation in eastern New York: Geological Society of America Bulletin, v. 87, p. 1443-1448.

FLOWER, R. H., 1964, The foreland sequence of the Fort Ann region, New York: Socorro, New Mexico Bureau of Mines and Mineral Resources Memoir 12, p. 153-161.

FORTEY, R. A., 1984, Global earlier Ordovician transgressions and regressions and their biological implications, in Bruton, D. L., ed., Aspects of the Ordovician System: Oslo, University of Oslo

Palaeontologic Contribution 295, p. 37-50.

GONZALEZ-BONORINO, G., 1990, Early development and flysch sedimentation in Ordovician Taconic foreland basin, west-central Newfoundland: Canadian Journal of Earth Science, v. 27, p. 1247-1257.

HALL, L. M., AND ROBERTS, D., 1988, Timing of Ordovician deformation in the Caledonian-Appalachian orogen, in Harris, A. L., and Fettes, D. J., eds., The Caledonide-Appalachian Orogen: London, Geological Society of London, Special Publication 38, p. 291-309.

HARDIE, L. A., 1989, Cyclic platform carbonates in the Cambro-Ordovician of the central Appalachians: American Geophysical Union, IGC Field Trip T161, p. 51-88.

HARLAND, W. B., ARMSTRONG, R. L., COX, A. V., CRAIG, L. E., SMITH, A. G., AND SMITH, D. G., 1990, A Geologic Time Scale 1989: Cambridge, Cambridge University Press, 263 p.

HOBSON, J. P., 1957, Lower Ordovician (Beekmantown) succession in Berks County, Pennsylvania: Bulletin of the American Association of Petroleum Geologists, v. 41, p. 2710-2722.

HOBSON, J. P., 1963, Stratigraphy of the Beekmantown group in southeastern Pennsylvania: Pennsylvania Geological Survey Bulletin G-37, 331 p.

JACOBI, R. D., 1981, Peripheral bulge— a causal mechanism for the Lower/Middle Ordovician unconformity along the western margin of the northern Appalachians: Earth and Planetary Science Letters, v. 56, p. 245-251.

JAMES, E., 1820, Observations on the geology of a part of the state of Vermont, the shores of Lake Champlain, St. Johns River, and Montreal: The Plough Boy, v. 1, p. 250.

JAMES, N. P., STEVENS, R. K., BARNES, C. R., AND KNIGHT, I., 1989, Evolution of a Lower Paleozoic continental-margin carbonate platform, northern Canadian Appalachians, in Crevello, P. D., and others, eds., Controls on Carbonate Platform and Basin Development: Tulsa, Society of Economic Paleontologists and Mineralogists Special Publication 44, p. 123-146.

KEITH, B. D., 1988, Regional facies of Upper Ordovician series of eastern North America, in Keith, B. D., ed., The Trenton Group (Upper Ordovician Series) of Eastern North America: Deposition, Diagenesis, and Petroleum: Tulsa, American Association of Petroleum Geologists, AAPG Studies in Geology 29, p. 1-16.

KNIGHT, I., JAMES, N. P., AND LANE, T. E., 1991, The Ordovician St. George unconformity, Northern Appalachians: the relationship of plate convergence at the St. Lawrence promontory to the Sauk/Tippecanoe sequence boundary: Geological Society of America Bulletin, v. 103, p. 1200-1225.

MAZZULLO, S. J., AGOSTINA, P., SEITZ, J. N., AND FISHER, D. M., 1978, Stratigraphy and depositional environments of the Upper Cambrian-Lower Ordovician sequence, Saratoga Springs, New York: Journal of Sedimentary Petrology, v. 48, p. 99-116.

MEHRTENS, C. J., 1985, The Cambrian platform in northwestern Vermont: Vermont Geology, v. 4, p. E1-E21.

MEHRTENS, C. J., 1988, Bioclastic turbidites in the Trenton limestone: significance and criteria for recognition, in Keith, B. D., ed., The Trenton Group (Upper Ordovician Series) of Eastern North America: Deposition, Diagenesis, and Petroleum: Tulsa, American Association of Petroleum Geologists Studies in Geology 29, p. 87-112.

MEHRTENS, C., 1989, Evolution of the Cambrian platform and platform margin in northwestern Vermont, in Colpron, M., and Doolan, B., eds., Proceedings of the Quebec-Vermont Appalachian Workshop: Burlington, University of Vermont, p. 8-10.

MEHRTENS, C., AND HILLMAN, D., 1988, The Rockledge formation: A Cambrian slope apron deposit in northwestern Vermont: Northeastern

Geology, v. 10, p. 287-299.

MULLINS, H. T., DOLAN, J., BREEN, N., ANDERSEN, B., GAYLORD, M., PETRUCCIONE, J. L., WELLNER, R. W., MELILLO, A. J., AND JURGENS, A. D., 1991, Retreat of carbonate platforms: response to tectonic processes: Geology, v. 19, p. 1089-1092.

MUSSMAN, W. J., MONTANEZ, I. P., AND READ, J. F., 1988, Ordovician Knox paleokarst unconformity, Appalachians, *in* James, N. P., and Choquette, P. W., eds., Paleokarst: New York, Springer-Verlag Publications, p. 211-228.

RANKIN, D. W., DRAKE, A. A., JR., GLOVER, L. III, GOLDSMITH, R., HALL, L. M., MURRAY, D. P., RATCLIFFE, N. M., READ, J. F., SECOR, D. T., JR., AND STANLEY, R. S., 1989, Pro-orogenic terranes, *in* Hatcher, R. D., and others, eds. The Appalachian-Ouachita Orogen in the United States: Boulder, Geological Society of America, DNAG, The Geology of North America, v. F-2, p. 7-100.

RAST, N., STURT, B. A., AND HARRIS, A. L., 1988, Early deformation in the Caledonian-Appalachian orogen, *in* Harris, A. L., and Fettes, D. J., eds., The Caledonide-Appalachian Orogen: London, Geological Society of London, Special Publication 38, p. 111-122.

READ, J. F., 1989, Controls on evolution of Cambro-Ordovician passive margin, U.S. Appalachians, *in* Crevello, P. D., and others, eds., Controls on Carbonate Platform and Basin Development: Tulsa, Society of Economic Paleontologists and Mineralogists Special Publication 44, p. 147-165.

RODGERS, J., 1970, The Tectonics of the Appalachians, New York, Wiley Interscience Publishers, 271 p.

RODGERS, J., 1971, The Taconic orogeny: Geological Society of America Bulletin, v. 82, p. 1141-1178.

ROWLEY, D. B., AND KIDD, W. S. F., 1981, Stratigraphic relationships and detrital composition of the medial Ordovician flysch of western New England: Implications for the tectonic evolution of the Taconic orogeny: Journal of Geology, v. 89, p. 199-218.

SCHWAB, F. L., NYSTUEN, J. P., AND GUNDERSON, L., 1988, Pre-Arenig evolution of the Appalachian-Caledonide orogen: Sedimentation and stratigraphy, *in* Harris, A. L., and Fettes, D. J., eds., The Caledonide-Appalachian Orogen: London, Geological Society of London, Special Publication 38, p. 75-91.

SHANMUGAM, G., AND WALKER, K. R., 1978, Tectonic significance of distal turbidites in the Middle Ordovician Blockhouse and Lower Sevier formations in east Tennessee: American Journal of Science, v. 278, p. 551-578.

SHANMUGAM, G., AND WALKER, K. R., 1980, Sedimentation, subsidence, and evolution of a foredeep basin in the Middle Ordovician, southern Appalachians: American Journal of Science, v. 280, p. 479-496.

SHAW, A., 1958, Stratigraphy and structure of the St. Albans area, northwestern Vermont: Geological Society of America Bulletin, v. 69, p. 519-567.

SLEEP, N. H., 1971, Thermal effects of the formation of Atlantic continental margins by continental breakup: Geophysical Journal of the Royal Astronomical Society, v. 24, p. 325-350.

STANLEY, R. S., AND RATCLIFFE, N. M., 1985, Tectonic synthesis of the Taconic orogeny in western New England: Geologial Society of America Bulletin, v. 96, p. 1227-1250.

THOMPSON, J. B., JR., 1967, Bedrock geology of the Pawlet quadrangle, Vermont: part II, eastern portion: Vermont Geological Survey, Bulletin 30, p. 61-98.

VOLLMER, F. W., AND BOSWORTH, W., 1984, Formation of melange in a foreland basin overthrust setting: Example from the Taconic orogen: Geological Society of America Special Paper 198, p. 53-70.

WASHINGTON, P. A., 1982, A revision of the stratigraphy of the upper limestone sequence near Middlebury, Vermont: Northeastern Geology, v. 4, p. 81-84.

WASHINGTON, P. A., 1985, Roof penetration and lock-up on the leading imbricate of the Shoreham, Vermont, duplex (abs.): Geological Society of America Abstracts with Programs, v. 17, p. 68.

WASHINGTON, P. A., 1987a, The thickness of the Middlebury limestone: Green Mountain Geologist, v. 14, no. 1, p. 12-15.

WASHINGTON, P. A., 1987b, Cleavage vs. folding vs. thrusting: Relative timing of structural events in the central Champlain Valley: Northfield, New England Intercollegiate Geological Conference, 1987 Guidebook, p. 351-368.

WASHINGTON, P. A., 1992a, Stratigraphic relations in the central Champlain Valley: Recurring estuarine deposits and a retreating shelf margin: Shoreham, Field Conference of Taconic Geologists, 1992 Guidebook, p. 1-47.

WASHINGTON, P. A., 1992b, Structural transitions: Southern Champlain thrust to northern Taconics, central Champlain Valley: Shoreham, Field Conference of Taconic Geologists, 1992 Guidebook. p. 48-80.

WASHINGTON, P. A., AND CHISICK, S. A., 1986, The Beekmantown Group in the central Champlain Valley: Vermont Geology, v. 5, p. F1-F17.

WASHINGTON, P. A., AND CHISICK, S. A., 1989, Medium duration lowstands of sea-level recorded in Cambrian and Ordovician strata of the Champlain Valley (VT & NY): Possible evidence of mantle plume activity (abs.): Geological Society of America Abstracts with Programs, v. 21, p. A-96.

ZEN, E., 1968, Nature of the Ordovician orogeny in the Taconic area, *in* Zen, E., and others, eds., Studies of Appalachian Geology: Northern and Maritime: New York, Wiley Interscience Publishers, p. 129-139.

ZERRAHN, G. J., 1978, Ordovician (Trenton to Richmond) depositional patterns of New York State, and their relation to the Taconic orogeny: Geological Society of America Bulletin, v. 89, p. 1751-1760.

TECTONIC CONTROL ON FORMATION AND CYCLICITY OF MAJOR APPALACHIAN UNCONFORMITIES AND ASSOCIATED STRATIGRAPHIC SEQUENCES

FRANK R. ETTENSOHN

Department of Geological Sciences, University of Kentucky, Lexington, KY 40506-0053

ABSTRACT: Recently developed flexural models suggest that lithospheric responses to early craton-margin orogenies should result in at least one major unconformity, both within and beyond the foreland basin, as well as a distinct sequence of lithologies largely restricted to the foreland basin. Consequently, interpretation of tectonic origin is based on (1) the presence of a distinctive overlying flexural sequence, (2) the coincidence of unconformity formation with the inception of established orogenies or tectophases therein, and (3) the distribution of unconformities relative to probable loci of tectonism.

Based on the above criteria, ten of the 13 major interregional and regional unconformities in the Appalachian basin appear to reflect major tectonic control, one is uncertain, one is largely eustatic in origin, and one probably reflects some combination of tectonic and eustatic control. Eight of the ten tectonically related surfaces are concurrent with the initiation of tectophases in the Taconian, Salinic and Acadian orogenies, whereas the other two reflect overlap of Mississippian Ouachita flexural events into extreme southern parts of the Appalachian basin. A widespread Early Pennsylvanian unconformity probably coincides with the initiation of the Alleghanian orogeny but lacks the anticipated overlying flexural stratigraphic sequence. Although this surface could reflect major eustatic influence, the differences in the accompanying sequence might just as likely result from the different style of tectonism accompanying this late-stage orogeny. The only certain, largely eustatically derived unconformity in the Appalachian basin appears to be that at the Ordovician-Silurian boundary, and even it bears some overprint of Taconian tectonic influence. However, a Middle Mississippian unconformity may represent some combination of eustatic lowering and relaxational bulge movement.

Inasmuch as typical Appalachian interregional or regional unconformities recur repeatedly in Paleozoic rocks both within and beyond the Appalachian basin, they must be considered cyclic. The cycles, however, are irregular and appear to be largely related to concurrent phases of tectonism.

INTRODUCTION

In the ongoing discussion about the predominance of tectonism vs. eustasy in explaining patterns of relative sea-level history as expressed in stratigraphy, sedimentology, and paleoecology, the significance of unconformities commonly seems to be minimized or ignored. This is surprising because unconformities are generally integral parts of cyclic sequences on all scales; yet it is the cyclic lithologies that have garnered more attention. If we understand the word "cyclic" to mean recurrence in the same order several times at more or less irregular intervals (Gary and others, 1972), then some unconformities and their bounded sediment packages are clearly cyclic.

The idea of recurrent unconformities found its first major expression in Sloss's work (1963) on cratonic sequences. Although Sloss largely dealt with the larger North American interregional unconformities, the theme of recurrence is also implicit in other work on regional unconformities (Ham and Wilson, 1967). Logically, the *relative* sea-level fluctuations that accompany the development of such unconformities must reflect control by tectonic or eustatic mechanisms or some combination thereof. The "combination" option is probably the best explanation, but most workers suggest that one of the mechanisms typically predominates.

The predominance of eustasy has been championed by many (e.g., Dennison and Head, 1975; Johnson and others, 1985; Boswell and Donaldson, 1988) and has gained new adherents with the resurgence of Milankovitch mechanisms. If the conclusions of Matthews (1987) are correct, continental glaciation has existed for at least 90 percent of Phanerozoic time and could have controlled most sea-level fluctuations. However, as Sloss (1984) has pointed out, the amplitudes of most vertical fluctuations required for unconformity formation "exceed the maxima attributable to any eustatic mechanism," and the durations of these eustatic fluctuations are too short to account for the intervals between major recurrent unconformities. Hence, other workers (Sloss and Speed, 1974; Ettensohn, 1985, 1991) have emphasized the likely predominance of tectonic control. Moreover, the fact that most interregional and regional unconformities formed concurrently with the inception of major tectonic events on craton margins (Ham and Wilson, 1967; Johnson, 1971; Sloss and Speed, 1974; Sloss, 1984) is another factor supporting the importance of tectonic control.

Possible mechanisms, however, by which craton-margin tectonism influenced the development of interregional and regional unconformities were largely uncertain until the advent of lithospheric-flexure modeling (Beaumont, 1981; Jordan, 1981; Quinlan and Beaumont, 1984). Although flexural modeling may have different lithospheric responses depending upon the relative rheology assumed for the lithosphere (Quinlan and Beaumont, 1984), comparison of model predictions with sedimentary sequences in the Appalachian basin seems to support best the loading and unloading of a temperature-dependent viscoelastic lithosphere as modeled by Quinlan and Beaumont (1984), Beaumont and others (1987, 1988) and Jamieson and Beaumont (1988).

These models suggest that an unconformity of predominantly tectonic origin is only one part of a predictable set of stratigraphic responses to a series of flexural events associated with an ongoing orogeny. Hence, if an unconformity can be shown to be part of such a stratigraphic response, then the unconformity is probably of largely tectonic origin. Logically, unconformity formation should also coincide with a period of orogeny, and the unconformity should exhibit a distribution pattern commensurate with the extent of flexural uplift and orogeny. Application of these three criteria to major Appalachian unconformities should then provide some indication as to the likelihood of tectonic origin.

Before examining the unconformities, however, it is important to understand their flexural origin and the origin of associated stratigraphic sequences. Basic models and their stratigraphic implications are outlined below.

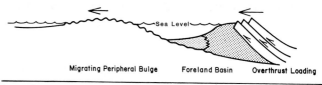

Fig. 1.—Schematic diagram showing development of a foreland basin, peripheral bulge, and bulge-related unconformity due to deformational loading in the orogen. Sediments in foreland basin contribute to loading and in distal parts of the basin during the initiation of deformation will largely be organic-rich muds (adapted from Quinlan and Beaumont, 1984).

BASIC MODELS

One major component of orogeny is overthrusting due to crustal foreshortening. Flexural models (Quinlan and Beaumont, 1984; Beaumont and others, 1988) suggest that once thrusting has commenced and thrust blocks begin moving cratonward, the stage is set for a sequence of four flexural events that will give rise to at least one major unconformity and an ensuing sedimentary sequence (Ettensohn, 1991). During early stages of an orogeny, surface (thrusts, nappes, folds) and subsurface (buried, obducted blocks, flakes) loads will accumulate on the cratonic margin severely loading the lithosphere. As a result of isostatic compensation by the lithosphere, the adjacent craton will deform into a downwarped flexural or retroarc foreland basin just cratonward of the orogen and an uplifted peripheral bulge on the distal (cratonward) margin of the basin (Fig. 1). Most of the loading is produced by stacking of thrust sheets, but a subordinate component is attributable to sediment loading (Beaumont, 1981; Tankard, 1986). As long as foreshortening continues and the thrust sheets continue to migrate cratonward, the peripheral bulge and foreland basin will also migrate cratonward. Although most of the loading and the accompanying basin-and-bulge migration will progress cratonward in a direction perpendicular to the strike of the orogenic belt, if the orogeny is diachronous along its length, deformational loading and attendant basin-and-bulge migration will also shift parallel to the strike of the orogenic belt (Ettensohn, 1987).

Bulge Movement and Unconformity Development

The first flexural event accompanying the inception of major loading is bulge moveout which results in widespread uplift of the foreland and generation of a regional unconformity (Fig. 1) (Quinlan and Beaumont, 1984). If we assume a constant sea level, erosion along the migrating bulge may reflect subaerial exposure or elevation into agitated wave base. According to Beaumont (pers. commun., 1991), the width of a bulge may be one to two times the width of its adjacent foreland basin, and the stresses responsible for bulge formation and migration may be

transmitted through continental lithosphere across distances of up to 1300 km and perhaps farther (Karner and Watts, 1983; Ziegler, 1987).

Clearly this kind of mechanism could easily account for many regional unconformities, but it may not be able to explain fully the continent-wide, interregional unconformities that bound the sequences of Sloss (1963). Each of these unconformities formed at the beginning of significant episodes of tectonism which apparently marked periods of major plate reorganization. The initial response of the continent at these times was apparently one of impedance to subduction or collision, and as a result much of the continent abruptly flexed upward. Sloss and Speed (1974) called these periods emergent episodes and suggested that they reflected time spans of a few million years. Dickinson (1974, p. 22) called such a response "braking" inasmuch as subduction is initially retarded or braked by the inception of convergence. Such unconformities may even develop throughout much of a foreland basin. Although the physics behind this mechanism is unclear, if these continents were largely surrounded by subduction zones as suggested by Scotese (1990), plate reorganization begun on one margin may have conditioned nearly synchronous tectonism and associated flexure on other margins as well.

Although the distribution of interregional unconformities may be nearly continent-wide (Sloss, 1963), the regional unconformities exhibit distributions more localized to parts of the craton and foreland basin adjacent to the locus of tectonism. Generally, the distribution of such unconformities will approximately parallel the strike of the associated orogeny. Moreover, because much Appalachian tectonism was apparently localized near continental promontories which were subject to greater shortening and resulting deformation (Dewey and Burke, 1974; Dewey and Kidd, 1974; Ettensohn, 1985, 1991), the distribution of both regional and interregional unconformities in the Appalachian basin is commonly asymmetric toward the involved promontories, even within a foreland basin (Ettensohn, 1990). However, in parts of the foreland basin most proximal to those promontories, neither interregional nor regional unconformities generally develop. Because deformation is so intense and persistent at the promontories, subsidence associated with

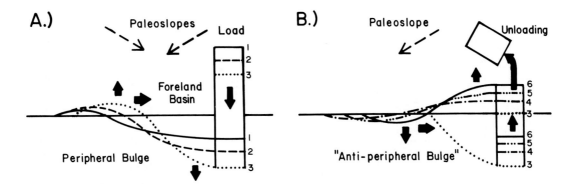

Fig. 2.—Schematic diagram showing two types of flexural response to lithospheric-stress relaxation (redrawn from Beaumont and others, 1988): (A) "loading-type" relaxation—thrust migration ceases and the resulting static gravitational load results in a deepening foreland basin and migration of peripheral bulge toward the load; (B) "unloading-type" relaxation—erosional unloading results in rebound near the unloaded area and an "anti-peripheral bulge" or peripheral sag that deepens and migrates toward the former load.

deformational loading in immediately adjacent parts of the foreland basin is commonly sufficient to offset the effects of bulge uplift or braking. As a result, times of major interregional-unconformity development (Early-Middle Ordovician, Early Devonian, and Early Pennsylvanian) are generally represented by conformable sequences (see Patchen and others, 1985a, b) in parts of foreland basins just cratonward of the promontories.

Foreland-Basin Formation and Black-Shale Deposition

Subsidence of the foreland basin immediately follows bulge moveout and represents an isostatic response to deformational loading in the adjacent orogeny. This loading generally represents a transfer of mass from one plate to another, but during the early Appalachian orogenies (Taconian through Acadian), much of this transfer apparently occurred along a steep basement ramp formed by the eastern rifted margin of the continent. It is likely that this ramp could have accommodated up to 20 km of vertical deformation without creating major subaerial topography or source areas (Jamieson and Beaumont, 1988), so that at least initially, much of this deformation must have occurred in the subsurface in the form of buried, obducted blocks and flakes (Karner and Watts, 1983; Jamieson and Beaumont, 1988). Because no major source of externally derived sediment is available during early phases of such orogeny, the adjacent foreland basin experiences sediment starvation. In the absence of major clastic influx, organic matter from the water column and suspended clays and silt compose most of the sediment, and because the foreland basin is undergoing rapid subsidence with which sedimentation cannot keep pace, the water column soon becomes stratified so that the organic matter is quickly preserved as dark or black shale in the resulting anoxic environments. Hence, the bulge-induced unconformity is typically overlain abruptly by black shales, although a transgressive carbonate or clastic sequence or a condensed succession may intervene. While dark shales predominate in central parts of the basin, transgressive-carbonate deposition may persist atop the unconformity in distal parts of the basin and on the bulge because of

reduced subsidence away from loading and elevation into agitated waters along the bulge (Quinlan and Beaumont, 1984).

Although dark-shale sedimentation may predominate in foreland basins during the initial phases of early continental-margin orogenies, it will not necessarily characterize foreland-basin deposits during initial phases of the later orogenies because of changing tectonic regimes. In fact, during later orogenies, the deformational front is likely to have surmounted the continental margin and developed sufficient topography to advance substantially onto the craton as a surficial load (Jamieson and Beaumont, 1988). Because these more expansive thrusts spread their load over a greater area, the resulting lithospheric flexure will generate a broader, shallower foreland basin (Karner and Watts, 1983), and this basin widening may be enhanced by the increasing flexural rigidity of the lithosphere with time (Watts and others, 1982). Because of the shallowness of such basins, their distance from the sea, and their proximity to sources of coarse clastic sediments, these basin fillings may consist largely of terrestrial clastic sediments, bearing little resemblance to the marine flexural sequence developed in the following parts of this "Basic Models" section.

Loading-type Relaxation and Deep-water Basin Infill

Dark shales predominate during the most active phases of tectonism, but the major influx of coarser clastic sediments typically does not begin until an ensuing episode of tectonic quiescence. During tectonic quiescence, active deformation and thrust migration cease, and the deformational load becomes static. The lithosphere responds to the static load by relaxing stress so that the foreland basin deepens and narrows while the peripheral bulge is uplifted and shifts toward the load (Fig. 2A). By this time, substantial relief has been generated by emplacement of a surface load (fold-thrust belt), and surface drainage nets have had adequate time to develop. As a result, coarser clastic debris is eroded and transported into the foreland basin in the form of deeper water deltaic deposits, turbidites and debris flows. Nearshore clastic sediments may also be redistributed as

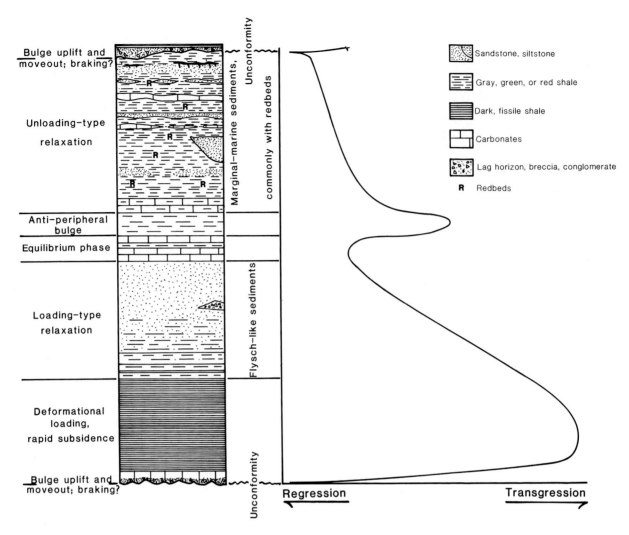

FIG. 3.—Generalized sequence of flexural events, accompanying flexural sequence of lithologies, and transgressive-regressive curve for early subduction-type orogenies on a continental margin. The four major parts of the sequence include unconformity formation due to bulge uplift and moveout, dark-shale deposition reflecting rapid subsidence accompanying active deformational loading, deposition of flysch-like clastics accompanying loading-type relaxation, and deposition of marginal-marine sediments with redbeds reflecting unloading-type relaxation. The sequence is bound by unconformities marking the inception of deformational events.

tempestites by storms. While these flysch-like sediments accumulate at increasing rates in the foreland basin, uplift and basinward movement of the adjacent bulge may generate a regressive carbonate sequence or another regional unconformity truncating previously deposited flysch-like sediments, depending upon the relative disposition of sea level (Fig. 2A). Overall, the basin sequence during this phase will be regressive.

Unloading-type Relaxation and Basin "Overflow"

Because the rate of clastic influx to the basin in the preceding phase eventually comes to exceed the rate of basin subsidence, the foreland basin fills with clastic sediments while the orogenic highlands undergo extensive lowering due to subsidence and erosion (Fig. 2B). As a result, a brief period of elevational equilibrium ensues between the filled basin and eroded high-

lands (Fig. 2B). If enough time lapses during this stage, an extensive blanket of shallow-water carbonates or mixed carbonates and shales may be deposited. Normally, this phase is short-lived because the area of the former orogen and foreland basin begin to rebound upward in isostatic response to the lost load, and a compensating "antiperipheral bulge" (Fig. 2B), which might be better called a peripheral sag, forms and moves toward the rebounding area (Beaumont and others, 1988). Although a short-lived transgressive sequence of shallow open-marine carbonates or shales may be deposited in the antiperipheral bulge or peripheral sag (Fig. 3), the overall sedimentary response is a cratonward prograding wedge of marginal-marine and terrestrial sediments which commonly contain redbeds. Because of rebound and the resulting progradation, it appears that the foreland basin has "overflowed" onto the craton. Yet, the foreland basin may experience cannibalization in this process resulting in an unconformity in proximal parts of the basin. Such an unconformity, though, could easily be destroyed or subsumed by erosion accompanying the bulge moveout of a succeeding tectophase or orogeny.

Local Unconformities

Interregional and regional unconformities are not the only ones to be affected by flexural movements. Bulge migration, even during times of tectonic quiescence, may reactivate basement structures (Bradley and Kusky, 1986) or cause faulting in surficial rocks (Knight and others, 1991) resulting in local uplift and the development of coincident unconformities and facies changes. In fact, because upper parts of the lithosphere are extremely viscous and only relax stress over extremely long periods of time (Quinlan and Beaumont, 1984), brittle deformation along basement zones of weakness may actually enhance the progress of flexure in the shallow crust.

Probable examples of this type of movement are present in the Ordovician rocks of eastern New York (Bradley and Kusky, 1986) and maritime Canada (Knight and others, 1991) and in Mississippian rocks near the Kentucky River Fault zone in eastern Kentucky (Dever and others, 1977; Ettensohn, 1981; Ettensohn and others, 1988a), near the Pocono dome in West Virginia (Yielding and Dennison, 1986; Dennison, 1986; Koehler and Smosna, 1989), and elsewhere in the Appalachian basin (Warne, 1990).

Summary

If regional and interregional unconformities in foreland basins and nearby cratonic areas are in fact controlled largely by concurrent tectonism, the suggested flexural scenario, at least for the early orogenies on a continental margin, requires that such an unconformity be succeeded by a distinct sequence of lithologies (Fig. 3). This flexural sequence begins with an unconformity, and is overlain in ascending order by transgressive carbonates or shallow-marine sands, dark shales, a flysch-like clastic sequence, and a sequence of marginal-marine clastic sediments with redbeds (Fig. 3). Moreover, because orogenies progress in pulses or tectophases on the order of five million years or less in duration (Jamieson and Beaumont, 1988), unconformities and their associated sedimentary sequences are generally cyclic during any one orogeny. Every cycle, however, may not exhibit the complete sequence of lithologies because a new tectophase may begin before the sedimentary expression of the previous one is complete or because erosion accompanying new bulge uplift and moveout destroys parts of the previous sedimentary expression. One consistent feature of these cycles is that the sedimentary expression of each successive tectophase cycle during an orogeny migrates farther cratonward than the previous one (Fig. 2A), reflecting the continued cratonward movement of deformation.

Although the distinctive, dark-shale-through-redbed, flexural sequence may be restricted to foreland basins where flexure is greatest, flexure is to some extent transmitted to adjacent parts of the craton as well. As a result, normal cratonic sedimentation may reflect patterns of transgression and regression clearly corresponding in nature and origin to major flexural events in foreland basins. As already mentioned, flexural stresses may be transmitted across distances in excess of 1300 km.

DEVELOPMENT OF MAJOR APPALACHIAN UNCONFORMITIES AND ATTENDANT SEQUENCES

Examination of any Appalachian correlation chart (Patchen and others, 1985a, b) shows that the Appalachian sedimentary sequence is rife with unconformities. A few are obviously interregional in nature, but most are regional and local. Of these, only the interregional and regional unconformities of substantial extent will be scrutinized in following sections. In an effort to determine relative importance of tectonism on unconformity development, each unconformity will be examined to determine if a flexurally related sequence of sedimentary units is associated with the unconformity, if its origin coincided with the inception of a tectonic event, and if unconformity distribution can be related to probable loci of tectonism.

Ordovician

Except for the unconformity generated at the base of the Sauk Sequence as latest Precambrian and Cambrian seas inundated long exposed and eroded Precambrian basement rocks, an interregional unconformity near the Lower-Middle Ordovician boundary (Canadian-Whiterockian; Arenig-Llanvirn) is the first major unconformity in the Appalachian basin. In maritime Canada it is called the St. George unconformity (Knight and others, 1991) and in the United States it has been called the post-Sauk unconformity by Sloss (1963), the Owl Creek Discontinuity by Wheeler (1963), unconformity C by Rodgers (1971), and the sub-Tippecanoe unconformity by Sloss (1988). Its formation broadly coincides with the beginning of what most workers call the Taconian orogeny, the Blountian orogeny or the Blountian phase of the Taconian orogeny (Zen, 1972; Drake and others, 1989).

The Taconian orogeny is generally thought to represent continental margin-island arc collision along an east-dipping subduction zone (Rodgers, 1987; Drake and others, 1989). Although subduction may have begun far east of the continental shelf as early as the Cambrian Period (Pavlides, 1989), actual collision and docking of a deformational load on the continental margin apparently did not occur until the Early-Middle Ordovician transition. When collision finally did occur, nearly the entire craton was uplifted in response. Although part of the resulting unconformity may be explained by bulge uplift and moveout (Jacobi, 1981; Quinlan and Beaumont, 1984), the craton-wide extent of the unconformity is difficult to explain in this way; some other form of craton reorganization such as continental braking or reciprocal tectonic movement on other continental margins, or perhaps eustasy (Mussman and Read,

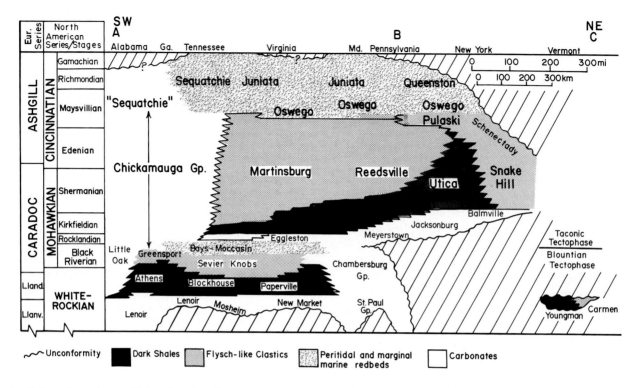

FIG. 4.—Schematic section paralleling the strike of the Appalachian basin showing the repetition and northeastward migration of flexural foreland-basin sequences reflecting the different loci and timing of the Blountian and Taconic tectophases. No vertical scale intended. See Figure 5 for location of this section (from Ettensohn, 1991).

1986) is necessary. In the Appalachian basin, however, development of this unconformity was locally associated with normal faulting (Thompson, 1967; Zen, 1967; Williams and Stevens, 1974; Bradley and Kusky, 1986; Knight and others, 1991) and folding (Thomas, 1968; Lowry, 1974), possible angular relationships (Shaw, 1970; Sloss, 1988), local restriction of erosion to anticlinal axes (Lowry, 1974), and a typical flexural sequence (Fig. 4) of overlying sedimentary units (Ettensohn, 1991), all of which suggest the predominance of a tectonic component.

The associated sedimentary sequence, moreover, is restricted to the Sevier basin and small remnant basins which occur just cratonward of the Virginia and Alabama promontories (Fig. 5) near the loci of the earliest or Blountian tectophase of the Taconian orogeny (Kay and Colbert, 1965; Drake and others, 1989). The underlying unconformity is far more extensive (Fig. 6A) than the resulting flexural sequence in the Sevier basin (Fig. 7), but the greatest relief on this unconformity (several hundred feet) occurs in parts of Virginia and Tennessee (Dennison, 1986) cratonward of the Virginia promontory where bulge uplift would have been greatest. In contrast, the unconformity is lacking in the easternmost, preserved parts of the Appalachian basin (Fig. 6A, arrows) at each of the three promontories where subsidence would have overwhelmed bulge uplift. The unconformity is also absent in central Pennsylvania, Maryland and northern Virginia. This absence is difficult to explain unless it is related to concurrent subsidence near the Fortieth Parallel lineament and related structures (see Dennison, 1982), which

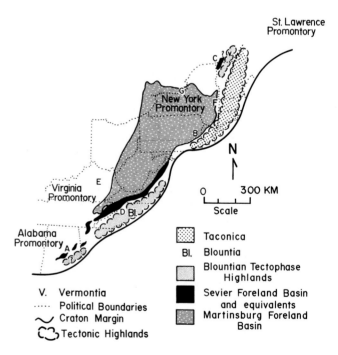

FIG. 5.—Schematic map of central and southern Appalachian area showing relative positions of Taconian tectonic highlands and black-shale foreland basins relative to continental promontories during Middle and Late Ordovician time. Note the northwestward and northeastward shift of the Martinsburg basin relative to the earlier Sevier basin and equivalents. Single letters refer to mid- and end-points of sections in Figures 4, 7, and 8 (from Ettensohn, 1991).

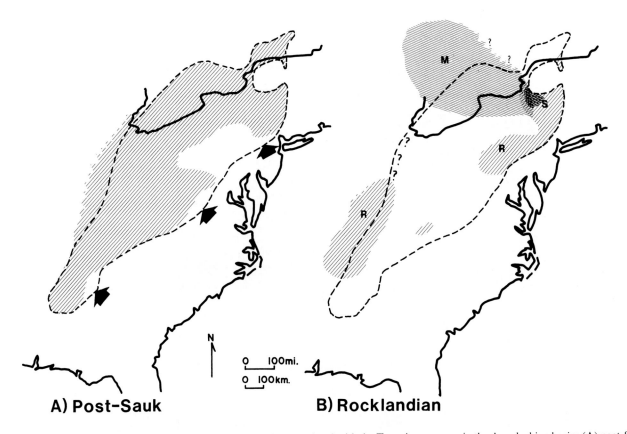

Fig. 6.—Approximate distribution of Ordovician unconformities associated with the Taconian orogeny in the Appalachian basin: (A) post-Sauk unconformity associated with the initiation of the Blountian tectophase; bold arrows point to areas in the basin most proximal to promontories where unconformities are absent; (B) Rocklandian unconformity marking the initiation of the Taconic tectophase; the three apparently related unconformities behind the New York promontory (R=Rocklandian; S=Shermanian; M=Maysvillian) appear to reflect the diachroneity of the bulge movement and the "skipping" of intervals possibly as a result of interactions with basement structural inhomogeneities. See Figure 8 for a sectional view of this area.

define the Pennsylvania reentrant.

The better known part of the Taconian orogeny is called the Taconic tectophase, and it represents the migration of convergence northward from the Virginia promontory and actual docking at the New York promontory (Fig. 5). This progression of events is indicated by the fact that foreland-basin subsidence reflected by the Martinsburg and equivalent black shales began in the Kirkfieldian Stage (late Caradocian) in northeastern Tennessee and advanced northeastward in time (Fig. 3). A Rocklandian (late Caradocian) unconformity marking the beginning of the tectophase is absent or poorly developed south of Pennsylvania (Fig. 6B), probably because sufficient residual loading from the Blountian tectophase remained in the area of the Virginia promontory to offset the effects of new bulge migration (Ettensohn, 1991). However, on the western margin of the Appalachian basin away from the effects of subsidence, the Rocklandian unconformity is again present (Fig. 6B), but the unconformity is best developed just cratonward of the New York promontory (Fig. 6B), the locus of the Taconic tectophase (Kay and Colbert, 1965; Rodgers, 1971; Drake and others, 1989). Here, bulge uplift and erosion were so severe that the Rocklandian unconformity has merged with the post-Sauk unconformity

forming a compound unconformity (Fig. 4). The initial unconformity-transgressive carbonate-black shale sequence is also well developed in the area of the New York promontory (Fig. 4).

As deformation migrated westward with time, so did various units in the flexural sequence (Fig. 7). Especially noticeable is the basinal Utica Shale which becomes younger to the west (Fig. 8). The underlying unconformity shows a similar pattern which probably reflects the time-transgressive nature of bulge migration. The area in central New York (Fisher, 1977) where the unconformity seemingly "skips" certain intervals (Figs. 6B and 8) may represent bulge interaction with basement structural inhomogeneities. The interaction was very localized, however, for the unconformity effectively remains at the same Maysvillian horizon throughout the remainder of its distribution (Fig. 6B) in western New York (Fisher, 1977), Ontario (Twenhofel and others, 1954) and parts of the Michigan basin (Shaver, 1985).

Again, the distribution of this unconformity and its association with a well developed flexural sequence strongly support the predominance of a tectonic component. Most compelling of all, however, is the primary localization of the unconformity near the locus of Taconic tectonism, the New York promontory.

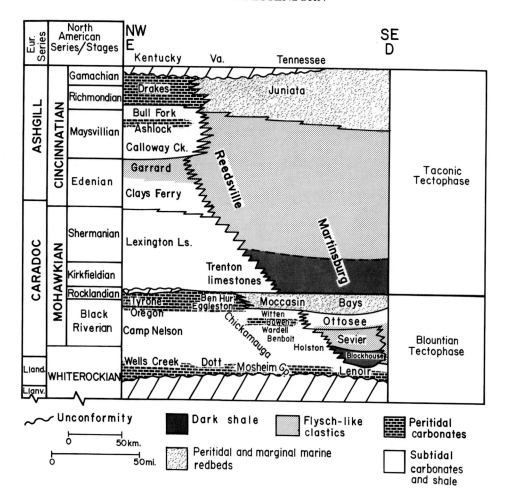

FIG. 7.—Schematic section perpendicular to the strike of the Appalachian basin showing the repetition and relative cratonward migration of flexural foreland-basin sequences during two successive Taconian tectophases. The sequences apparently represent the passage of peripheral and anti-peripheral bulges on the distal margins of the foreland basins during respective tectophases. Crosshatched areas associated with unconformities represent missing section. No vertical scale intended. See Figure 5 for location of this section (from Ettensohn, 1991).

Other lines of evidence supporting the importance of Taconian tectonism in the development of the Ordovician stratigraphic sequence in the Appalachian basin are discussed in detail by Ettensohn (1991).

Ordovician-Silurian Transition

The Ordovician System and the sedimentary sequence resulting from the Taconic tectophase end with the typical marginal-marine redbed sequence in the Appalachian basin (Queenston-Juniata-Sequatchie; Figs. 3, 7, 8), but the sequence is truncated in parts of the Appalachian basin and on adjacent parts of the craton (Figs. 4, 7) by a widespread unconformity called the Cherokee Discontinuity by Dennison and Head (1975). Associated with the unconformity is an underlying redbed complex that is far more extensive than any other Appalachian basin, tectono-stratigraphic redbed sequence (Dennison, 1976, Fig. 4), suggesting widespread sea-level lowering. Moreover, this lowering culminates in a widespread interregional uncon-

formity near the Ordovician-Silurian boundary. The worldwide nature of this lowering (McKerrow, 1979) and its coincidence with glacial deposits in Gondwana, Europe, and North America (Hambrey, 1985; Caputo and Crowell, 1985) strongly indicate glacio-eustatic origin. Although this Late Ordovician-to-Middle Silurian (Caradocian-Wenlockian) glaciation may have lasted between 20 and 35 million years (Hambrey, 1985; Bjorlykke, 1985), the peak occurred during a two-million-year interval in latest Ordovician time (late Richmondian-Gamachian; late Ashgill) when much of the unconformity was generated. The truncation of Ordovician redbeds in the Appalachian basin during and after this event may have been the source for the redbeds and remobilized iron during subsequent Early Silurian (Llandovery) transgressions resulting in the Clinton Group (Ziegler and McKerrow, 1975).

The aerial extent of the Cherokee Discontinuity in and near the Appalachian basin has a distribution (Fig. 9) that suggests some tectonic influence as well. The unconformity distribution shows three prominent eastward salients just cratonward of the

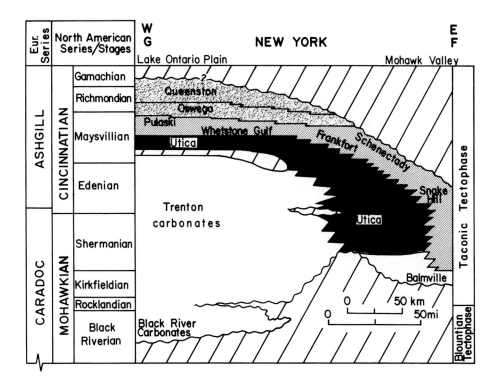

Fig. 8.—Schematic section perpendicular to the strike of the Appalachian basin in New York showing the westward extension of the Martinsburg basin reflected in the Utica Shale and its westward migration in time (after Fisher, 1977). The diachroneity of the Utica and underlying unconformities most likely reflects the diachroneity of basin-and-bulge movement during the Taconic tectophase. No vertical scale is intended; legend same as that on Figure 7. See Figure 5 for location of this section (from Ettensohn, 1991).

New York, Virginia, and Alabama promontories (Fig. 9). Inasmuch as these promontories were the probable loci of Taconian tectophases (Ettensohn, 1991), these are locations where major rebound and concomitant unconformity formation would be expected in the Appalachian basin during unloading-type relaxation (Fig. 2B).

Moreover, the nature of the unconformity at the three salients also supports the likelihood of tectonic influence, at least at the northern two salients. In the eastwardly projecting unconformity salient at the New York promontory, the Ordovician-Silurian boundary is marked by an angular unconformity in the north which becomes a disconformity to the south where Silurian rocks rest on Ordovician rocks as old as Trentonian (Rodgers, 1971; Stephens and Wright, 1981; Liebling and Scherp, 1982). In contrast, at the two isolated unconformity outliers near the Virginia promontory (Fig. 9), erosion apparently truncated all the Upper Ordovician redbeds so that Silurian rocks rest on top of Maysvillian parts of the Martinsburg Formation (Dennison, 1976; Patchen and others, 1985a). In the Alabama salient (Fig. 9), however, the unconformity is generally either a paraconformity or a subtle disconformity involving truncation of Richmondian rocks, not unlike the situation in western parts of the Appalachian basin and on the adjacent craton (Dennison, 1976; Patchen and others, 1985b; Shaver, 1985). The overall pattern in these eastwardly projecting unconformity salients is one of increasing erosional truncation and deformation toward the New York

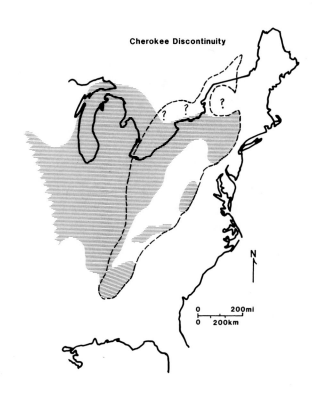

Fig. 9.—Approximate distribution of the Cherokee (Ordovician-Silurian) Discontinuity in the Appalachian basin and adjacent areas.

promontory, where the last and apparently most intense, of the Taconian tectophases had just occurred (Drake and others, 1989; Ettensohn, 1991).

The only place in the study area where the unconformity is not present is in central and south-central parts of the Appalachian basin (Fig. 9) where subsidence outpaced sea-level lowering, possibly because of anti-peripheral-bulge formation (Fig. 2B).

Although distribution patterns, magnitude of erosion and the angularity of the Cherokee Discontinuity in eastern parts of the Appalachian basin (Fig. 9) suggest late-stage, tectonic (rebound?) influence (Quinlan and Beaumont, 1984; Tankard, 1986), elsewhere the subtle nature of the discontinuity, its lack of angularity, and the absence of an overlying Silurian flexural sequence strongly support a glacio-eustatic origin for this discontinuity (Dennison and Head, 1975).

Silurian

In the Appalachian region, the Silurian Period is typically viewed as a time of tectonic quiescence between the Taconian and Acadian orogenies. The stratigraphic record of this period, especially in north and north-central parts of the Appalachian basin, is divided into a number of seemingly cyclic sequences by unconformities (Rickard, 1975; Brett and others, 1990), many of which are commonly attributed to eustasy (Dennison and Head, 1975; Johnson and others, 1985; Dennison, 1986; Brett and others, 1990). The apparently global nature of Early Silurian (Llandoverian) cycles and their coincidence with Gondwana glaciation certainly suggests eustatic control (Johnson and others, 1985; Johnson, 1987). However, a growing body of evidence in the form of volcanism, plutonism, and deformation from the northern and central Appalachians (Laird, 1988; Thirlwall, 1988; Wones and Sinha, 1988; Bevier and Whalen, 1990) suggests that probably by mid-Clinton time (latest Llandoverian), and certainly by late-Clinton time (earliest Wenlockian), some type of Silurian tectonic activity was ongoing. This activity may have coincided with a period of weak Silurian tectonism called the Salinic disturbance (Boucot, 1962; Rodgers, 1970, 1987; Fairbairn, 1971). The disturbance probably represents a weaker, southern phase of the Caledonian orogeny involving convergence between a southern prong of Baltica or separated Avalonian terranes and the New England-Maritime province area (McKerrow and Ziegler, 1972; Scotese and McKerrow, 1990). According to Rodgers (1987), the Salinic disturbance was concentrated near the St. Lawrence promontory, but evidence from the Appalachian basin indicates that the effects of convergence were also felt near the New York and Virginia promontories.

The earliest line of evidence is a regional angular unconformity of mid-Clinton or of late Llandoverian age in New York, central Pennsylvanian and southern Ontario (Rickard, 1975; Brett and others, 1990) and an equivalent disconformity on the adjacent craton (Fig. 10) throughout much of east-central United States (Ham and Wilson, 1967; Shaver, 1985; Patchen and others,

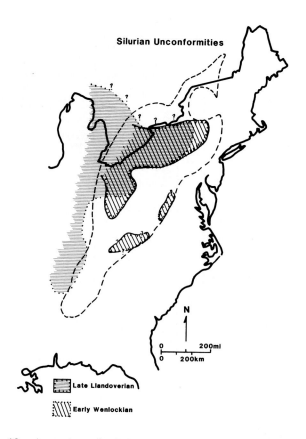

FIG. 10.—Approximate distributions of two Middle Silurian (Clintonian; late Llandoverian and early Wenlockian) unconformities probably related to the Salinic disturbance in the Appalachian basin and adjacent areas. Note how the lobate distribution patterns roughly correspond to promontory positions.

1985b). Not only the angularity and distribution of the unconformity relative to the New York and Virginia promontories (Fig. 10), but also the presence of a partial, overlying flexural sequence (Fig. 11), support the predominance of tectonic influence in the formation of this unconformity. The unconformity is overlain by a condensed section or transgressive limestone in northwestern New York and adjacent parts of Ontario, which in turn grades upward into the dark, laminated Williamson Shale and equivalent units (Eckert and Brett, 1989; Brett and others, 1990, 1991). According to Brett and others (1990), the Williamson represents the deepest water Silurian deposits in the Appalachian basin.

The dark Williamson Shale grades upward into a greenish-gray flysch-like sequence with graptolites and then into an argillaceous, offshore carbonate unit, the Rockway Member of the Irondequoit Limestone (Fig. 11) with a condensed section at its base (Lin and Brett, 1988); the Rockway is unconformably overlain by the upper part of the Irondequoit Limestone (Brett and others, 1990, 1991). The appearance of greenish-gray shales apparently reflects the initiation of loading-type relaxation, whereas the abrupt appearance of the Rockway Member and its underlying condensation zone is most likely related to

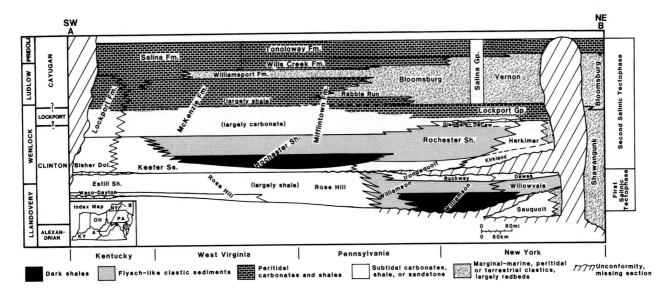

F ɪ ɢ. 11.—Schematic section partly paralleling the strike of the Appalachian basin showing the repetition and southward migration of flexural foreland-basin sequences and accompanying unconformities probably related to the Salinic disturbance.

eastward bulge movement and uplift accompanying loading-type relaxation (Fig. 2A). The concomitant deepening of the foreland basin (Fig. 2A) and its rapid infilling with more "flysch-like" clastics is probably represented by the much thicker, equivalent, gray sandy shales of the Willowvale Formation (Fig. 11) to the east (Lin and Brett, 1988).

Although the main parts of the Salinic disturbance were apparently early Cayugan (Ludlovian) in age (Boucot, 1962; Fairbairn, 1971), this mid-Clinton (late Llandoverian) sequence seems to reflect some sort of early Salinic tectonism, perhaps a tectophase concentrated near the New York promontory. Elsewhere in the northern and central Appalachians, the presence of tectonism at this time is supported by approximately concurrent volcanism, plutonism, and deformation (Laird, 1988; Thirlwall, 1988; Bevier and Whalen, 1990). This partial mid-Clinton (late Llandoverian) sequence and its truncation by a subsequent unconformity (Fig. 11) indicates, however, that complete relaxation was never attained before renewed tectonism intervened. Typically, this late Llandoverian sequence is associated with the fourth of Johnson and others' (1985) Llandoverian eustatic sea-level rises (Lin and Brett, 1988; Eckert and Brett, 1989), but the evidence presented here suggests the sequence and the underlying unconformity are predominantly of tectonic origin, at least in the area behind the New York promontory.

A second unconformity of regional extent developed shortly thereafter in northwestern parts of the Appalachian basin (Fig. 10). This subtle, late-Clinton or early Wenlockian unconformity occurs at the base of the upper Irondequoit and Keefer Formations in central and western New York, central Pennsylvania, southern Ontario and in some places on the adjacent craton (Brett, 1983; Shaver, 1985; Brett and others, 1990, 1991). The unconformity distribution is strongly asymmetric toward the New York and Virginia promontories, especially the New York

promontory (Fig. 10), suggesting that the unconformity was probably related to bulge uplift and moveout emanating from the promontories. Two small unconformity "outliers" behind the Virginia promontory (Fig. 10) nearly coincide with earlier Cherokee outliers (Fig. 9). However, these outliers are post-Keefer (Lockportian or late Wenlock-early Ludlow) and therefore slightly younger (Patchen and others, 1985a, b) than the late Clinton or early Wenlockian age of the unconformity elsewhere. If these outliers are part of the same unconformity, they may merely reflect the along-strike diachroneity of the disturbance (see Ettensohn, 1987) or peripheral-bulge return during loading-type relaxation (Fig. 2A). Nonetheless, the fact that they have persisted in the same areas as earlier surfaces may reflect reactivation of major basement structures in these areas (see Linn and others, 1990).

Unconformity development was immediately followed by transgression reflecting basin subsidence. The basin is now largely defined by the distribution of the Rochester Shale behind the New York and Virginia promontories (see Brett, 1983, Fig. 13). The basinal sequence begins with shallow-water, transgressive carbonates of the upper Irondequoit Limestone (Brett, 1983; Brett and others, 1990) in central and western New York and with the transgressive shoreface and shallow-marine shelf sands of the Keefer Sandstone and its equivalents (Smosna, 1983; Meyer and others, 1992) in east-central New York and Pennsylvania, western Virginia, and West Virginia. These units grade upward into the deeper water, dark gray shales and mudstones of the Rochester Shale (Fig. 11). Locally, the Rochester contains black, fissile, pyritic shales which have been interpreted as lagoonal deposits behind Keefer sand bars (Folk, 1962; Brett, 1983), but Luttrell (1968) and Smosna and Patchen (1978) rejected this interpretation based on paleontology and regional relationships. These areas of fissile black shale may reflect little more than loci of maximum subsidence behind promontories.

Subsidence throughout most of the Rochester basin, however, was apparently not as great or as long in duration as that characterizing the other dark-shale foreland basins in this study. In contrast, most of the Rochester basin was characterized by depths only slightly greater than storm wave base, and the greatest depths occurred early during the first half of Rochester deposition (Brett, 1983). Hence, early Rochester deposition probably represented the phase of maximum basin subsidence resulting from deformational loading (Figs. 1 and 3) during the second Salinic tectophase (Fig. 1).

The upper half of Rochester deposition, however, is a regressive sequence with an upward increasing frequency of calcareous tempestites (Brett, 1983; Brett and others, 1990). This part of the Rochester basin apparently reflects loading-type relaxation and south-to-southeasterly bulge migration toward the orogen (Fig. 2A). This bulge migration, moreover, is not only reflected by upward shallowing, regression, and the southward migration of facies, but also by an unconformity on the basin margins (Brett and others, 1990) as predicted by flexural models (Fig. 11). Although flexural models would predict a foreland basin infilling rapidly with flysch-like clastics at this point in time (Ettensohn, 1991), the overall weakness of the disturbance apparently did not produce source areas of sufficient relief to provide the necessary coarser clastic sediment. Instead, the basin filled with muds, silts, and fossil-rich tempestites.

By Lockportian time (late Wenlockian-early Ludlovian), very shallow-water carbonates of the Lockport Group and equivalents in the lower McKenzie and Mifflintown Formations were deposited throughout most of the northern and central Appalachian basin (Zenger, 1965; Rickard, 1975; Patchen and Smosna, 1975; Brett and others, 1990). The widespread distribution and abrupt appearance of these shallow-water carbonates apparently reflects the period of elevational equilibrium between the filled foreland basin and the eroded load (Fig. 2B, no. 3), during which very shallow seas spread throughout the former foreland basin. However, the increased abundance of slightly deeper water, darker marine shales in eastern parts of the McKenzie distribution (Patchen and Smosna, 1975) may reflect the formation of an antiperipheral bulge or peripheral sag and the beginning of unloading-type of relaxation (Fig. 11).

The hallmark of unloading-type relaxation is the cratonward progradation of a clastic wedge dominated by marginal-marine redbeds and related lithologies derived from rebounding parts of the orogen and foreland basin (Fig. 2B). In this case, the flexural sequence is terminated by the Bloomsburg and Vernon Formations, which represent a westwardly prograding wedge of marginal-marine red mudstones, dolostones, and greenish-gray to black shales (Dennison, 1986; Brett and others, 1990). Continuing rebound, in eastern New York, however, was apparently substantial enough that parts of the Vernon Shale itself were eventually uplifted and truncated (Brett and others, 1990).

By the time major Salinic rebound had ended, eastern parts of the north and central Appalachian basin, behind the New York and Virginian promontories, were locally uplifted or flooded with marginal-marine redbeds. Although rebound was not as intense in western parts of the basin, resulting uplift and basin isolation, combined with paleogeographic location in the evaporative trade-wind belt (Scotese and McKerrow, 1990) and a possible rain shadow (Grabau, 1913), generated conditions conducive to the deposition of restricted, shallow-marine carbonates and evaporites. These lithologies are represented by the Salina Group and its equivalents (Fig. 11).

Despite the poor to moderately developed nature of the Silurian flexural sequence in the Appalachian basin, the sequence is important as an example of response to a relatively weak orogenic event. The absence or poor development of dark, anaerobic basin facies during subsidence stages and the reduced input and extent of coarser clastic sediments during subsequent relaxation stages are typical of such a response. Moreover, the sequence is important because it suggests the likelihood of significant tectonic influence on the deposition of rocks, whose formation was previously ascribed largely to eustasy (Dennison and Head, 1975; Brett, 1983; Brett and others, 1990). Although tectonism has been called on at times in order to create sources for prograding clastic wedges in the sequence, the tectonism has been invariably identified as late recurrences of Taconian uplift (Folk, 1962; Brett, 1983). However, the fact that the late Clinton to Cayugan age (Wenlockian-earliest Pridolian) of the main flexural sequence overlaps early Cayugan (Ludlovian) dates (Boucot, 1962; Fairbairn, 1971) for the main part of the nearby Salinic disturbance, suggests that interpretations of late-stage Taconic uplift are unnecessary.

Silurian-Devonian Transition

The Silurian-Devonian transition in the Appalachian basin is represented by the Upper Silurian-Lower Devonian (Pridolian-Lochkovian) Helderberg Group. Although some evidence of latest Silurian tectonism has been suggested in basin areas just behind the New York (Salkind, 1979) and Virginia (Linn and others, 1990) promontories, no evidence of major regional unconformities or accompanying flexural stratigraphic sequences is present in Helderbergian rocks. Local instances of tectonism probably reflect reactivation of local structures by Salinic rebound, and Helderbergian lithofacies patterns in large part apparently reflect eustatic sea-level changes, differential basin subsidence and varying sedimentation rates (Dorobek and Read, 1986).

Devonian-Early Mississippian

Latest Early Devonian-through-Early Mississippian time in the Appalachian basin saw major episodes of unconformity formation and deposition of associated flexural stratigraphic sequences related to the Acadian orogeny. This sedimentary record is divided into four distinct, unconformity-bound sedimentary sequences that correspond to Acadian tectophases or phases of major intense deformation (Boucot and others, 1964;

Rodgers, 1970, p. 136; Johnson, 1971; Ettensohn, 1985, 1987). Hence, the tectophases are interpreted to have resulted from the collision of displaced Avalonian terranes, which approached North America with Baltica during the Caledonian orogeny, with consecutive promontories. The first tectophase (Early-Middle Devonian; Emsian-Eifelian) probably reflects collision with the St. Lawrence promontory; the second tectophase (Middle Devonian; Eifelian-early Givetian) with the New York promontory; the third tectophase (Late Middle Devonian-latest Devonian; late Givetian-Famennian), southward migration of oblique convergence between the New York and Virginia promontories; and the fourth tectophase (Early Mississippian; Tournaisian), convergence with the Virginia promontory and southward migration of deformation. The above scenario parallels Rodger's (1967) early observations of a northeastwardly to southwestwardly transgressing orogeny.

Each flexural sequence corresponding to a tectophase exhibits the same generalized four-part cycle (Fig. 3). Each cycle is bound at the base by an unconformity marking bulge moveout or continental braking, and the beginning of the Acadian orogeny is marked by the Emsian, pre-Oriskany or Wallbridge Discontinuity. This and each successive Acadian unconformity generally show a tripartite distribution behind the promontories (Fig. 12).

Each unconformity is typically overlain by a transgressive carbonate or calcareous sandstone (Fig. 13) indicating initiation of basin subsidence and marking the inception of the second part of the cycle. Only in the Sunbury cycle is this part of the sequence absent, apparently reflecting the deeper, basinward position (Fig. 14D) at which the cycle began. Where present, however, the basal carbonates, directly give way to black shales as subsidence and deepening increased in response to deformational loading. Moreover, black-shale basins are asymmetric toward the promontories, and more importantly, each successive basin moved progressively more cratonward in time (Figs. 13, 14) clearly indicating flexural response to the continuing cratonward migration of deformation. In addition, the numerous third-tectophase, black-shale basins not only reflect a pulsatory type of tectonism, but also indicate a prominent along-strike or transpressive component of tectonism (Ettensohn, 1987) by their progressive southwestward movement (Fig. 14C).

The third part of the preserved cycles is a clastic wedge representing major infilling of foreland basins with coarser clastic sediments of turbiditic, deltaic, and storm-related origins. Much of the "Catskill Delta" and parts of the Hamilton Group represent this part of the cycles. These sediments reflect the filling of deep basins during loading-type relaxation.

The fourth part of the cycles, reflecting a brief period of basin equilibrium followed by unloading-type relaxation and a cratonward progradation of marginal-marine sediments with redbeds (Fig. 3), is poorly developed or absent in all the Acadian cycles except for the last one. Only in the sedimentary expression of the final Acadian tectophase (Sunbury cycle) during later parts of the Mississippian (Fig. 15) is a major progradational

wedge of marginal-marine redbeds present. Although the Catskill Delta shows evidence of redbed facies developed on the eastern margin of the Appalachian basin (Rickard, 1975), in no one subcycle or cycle, except in the final one, did they prograde far into the basin. Apparently, Acadian tectophases succeeded each other so rapidly that complete flexural and sedimentary responses never developed. As a result, clastic wedges in each of the lower three flexural sequences only succeeded to the third part of the cycle before being abruptly uplifted and truncated by unconformities marking inception of subsequent tectophases (Fig. 13).

Mississippian

Only one major regional unconformity is present in the Mississippian of the Appalachian basin, although on and near major structures in the basin, local unconformities with substantial erosion (Ettensohn, 1980, 1981; Yielding and Dennison, 1986; Ettensohn and others, 1988a) apparently developed because of bulge-induced reactivation. In general, however, Mississippian time in the Appalachian basin was one of tectonic relaxation after the final tectophase (unconformity and Sunbury Shale in Figs. 13, 15A) of the Acadian orogeny—both loading-type and unloading-type relaxation. As interpretations in Figure 15 suggest, Lower and Middle Mississippian (Kinderhookian and Valmeyeran; Tournaisian and Visean) clastics and some of the lower parts of overlying carbonates represent a period of loading-type relaxation (Fig. 15B).

A regional Middle Mississippian or mid-Valmeyeran (Osage-Meramec; early Visean) unconformity and associated shallow-marine sediments (Warsaw, Renfro Member of the Slade, Maccrady) overlie the Borden-Price-Pocono-Grainger clastic sequence and intervene at such a time and place as to suggest association with concomitant eastward bulge movement accompanying relaxation (Figs. 2A, 15B). Although locally subtle, the unconformity is widespread in the Appalachian basin and has the typical tripartite distribution (Fig 16) relative to promontories (Weller and others, 1948; Youse, 1963; Patchen and others, 1985a, 1985b; Warne, 1990) that characterizes the results of previous Appalachian flexural movements (Figs. 6, 9, 10, 12). Moreover, units just below the unconformity typically represent very shallow subtidal to supratidal environments. The position and timing of these units in the flexural sequence almost certainly reflect relaxational bulge movement and uplift as is apparent in other flexural sequences (see Bowen and upper Ashlock Formations, Fig. 7; Rockway-Dawes and Glenmark-DeCew Formations, Fig. 11). However, the prominence and widespread nature of the overlying unconformity when compared with the Ordovician and Silurian examples above (Figs. 7, 11), seem to necessitate some additional mechanism of relative sea-level lowering, and that was mostly likely related to eustasy (Warne, 1990).

The area of greatest relative uplift and eustatic lowering coincided with eastern and central parts of the basin where aggrading delta sediments were already comparatively close to

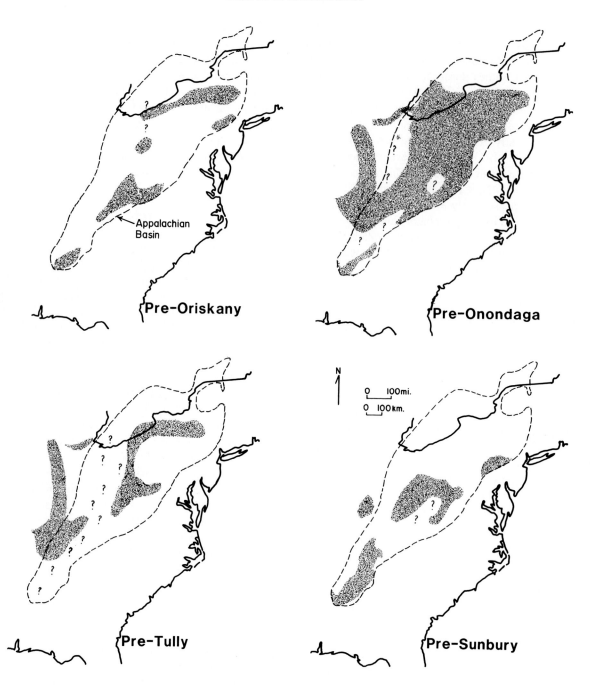

FIG. 12.—Approximate preserved distribution of four Acadian unconformities showing their preferential development just behind promontories. Even through the major brunt of any one tectophase was apparently focused largely on a single promontory, the presence of unconformities near other promontories suggests concurrent episodes of convergence at all promontories during each tectophase.

sea level (Figs. 15A, 15B). In this area especially, the intertidal and supratidal redbeds and evaporites of the Maccrady Formation (Warne, 1990) were deposited (Fig. 15B). Unusually thick deposits of Maccrady evaporites apparently reflect local structural basins (Warne, 1990) reactivated by bulge movement.

After regional uplift accompanying passage of the bulge, shallow-water carbonate deposition, probably accompanied by some eustatic sea-level rise, commenced throughout the basin

and is represented by the Greenbrier, Newman, Slade, and equivalent formations. The uppermost, high-energy carbonates in these formations (Chesterian; late Visean) apparently represent the time of elevational equilibrium between the filled foreland basin and the beveled Acadian highlands (Fig. 15C). In contrast, the overlying clastic sediments of the upper Newman, Bluefield and Pennington Formations (Chesterian; Serpukhovian or Namurian A) apparently represent a westwardly prograding

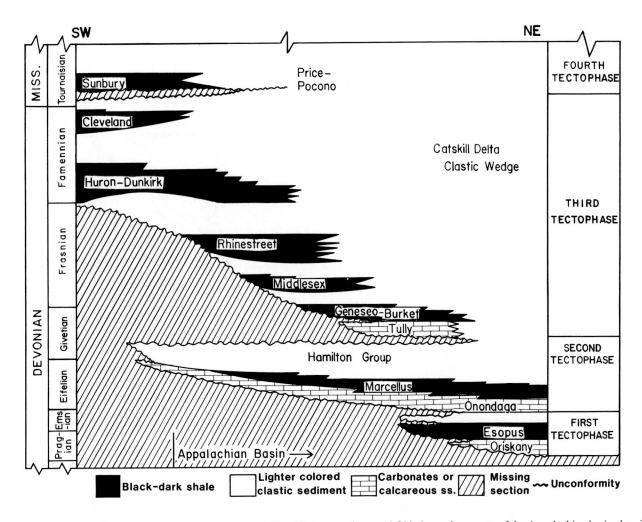

Fig. 13.—Composite stratigraphic section from east-central New York to north-central Ohio in northern parts of the Appalachian basin showing the distribution in time of pre-tectophase unconformities (see Fig. 12) and unconformity-bound flexural sequences of black shales and coarser clastic sediments attributed to four Acadian tectophases (adapted from Ettensohn, 1987). Note the progressive westward migration of successive black-shale basins (see Fig. 14). Of the four tectophases, only the flexural sequence of the final one developed to completion (see Fig. 15).

marginal-marine, clastic wedge formed during unloading-type relaxation (Fig. 15D). This type of marginal-marine sedimentation continued into earliest Pennsylvanian time (Pocahontas; Morrowan or Bashkirian) when it was abruptly ended by a major period of Early Pennsylvanian uplift and erosion probably related to the inception of the Alleghanian orogeny (Fig. 15E).

These patterns, however, are not as clear in southern parts of the Appalachian basin where lower parts of the Mississippian sequence are interrupted by two regional unconformities (Fig. 17). The distribution patterns of both unconformities strike more or less east-west, parallel to the Ouachita orogen (Fig. 18), but nearly perpendicular to the Appalachian orogen. The age of these unconformities and the fact that they approximately outline the Ouachita foreland basin suggest that they are related to the Ouachita orogeny.

The oldest of the unconformities formed near the Kinderhook-Valmeyer (Tournaisian) boundary (Fig. 17) and is prominent throughout central and west-central United States (Ham and

Wilson, 1967). Although continent-continent collision did not occur until Pennsylvanian time, according to Arbenz (1989) major Ouachita convergence appears to have begun at least by the end of Kinderhookian time (middle Tournaisian), a timing which coincides with development of the unconformity. Although the southern Appalachian basin was too distant from Ouachita tectonism (see Thomas, 1989, Fig. 3C) during Middle Mississippian time to develop a typical flexural sequence, the unconformity is present in southern parts of the Appalachian basin (Figs. 17, 18), and dark shales and turbidites in the Tesnus, Caney, Barnett, and Stanley Formations farther to the west apparently represent inception of a flexural sequence there.

The second Mississippian unconformity in the southern Appalachian basin formed near the Valmeyer-Chester (Visean) transition (Fig. 17). The fact that the unconformity is overlain by a flexural sequence in northwest-central Alabama (Fig. 17) suggests that the unconformity reflects the inception of a second Ouachita tectophase. Moreover, the presence of this unconformity

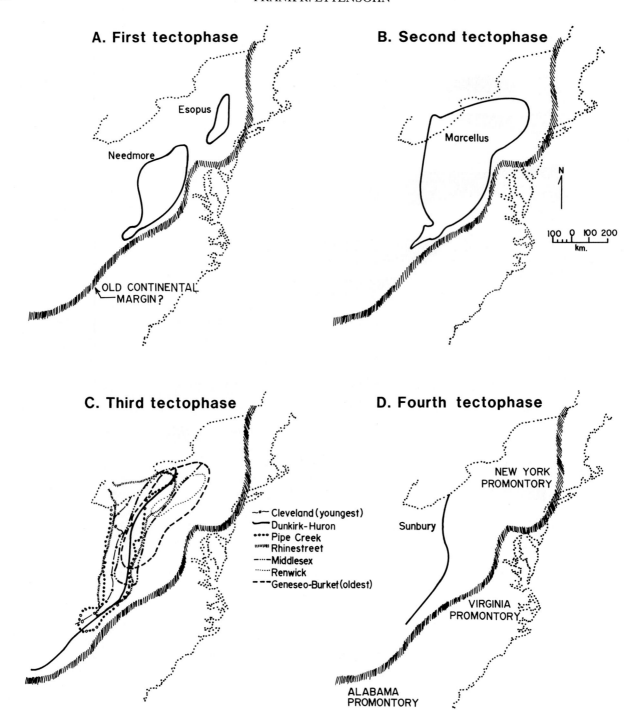

FIG. 14.—Approximate aerial distribution of successive black-shale basins by Acadian tectophase: (A) first-tectophase basins; (B) second-tectophase basin; (C) third-tectophase basins; and (D) proximal margin of the fourth-tectophase basin. Note the progressive westward migration of basins from tectophase to tectophase and within the third tectophase (compare with Figure 13).

in parts of the Appalachian basin (Fig. 18) and its accompanying flexural sequence in adjacent parts of the Black Warrior basin indicate that by this time, Ouachita loading had apparently migrated far enough cratonward to involve extreme southwestern parts of the Appalachian basin in its foreland-basin sedimentation. The unconformity truncates the Valmeyeran (early Meramecian; Visean) Tuscumbia Limestone and is overlain by Chesterian (latest Visean-Serpukhovian) transgressive limestones and sandstones of the Lewis sequence and equivalents in the Pride Mountain and Floyd Formations (Pashin and others, 1991; Pashin and Rindsburg, 1993a, b). Apparently most of the Genevievian Stage (late Meramecian; Visean) is absent or

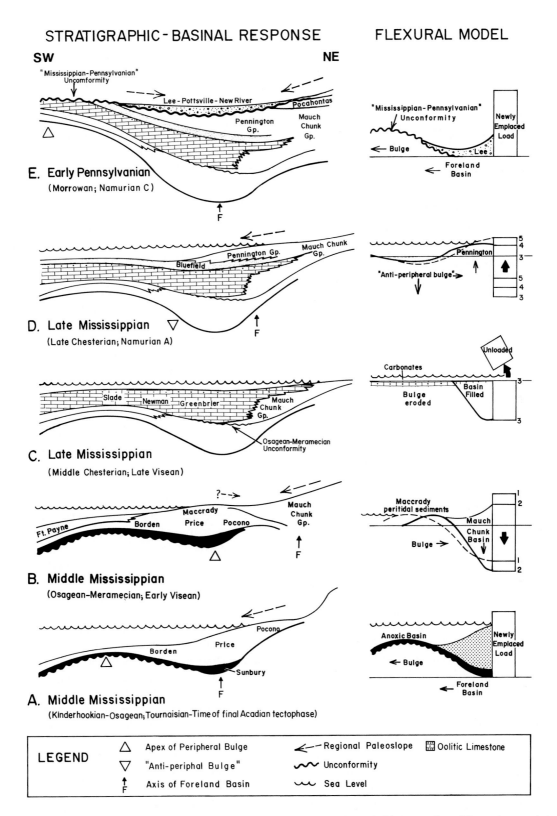

Fig. 15.—Schematic east-west section across the central Appalachian basin showing the probable succession of flexural events between the last tectophase of the Acadian orogeny (A) and the inception of the Alleghanian orogeny (E). In this scenario, the Sunbury black shale and sub-Sunbury unconformity represent rapid foreland-basin subsidence and bulge movement respectively accompanying the final Acadian tectophase, whereas remaining parts of the Mississippian section reflect subsequent Acadian relaxation. Lithologic symbols as on Figure 13 (adapted from Ettensohn and Chesnut, 1989).

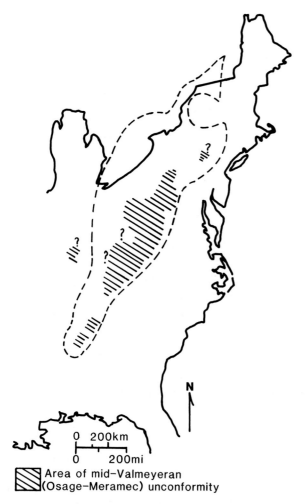

Area of mid-Valmeyeran
(Osage-Meramec) unconformity

FIG. 16.—Approximate aerial distribution of the Middle Mississippian (mid-Valmeyeran; Osage-Meramec) unconformity in the Appalachian basin.

condensed along this surface (Waters and Maples, 1992). Deposition of the dark resistive Floyd Shale continued to the southwest in deeper parts of the Ouachita foreland basin, while on the northeastern margin of the basin in the Appalachian "sphere of influence" upper parts of the Pride Mountain, the Hartselle, and the lower Bangor Limestone (Fig. 17) apparently represent shallow-water, shelf-edge deposition (Thomas, 1972; Thomas and Mack, 1982; Pashin and others, 1991) during the inception of the Appalachian equilibrium phase (Fig. 15C) mentioned earlier (Ettensohn and Pashin, 1993).

In deeper parts of the Ouachita foreland basin, the loading and subsidence phase represented by the Floyd Shale and equivalent parts of the Bangor Limestone is followed by the interbedded sands and shales of the lower and middle Parkwood Formation, which largely reflect gravity-driven deposition during loading-type relaxation. This phase of relaxation was ended by a period of elevational equilibrium between the filled basin and eroded uplands during which time the shallow-water, oolitic *Millerella* Limestone, abruptly developed throughout large parts of the known foreland basin (Fig. 17). The *Millerella* Limestone

is overlain by the upper Parkwood Formation, a northeastwardly prograding wedge of deltaic to marginal-marine clastics containing gray to black shales, sandstones, discontinuous coals, limestones and red shale. It is interpreted to represent deposition during unloading-type relaxation in the Ouachita orogen (Ettensohn and Pashin, 1993). As with the relaxational sequence in the Appalachian basin, upper Parkwood deposition continued into Early Pennsylvanian time (Butts, 1926; Jennings and Thomas, 1987) when it was locally truncated by an Early Pennsylvanian unconformity (Pashin and others, 1991; Ettensohn and Pashin, 1993) probably reflecting initiation of the Alleghanian orogeny.

Pennsylvanian

The so-called "Mississippian-Pennsylvanian" unconformity is one of the most prominent and widespread of Sloss's (1963) interregional unconformities. Despite the Mississippian-Pennsylvanian appellation, the unconformity is of Early Pennsylvanian age throughout the Appalachian basin (Ettensohn and Chesnut, 1989; Chesnut, 1989, 1992). Moreover, in three Appalachian areas (Fig. 19), east-central Pennsylvania (Edmunds and others, 1979), western Virginia and southern West Virginia (Englund and others, 1979), and northern Alabama (Butts, 1926; Jennings and Thomas, 1987; Pashin and others, 1991), Mississippian and Pennsylvanian rocks are apparently both vertically and laterally gradational, and rocks near the boundary reflect marginal-marine to terrestrial conditions (see Englund and others, 1979), indicating the continuation of unloading-type relaxation into the earliest Pennsylvanian (Fig. 15E). Englund and others (1979) and Pashin and others (1991) have clearly demonstrated that these Lower Pennsylvanian rocks near the boundary were truncated by a younger Early Pennsylvanian surface which cuts progressively deeper through Lower Pennsylvanian and Upper Mississippian rocks to the west and northwest (Fig. 20). The distribution of the unconformity relative to promontories, modeling (Quinlan and Beaumont, 1984; Beaumont and others, 1987, 1988), and the coincidence of unconformity formation with early phases of the Alleghanian orogeny suggest that the unconformity is probably related to northwestward peripheral-bulge migration accompanying inception of the Alleghanian orogeny. This phase of orogeny probably reflects initiation of convergence between Gondwanaland and the southeast margin of Laurussia and the subsequent development of a magmatic arc (Sinha and Zietz, 1982). The timing of pluton emplacement in the arc (320-300 Ma; Dallmeyer, 1986), moreover, overlaps the period of unconformity development indicating that early phases of the orogeny and unconformity formation were probably contemporaneous. However, the fact that the unconformity is widespread beyond the limits of the Appalachian area (Fig. 19) may merely reflect the presence of approximately coeval orogenies on the southern and western margins of the continent as well, or the possibility of a major eustatic component.

Throughout most of the Appalachian basin, this unconformity deeply truncates Mississippian rocks so that an intersystemic,

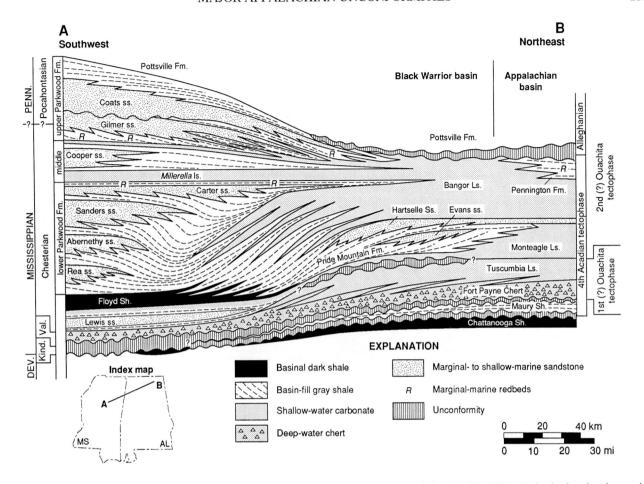

FIG. 17.—Schematic northeast-southwest section from the southern end of the Appalachian basin into the Black Warrior basin showing the overlap of probable Ouachita (Kinderhook-Valmeyer and Valmeyer-Chester) unconformities into the Appalachian basin. The Lewis-through-Parkwood section in the Black Warrior basin is an ideal flexural sequence not unlike that in the Appalachian basin (Fig. 15); however, it is younger than the Appalachian sequence (from Ettensohn and Pashin, 1993).

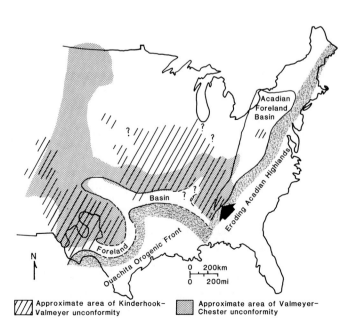

Mississippian-Pennsylvanian unconformity is apparent (Figs. 19, 20). In parts of the basin immediately behind the three promontories, however, the unconformity becomes an intra-Pennsylvanian surface, and most proximal to the promontory, a conformable sequence (Figs. 19, 20). The localization of areas of conformity immediately behind the promontories apparently reflects sufficient residual loading and resulting subsidence at these loci in the basin to offset the effects of bulge uplift.

Unlike earlier Appalachian foreland basins, the Alleghanian basin was broader and its initial sediments were largely fluviatile, orthoquartzitic sandstones and conglomerates (Pottsville Group) overlain by a finer grained sequence of cyclic marginal-

FIG. 18.—Map of eastern and central United States showing the approximate aerial distributions of two Mississippian (Kinderhook-Valmeyer and Valmeyer-Chester) unconformities probably associated with the Ouachita orogeny, which overlap into the Appalachian basin (arrow). Both unconformities generally parallel Ouachita orogen, although the prominent north-south extension of the Valmeyer-Chester unconformity in west-central United States parallels the Antler orogen and may reflect a coeval phase of Antler orogeny and bulge movement.

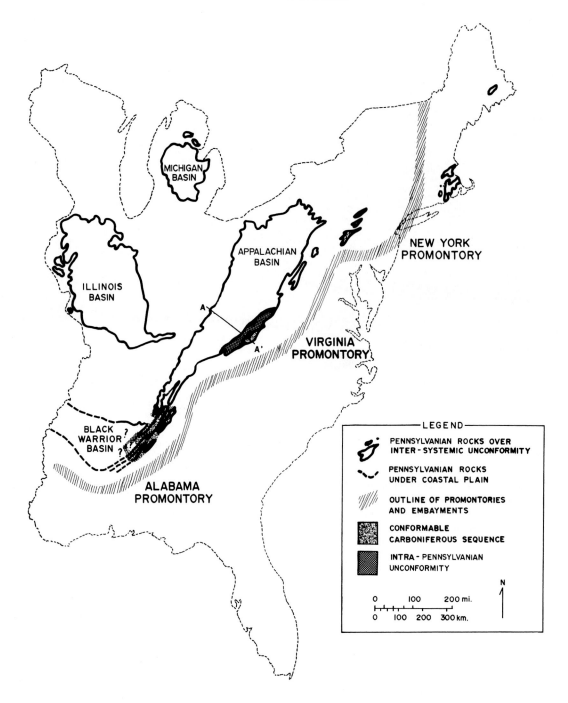

Fɪɢ. 19.—Nature of the so-called Mississippian-Pennsylvanian unconformity in the Appalachian basin and adjacent areas in eastern United States. In proximal parts of the Appalachian basin immediately behind the promontories the sequence is wholly conformable. Cratonward away from these areas, the unconformity becomes an intra-systemic (Pennsylvanian) unconformity and then an inter-systemic (Mississippian-Pennsylvanian) unconformity.

marine clastic sediments (Chesnut, 1989, 1992). Moreover, in a fashion typical of foreland basins (Fig. 13), a series of Pottsville (Lee) basins developed, and each progressively shifted more cratonward in time (Chesnut, 1989, 1992) apparently reflecting the cratonward migration of deformation. The otherwise anomalous nature of the Alleghanian foreland basin and its sedimentary sequence are probably related to the wholly surficial nature

of the deformational load (Karner and Watts, 1983) and to the fact that the load had by that time surmounted the former continental margin and moved far westward onto the craton (Beaumont and others, 1987; Jamieson and Beaumont, 1988). Although the Alleghanian orogeny probably progressed in a series of tectophases (Geiser and Engelder, 1983), our incomplete understanding of flexural response to such "thin-skinned"

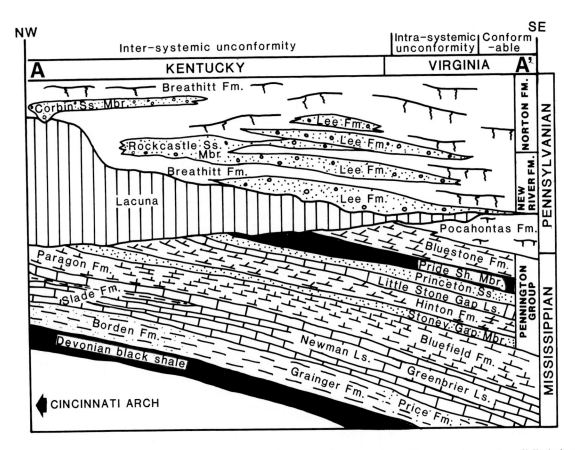

FIG. 20.—Schematic cross section perpendicular to the Appalachian basin showing the progressive northwestward truncation of Mississippian rocks below the Early Pennsylvanian ("Mississippian-Pennsylvanian") unconformity, which is probably associated with the inception of the Alleghanian orogeny (see Fig. 15E). Location of this section is shown on Figure 19.

orogenic events and our poor knowledge of detailed Pennsylvanian-Permian stratigraphy at present prevent the recognition of these tectophases in the stratigraphic record.

DISCUSSION AND CONCLUSIONS

The preceding review of the Paleozoic stratigraphic record in the Appalachian basin indicates that 13 interregional and regional unconformities, judged to be major based on their extent within and beyond the basin, are present (Fig. 21). Irrespective of possible tectonic or eustatic origins, the recurrence of unconformities, many accompanied by similar overlying sedimentary sequences, strongly suggests some type of cyclicity, even though recurrence intervals are irregular.

Ten of these unconformities, two in the Middle Ordovician, two in the Silurian, three in the Devonian, one at the Devonian-Mississippian boundary, and two in the Middle Mississippian, (Fig. 21), are interpreted to be primarily tectonic in origin. Interpretation of tectonic origin is based on the presence of a distinctive, overlying, flexural stratigraphic sequence, the coincidence of unconformity formation with the inception of established orogenies or tectophases therein, and the distribution of unconformities relative to probable loci of tectonism. In contrast, the absence of an overlying flexural sequence and no

coincidence with orogeny suggest that unconformities were predominantly eustatic in origin. Only the unconformity at the Ordovician-Silurian boundary seems to be of this type.

Of the criteria for a tectonic origin, the presence of a flexural stratigraphic sequence overlying an unconformity is probably most indicative of tectonic origin. This sequence, including a basal transgressive carbonate or clastic unit followed by dark shales, a coarser clastic flysch-like basinal unit, and a final marginal-marine clastic or redbed unit, is so distinctive because each lithology and its position in the sequence reflect a specific flexural response of the basin to deformational loading and subsequent relaxation during orogenesis. This flexural successor sequence, moreover, is primarily marine and subsidence dominant. How such a specific sequence of lithologies could form repeatedly over intervals of a few million to a few tens of millions of years as a result of eustasy is difficult to imagine.

Such a sequence, however, is apparently only typical during the early orogenic phases of an orogen when thick deformational loads pile up along a narrow belt as they attempt to surmount an initially steep continental margin. The thick, concentrated loads result in rapidly subsiding, narrow foreland basins dominated by marine sediments which were typical of Appalachian foreland basins during the Taconian (Middle-Late Ordovician), Salinic (Middle-Late Silurian) and Acadian (Early Devonian-Middle

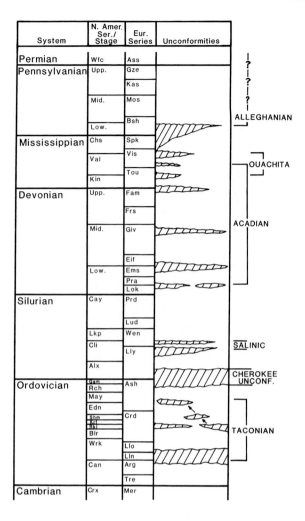

System	N. Amer. Ser./Stage	Eur. Series	Unconformities
Permian	Wfc	Ass	
Pennsylvanian	Upp.	Gze	
		Kas	
	Mid.	Mos	
		Bsh	ALLEGHANIAN
	Low.		
Mississippian	Chs	Spk	
	Val	Vis	OUACHITA
		Tou	
	Kin		
Devonian	Upp.	Fam	
		Frs	ACADIAN
	Mid.	Giv	
		Eif	
	Low.	Ems	
		Pra	
		Lok	
Silurian	Cay	Prd	
		Lud	
	Lkp	Wen	
	Cli	Lly	SALINIC
	Alx		CHEROKEE UNCONF.
Ordovician	Gam	Ash	
	Rch		
	May		
	Edn		
	Shm	Crd	
	Krf		
	Rkl		TACONIAN
	Blr		
	Wrk	Llo	
		Lln	
	Can	Arg	
		Tre	
Cambrian	Crx	Mer	

Fig. 21.—Distribution in time of the thirteen major regional and interregional unconformities found in the Appalachian basin and the probable orogenic events with which most are associated. Question marks above the initial Alleghanian unconformity reflect our incomplete understanding of Pennsylvanian-Permian stratigraphy and the possibility of other tectonically related unconformities.

Mississippian) orogenies. Formation of the two Ordovician, the two Silurian, and the five Devonian-Mississippian unconformities, as well as their overlying, dominantly marine flexural sequences, clearly reflect these circumstances. Moreover, each, with the exception of the mid-Valmeyeran unconformity, apparently coincides with inception of a tectophase in their respective orogenies. The mid-Valmeyeran unconformity, in contrast, formed during a tectophase and seems to reflect the coincidence of relaxational bulge uplift and movement with sea-level lowering.

The unconformities are generally more widespread than their overlying flexural sequences. This is especially true of the interregional unconformities (Early-Middle Ordovician; Early Devonian; Early Pennsylvanian), each of which represents the inception of a major, new orogeny and a new phase of cratonic reorganization. The overlying flexural sequences, in contrast,

are typically best developed in more proximal, rapidly subsiding parts of the foreland basin. Consequently, flexural sequences may be absent above overlying unconformities in distal parts of the foreland basin and on the adjacent craton even though the transgressive-regressive regimes inherent in the flexural sequence (Fig. 3) are still present. Such a situation is apparent in the extreme southern parts of the Appalachian basin which was distal to flexural responses from the nearby Ouachita orogen during the Mississippian. In this area, a prominent Kinderhook-Valmeyer unconformity apparently represents initiation of the Ouachita orogeny, but that part of the Appalachian basin was far enough removed from the locus of initial loading and subsidence that no flexural sequence is present. By the Valmeyer-Chester transition, however, deformation during the next Ouachita tectophase had apparently migrated far enough cratonward that a moderately developed flexural sequence accompanied the unconformity in extreme southern parts of the Appalachian basin and in adjacent parts of the Black Warrior basin. This part of the basin, however, is rather complex because of the overlapping effects of Acadian, Ouachita, and Alleghanian flexure.

The most widespread, but latest of the major Appalachian unconformities developed in the Early Pennsylvanian, late in the history of the orogen. Unconformity formation at this time appears to have coincided with inception of the Alleghanian orogeny, but the typical marine flexural sequence is absent. Although eustatic causes cannot be ruled out, the predominance of largely terrestrial clastic sediments in a broad, shallow foreland basin suggests a more "thin-skinned," dispersive type of loading characteristic of late-stage orogenies.

Mapping the extent of unconformities can also help to elucidate their origin, but this kind of evidence may not be as diagnostic as the previous two criteria. The distribution of unconformities of interpreted tectonic origin on the adjacent craton should strike more or less parallel to the trends of the involved orogenies; as predicted, Taconian, Salinic, and Acadian unconformities generally strike northeast-southwest (Figs. 6, 10, 12, 16), whereas the two southern Appalachian unconformities apparently related to the Ouachita orogeny generally strike east-west (Fig. 18). The Alleghanian surface, in contrast, is nearly ubiquitous (Fig. 19) suggesting a widespread coeval eustatic drawdown or the near contemporaneity of orogenies on three margins of the continent.

Moreover, within foreland basins, tectonically related unconformities generally show an asymmetry toward the continental promontories most involved in the concurrent deformation. It is of note, however, that the same pattern is apparent in the distribution of the Ordovician-Silurian unconformity (Fig. 6) which lacks a flexural sequence and does not coincide with an orogeny. Although this unconformity is widely accepted as having a glacio-eustatic origin, the peculiar distribution pattern suggests continued tectonic influence (probably late Taconian rebound) and the need for caution in using distribution patterns.

Hence, the recurrence of unconformities in the Appalachian basin suggests the presence of an irregular second- or third-order cyclicity, and lines of evidence discussed above indicate that much of this cyclicity must reflect episodic tectonic activity. Although tectonically related cyclicity appears to predominate at this scale, a major eustatic component is probable for the Ordovician-Silurian unconformity, the mid-Valmeyeran Mississippian unconformity, and may have contributed to the widespread development of the Early Pennsylvanian unconformity.

Before the realization of far-field tectonics and its flexural underpinnings, there was little reason to relate unconformity development to any process other than eustasy. However, it is now apparent that flexural perturbations of the lithosphere affect overlying sedimentary sequences and may extend over 1000 km away from the originating orogeny. Consequently, in view of the spatial and temporal proximity of the Appalachian basin to a number of Paleozoic orogenies, it is probably unrealistic, except during times of significant glaciation, to relate major unconformity formation and accompanying second- and third-order sedimentary cycles to the predominance of any cause except tectonism. This does not preclude, however, the probable significance of eustasy in generating the many fourth- and fifth-order cycles that abound throughout the Appalachian basin.

ACKNOWLEDGMENTS

I wish to thank John Dennison, Kevin Stewart and Andrew Warne who reviewed the paper and provided many helpful suggestions. I also want to thank Judianne Lesniewski, Tina Buster, and Margie Palmer for their help with manuscript preparation.

REFERENCES

ARBENZ, J. K., 1989, The Ouachita system, in Balley, A. W., and Palmer, A. R., eds., The Geology of North America— An Overview: Boulder, Geological Society of America, The Geology of North America, v. A, p. 371-396.

BEAUMONT, C., 1981, Foreland basins: Geophysical Journal of the Royal Astronomical Society, v. 65, p. 291-329.

BEAUMONT, C., QUINLAN, G. M., AND HAMILTON, J., 1987, The Alleghanian orogeny and its relationship to the evolution of the eastern interior, North America, in Beaumont, C., and Tankard, A. J., eds., Sedimentary Basins and Basin-Forming Mechanisms: Calgary, Canadian Society of Petroleum Geologists Memoir 12, p. 425-445.

BEAUMONT, C., QUINLAN, G. M., AND HAMILTON, J., 1988, Orogeny and stratigraphy: Numerical models of the Paleozoic in the eastern interior of North America: Tectonics, v. 7, p. 389-416.

BEVIER, M. L., AND WHALEN, J. B., 1990, Tectonic significance of Silurian magmatism in the Canadian Appalachians: Geology, v. 18, p. 411-414.

BJORLYKKE, K., 1985, Glaciations, preservation of their sedimentary record and sea level changes—a discussion based on the Late Precambrian and Lower Paleozoic sequence in Norway: Palaeogeography, Palaeoclimatology, and Palaeoecology, v. 51, p. 197-207.

BOSWELL, R. M., AND DONALDSON, A. C., 1988, Depositional architecture of the Upper Devonian Catskill Delta complex: Central Appalachian basin, U.S.A., in McMillan, N. J., Embry, A. F., and Glass, D. J., eds., Devonian of the World, Proceedings of the Second International Symposium on the Devonian System: Calgary, Canadian Society of

Petroleum Geologists Memoir 14, v. 2, p. 65-84.

BOUCOT, A. J., 1962, Chapter 10, Appalachian Siluro-Devonian, in Coe, K., ed., Some Aspects of the Variscan Fold Belt: Manchester, Manchester University Press, p. 155-163.

BOUCOT, A. J., FIELD, M. T., FLETCHER, R., FORBES, W. H., NAYLOR, R. S., AND PAVLIDES, L., 1964, Reconnaissance bedrock geology of the Presque Isle Quadrangle, Maine: Augusta, Maine Geological Survey Quadrangle Mapping Series No. 2, 123 p.

BRADLEY, D. C., AND KUSKY, T. M., 1986, Geologic evidence for rate of plate convergence collision: Journal of Geology, v. 94, p. 667-681.

BRETT, C. E., 1983, Sedimentology, facies, and depositional environments of the Rochester Shale (Silurian; Wenlockian) in western New York and Ontario: Journal of Sedimentary Petrology, v. 53, p. 947-971.

BRETT, C. E., GOODMAN, W. M., AND LO DUCA, S. T., 1990, Sequence stratigraphy of the type Niagaran Series (Silurian) of western New York and Ontario, in Lash, G. G., ed., Field Trip Guidebook, Western New York and Ontario: Fredonia, New York State Geological Association, p. C1-C71.

BRETT, C. E., GOODMAN, W. M., AND LO DUCA, S. T., 1991., Part 2: Silurian sequences of the Niagara Peninsula, in Cheel, R. J., ed., Sedimentology and Depositional Environments of Silurian Strata of the Niagara Escarpment, Ontario and New York, Field Trip B4: Guidebook: Toronto, Geological Association of Canada, p. 2-99.

BUTTS, C., 1926, The Paleozoic rocks, in Adams, G. I., Butts, C., Stephenson, L. W., and Cooke, C. W., Geology of Alabama: University, Alabama Geological Survey Special Report 14, p. 41-230.

CAPUTO, M. V., AND CROWELL, J. C., 1985, Migration of glacial centers across Gondwana during Paleozoic Era: Geological Society of America Bulletin, v. 96, p. 1020-1036.

CHESNUT, D. R., JR., 1989, Stratigraphic framework of the Pennsylvanian-age rocks of the central Appalachian basin, eastern U.S.A., in Yugan, J., and Chun, L., eds., Compte Rendu, Onzième Congrès International de Stratigraphie et de Géologie du Carbonifère: Nanjing, Nanjing University Press, v. 2, p. 1-19.

CHESNUT, D. R., JR., 1992, Stratigraphic and structural framework of the Carboniferous rocks of the central Appalachian basin in Kentucky: Lexington, Kentucky Geological Survey, Bulletin 3, Series XI, 42 p.

DALLMEYER, R. D., 1986, Polyphase terrane accretion in the southern Appalachians: Geological Society of America, Abstracts with Programs, v. 18, p. 579.

DENNISON, J. M., 1976, Appalachian Queenston delta related to eustatic sea-level drop accompanying Late Ordovician glaciation centered in Africa, in Basset, M. G., ed., The Ordovician System: Proceedings of a Palaeontological Association Symposium: Cardiff, University of Wales Press and National Museum of Wales, p. 107-120.

DENNISON, J. M., 1982, Geologic history of the 40th Parallel lineament in Pennsylvania: Spring Program, Appalachian Basin Industrial Associates, v. 2, p. 119-137.

DENNISON, J. M., 1986, Sedimentary tectonics of the Appalachian basin: Raleigh, Society of Economic Paleontologists and Mineralogists Mid-Year Meeting, Short Course Notes, 148 p.

DENNISON, J. M., AND HEAD, J. W., 1975, Sea level variations interpreted from Appalachian basin Silurian and Devonian: American Journal of Science, v. 275, p. 1089-1120.

DEVER, G. R., JR., HOGE, H. P., HESTER, N. C., AND ETTENSOHN, F. R., 1977, Stratigraphic evidence for late Paleozoic tectonism in northeastern Kentucky: Lexington, Field Trip Guidebook, Eastern Section, American Association of Petroleum Geologists, Kentucky

Geological Survey, 80 p.

DEWEY, J. F., AND BURKE, K. C. A., 1974, Hot spots and continental breakup: Implications for collisional orogeny: Geology, v. 2, p. 57-60.

DEWEY, J. F., AND KIDD, W. S. F., 1974, Continental collisions in the Appalachian-Caledonian belt: Variations related to complete and incomplete suturing: Geology, v. 2, p. 543-546.

DICKINSON, W. R., 1974, Plate tectonics and sedimentation, in Dickinson, W. R., ed., Tectonics and Sedimentation: Tulsa, Society of Economic Paleontologists and Mineralogists Special Publication 22, p. 1-27.

DOROBEK, S. L., AND READ, J. F., 1986, Sedimentology and basin evolution of the Siluro-Devonian Helderberg Group, central Appalachians: Journal of Sedimentary Petrology, v. 56, p. 601-613.

DRAKE, A. A., SINHA, A. K., LAIRD, J., AND GUY, R. E., 1989, The Taconic orogen, in Hatcher, R. D., Jr., Thomas, W. A., and Viele, G. W., eds., The Appalachian-Ouachita orogen in the United States: Boulder, Geological Society of America, The Geology of North America, v. F-2, p. 101-177.

ECKERT, B.-Y., AND BRETT, C. E., 1989, Bathymetry and paleoecology of Silurian benthic assemblages, late Llandoverian, New York State: Palaeogeography, Palaeoclimatology, Palaeoecology, v. 74, p. 297-326.

EDMUNDS, W. E., BERG, T. M., SEVON, W. D., PIOTROWSKI, R. C., HEYMAN, L., AND RICKARD, L. V., 1979, The Mississippian and Pennsylvanian (Carboniferous) systems in the United States— Pennsylvania and New York: Washington, D. C., United States Geological Survey Professional Paper 1110-B, p. B1-B33.

ENGLUND, K. J., ARNDT, H. H., AND HENRY, T. W., 1979, IX ICC Field Trip No. 1, Proposed Pennsylvanian stratotype, Virginia and West Virginia: Falls Church, American Geological Institute, American Geological Institute Selected Guidebook Series 1, 136 p.

ETTENSOHN, F. R., 1980, An alternative to the barrier-shoreline model for deposition of Mississippian and Pennsylvanian rocks in northeastern Kentucky: Geological Society of America Bulletin, v. 91, pt. 1, p. 130-135; pt. 2, p. 934-1056.

ETTENSOHN, F. R., 1981, Mississippian-Pennsylvanian boundary in northeastern Kentucky, in Roberts, T. G., ed., Stratigraphy, Sedimentation: Falls Church, Geological Society of America, Cincinnati 1981 Field Trip Guidebook, American Geological Institute, p. 195-257.

ETTENSOHN, F. R., 1985, The Catskill Delta complex and the Acadian orogeny, in Woodrow, D. W., and Sevon, W. D., eds., The Catskill Delta: Boulder, Geological Society of America Special Paper 201, p. 39-49.

ETTENSOHN, F. R., 1987, Rates of relative plate motion during the Acadian orogeny based on the spatial distribution of black shales: Journal of Geology, v. 95, p. 572-582.

ETTENSOHN, F. R., 1990, Estimating absolute depths for Devonian-Mississippian black shales in eastern United States: Implications for eustatic vs. tectonic control of sea level: Fall Program, Appalachian Basin Industrial Associates, v. 17, p. 40-139.

ETTENSOHN, F. R., 1991, Flexural interpretation of relationships between Ordovician tectonism and stratigraphic sequences, central and southern Appalachians, U.S.A., in Barnes, C. R., and Williams, S. H., eds., Advances in Ordovician Geology: Ottawa, Geological Survey of Canada Paper 90-9, p. 213-224.

ETTENSOHN, F. R., AND CHESNUT, D. R., JR., 1989, Nature and probable origin of the Mississippian-Pennsylvanian unconformity in the eastern United States, in Yugan, J., and Chun, L., eds., Compte Rendu, Onzième Congrès International de Stratigraphie et de

Géologie du Carbonifère: Nanjing, Nanjing University Press, v. 4, p. 145-159.

ETTENSOHN, F. R, DEVER, G. R., JR., AND GROW, J. S., 1988a, A paleosol interpretation for profiles exhibiting subaerial exposure "crusts" from the Mississippian of the Appalachian basin, in Reinhardt, J., and Sigleo, W. R., eds., Paleosols and Weathering Through Geologic Time: Principles and Applications: Boulder, Geological Society of America Special Paper 216, p. 49-79.

ETTENSOHN, F. R., MILLER, M. L., DILLMAN, S. B., ELAM, T. D., GELLER, K. L., SWAGER, D. R., MARKOWITZ, G., WOOCK, R. D., AND BARRON, L. S., 1988b, Characterization and implications of the Devonian-Mississippian black-shale sequence, eastern and central Kentucky, U.S.A.: Pycnoclines, transgression, regression, and tectonism, in McMillan, N. J., Embry, A. F., and Glass, D. J., eds., Devonian of the World, Proceedings of the Second International Symposium of the Devonian System: Calgary, Canadian Society of Petroleum Geologists Memoir 14, v. 2, p. 323-345.

ETTENSOHN, F. R., AND PASHIN, J. C., 1993, Mississippian stratigraphy of the Black Warrior basin and adjacent parts of the Appalachian basin: Evidence for flexural interaction between two foreland basins, in Pashin, J. C., ed., New Perspectives on the Mississippian System of Alabama: Tuscaloosa, 30th Annual Field Trip, Alabama Geological Society, p. 29-40.

FAIRBAIRN, H. W., 1971, Radiometric age of mid-Paleozoic intrusives in the Appalachian-Caledonide mobile belt: American Journal of Science, v. 220, p. 203-217.

FISHER, D. W., 1977, Correlation of the Hadrynian, Cambrian, and Ordovician rocks in New York State: New York State Museum Map and Chart Series 25.

FOLK, R. L., 1962, Petrography and origin of the Silurian Rochester and McKenzie shales, Morgan County, West Virginia: Journal of Sedimentary Petrology, v. 32, p. 539-578.

GARY, M., McAFEE, R., JR., AND WOLF, C. L., 1972, Glossary of Geology: Washington, D. C., American Geological Institute, 858 p.

GEISER, P., AND ENGELDER, T., 1983, The distribution of layer parallel shortening fabrics in the Appalachian foreland of New York and Pennsylvania: Evidence for two non-coaxial phases of the Alleghanian orogeny, in Hatcher, R. D., Jr., Williams, H., and Zietz, I., eds., Contributions to the Tectonics and Geophysics of Mountain Chains: Boulder, Geological Society of America Memoir 158, p. 161-175.

GRABAU, A. W., 1913, Early Paleozoic delta deposits of North America: Geological Society of America Bulletin, v. 24, p. 399-512.

HAM, W. E., AND WILSON, J. L., 1967, Paleozoic epeirogeny and orogeny in the central United States: American Journal of Science, v. 265, p. 332-407.

HAMBREY, M. J., 1985, The Late Ordovician-Early Silurian glacial period: Palaeogeography, Palaeoclimatology, Palaeoecology, v. 51, p. 273-289.

JACOBI, R. D., 1981, Peripheral bulge—a causal mechanism for the Lower/Middle Ordovician disconformity along the western margin of the northern Appalachians: Earth and Planetary Science Letters, v. 56, p. 245-251.

JAMIESON, R. A., AND BEAUMONT, C., 1988, Orogeny and metamorphism: A model for deformation and pressure-temperature-time paths with applications to the central and southern Appalachians: Tectonics, v. 7, p. 417-445.

JENNINGS, J. R., AND THOMAS, W. A., 1987, Fossil plants from Mississippian-Pennsylvanian transition strata in the southern Appalachians: Southeastern Geology, v. 27, p. 207-217.

JOHNSON, J. G., 1971, Timing and coordination of orogenic, epeirogenic, and eustatic events: Geological Society of America Bulletin, v. 82, p. 3263-3298.

JOHNSON, M. E., 1987, Extent and bathymetry of North American platform seas in the Early Silurian: Paleooceanography, v. 2, p. 185-211.

JOHNSON, M. E., RONG, J., AND YANG, X., 1985, Intercontinental correlation by sea-level events in the Early Silurian of North America and China (Yangtze platform): Geological Society of America Bulletin, v. 96, p. 1384-1397.

JORDAN, T. E., 1981, Thrust loads and foreland basin evolution, Cretaceous, western United States: American Association of Petroleum Geologists Bulletin, v. 65, p. 2506-2520.

KARNER, G. D., AND WATTS, A. B., 1983, Gravity anomalies and flexure of the lithosphere at mountain ranges: Journal of Geophysical Research, v. 88, no. B12, p. 10449-10477.

KAY, M., AND COLBERT, E. H., 1965, Stratigraphy and Life History: New York, John Wiley, 736 p.

KNIGHT, I., JAMES, N. P., AND LANE, T. E., 1991, The Ordovician St. George unconformity, northern Appalachians: The relationship of plate convergence at the St. Lawrence Promontory to the Sauk/Tippecanoe sequence boundary: Geological Society of America Bulletin, v. 103, p. 1200-1225.

KOEHLER, B., AND SMOSNA, R., 1989, Mississippian oolites on the West Virginia dome (abs.): American Association of Petroleum Geologists Bulletin, v. 73, p. 1035.

LAIRD, J., 1988, Arenig to Wenlock age metamorphism in the Appalachians, in Harris, A. L., and Fettes, D. J., eds., The Caledonian-Appalachian Orogen: Oxford, Geological Society Special Publication No. 38, p. 311-345.

LIEBLING, R. S., AND SCHERP, H. S., 1982, Late Ordovician/Early Silurian hiatus at the Ordovician-Silurian boundary in eastern Pennsylvania: Northeastern Geology, v. 4, p. 17-19.

LIN, B.-Y., AND BRETT, C. E., 1988, Stratigraphy and disconformable contacts of the Williamson: Revised correlation of the Late Llandoverian in New York State: Northeastern Geology, v. 10, p. 241-253.

LINN, F. M., TEXTORIS, D. A., AND DENNISON, J. M., 1990, Syn-depositional tectonic influence on lithofacies of the Helderberg Group (Silurian-Devonian) of the central Appalachians: Fall Program, Appalachian Basin Industrial Associates, v. 17, p. 190-213.

LOWRY, W. D., 1974, North American geosynclines: Test of continental drift theory: American Association of Petroleum Geologists Bulletin, v. 58, p. 575-620.

LUTTRELL, E. R., 1968, An analysis of the Silurian Keefer Sandstone of Pennsylvania: Unpublished Ph.D. Dissertation, Princeton University, Princeton, 111 p.

MATTHEWS, R. K., 1987, Eustatic controls on near-surface carbonate diagenesis: Colorado School of Mines Quarterly, v. 82, p. 17-40.

MCKERROW, W. S., 1979, Ordovician and Silurian changes in sea level: Journal of the Geological Society of London, v. 16, p. 17-145.

MCKERROW, W. S., AND ZIEGLER, A. M., 1972, Paleozoic oceans: Nature, Physical Sciences, v. 240, p. 92-94.

MEYER, S. C., TEXTORIS, D. A., AND DENNISON, J. M., 1992, Lithofacies of the Silurian Keefer Sandstone, east-central Appalachian basin, USA: Sedimentary Geology, v. 76, p. 187-206.

MUSSMAN, W. J., AND READ, J. F., 1986, Sedimentology and development of a passive- to convergent-margin unconformity: Middle Ordovician Knox unconformity, Virginia Appalachians: Geological Society of America Bulletin, v. 97, p. 282-295.

PASHIN, J. C., OSBORNE, W. E., AND RINDSBERG, A. K., 1991, Outcrop characterization of sandstone heterogeneity in Carboniferous reservoirs, Black Warrior basin, Alabama: Bartlesville, Bartlesville Project Office, United States Department of Energy, DOE/BC/14448-6, 126 p.

PASHIN, J. C., AND RINDSBURG, A. K., 1993a, Tectonic and paleotopographic control of basal Chesterian sedimentation in the Black Warrior Basin: Gulf Coast Association of Geological Societies Transactions, v. 43, p. 291-304.

PASHIN, J. C., AND RINDSBURG, A. K., 1993b, Origin of the carbonate-siliciclastic Lewis Cycle (Upper Mississippian) in the Black Warrior basin of Alabama: Geological Survey of Alabama Bulletin 157, 54 p.

PATCHEN, D. G., AVARY, K. L., AND ERWIN, R. B., 1985a, Southern Appalachian region, in Lindberg, F. A., ed., Correlation of Stratigraphic Units of North America (COSUNA) Project: Tulsa, American Association of Petroleum Geologists COSUNA Chart SAP.

PATCHEN, D. G., AVARY, K. L., AND ERWIN, R. B., 1985b, Northern Appalachian region, in Lindberg, F. A., ed., Correlation of Stratigraphic Units of North America (COSUNA) Project: Tulsa, American Association of Petroleum Geologists COSUNA Chart NAP.

PATCHEN, D. G., AND SMOSNA, R. A., 1975, Stratigraphy and petrology of Middle Silurian McKenzie Formation in West Virginia: American Association of Petroleum Geologists Bulletin, v. 59, p. 2266-2287.

PAVLIDES, L., 1989, Early Paleozoic composite melange terrane and its origin, central Appalachian Piedmont Belt, Virginia and Maryland, in Horton, J. W., Jr., and Rast, N., eds., Melanges and Olistostromes of the U.S. Appalachians: Boulder, Geological Society of America Special Paper 228, p. 137-193.

QUINLAN, G. M., AND BEAUMONT, C., 1984, Appalachian thrusting, lithospheric flexure, and the Paleozoic stratigraphy of the eastern interior of North America: Canadian Journal of Earth Science, v. 21, p. 973-996.

RICKARD, L. V., 1975, Correlation of the Silurian and Devonian rocks in New York State: Albany, New York State Museum and Science Service Map and Chart Series 24.

RODGERS, J., 1967, Chronology of tectonic movements in the Appalachian region of eastern North America: American Journal of Science, v. 265, p. 408-427.

RODGERS, J., 1970, The Tectonics of the Appalachians: New York, Wiley-Interscience, 271 p.

RODGERS, J., 1971, The Taconic orogeny: Geological Society of America Bulletin, v. 82, p. 1141-1178.

RODGERS, J., 1987, The Appalachian-Ouachita orogenic belt: Episodes, v. 10, p. 259-266.

SALKIND, M., 1979, Silurian tectonic activity in southeastern New York: Northeastern Geology, v. 1, p. 48-59.

SCOTESE, C. R., 1990, Atlas of Phanerozoic plate tectonic reconstructions, International Lithosphere Program (IUGG-IUGS): Arlington, Paleomap Project Technical Report No. 10-90-1.

SCOTESE, C. R., AND MCKERROW, W. S., 1990, Revised world maps and introduction, in McKerrow, W. S., and Scotese, C. R., eds., Paleozoic Palaeogeography and Biogeography: London, Geological Society Memoir 12, p. 1-21.

SHAVER, R. H., 1985, Midwestern basins and arches region, in Lindberg, F. A., ed., Correlation of Stratigraphic Units of North America (COSUNA) Project: Tulsa, American Association of Petroleum Geologists COSUNA Chart MBA.

SHAW, C. E., JR., 1970, Age and stratigraphic relations of the Talladega Slate: Evidence of pre-Middle Ordovician tectonism in central Alabama: Southeastern Geology, v. 11, p. 253-267.

SINHA, A. K., AND ZIETZ, I., 1982, Geophysical and geochemical evidence for a Hercynian magmatic arc, Maryland to Georgia: Geology, v. 10, p. 593-596.

SLOSS, L. L., 1963, Sequences in the cratonic interior of North America: Geological Society of America Bulletin, v. 74, p. 93-113.

SLOSS, L. L., 1984, Comparative anatomy of cratonic unconformities, in Schee, J. S., ed., Interregional Unconformities and Hydrocarbon Accumulation: Tulsa, American Association of Petroleum Geologists Memoir 36, p. 1-6.

SLOSS, L. L., 1988, Tectonic evolution of the craton in Phanerozoic time, in Sloss, L. L., ed., Sedimentary Cover—North American Craton: U.S.: Boulder, Geological Society of America, The Geology of North America, v. D2, p. 25-51.

SLOSS, L. L., AND SPEED, R. C., 1974, Relationship of cratonic and continental-margin tectonic episodes, in Dickinson, W. R., ed., Tectonics and Sedimentation: Tulsa, Society of Economic Paleontologists and Mineralogists Special Publication 22, p. 38-55.

SMOSNA, R., 1983, Diagenetic history of the Silurian Keefer Sandstone in West Virginia and Kentucky: Journal of Sedimentary Petrology, v. 53, p. 1319-1329.

SMOSNA, R., AND PATCHEN, D., 1978, Silurian evolution of central Appalachian basin: American Association of Petroleum Geologists Bulletin, v. 62, p. 2308-2328.

STEPHENS, G. C., AND WRIGHT, T. G., 1981, Stratigraphy of the Martinsburg Formation of Harrisburg in the Great Valley of Pennsylvania: American Journal of Science, v. 281, p. 1009-1020.

TANKARD, A. J., 1986, On the depositional response to thrusting and lithospheric flexure: Examples from the Appalachian and Rocky Mountain basins, in Allen, P. A., and Homewood, P., eds., Foreland Basins: Oxford, International Association of Sedimentologists Special Publication 8, p. 369-392.

THIRLWALL, M. F., 1988, Wenlock to mid-Devonian Volcanism of the Caledonian-Appalachian orogen, in Harris, A. L., and Fettes, D. J., eds., The Caledonian-Appalachian Orogen: Oxford, Geological Society Special Publication 38, p. 415-428.

THOMAS, W. A., 1968, Contemporaneous normal faults on flanks of Birmingham anticlinorium: American Association of Petroleum Geologists Bulletin, v. 52, p. 2123-2136.

THOMAS, W. A., 1972, Mississippian stratigraphy of Alabama: Tuscaloosa, Alabama Geological Survey Monograph 12, p. 44-57.

THOMAS, W. A., 1988, The Black-Warrior basin, in Sloss, L. L., ed., Sedimentary Cover—North American Craton: Boulder, Geological Society of America, The Geology of North America, v. D2, p. 471-492.

THOMAS, W. A., 1989, The Appalachian-Ouachita orogen beneath the Gulf Coastal Plain between the outcrops in the Appalachian and Ouachita mountains, in Hatcher, R. D., Jr., Thomas, W. A., and Viele, G. W., eds., The Appalachian-Ouachita Orogen in the United States: Boulder, Geological Society of America, The Geology of North America, v. F2, p. 537-553.

THOMAS, W. A., AND MACK, G. H., 1982, Paleogeographic relationships of a Mississippian barrier-island shelf-bar system (Hartselle Sandstone) in Alabama to the Appalachian-Ouachita orogenic belt: Geological Society of America Bulletin, v. 93, p. 6-19.

THOMPSON, J. B., JR., 1967, Bedrock geology of the Pawlet quadrangle, Vermont, part II, eastern portions: Vermont Geological Survey Bulletin 30, p. 61-98.

TWENHOFEL, W. H., AND OTHERS, 1954, Correlation of the Ordovician formations of North America: Geological Society of America Bulletin, v. 65, p. 247-298.

WARNE, A. G., 1990, Regional stratigraphic analysis of the Lower Mississippian Maccrady Formation of the central Appalachians: Unpublished Ph.D. dissertation, University of North Carolina, Chapel Hill, 493 p.

WATERS, J. A., AND MAPLES, C. G., 1992, Onset of basin development in the Black Warrior basin: Evidence from echinoderm biostratigraphy (abs.): Geological Society of America, Abstracts with Programs, v. 24, p. A154.

WATTS, A. B., KARNER, G. D., AND STECKLER, M. S., 1982, Lithospheric flexure and the evolution of sedimentary basins: Philosophical Transactions of the Royal Society of London, v. A305, p. 249-281.

WELLER, J. M., AND OTHERS, 1948, Correlation of the Mississippian formations of North America: Geological Society of America Bulletin, v. 59, p. 91-196.

WHEELER, H. E., 1963, Post-Sauk and pre-Absaroka stratigraphic patterns in North America: American Association of Petroleum Geologists Bulletin, v. 47, p. 1497-1526.

WILLIAMS, H., AND STEVENS, R. K., 1974, The ancient continental margin of eastern North America, in Burk, C. A., and Drake, C. L., eds., Geology of Continental Margins: New York, Springer-Verlag, p. 781-796.

WONES, D. R, AND SINHA, A. K., 1988, A brief review of Early Ordovician to Devonian plutonism in the North American Caledonides, in Harris, A. L., and Fettes, D. J., eds., The Caledonian-Appalachian Orogen: Oxford, Geological Society Special Publication 38, p. 381-388.

YIELDING, C. A., AND DENNISON, J. M., 1986, Sedimentary response to Mississippian tectonic activity at the east end of the 38th parallel fracture zone: Geology, v. 14, p. 621-624.

YOUSE, A. C., 1963, Gas producing zones of Greenbrier (Mississippian) Limestone, southern West Virginia and eastern Virginia: American Association of Petroleum Geologists Bulletin, v. 48, p. 465-486.

ZEN, E., 1967, Time and space relationships of the Taconic allochthon and autochthon: Boulder, Geological Society of America Special Paper 87, 107 p.

ZEN, E., 1972, The Taconide zone and the Taconic orogeny in the western part of the northern Appalachian orogen: Geological Society of America Special Paper 135, 72 p.

ZENGER, D. H., 1965, Stratigraphy of the Lockport Formation (Middle Silurian) in New York State: Albany, New York State Museum and Science Service Bulletin 404, 210 p.

ZIEGLER, A. M., AND MCKERROW, W. S., 1975, Silurian marine red beds: American Journal of Science, v. 275, p. 31-56.

ZIEGLER, P. A., 1987, Late Cretaceous and Cenozoic intra-plate compressional deformations in the Alpine foreland—a geodynamic model: Tectonophysics, v. 137, p. 389-420.

ABSTRACTS

Some abstracts were not developed into full-length papers. For others, arrangements were made to publish elsewhere. The following abstracts complete the full listing of presentations offered for the symposium.

SYNTHETIC FORELAND-BASIN STRATIGRAPHY ASSOCIATED WITH CONSTRUCTIVE, STEADY STATE, AND DESTRUCTIVE OROGENS

DAVID D. JOHNSON AND CHRISTOPHER BEAUMONT

Oceanography Department, Dalhousie University, Halifax, Nova Scotia, B3H-4J1

Foreland-basin stratigraphy and orogen state are determined by the rate of mass accretion to an orogen by thrust tectonics, the efficiency of mass redistribution by surface processes, and lithospheric flexure. Orogen state can be characterized as constructive, steady, or destructive depending on the mass net balance in the orogen (Jamieson and Beaumont, 1988, Tectonics, v. 7, pp. 417-445). We have constructed a kinematic planform foreland-basin model to look for stratigraphic relationships between synthetic foreland basin stratigraphy and orogen state.

The foreland-basin model links thin-skinned tectonic development of an orogen, lithospheric flexure (Beaumont and others, 1988, Tectonics, v. 7, p. 389-416) and mass redistribution by surface processes (Beaumont and others, 1992, Thrust Tectonics, p. 1-18). The tectonic model uses critical wedge principles to construct a two-sided wedge-shaped orogen. Sediments are accreted to the toe of each wedge at a rate proportional to the convergence rate of each leading slip line with the adjacent autochthon. The wedges, which need not be symmetric, grow in proportion to the net rate of mass influx. Their geometry is consistent with flexural adjustment of the lithosphere, conservation of mass, the criticality of each Coulomb wedge and match of wedge heights at their interface. The lithospheric-flexure model includes elastic or thermally activated linear viscoelastic rheologies. The surface process model couples climatic, hillslope (mass diffusion) and fluvial (mass transport) processes to erode, redistribute, and deposit mass across the orogen, its foreland basin and peripheral bulge.

Synthetic stratigraphic assemblages are constructed for a range of tectonic, lithospheric, and surface process model parameters, to determine under what circumstances an assemblage can be considered diagnostic of an orogens's state, or change in state.

SYNTHETIC MODELS OF UNCONFORMITY DEVELOPMENT IN FORELAND BASINS

CHRISTOPHER BEAUMONT AND DAVID D. JOHNSON

Oceanography Department, Dalhousie University, Halifax, Nova Scotia, B3H-4J1

Two competing hypotheses can explain distal unconformities in foreland basin stratigraphy: peripheral-bulge migration with stress relaxation in the lithosphere (Quinlan and Beaumont, 1984, Canadian Journal of Earth Sciences, v. 21, p. 973) or orogen tectonics and basin-filling mechanisms (Flemmings and Jordan, 1990, Geology, v. 18, p. 335; Sinclair and others, in review). We investigate synthetically the origin of unconformities using the planform model (Johnson and Beaumont, abstract above). The figure shows an unconformity bounded sequence for one tectonic cycle, on a stress-relaxing lithosphere. I-type erosion occurs when constructive orogenic loading outstrips basin-filling and lithospheric relaxation. F-type erosion occurs when peripheral bulge migration with lithospheric-stress relaxation dominates. I- and F-type unconformities remain distinct when intervening sediments are preserved, otherwise the composite unconformity reflects the superposition of tectonic and relaxation dominated phases that may span several tectonic cycles.

Does synthetic basin stratigraphy provide the evidence to distinguish I- from F-type erosion and to determine which dominates?

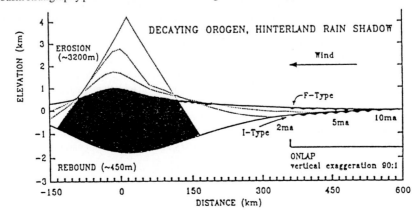

TECTONISM, SEA-LEVEL CHANGE, AND PALEOCLIMATE: EFFECTS ON ATLANTIC PASSIVE MARGIN SEDIMENTATION

C. WYLIE POAG

United States Geological Survey, Woods Hole, MA 02543

I have used >10,000 line-km of multichannel seismic-reflection profiles and 88 key boreholes to map 23 postrift Mesozoic and Cenozoic depositional sequences of the U. S. Middle Atlantic margin. From these data I infer that tectonic uplift was consistently a dominant force in determining the architecture and distribution of the sequences. Relative uplift among three primary source terrains directly determined the location of siliciclastic dispersal routes, rates of sediment accumulation, and the latitudinal position of associated offshore depocenters.

The principal role of sea-level change appears to have been to distribute and redistribute sediments once they reached the basin complex. Sea-level change was particularly effective during long-term rises and short-term falls, when it determined the bathymetric position of depocenters. A marked increase in sediment supply (triggered by source-terrain uplift), however, could mask the effects of short-term sea-level rise. Moreover, in the absence of tectonic uplift, major sea-level falls generally did not accelerate sediment accumulation.

Paleoclimatic shifts appear to have influenced deposition most effectively when associated with extreme conditions, such as extensive aridity or the buildup of continental ice sheets. But the relative amount of carbonate production on the continental shelves also was responsive to paleoclimatic change.

TECTONIC CONTROLS ON THE SEDIMENTARY RECORD OF THE EARLY MESOZOIC NEWARK SUPERGROUP, EASTERN NORTH AMERICA

JOSEPH P. SMOOT

United States Geological Survey, Box 25046, MS 939, Denver Federal Center, Denver, CO 80225

The Newark Supergroup consists of continental sedimentary and igneous rocks of Late Triassic to Early Jurassic age that fill a series of exposed half-graben basins along the eastern coast of North America. The rift basins formed along reactivated Paleozoic faults during extension that later led to the opening of the Atlantic Ocean. The basins had internal drainage systems, with base level largely independent of sea level. In general, each basin exhibits a similar vertical succession of sedimentary environments, although the ages of the deposits may differ from basin to basin: 1) Thin (<200 m), discontinuous immature fluvial conglomerates unconformably overlie Paleozoic or Precambrian rocks; these deposits reflect local provenance and small drainages. 2) Moderately thick sequence (500-1000 m) of conglomerate, crossbedded sandstone, and siltstone reflecting braided-river deposition with greater maturity than the basal conglomerates which they abruptly, and possibly unconformably, overlie. 3) Thick sequence (500-2000 m) of medium to fine sandstone and siltstone reflecting deposits of meandering streams and vegetated muddy plains. 4) Very thick sequence (1000-6000 m) of lacustrine mudstone and siltstone, in moist basins forming cyclic patterns of deep-water to subaerial deposits. In addition to this succession, alluvial-fan deposits intertongue with lacustrine and fluvial deposits along the faulted basin margins.

The vertical succession of depositional environments reflects: 1) the initial erosion surface prior to rifting; 2) basin subsidence and the localization and capture of regional drainages; 3) progressive loss of stream power as fault-related uplift constricted outlets and sediment aggradation lowered gradients; 4) partial or complete hydrographic closure of the basins, as faulting restricted outlets. Fault-related uplift created highlands producing alluvial fans sometime before the end of the third episode of deposition. The timing of these events for each basin probably depends on the relationship of the regional extensional field with the orientation and distribution of Paleozoic faults and preexisting drainages. A progressive vertical decrease in the abundance and thickness of deep-lake laminites within cyclic lacustrine strata may reflect an increase in basin area during subsidence. Three volcanic events extruding large volumes of flood basalt occurred apparently simultaneously in at least three basins and may have been associated with periods of increased subsidence. In several basins, the final stage of sedimentation is fluvial, suggesting a decrease in subsidence or a lowering of the outlet. Tectonic activity may also be reflected by decimeter-scale sequences of grain-size changes in alluvial-fan deposits and by radical shifts in the locations of marginal fluvial deposits. Climatic fluctuation, however, is the most important contributor to development of lake cycles and may also account for some fluvial variability.

CLIMATE CONTROLS ON CYCLIC SEDIMENTATION: CLIMATOSTRATIGRAPHY

C. BLAINE CECIL, N. TERENCE EDGAR, THOMAS S. AHLBRANDT
United States Geological Survey, Reston, VA 22092

Cyclic sedimentation is generally attributed to tectonic and (or) eustatic controls. The stratigraphy of chemical and siliciclastic sedimentary rocks cannot, however, be explained on the basis of these physical processes alone. As an example, the Pennsylvanian System of the United States contains transgressive-regressive cycles (cyclothems) that appear to be eustatically driven; also, basins were tectonically active as subsidence was necessary to provide accommodation space. On a basin scale, however, stratigraphic repetition of chemical rocks (coal beds, paleosols, and marine and nonmarine limestone) and siliciclastic rocks is indicative of paleoclimatic cycles as well as baselevel change induced by tectonics and (or) eustasy. Such climate cycles (changes in rainfall patterns) are recorded in stratigraphic sequences by 1) changes in paleosediment flux, 2) laterally extensive paleosols, whose characteristics range from modern aridisols (arid-climate soils) to vertisols (seasonal-climate soils) to oxisols (everwet climate soils) , and 3) paleobotanical changes that have long been attributed to changes in paleoclimate. On a continental scale, zonal-circulation paleoclimates are also recorded in addition to paleoclimate cycles.

The paleogeography of North America during the Pennsylvanian was such that the Appalachian basin was equatorial relative to the midcontinent and western United States. The resulting paleoclimate gradient from the Appalachian basin to the western United States (wetter to drier, respectively) is indicated by the development of coal beds in the Appalachian basin while evaporites were being deposited in the Paradox basin contemporaneously with alluvial fans and eolianites in the Rocky Mountain region. Paleoclimate, therefore, appears to be on a par with tectonic and eustatic changes as a primary control on sedimentation and stratigraphy. Thus, sedimentary sequences that exhibit the effects of climate as a primary control on sedimentation may be appropriately classified as climatostratigraphic units.

CAUSES OF PENNSYLVANIAN CYCLICITY IN THE APPALACHIAN BASIN

ALAN C. DONALDSON
Geology and Geography Department, West Virginia University, Morgantown, WV 26506

Three types of cycles are recognized for the Pennsylvanian rocks of the Appalachian basin: 1) autocycles (river avulsion and shifts in supply); 2) sub-regional allocycles (tectonism within the basin or uplift of parts of the orogenic belt; and 3) regional and eustatic allocycles (regional tectonism or eustasy). Autocycles occur within minor allocycles and are interpreted as sediment-supply shifts accompanying river avulsions within relatively fixed-drainage basins. Minor allocycles of coal measures commonly are 18 to 30 meters in thickness and contain extensive paleosols, coal, marine and/or freshwater limestone beds. Current estimates of allocycle durations vary greatly depending on the time scale used. Minor allocycles probably are glacio-eustatic and shoreline T-R shifts range from 32 to >800 km. Intermediate allocycles consist of several minor allocycles either in progradational, aggradational or retrogradational sets with relatively more transgressed units serving as the boundary units and are 90 to 115 m thick. Major allocycles are bounded by extensive transgressive units, are about 300 m thick, and approximate the Lower, Middle, and Upper Series of Pennsylvanian. These major allocycles are subregional, reflect basin tectonism shown by shifts in the seaway during the Pennsylvanian, and do not correlate with worldwide sea level curves, suggesting the basin's response to thrust-sheet loading during the Alleghanian orogeny. Cycles in paleoclimate were significant in influencing the lithic response within the allocycles (Cecil, 1990).

COMPARISON OF CARBONATE CYCLES IN THE CORDILLERAN REGION WITH MIDCONTINENTAL CYCLOTHEMS SUGGESTS A COMMON EUSTATIC ORIGIN

RALPH L. LANGENHEIM, JR.

Museum of Natural History, University of Illinois, 245 NHB, 1301 W. Green Street, Urbana , IL 61801

Cyclic carbonates from the middle of zone 21 through the Zone of *Fusulina* at Arrow Canyon, Nevada are asymmetric. They begin with an abrupt initiation of deeper-water deposition and a relatively restricted benthonic fauna. These sediments are gradationally followed by more coarsely grained sediment containing a more varied fauna. Some of the cycles shallow to subaerial exposure. Using current 'best' radiogenic dates for the base of the Bashkirian, Moscovian, and Kasimovian (Harland and others, 1989), calculated cycle duration ranges from 224,000 to 328,000 years. Carbonate rocks below zones 19, 20, and the lower half of 21, are less uniformly cyclical and the cycles reflect a more or less symmetrical rise and fall of sea level. The change to asymmetrical cycles occurs in the near vicinity of the onset of southern hemisphere glaciation, about 315 my (Harland and others, 1989). Asymmetric cycles, similar in character and length to those at Arrow Canyon characterize contemporaneous passive margin deposits of central Utah, through southern Arizona, into New Mexico.

The Arrow Canyon cycles are compatible in asymmetry and apparent duration with the well-documented cyclothems of the Midcontinent. Although it is impossible to biostratigraphically distinguish and correlate western and middlewestern cycles, their occurrence in similar numbers within zonal units and their similar length strongly suggests contemporaneity of individual cycles, continuity within sedimentary basins and a common, eustatic causal process.

DISTINGUISHING TECTONIC AND EUSTATIC SIGNALS IN A CYCLIC STRATIGRAPHIC RECORD

P. W. GOODWIN AND E. J. ANDERSON

Geology Department, Temple University, Philadelphia, PA 19122

In foreland basins tectonic processes tend to act continuously and leave a record of gradual facies changes (both laterally and vertically). In marked contrast, eustatic processes appear to be organized in a hierarchy in which sea-level rise in the shortest term cycle (the precessional cycle) is so rapid that it is recorded as a stratigraphic surface. Such surfaces mark the boundaries of meter-scale cycles or PACs (6th order sequences), the basic building blocks of the stratigraphic record. Indeed, all actual surfaces at cycle boundaries that are not unconformities (including the boundaries of the 5th or 4th order sequences driven by eccentricity cycles) are a product of this process. Eccentricity (at both scales) produces trends by modifying the magnitude of sea-level change in the basic 20-ky cycle and by producing longer term progressive patterns of deepening and shallowing. However, these larger-scale processes are not directly responsible for surfaces at cycle boundaries.

While eustatic processes seem to be responsible for small-scale cyclic packaging of the stratigraphic record, tectonic processes play different roles. Tectonic subsidence provides the long-term accommodation space for the stratigraphic record and, on a large scale, determines its general composition and degree of completeness. Also tectonically driven differential subsidence leads to lateral thickness and depth variations in facies. Correlations of the 6th and 5th order sequences in the Silurian-Devonian carbonates of the Appalachian basin indicate that differential subsidence is a significant contributor at the scale of 4th order sequences and perhaps at the 5th order scale. These observations suggest that eustatic and tectonic processes both play significant roles in the development of the stratigraphic record, but that these roles are noncompeting.

TECTONIC VS. EUSTATIC UNCONFORMITIES: EXAMPLES FROM UPPER ORDOVICIAN AND LOWER SILURIAN ROCKS, SOUTHERN APPALACHIANS

JOACHIM DORSCH[1], STEVEN G. DRIESE[1], ANNE R. GOGOLA[1], J. C. BOLTON[1], RICHARD K. BAMBACH[2]

[1]*Department of Geological Sciences, University of Tennessee, Knoxville, TN 37996*
[2]*Department of Geological Sciences, Virginia Polytechnic Institute and State University, Blacksburg, VA 24061*

The Martinsburg-Reedsville, Juniata-Sequatchie, and Tuscarora-Clinch-Rockwood-Red Mountain Formations, (Upper Ordovician to Lower Silurian) comprise the major part of the Taconic clastic wedge in southwestern Virginia, eastern Tennessee, northeastern Alabama, and northwestern Georgia. This clastic wedge constitutes the fill of a peripheral foreland basin which formed in response to thrust loading. The well-known disconformity at the base of the Tuscarora Fm. in the eastern Valley and Ridge of Virginia was recently traced as a paraconformity to the northwest (i.e., basinward). A white, thick-bedded "upper" Tuscarora abruptly overlies the Martinsburg or Juniata Formations. with a distinctive basal conglomerate in the southeastern Valley and Ridge; a yellowish, thin-bedded, more mudstone-rich "lower" Tuscarora interpreted as facies-equivalent for parts of the Juniata Formation, occurs beneath the "upper" Tuscarora in the northwestern Valley and Ridge. This regional unconformity, with maximum lacuna towards the southeast, divides the clastic wedge into two parts. The nature of the unconformity along and across structural strike, its regional extent and disappearance towards Tennessee, and the overall tectonic setting all favor a tectonic origin for this unconformity. Basin rebound following the cessation of thrusting, concomitant with erosional/tectonic unloading of the Taconic orogen farther to the southeast, is suggested as the causal mechanism for the erosional unconformity. This interpretation contrasts markedly with stratigraphic relationships observed in eastern Tennessee, northeastern Alabama and northwestern Georgia, where the equivalent stratigraphic interval exhibits a small lacuna between the Juniata-Sequatchie and Clinch-Rockwood-Red Mountain Fms. Basin development was not related to collisional tectonism, but represents a "left-over moat" of the Blount-phase. Paleosols occur within the upper Juniata Fm. in the eastern Valley and Ridge, sharply overlain by a 0.6- to 3-m thick, very fine- to medium-grained and phosphatic, highly bioturbated, transgressive sandstone of marine shoreface origin that occurs in the basal Clinch, Rockwood and Red Mountain Fms. This regionally correlative sandstone, informally known as the "basal transgressive sandstone" (bts), was deposited along ravinement surfaces during erosional shoreface retreat; it therefore documents a lowstand followed by a eustatic rise. To the west, the bts overlies open-marine shale and limestone deposits containing a typical Ashgillian brachiopod fauna; shelf strata above the bts contain a Lower Llandovery (A1-2) brachiopod fauna. Hence, the magnitude of the lacuna appears decrease in a northwesterly direction, where the Ordovician to Silurian boundary is nearly conformable.

SEA-LEVEL DROP CONTRASTED WITH PERIPHERAL BULGE MODEL FOR APPALACHIAN BASIN DURING MID-ORDOVICIAN

JOHN M. DENNISON
Geology Department, University of North Carolina at Chapel Hill, Chapel Hill, NC 27599-3315

Middle Ordovician strata record the early Taconic orogeny in the Appalachian basin and also exhibit an imprint of sea-level history. The Bays-Moccasin-Bowen redbeds and the Colvin Mountain-Greensport Formations result from the Blount Phase of the Taconic orogeny. Geographic distribution, geometry, and age relationships of these strata and their equivalents indicate also that a sea-level drop occurred in the Appalachian basin in latest Wilderness time, and its effects were superimposed on the longer lasting accumulation of the siliciclastic wedge of the Blount delta.

Evidence for brief sea-level drop then is present from Alabama to Virginia and Kentucky, and also in New York. Lithostratigraphic evidence for sea-level drop includes thin redbeds, sun-cracked shale sandwiched between marine carbonates, dolomite sandwiched between deeper water limestones, erosional truncation of bentonites on the Nashville dome, and Fincastle Conglomerate transported into deep-water shales in Virginia. Associated bentonite chronology and conodont biostratigraphy provide the best means for measuring time precisely.

Time constraints for this event make it sufficiently contemporaneous that migration of a peripheral bulge is precluded as an explanation. The geographic extent is too large, and it does not migrate through time, as would be expected from an orogenic mechanism.

Details of the Bays-Moccasin-Bowen redbeds suggest two sea-level minima in a short time span, recorded by the distal end of the Moccasin and Bowen redbeds and with the later one representing the greatest drop in sea level.

In northern West Virginia, Maryland, Pennsylvania, and Ohio there is not such clear evidence for this sea-level drop. This may be because deeper carbonate facies accumulated there, or simply because these strata have not been studied in sufficient lithologic detail.

INDEX

INDEX

258   INDEX

133, 184-186, 189-191, 194-195, 197, 204, 212, 217, 221, 225, 228, 234, 237, 246-249
Hensley Member, 55
Herkimer Formation, 153, 157, 159, 164
Hiatuses, 171, 185, 203, 208, 210
Highstand deposits, 69, 148, 152, 156-157, 159, 184
Highstand systems tract, 8, 32-33, 56-57, 59, 61-62, 69-74, 77-79, 147-149, 152-159, 165-166, 181, 184, 189-190, 192, 205
Hill Formation, 149, 152
Hillier Formation, 187-188, 196
Hindeodus, 71-72
Holocene, 2, 36, 39, 75, 122
Homerian, 151, 154
Hudson Bay lowlands, 197
Hugoton embayment, 81
Huron Black Shales, 127, 133, 136, 138, 141-142
Hushpuckney Shale, 73
Hyden Formation, 54

I

I-type erosion, 245
Iapetus Ocean, 157, 184
Iatan Formation, 66, 76
Ice Sheets, 25, 77, 80, 83, 121, 246
Idiognathodus, 71-72
Idioprioniodus, 71
Illinois, 57, 63, 65, 70, 72-76, 82-84, 248
Illinois basin, 57, 63, 65, 72-74, 83
Indiana, 203
Individual cycle method, 41
Indonesia, 56, 102
Interregional unconformities, 217-218, 221, 234, 237-238
Intraplate stress, 27, 30, 43, 47, 103
Iola cyclothem, 37-38, 69
Iowa, 39-40, 42-46, 65-66, 72, 75, 77, 81-82, 197
Iowa Platform, 40, 42-46
Irish Valley Member, 123-124
Irondequoit Formation, 149-150, 157, 160-161, 226-227
Island Formation, 153, 209
Isostatic uplift, 116, 162, 164-166, 186

J

Java Formation, 79, 127, 134-136, 138-140, 142
Java Sea, 79
Java Village, 136
Jeffersonian Stage, 208-209, 211-214
Joana Limestone, 31-32
Johnstown, 107
Jonesboro fault system, 5, 7, 18, 20
Jordanville Sandstone, 152
Joslin Hill, 153
Juniata Formation, 171-172, 174, 176-178, 186, 212-213, 249
Jurassic Period, 1, 5-8, 15, 17, 32, 38-39, 46-47, 53, 57-59, 63, 66, 69, 71, 73, 75, 77, 79, 81-82, 107-108, 111, 114, 116-117, 121-122, 124-127, 133, 136, 142, 147-149, 151-159, 162, 164-166, 171-172, 177-178, 181-182, 185, 188-189, 195, 197, 203-204, 206-214, 217-221, 224, 226-229, 231-232, 234, 237-239, 246, 249

K

K-bentonites, 3, 195
Kanawha Formation, 82, 110

Kansas, 29, 38, 40, 45, 65-66, 69, 72, 75, 80-81
Kansas City Group, 38, 45, 69, 75
Karner, 30, 43, 196, 218-219, 236
Kasimovian, 248
Kaskaskia, 107
Kayaderosseras, 210, 212
Kayaderosseras stratigraphic break, 210
Keefer Formations, 152, 161, 164-165, 227
Kendrick Shale Member, 55
Kentucky, 51, 53-54, 57, 77, 82, 107, 111-112, 115, 217, 249
Kentucky River Fault Zone, 221
Kinderhookian Stage, 229, 231
Kinderscoutian Stage, 115-116
Kinematic planform foreland basin model, 245
Kings Falls Formation, 195
Kirkfieldian Stage, 171, 179, 182, 197, 223
Kirkland Hematite, 151-152, 157
Knox Group, 171, 178, 210-212, 214
Knoxville, 249
Kope Formation, 197

L

Lacustrine
  cycles, 5, 7-8, 13-20, 25-26, 28-33, 38, 40, 42, 44-45, 51-52, 54-56, 63, 65-66, 71-73, 75, 77, 80-84, 89-95, 97-103, 121-129, 133-143, 147-148, 150, 154, 157-164, 166, 171-172, 174, 177-178, 188, 217, 221, 226, 229, 239, 245-248
  facies, 1-2, 5-8, 11, 15, 18, 25, 28, 31, 35-36, 45, 53, 66, 69-77, 79-80, 82, 84, 89, 92, 110, 123-125, 127, 129, 133-135, 137-143, 147-154, 156-158, 160-162, 164-166, 174, 178, 181-182, 192, 195, 197, 221, 228-229, 248-249
  lithofacies, 5, 8, 11-12, 18, 21, 35, 37, 39, 123, 141, 186-187, 198, 212, 228
Lagoonal deposits, 227
Lake District, 157
Lake Erie, 136
Lake Ontario, 182
Large scale sequences, 148
Laurentian terrane, 157, 211
Laurussia, 234
Leatham Member, 32
Lecompton Cyclothem, 38
Lee Formation, 53, 82, 114
Lemon Fair carbonates, 209
Lewis sequence, 232
Lewiston Member, 152, 157, 159-160
Lillydale Shale, 110
Lindsay Formation, 195
Lithospheric flexure, 1-2, 25-27, 30-31, 65, 103, 147, 162, 166-167, 179, 198, 218-219, 221, 238, 245
Little Falls, 195
Lizard Creek Member, 152, 157
Llandoverian, 148-149, 151, 154, 157, 162-166, 224, 226-227, 249
Loading-type relaxation, 219-220, 226-229, 234
Lockatong Formation, 7-8
Lockport Group, 152-154, 157, 164-165, 228
Lockportian Stage, 227-228
Long-term climate, 35, 38, 45-48
Lost City Limestone, 80
Lowstand deposits, 32-33, 56-58, 60, 66, 69, 71-72, 107, 117, 141-142, 148, 152, 154-159, 159, 166, 181-182, 184, 186, 203, 206, 208, 210-

264 INDEX